Agricultural Crop Improvement

Larger amounts of nutritious food are required to feed the growing world population, which is a great challenge due to water shortages and reduced crop plant yield. To overcome this issue, there must be improvement in crop plant production systems. *Agricultural Crop Improvement: Plant and Soil Relationships* addresses key issues of crop plant yield and production, with molecular and physiological interventions to evolve future strategies that will overcome these challenges faced by the agricultural sector.

Features:

- Investigates modern and traditional agricultural techniques, including nanomaterials, nanosensors, genetic engineering, molecular breeding, nutrient and plant hormone interactions, microbiome engineering, soil enzymes, biostimulants, and intercropping
- Explores how technologies and scientific advancements from biotechnology, information technology, and environmental science are revolutionizing sustainable farming practices

Agricultural Crop Improvement: Plant and Soil Relationships is a valuable source of information for researchers and extension workers involved in sustainable crop plants improvement and practices, and a useful resource for students and scientists working on aspects of agriculture, botany, plant science, crop physiology, crop biotechnology, and microbiology.

Agricultural Crop Improvement

Plant and Soil Relationships

Edited by
Azamal Husen

CRC Press
Taylor & Francis Group
Boca Raton London New York

CRC Press is an imprint of the
Taylor & Francis Group, an **informa** business

Designed cover image: Shutterstock

First edition published 2025
by CRC Press
2385 NW Executive Center Drive, Suite 320, Boca Raton FL 33431

and by CRC Press
4 Park Square, Milton Park, Abingdon, Oxon, OX14 4RN

CRC Press is an imprint of Taylor & Francis Group, LLC

ISBN: 9781032616261 (hbk)
ISBN: 9781032630359 (pbk)
ISBN: 9781032630366 (ebk)

DOI: 10.1201/9781032630366

Typeset in Times
by KnowledgeWorks Global Ltd.

Contents

About the Editor

Azamal Husen is serving as Full-term Professor and Head of the Department of Biology, University of Gondar, Ethiopia, and currently working as a Foreign Delegate at Wolaita Sodo University, Wolaita, Ethiopia. He is also working as an Adjunct Professor at Graphic Era (Deemed to be University), Dehradun, Uttarakhand, India. He also worked as Full Professor at Sankalchand Patel University, Visnagar, India, and Visiting Faculty of the Forest Research Institute and the Doon College of Agriculture and Forest at Dehradun, India. His research and teaching experience of 25 years encompasses Nanobiotechnology, Plant Physiology, Environmental Botany, Medicinal Plants; and Clonal Propagation. He has conducted research sponsored by the World Bank, the National Agricultural Technology Project, the Indian Council of Agriculture Research, the Indian Council of Forest Research Education, and the Japan Bank for International Cooperation. Husen has published extensively (over 300, which includes research papers, books, book chapters, etc.) and served on the Editorial Board and as reviewer of reputed journals published by Elsevier, Frontiers Media, Taylor & Francis, Springer Nature, RSC, Oxford University Press, Sciendo, the Royal Society, CSIRO, PLOS, MDPI, John Wiley & Sons, and UPM Journals. He is on the advisory board of Cambridge Scholars Publishing, UK. He is a fellow of the Plantae group of the American Society of Plant Biologists, and a member of the International Society of Root Research, Asian Council of Science Editors, and International Natural Product Sciences. He has given numerous talks and presentations in many conferences and workshops. Husen has also given various orientation courses for academic staff, prepared different modules, developed BSc, MSc, and PhD curriculum. Moreover, he has received a recognition award from University of Gondar, Ethiopia for a decade of service in teaching, research, and community services. He is Editor-in-Chief of the *American Journal of Plant Physiology*, and a Series Editor of *Exploring Medicinal Plants* (Taylor & Francis, USA); *Plant Biology, Sustainability, and Climate Change* (Elsevier, USA); and *Smart Nanomaterials Technology* (Springer Nature, Singapore). He has achieved the distinguished honour of being recognized as one of the 'World's Top 2% Scientists' for the years 2022, 2023, and 2024 by Stanford University, USA. This recognition has also been prominently featured in the Elsevier Data Repository. Moreover, he has also worked as a special issue guest editor on various topics such as Advances in Plant Metabolomics: Insights, Applications, and Integration (Plant Physiology and Biochemistry), Exploring Nanotechnology in Medicinal Plants (*Journal of Herbal Medicine*), and Integrating Nanotechnology with Plant Compounds: Enhancing Abiotic Stress Resilience and Crop Improvement via Signaling Molecules (*Frontiers in Plant Science*).

Contributors

Altaf Ahmad
Department of Botany
Aligarh Muslim University
Aligarh, India

Naeem Ahmad
College of Agronomy
Northwest A&F University
Yangling, China
and
Key Laboratory of Crop Physi-ecology and Tillage Science
Northwestern Loess Plateau, Ministry of Agriculture
Yangling, China

Sami Ul-Allah
Department of Plant Breeding and Genetics
Bahauddin Zakariya University
Multan, Pakistan

Mohamed A. Amin
Department of Botany and Microbiology
Faculty of Science
Al-Azhar University
Cairo, Egypt

Ana Cristina Ramírez Anguiano
Chemistry Department
School of Exact Sciences and Engineering (CUCEI)
University of Guadalajara
Zapopan, Mexico

Iqra Arshad
Nuclear Institute of Agriculture and Biology College (NIAB-C)
Pakistan Institute of Engineering and Applied Sciences
Faisalabad, Pakistan

Asad Azeem
Department of Plant Breeding and Genetics
University of Agriculture
Faisalabad, Pakistan
and
College of Agriculture
The University of Layyah
Dera Ghazi Khan, Pakistan

Sangita Banga
Department of Sciences
Manav Rachna University
Faridabad, India

Megha Barot
Center for Research and Innovation
Sakalchand Patel University
Visnagar, India

Aaliya Batool
Department of Botany
University of Agriculture
Faisalabad, Pakistan

Mohammed Berrada
Department of Chemistry
Faculty of Sciences, Ben M'Sick
University Hassan II of Casablanca
Casablanca, Morocco

Farheen Bhatti
Department of Botany
Government College Women University
Faisalabad, Pakistan

Ritu Chaudhary
Department of Botany
Aligarh Muslim University
Aligarh, India

Ramachandran Chelliah
Department of Food Science and Biotechnology
College of Agriculture and Life Sciences
Kangwon National University
Gangwon-do, South Korea
and
Saveetha School of Engineering (SIMATS) University
Chennai, India

Gowardhan Kumar Chouhan
Plant Microbe Interaction Lab
Institute of Environment and Sustainable Development
Banaras Hindu University
Varanasi, India

Fouad Damiri
Department of Chemistry
Faculty of Sciences, Ben M'Sick
University Hassan II of Casablanca
Casablanca, Morocco

Milagros Melissa Flores Fonseca
Medical Science
University of Colima, Avenida Universidad
Colima, Mexico

Hossam M. Fouda
Department of Botany and Microbiology
Faculty of Science
Al-Azhar University
Cairo, Egypt

Anand Kumar Gaurav
Plant Microbe Interaction Lab
Institute of Environment and Sustainable Development
Banaras Hindu University
Varanasi, India

Anindya Ghosh
Department of Biotechnology
St. Xavier's College (Autonomous)
Kolkata, India

Pronobesh Ghosh
Department of Sericulture
Raiganj University
Raiganj, India

Tushar Goyal
Plant Microbe Interaction Lab
Institute of Environment and Sustainable Development
Banaras Hindu University
Varanasi, India

Eksha Guliani
Department of Sciences
Manav Rachna University
Faridabad, India

Gayathri Gururajan
Department of Microbiology
Vels Institute of Science, Technology & Advanced Studies (VISTAS)
Chennai, India

Amina Hafeez
Department of Plant Breeding and Genetics
University of Agriculture
Faisalabad, Pakistan

Muhammad Bilal Hafeez
College of Agronomy
Northwest A&F University
and
Key Laboratory of Crop Physi-ecology and Tillage Science
Northwestern Loess Plateau, Ministry of Agriculture
Yangling, China

Muhammad Sheraz Hassan
Department of Plant Breeding and Genetics
Ghazi University
Dera Ghazi Khan, Pakistan

Azamal Husen
Department of Biotechnology
Smt. S. S. Patel Nootan Science & Commerce College
Sankalchand Patel University
Visnagar, India
and
Department of Biotechnology,
Graphic Era (Deemed to be University)
India
and
Wolaita Sodo University
Wolaita Sodo, Ethiopia

Abdullah A. Al-Kahtani
Department of Chemistry
College of Science
King Saud University
Riyadh, Saudi Arabia

Divya Kapoor
Department of Microbiology
CCS Haryana Agricultural University
Hisar, India

Murugan Karuvelan
Department of Microbiology
Vels Institute of Science, Technology & Advanced Studies (VISTAS)
Chennai, India

Abida Kausar
Department of Botany
Government College Women University
Faisalabad, Pakistan

Sana Khalid
Department of Environmental Sciences
COMSATS University
Islamabad, Pakistan

Noopur Khare
Department of Food Technology
Bhai Gurdas Institute of Engineering and Technology
Sangrur, India

Pragati Khare
Department of Pharmacy
Bhagwant University
Ajmer, India

Deepak Kumar
Plant Microbe Interaction Lab
Institute of Environment and Sustainable Development
Banaras Hindu University
Varanasi, India

Jun Li
College of Agronomy
Northwest A&F University
and
Key Laboratory of Crop Physi-ecology and Tillage Science
Northwestern Loess Plateau, Ministry of Agriculture
Yangling, China

Atul Loyal
Department of Genetics and Plant Breeding
CCS Haryana Agricultural University
Hisar, India

Arvind Malik
Department of Genetics and Plant Breeding
CCS Haryana Agricultural University
Hisar, India

Amit Kumar Mandal
Department of Sericulture
Raiganj University
Raiganj, India

Pankaj Mandal
Department of Sericulture
Raiganj University
Raiganj, India

Sayanti Mandal
Department of Chemistry and Biochemistry
School of Basic Sciences & Research
Sharda University
Greater Noida, India

Arpan Modi
Axenic Agritech Private Limited
Ahmedabad, India

Rittick Mondal
Department of Sericulture
Raiganj University
Raiganj, India

Martha Isabel Torres-Morán
Department of Agricultural Production
University Center for Biological and Agricultural Sciences (CUCBA)
University of Guadalajara
Zapopan, Mexico

Swayambhik Mukherjee
Department of Biotechnology
St. Xavier's College (Autonomous)
Kolkata, India

Arpan Mukherjee
Plant Microbe Interaction Lab
Institute of Environment and Sustainable Development
Banaras Hindu University
Varanasi, India

Behzad Murtaza
Department of Environmental Sciences
COMSATS University
Islamabad, Pakistan

Yalavarthi Nagaraju
Central Silk Research and Training Institute
Berhampore, India

Nabeel Khan Niazi
Institute of Soil and Environmental Sciences
University of Agriculture
Faisalabad, Pakistan

Deog Hwan Oh
Department of Food Science and Biotechnology
College of Agriculture and Life Sciences
Kangwon National University
Gangwon-do, South Korea

Mahmoud S. Osman
Department of Botany and Microbiology
Faculty of Science
Al-Azhar University
Cairo, Egypt

Priti Pal
Department of Environmental Engineering
Shri Ramswaroop Memorial College of Engineering & Management
Lucknow, India

Armando Rafael Hernández-Pérez
Department of Agricultural Production,
University Center for Biological and Agricultural Sciences (CUCBA)
University of Guadalajara
Zapopan, Mexico

Shradha Raj
Department of Microbiology
Vels Institute of Science, Technology & Advanced Studies (VISTAS)
Pallavaram, India

Ana Paulina Velasco-Ramírez
Department of Agricultural Production
University Center for Biological and Agricultural Sciences (CUCBA)
University of Guadalajara
Zapopan, Mexico

Sandra Fabiola Velasco-Ramírez
Department of Chemistry
University Center of Exact Sciences and Engineering (CUCEI)
University of Guadalajara
Zapopan, Mexico

Alejandro Velasco-Ramírez
Department of Agricultural Production
University Center for Biological and Agricultural Sciences (CUCBA)
University of Guadalajara
Zapopan, Mexico

P. Gangadhara Rao
College of Horticulture
Dr. Y.S.R. Horticultural University
Venkataramannagudem, India

Faisal Rasheed
Department of Botany
Aligarh Muslim University
Aligarh, India

Aryadeep Roychoudhury
Discipline of Life Sciences, School of Sciences
Indira Gandhi National Open University
New Delhi, India

Momna Rubab
Department of Food Engineering and Biotechnology
University of Engineering and Technology
Lahore, Pakistan

Arpit Sand
Department of Sciences
Manav Rachna University
Faridabad, Haryana, India

Manasa S. G.
Department of Agricultural Microbiology
COH
Bidar, Karnataka, India

Avinash S. H.
Department of Agricultural Microbiology
S. Nijalingappa Sugar Institute
Belagavi, India

Yousuf Shafiq
College of Agronomy
Northwest A&F University
Yangling, China

Muhammad Shahid
Department of Environmental Sciences
COMSATS University
Islamabad, Pakistan

Harshita Shand
Department of Sericulture
Raiganj University
Raiganj, India

Mayur Mukut Murlidhar Sharma
Department of Agriculture and Industry
Kangwon National University
Chuncheon, Korea

Muhammad Rizwan Shoukat
State College of Water Resources and Civil Engineering
China Agricultural University
Beijing, China

Akhilesh Kumar Singh
Departments of Biotechnology
School of Life Sciences, Mahatma Gandhi Central University
Motihari, India

Sachidanand Singh
Department of Biotechnology
School of Energy and Technology
Pandit Deendayal Energy University, Knowledge Corridor
Gandhinagar, India

Ghazala Sultan
Department of Computer Science
Faculty of Science
Aligarh Muslim University
Aligarh, India

Tasveer Zahra Tariq
Department of Botany
Bahauddin Zakariya University
Multan, Pakistan

Jaya Tuteja
School of Basic Sciences
Galgotias University
Greater Noida, India

Jay Prakash Verma
Plant Microbe Interaction Lab
Institute of Environment and Sustainable Development
Banaras Hindu University
Varanasi, India

Xiaoli Wang
College of Agronomy
Northwest A&F University
Yangling, China
and
Key Laboratory of Crop Physi-ecology and Tillage Science
Northwestern Loess Plateau, Ministry of Agriculture
Yangling, China

Mithilesh Yadav
Department of Chemistry
Veer Bahadur Singh Purvanchal University
Jaunpur, India

Ahmad Latif Virk
State Key Laboratory of Subtropical Silviculture
Zhejiang A&F University
Zhejiang, China

Sheetal Yadav
Department of Microbiology
CCS Haryana Agricultural University
Hisar, India

Noreen Zahra
Department of Botany
Government College Women University
Faisalabad, Pakistan
and
Department of Botany
University of Agriculture
Faisalabad, Pakistan

Preface

The ever-increasing global human population confronts unprecedented challenges in the agricultural sector today. Higher amounts of nutritional food are required to feed the overall world population. This is very challenging due to water shortages and the consequent reduced crop plant yield in many parts of the world. Thus, to overcome these and other unseen problems, the overall crop plant production systems need to be improved. Additionally, modern sustainable agriculture needs to strengthen, review, and broaden its practices and business models, by integrating opportunities from different allied sectors and disciplines of science and technology. Therefore, searching for new technologies and approaches to accelerate crop plant yield and production under optimal and suboptimal conditions, and improving resource-use efficiency are crucial to ensure food security.

Sustainable crop yield refers to agricultural production in a way that does not cause any adverse impact to the environment, biodiversity, and overall, the quality of agricultural crop plants. In general, higher production and yield of crop plants is based mainly on the greater use of irrigation, synthetic fertilizer, and synthetic pesticides. Although these practices increase crop production, quite often they lead to soil erosion, decreased soil fertility, and several other adverse effects on ground water and the lake, river, and coastal ecosystems. Furthermore, more sustainable cropping systems for higher crop yield and production need be developed to protect and improve the soil and water resources. It is believed that the next generation of cropping systems may ensure soil health and accelerate crop plant production and minimize the inputs of synthetic fertilizers, pesticides, and water in a sustainable manner.

In this context, there is a pressing demand for redesigning agriculture systems to achieve sustainability by utilizing modern and advanced tools. This necessitates understanding the role of nanomaterials; nanosensors; molecular, physiological, and biochemical aspects; genetic engineering and micropropagation; molecular breeding; significance of macro- and micronutrients; interaction of mineral nutrients and plant hormones; nutrient sensing and signalling; rhizophore or microbiome engineering; soil enzymes; biostimulants or PGPR; intercropping procedures; and some traditional agricultural practices. Taken together, this book addresses key issues of crop plant yield and/or production and practices with molecular and physiological interventions and evolve future strategies to overcome challenges faced by the agricultural sector. The topics covered in this book are written by eminent and rising scholars. This book is a rich source of information in one place. This is a useful resource for researchers and extension workers involved in sustainable crop plant improvement and practices and related disciplines.

I am grateful to all the contributors for readily accepting my invitation, sharing their knowledge in specialized areas of research and readily adjusting with suggestions for improving their contributions. With great pleasure, I also extend my sincere thanks to Randy Brehm, Tom Connelly, and all associates at Taylor & Francis Group, LLC/CRC Press for their active cooperation. Finally, my special thanks go to Shagufta, Zaara, Mehwish, and Huzaifa for providing their time and overall support to put everything together.

Azamal Husen

Preface

1 Perspectives of Nanomaterials for Crop Plants Improvement and Practices

Murugan Karuvelan, Shradha Raj, Gayathri Gururajan, Ramachandran Chelliah, Ghazala Sultan, Momna Rubab, and Deog Hwan Oh

1.1 INTRODUCTION

Nanotechnology, a field that melds the infinitesimal with the innovative, has emerged as a cornerstone of advancement across a broad spectrum of disciplines, from medicine and energy to materials science and plant pathology. Esteemed research by scholars such as (Bachheti, Bachheti and Husen, 2024; Husen, 2020, 2022, 2023a,b,c; Husen and Siddiqi, 2023; Madhusudhan *et al.*, 2024; Mazzaglia *et al.*, 2017), underscores its pervasive influence. The scope of nanotechnology spans the characterization, fabrication, and manipulation of materials, ushering in an era of groundbreaking developments in chemical, biological, biomedical, pharmaceutical, agricultural, electronic, and bioengineering applications. The concept of nanotechnology, defined by Norio Taniguchi as "the study of matter on the atomic or molecular scale," has laid the foundational stone for this scientific renaissance (Agrawal and Rathore, 2014). In the agricultural sector, the application of nanotechnology is revolutionizing practices by introducing nanoformulated agrochemicals that enhance the efficiency of pesticides and fertilizers, nanosensors for precise disease identification and residue detection, and nanodevices for plant genetic engineering. These innovations extend to diagnostic tools for plant diseases, advancements in animal health and breeding, poultry production, and improved postharvest management, paving the way for enhanced crop productivity and sustainability. As the demand for sustainable agricultural practices and the development of disease-resistant crop varieties intensifies, the limitations of conventional methodologies become apparent.

Nanoparticles have surfaced as a groundbreaking solution, propelling agricultural research into new territories with their potential for disease diagnosis, yield enhancement, and the breeding of disease-resistant varieties. Notably, the biological synthesis of nanoparticles presents a promising avenue for creating nanofertilizers, nanoinsecticides, and nanopesticides, simultaneously addressing environmental concerns by reducing toxic pollutants, a testament to the environmentally friendly and biodegradable nature of these innovations (Shang *et al.*, 2019). The intricate dance of nanotechnology within agriculture outlines not just a future of potential but a present of progress, as evidenced by the detailed application of nanoparticles in crop protection detailed in Table 1.1. This introduction not only sets the stage for a deep dive into the revolutionary impacts of nanotechnology on agriculture but also integrates critical references to establish a comprehensive backdrop for the ensuing discussion.

In the ever-evolving landscape of nanotechnology, nanomaterials emerge as a frontier of innovation, with their sizes ranging from 1 to100 nanometers. These materials are distinguished by their exceptional surface-to-volume ratio, unlocking a realm of distinctive physicochemical properties. Their unique attributes, including vibrant colors, remarkable strength, enhanced diffusion, and specific toxicity profiles (Gupta *et al.*, 2016), along with their high catalytic reactivity, solubility, and biochemical activities (Sudha *et al.*, 2018), have catapulted nanomaterials into the spotlight

DOI: 10.1201/9781032630366-1

TABLE 1.1
Overview of Nanoparticles' Role in Enhancing Crop Production, Applications, Mechanisms, and Environmental Considerations

S No	Nanoparticles as Nanofertilizers	Nanoparticles as Nanobiosensors	Nanoparticles as Transfer Vehicle	Nanoparticles as Nanoinsecticides and Nanopesticides
1	Enhanced growth	Early detection of pathogens	Drug delivery	Disease suppression
2	Promote seed germination	Real-time monitoring of cultivated fields	Enabled genome editing	Eradicates insects and pests
3	Stimulate nutrients supply	Indicates and eliminates environmental pollutants	Stress gene expression	Increases disease tolerance
4	Increased productivity		Acclimatization Pathogen related gene expression	

across various commercial sectors. From scratch-resistant paints to sophisticated electronics, eco-friendly cosmetics, advanced environmental cleanup solutions, cutting-edge sports equipment, sensitive sensors, and revolutionary energy storage devices (Sharifi *et al.*, 2012), nanomaterials have marked their presence. In the agricultural sector, nanotechnology's footprint expands with the advent of nanofertilizers, nanoherbicides, and nanopesticides, which promise targeted action and precision delivery of active ingredients. Additionally, nanosensors stand at the vanguard of agricultural innovation, enabling the detection of environmental shifts and precise nutrient management, while also acting as catalysts in soil and groundwater remediation efforts. Despite the essential role of fertilizers in boosting crop yields, their usage poses a threat to the delicate soil mineral balance, potentially diminishing soil fertility over time (Solanki *et al.*, 2015). The conventional manufacturing of pesticides and fertilizers bears a hefty price tag, urging the need for a more efficient, cost-effective solution. Herein, nanomaterials, particularly metals and metallic oxide nanoparticles, offer a promising avenue in agriculture to curtail nutrient loss, amplify yields, minimize reliance on traditional plant protection products, and slash production costs, thereby optimizing output (Usman *et al.*, 2020).

This cutting-edge approach to crop protection leverages nanomaterials through genetic modification via protein or genetic material transfer (DNA or RNA), and sophisticated disease detection facilitated by nanosensors (Khandelwal and Joshi, 2018). Such nanosensors represent a paradigm shift in agriculture, enabling comprehensive monitoring and quantification of vital factors like crop production, disease management, environmental pollution, soil quality, and the intricate interactions between pathogens and crops. Nanomaterials' integration into crop improvement and sustainable agriculture practices spans from naturally occurring to intentionally added components, synthesized through innovative physical, chemical, or biologically driven (green synthesis) methods (Iravani, 2014; Jamkhande *et al.*, 2019; Mughal and Hassan, 2022). These synthesis methods are broadly categorized into bottom-up approaches, where small nanostructures are assembled into larger nanomaterials, and top-down approaches, crafting nanomaterials from larger macromolecules (Khandelwal and Joshi, 2018; Singh *et al.*, 2015). The green synthesis of nanoparticles, employing plants and microorganisms, is particularly celebrated for its cost-effectiveness, simplicity, and environmental friendliness. This comprehensive exploration is dedicated to unraveling the multifaceted applications of nanomaterials in agriculture, spotlighting their role in crop enhancement, disease management, and the exciting future prospects of nanotechnology in elevating agricultural practices to new heights.

1.2 TYPES OF NANOMATERIALS FOR CROP PROTECTION

In the dynamic field of plant pathology, an expansive selection of nanomaterials is pioneering the fight against a broad spectrum of plant diseases. This innovative array, which spans metals, metal oxides, non-metals, and various other categories, marks a significant leap forward in our approach to agricultural health. Among these materials, the use of silver, zinc, copper, and carbon-based nanomaterials stands out for their potent antimicrobial properties. Additionally, biocompatible substances like chitosan, versatile liposomes, highly precise quantum dots, and structurally complex dendrimers are being harnessed to offer new dimensions in disease management strategies.

Esteemed researchers in the field, including Elmer and White (2018) and Bayoumy *et al.* (2020), have contributed valuable insights into the effectiveness of these nanomaterials. Their work sheds light on how these nanotechnological advances can be strategically deployed to safeguard crops from the ravages of pathogens. The applications of these nanomaterials in crop disease management, as we will discuss, represent a confluence of cutting-edge science and practical agricultural practices. By integrating these nanotechnologies into our plant pathology toolkit, we stand on the brink of revolutionizing how we protect and enhance crop yield, ensuring food security in an increasingly unpredictable world.

1.2.1 Silver (Ag) Nanomaterials in Crop Protection

Silver (Ag) nanomaterials stand at the forefront of agricultural innovation, particularly in the realm of crop disease management. Recognized as one of the earliest nanomaterials to be applied in agriculture, their utilization marks a significant advancement in our ongoing battle against plant pathogens. The antimicrobial prowess of Ag nanomaterials has been extensively documented, with research (Keshvadi *et al.*, 2019; Singh *et al.*, 2015) highlighting their role as a well-studied, well-characterized cornerstone in nanotechnology-based plant protection strategies.

The application of Ag nanomaterials transcends basic antimicrobial activity; they are instrumental in bolstering plant defense mechanisms. This includes the promotion of lignin deposition within vascular bundles, a critical factor in enhancing plant resilience against pathogens (Moussa *et al.*, 2013). Such mechanisms are vital for the structural integrity and defense capabilities of plants, providing a robust barrier against invasive pathogens.

Further emphasizing their potential, a groundbreaking study by Jagana, Hegde and Lella (2017) revealed that green-synthesized nano-Ag could dramatically reduce disease progression. In their research, nano-Ag was shown to suppress disease progression by nearly 70% in banana fruit when challenged with the pathogen *Colletotrichum musae,* a common antagonist in the agricultural sector. This evidence underscores the transformative impact of Ag nanomaterials in crop protection, offering a promising tool for enhancing agricultural productivity and sustainability through advanced nanotechnological interventions.

1.2.2 Zinc-Nanomaterials (Zn)

Zinc nanomaterials (Zn) stand out in the realm of plant pathology as a versatile and powerful ally against a wide range of plant diseases. Renowned for their antimicrobial properties, Zn nanomaterials have been the subject of numerous studies that underscore their efficacy in combating formidable plant pathogens. Research led by He et al. (2011) and Wani and Shah (2012), along with Sardella, Gatt and Valdramidis (2017), has illuminated the broad spectrum of Zn nanomaterials' antimicrobial activity. These investigations have demonstrated that Zn nanomaterials are exceptionally potent against a variety of pathogenic fungi that pose serious threats to crops. Notable among these are *Mucor plumbeus*, *Alternaria alternata*, *Fusarium oxysporum*, *Sclerotinia sclerotiorum*, *Rhizoctonia solani*, and *Rhizopus stolonifera*, all of which can cause devastating losses in agricultural production.

Beyond their antifungal capabilities, Zn nanomaterials have also shown promise in tackling bacterial pathogens, showcasing their multifaceted utility in plant disease management. This ability to target both fungal and bacterial diseases makes Zn nanomaterials an invaluable resource in the ongoing effort to safeguard crops from a myriad of threats. Their application in agriculture not only enhances crop resilience but also supports sustainable farming practices by reducing the reliance on traditional chemical pesticides. As the agricultural sector continues to face challenges from evolving pathogens, the role of Zn nanomaterials in disease management and crop protection becomes increasingly crucial, offering hope for more robust and healthy crop yields.

1.2.3 Copper Nanomaterials: Enhancing Plant Disease Resistance with Antimicrobial Properties

Copper nanomaterials stand out in the realm of agricultural nanotechnology for their potent antimicrobial capabilities, making them an invaluable asset in the diagnosis and management of plant diseases. These nanoscale copper-based agents serve as a cornerstone in developing advanced pesticides and fertilizers, tailored to bolster disease resistance across a diverse range of crops. The effectiveness of copper nanomaterials is not merely theoretical; it is grounded in empirical evidence showcasing their capacity to tackle formidable plant pathogens.

A notable study by Ponmurugan *et al.* (2016) demonstrated the remarkable potential of biologically synthesized copper nanomaterials, utilizing the bacterium *Streptomyces griseus* in combating root rot disease in tea plants, a disease instigated by the formidable pathogen *Poria hypolateritia.* This groundbreaking research highlights the role of nanoscale copper in disrupting the lifecycle of pathogens, thereby safeguarding plants from devastating diseases.

Furthermore, research has illuminated how applications of nano-copper oxide (CuO) precipitate a surge in the synthesis of defense-related enzymes within plants, such as polyphenol oxidase. This enzyme plays a pivotal role in the plant's innate defense mechanism, enhancing its ability to fend off attacks from various pathogens. Studies (Sarkar *et al.*, 2020; Wang, Liu and Han, 2013) have contributed to our understanding of the molecular pathways through which copper nanomaterials invigorate plant defenses, marking a significant advance in our approach to sustainable crop protection. In essence, copper nanomaterials encapsulate a promising frontier in plant pathology, offering a synergistic blend of disease control and enhancement of plant defense mechanisms. Through targeted applications, these nanomaterials are setting new standards in agricultural practices, leading the charge toward healthier, more resilient crops.

1.2.4 Harnessing Carbon Nanomaterials for Plant Health and Disease Management

Carbon nanomaterials, renowned for their exceptional physical properties and versatile functions across diverse research arenas, are emerging as a novel contender in the domain of plant disease management. Despite the relative paucity of detailed studies elucidating their role in this specific field, the available evidence points to a promising potential that carbon-based nanotechnologies hold for agriculture. Among the carbon nanomaterials, three main types have been identified as key players in combating plant diseases: fullerenes, graphene oxides, and carbon nanotubes. Graphene oxide, in particular, has demonstrated remarkable capabilities in enhancing plant resistance against bacterial invasions. A significant study revealed that graphene oxide could induce an impressive 95% disease resistance in plants against *Xanthomonas oryzae*, a notorious bacterial pathogen. This finding underscores the potent antimicrobial properties of graphene oxide, positioning it as a powerful tool in the arsenal against crop diseases.

Further research into carbon nanomaterials has unveiled their broader implications for plant health. Notably, these studies suggest that beyond their direct antimicrobial effects, carbon nanomaterials may play a crucial role in promoting plant growth and facilitating the eradication of pathogens. For instance, Chen, Wang and Han (2013) have contributed valuable insights into how carbon nanotechnologies can significantly bolster disease resistance in plants, offering a dual benefit of enhanced growth and pathogen resistance.

In light of these findings, carbon nanomaterials are poised to redefine the landscape of plant disease management. By leveraging the unique properties of fullerenes, graphene oxides, and carbon nanotubes, researchers and agriculturists alike can tap into a new paradigm of crop protection that not only wards off diseases but also supports robust plant growth. As we delve deeper into the capabilities of these nanomaterials, the future of sustainable agriculture looks increasingly promising, heralding an era where carbon nanotechnology plays a pivotal role in fostering healthier, more resilient crops.

1.2.5 Non-Metals, Metalloids, and Metallic Oxides in Nanoparticles

The exploration of nanoparticles composed of non-metals, metalloids, and metallic oxides presents an intriguing frontier in the fight against plant diseases. While comprehensive data on their efficacy remains somewhat limited, the burgeoning body of research hints at their significant promise in bolstering disease resistance within crops. Among the array of non-metal nanoparticles, nanosulfur and nanotitanium emerge as the subjects of particular scholarly focus. Nanosulfur has gained prominence for its integration into advanced fungicidal formulations, a development underscored by the work of Rao and Paria (2013), who highlighted its efficacy. On the other hand, nano-titanium is lauded for its autocatalytic properties that render it a potent antifungal agent, a finding supported by Boxi, Mukherjee and Paria (2016).

Diving deeper into the realm of metalloids, silicon-based nanomaterials (Si) have been identified as having multifaceted impacts on disease management strategies. A notable application includes the use of nano-Si particles combined with Ag to effectively curb the spread of powdery mildew, as documented by Park *et al.* (2006). Furthermore, manganese oxide (MnO), a metallic oxide nanoparticle, has been proven to control wilt disease in both eggplants and watermelons, an advancement chronicled by Elmer and White (2016). Additional metallic oxide nanoparticles, such as CuO, titanium dioxide (TiO_2), iron oxide (Fe_2O_3), and magnesium oxide (MgO), have also carved out roles in plant pathology, each bringing unique properties and mechanisms to the forefront of disease management. Thus, the diverse and specialized functionalities of these nanomaterials underscore their pivotal contributions to the science of plant pathology. By offering innovative, nuanced solutions, they represent a significant leap forward in our ongoing quest to safeguard and enhance crop health through strategic management and control of plant diseases.

1.3 INTERACTION OF NANOMATERIALS WITH PLANTS

Delving into the intricate dynamics of how nanomaterials interact with plant life is a fascinating and critical aspect of advancing agricultural technologies. This understanding is paramount for leveraging the potential benefits these tiny particles can offer in enhancing plant growth, disease resistance, and stress tolerance. At the heart of these interactions lies the journey of nanomaterials from their entry points into the internal reaches of the plant system. Nanoparticles typically find their way into plants through various natural gateways, including stomata (pores on the leaf surface), hydathodes (pore-like structures at leaf margins), stigmas (the pollen-receiving part of flowers), and wounds. This entry mechanism is pivotal for the transport of nanomaterials within plant structures, a process meticulously outlined by Sanzari, Leone and Ambrosone (2019).

To comprehend the full spectrum of these interactions, it's crucial to consider both the physiological and anatomical frameworks of plants. The outermost layer of a plant leaf, the epidermis, features a cuticle membrane as a protective layer that is particularly thick on the upper surfaces of vegetative organs. This barrier plays a significant role in mediating the interaction between nanomaterials and the plant, often serving as an obstacle to the penetration of nanoparticles. Within the epidermis, the presence of stomata, each flanked by two guard cells, provides a critical pathway for the entry of nanomaterials. These particles can settle on the cell wall of the substomatal cavity, marking the beginning of their journey through the plant's internal structure. Once inside, nanomaterials navigate the plant's complex system via two main routes: the symplastic pathway and the apoplastic pathway. The symplastic pathway facilitates movement through the cytoplasm of adjacent cells, a process detailed by Roberts and Oparka (2003), while the apoplastic pathway permits travel through the extracellular spaces, including cell walls and xylem vessels, as described by Sattelmacher (2001). These pathways collectively enable the distribution of nanomaterials throughout the plant. For nanomaterials to access the symplastic pathway, they must first internalize within the plant cell, crossing the plasma membrane barrier. This step involves several mechanisms that facilitate the entry and integration of nanomaterials into plant cells, a topic extensively explored by researchers such as (Rico *et al.*, 2011; Schwab *et al.*, 2015). Understanding these mechanisms is essential for developing strategies that enhance the beneficial impacts of nanomaterials on plant health and productivity. In summary, the interaction of nanomaterials with plants is a multifaceted process that involves entry through natural openings, navigation through complex pathways within the plant, and the overcoming of physiological barriers. This intricate dance between nanomaterials and plant structures holds the key to unlocking new frontiers in agricultural science and technology, offering promising pathways for the development of more resilient and productive crops in (Figure 1.1).

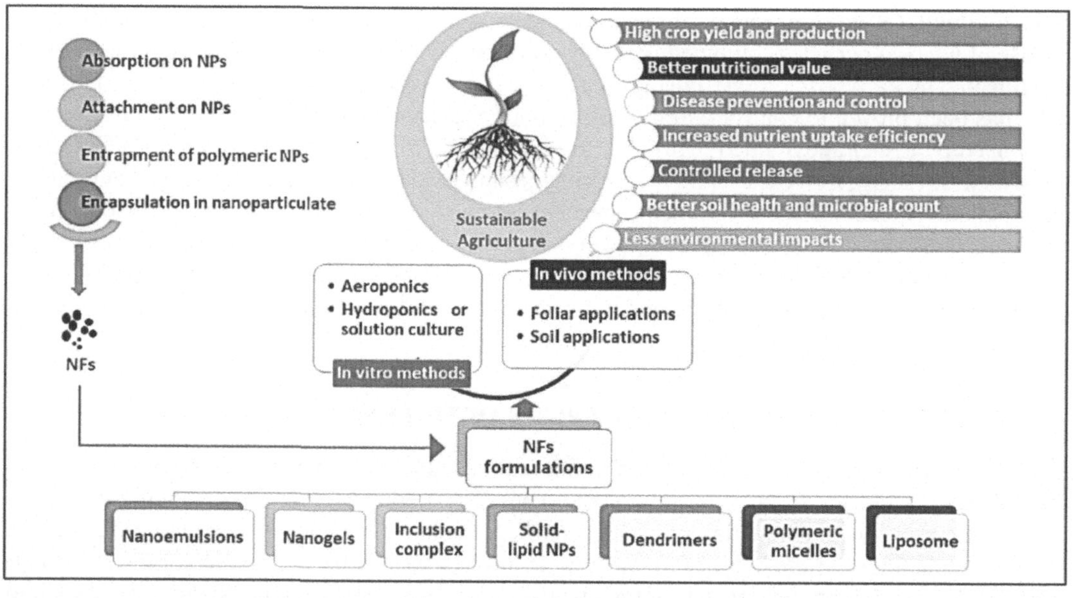

FIGURE 1.1 This figure depicts a detailed illustration of the fundamental mechanisms through which nanofertilizers enhance nutrient uptake in plants. This figure meticulously outlines the process starting from the application of nanofertilizers to their interaction with plant root systems. It highlights how these tiny, engineered particles, smaller than conventional fertilizers, penetrate more effectively into plant tissues. The diagram shows the nanoscale interactions between these fertilizers and plant cells, demonstrating how their size and surface properties allow for more efficient absorption and utilization of nutrients by the plants. Additionally, the figure elucidates the role of nanofertilizers in promoting root growth and increasing the surface area for nutrient absorption. It also discusses the slow-release properties of nanofertilizers, which ensure a steady supply of nutrients, minimizing waste and environmental impact. This visual representation serves as an educational tool to understand the advanced technology behind nanofertilizers and their potential to revolutionize agricultural practices by boosting plant health and crop yield.

1.3.1 Endocytosis

Endocytosis stands as a pivotal cellular process, marking the gateway through which nanomaterials enter the interior realms of plant cells. This sophisticated mechanism involves the delicate invagination of the plasma membrane, a process meticulously orchestrated to encapsulate nanomaterials within a newly formed vesicle. Once encapsulated, these vesicles embark on a journey through the cellular landscape, destined for various compartments within the cell where they can exert their influence. The phenomenon of endocytosis in plant cells, detailed by Etxeberria *et al.* (2006), highlights the nuanced interplay between nanomaterials and the cellular machinery. The plasma membrane, acting as both barrier and gateway, carefully folds inward, engulfing the nanomaterials in a process akin to the cell swallowing these tiny particles. This action results in the formation of a vesicle, a small, membrane-bound compartment that serves as a vessel for the safe transport of the nanomaterials into the cell's interior. Once inside, the journey of these vesicles is far from over. They navigate the complex cellular environment, potentially merging with other membrane-bound compartments such as endosomes, lysosomes, or the Golgi apparatus. These interactions are crucial, as they determine the fate of the nanomaterials, whether they will be processed, stored, or utilized within the cell. Understanding endocytosis and its role in the internalization of nanomaterials into plant cells opens new avenues for research and application in agriculture and biotechnology. By manipulating this process, scientists can enhance the delivery of beneficial nanoparticles to plant cells, improving disease resistance, nutrient uptake, and overall plant health. This intricate process underscores the incredible potential of nanotechnology in transforming our approach to crop management and agricultural sustainability.

1.3.2 Pore Formation

Pore formation is a remarkable yet disruptive mechanism through which some nanomaterials gain entry into plant cells, circumventing traditional cellular entry pathways. This process involves the direct interaction of nanoparticles with the plasma membrane, leading to the creation of pores or openings through which these materials can infiltrate the cell. Unlike endocytosis, where nanomaterials are encapsulated within vesicles, pore formation allows nanomaterials to pass directly into the cytosol, the liquid matrix within cells, without being enclosed in any cellular organelle.

The research conducted by Wong *et al.* (2016) sheds light on how certain nanomaterials can perturb the structural integrity of the plasma membrane, inducing the formation of these pores. The mechanism by which nanoparticles disrupt the membrane is complex and can vary depending on the characteristics of the nanoparticles, including their size, shape, and chemical composition. Once these pores are formed, they provide a direct gateway for nanomaterials to access the cytosol, offering a path that bypasses the cell's endocytic machinery. Further elucidating this mechanism, Serag *et al.* (2011) highlighted the significance of pore formation as a means for nanoparticles to directly reach the cytosol without being sequestered within organelles. This direct access to the cytosol can have profound implications for the delivery of substances designed to modulate cellular functions, including gene expression, enzymatic activities, and intracellular signaling pathways.

The phenomenon of pore formation by nanomaterials represents a double-edged sword. On one hand, it opens up innovative avenues for the targeted delivery of therapeutic agents, nutrients, or genetic material directly into the cell's interior, potentially enhancing plant health and productivity. On the other hand, the unregulated formation of pores can pose risks to cellular integrity and function, necessitating careful consideration of the potential impacts of nanoparticle exposure on plant cells. The insight on the dynamics of pore formation and its implications for cellular entry of nanomaterials is crucial for the development of safe and effective nanoparticle-based applications in agriculture and beyond. This knowledge enables the design of nanomaterials that harness the benefits of direct cytosol access while minimizing potential adverse effects, paving the way for advanced strategies in plant science and biotechnology.

1.3.3 Carrier Proteins

Carrier proteins play a crucial role in the nuanced ballet of interactions between nanomaterials and plant cells, offering a sophisticated mode of entry for nanoparticles seeking passage into the cellular interior. This mechanism, involving the binding of nanomaterials to specific proteins on the cell membrane, underscores the complexity and specificity of cellular uptake processes. Nel *et al.* (2009) shed light on this fascinating process, revealing how nanoparticles can harness the natural function of carrier proteins to facilitate their internalization. Carrier proteins, integral components of the cell membrane, are typically involved in the transport of molecules across the cell's boundary, selectively allowing substances to enter or exit the cell based on cellular needs and environmental cues. When nanomaterials bind to these proteins, they effectively "hitch a ride" into the cell, utilizing the proteins' transport capabilities to cross the membrane barrier. This binding can be influenced by various factors, including the size, shape, and surface chemistry of the nanomaterials, as well as the specificity and affinity of the carrier proteins for their cargo.

The interaction between nanomaterials and carrier proteins represents a targeted approach to cellular entry, potentially allowing for the precise delivery of nanoparticles designed to elicit specific responses within the cell. For example, nanoparticles could be engineered to bind to carrier proteins associated with nutrient uptake, thereby facilitating the delivery of nutrient-loaded nanoparticles to enhance plant growth or stress tolerance. Moreover, this mode of entry opens up possibilities for the development of nanoparticle-based therapeutics and bioactive compounds in plant science, enabling the targeted delivery of substances to modulate gene expression, enzyme activity, or signaling pathways within plant cells. Understanding the role of carrier proteins in the internalization of nanomaterials is essential for advancing the design and application of nanotechnology in agriculture. By elucidating the mechanisms underpinning nanoparticle-protein interactions, researchers can develop more effective and efficient strategies for enhancing plant health and productivity through nanotechnology, paving the way for innovative solutions to global agricultural challenges.

1.3.4 Role of Plasmodesmata

Plasmodesmata represent a fascinating and distinctive pathway through which nanomaterials can navigate into plant cells, showcasing the ingenuity of cellular architecture in facilitating intercellular communication and transport. These specialized structures, essentially microscopic channels that traverse the cell walls, serve as vital conduits for the movement of substances between plant cells, highlighting a marvel of biological engineering. One study (Roberts and Oparka, 2003) brought to light the critical role of plasmodesmata as alternative gateways for nanomaterials, providing a route that bypasses the cell's external membrane. This mode of entry leverages the plant's inherent systems for transport and communication, enabling nanoparticles to move seamlessly from one cell to another within the plant's vast cellular network. The significance of plasmodesmata extends beyond mere transport; as shown in Zhai *et al.* (2014), these structures are integral to the process of translocation through the phloem, the plant's nutrient distribution system. This insight underscores the importance of plasmodesmata in the overall health and functionality of plants, facilitating not just the movement of nanomaterials, but also the essential distribution of water, sugars, and other nutrients critical to plant growth and development.

In the context of nanotechnology's applications in agriculture and plant sciences, understanding the role of plasmodesmata opens new horizons for targeted delivery systems and intercellular communication strategies. By harnessing these natural pathways, scientists can develop innovative approaches to enhance plant resilience, optimize nutrient distribution, and improve crop yields, all while minimizing external environmental impacts. The exploration of plasmodesmata as conduits for nanomaterial entry into plant cells not only enriches our understanding of plant biology but also paves the way for advanced agricultural practices. This knowledge empowers researchers to design nanomaterials that can effectively navigate these cellular highways, offering promising solutions to some of the most pressing challenges in plant science and agriculture today.

1.3.5 Ion Channels

Ion channels, intricately woven into the fabric of cellular membranes, have been identified as potential gateways for nanomaterials seeking entry into cells. This notion, put forward by researchers (Rico *et al.*, 2011; Schwab *et al.*, 2015), posits that these channels could serve as conduits for the transport of nanoparticles into the internal cellular environment. Given the structural precision of ion channels, which typically feature openings around 1 nm in diameter, they present a fascinating yet challenging pathway for the ingress of nanomaterials. The dimensional constraints of ion channels naturally limit the passage of larger nanoparticles, suggesting that only those of a suitable size or those that have undergone specific modifications can traverse these cellular gateways. This size limitation underscores the complexity of cellular entry mechanisms and highlights the need for innovative approaches to tailor nanomaterials for successful internalization.

Beyond the cellular membrane, the journey of nanomaterials within plant systems encompasses the apoplastic translocation process, as detailed by Sanzari *et al.* (2019). This mechanism involves the movement of nanoparticles through the extracellular spaces and cell walls, playing a pivotal role in the radial transport of these materials into the root vascular tissue. From there, nanomaterials are able to ascend toward the aerial parts of the plant, an upward journey that is essential for the systemic distribution of nanoparticles within the plant architecture. The capacity of nanomaterials to effectively navigate through these diverse entry points and translocation pathways underscores their vast potential in revolutionizing agricultural practices. By facilitating targeted delivery and distribution within plants, nanomaterials open up new avenues for enhancing crop physiology, improving nutrient uptake efficiency, and fortifying plants against diseases and pests. As research in this domain continues to unfold, the strategic manipulation of nanomaterials to optimize their compatibility with ion channels and apoplastic translocation processes will be key in harnessing their full potential. The promise held by these advanced materials in agricultural sciences not only points toward improved crop health and yield but also marks a significant leap forward in sustainable farming practices, offering a glimpse into the future of precision agriculture.

1.4 ROLE OF NANOMATERIALS IN SEED GERMINATION, CROP GROWTH, AND QUALITY ENRICHMENT

Nanoscience stands at the forefront of agricultural innovation, offering a suite of nanotechnological solutions designed to enhance the fundamental processes of plant life, from the critical initial phase of seed germination through to growth, development, and environmental adaptation. This emerging field has the potential to revolutionize traditional farming practices by introducing cost-effective, efficient methods to boost agricultural productivity and sustainability. Seed germination plays a pivotal role in the life cycle of plants, setting the stage for successful seedling development, survival, and ultimately, the dynamics of plant populations. It's a process intricately influenced by a complex interplay of factors, such as environmental conditions, genetic predispositions, moisture levels, and soil fertility. The work of Manjaiah *et al.* (2019) underscores the delicate balance these factors maintain in determining the fate of a germinating seed.

Emerging research points to the significant impact nanomaterials have on enhancing seed germination and fostering plant growth and development. Yet, the precise mechanisms through which nanomaterials exert their beneficial effects on germination remain an area ripe for exploration. Early studies, including those by Lu *et al.* (2002) and later by Banerjee and Kole (2016), provide insights into the potential interactions between nanomaterials and plant seeds. These interactions suggest that nanomaterials may facilitate the penetration of the seed coat, thereby improving water absorption and utilization. This enhanced hydration could be the key to activating the seed's enzymatic systems, leading to improved germination rates and more robust seedling growth. However, the detailed mechanism behind nanomaterial-induced water uptake within seeds remains elusive, sparking curiosity and further investigation within the scientific community. The hypothesis is that nanomaterials might modify the physical properties of the seed coat or interact with biochemical pathways in a way that enhances the efficiency of water transport and usage within the seed. Understanding this mechanism could unlock new avenues for optimizing seed germination and plant development, particularly in challenging environmental conditions or poor soil fertility scenarios. Nanoscience's promise in agriculture extends beyond germination; it offers a holistic approach to enhancing plant physiology and crop management practices. As research continues to unravel the mysteries of nanomaterial interactions with plant systems, the potential for these tiny particles to contribute to sustainable agriculture and food security becomes increasingly apparent. The journey from seed to plant, under the influence of nanotechnology, represents a fascinating intersection of biology, chemistry, and engineering, opening up a world of possibilities for the future of farming.

The field of nanotechnology is rapidly emerging as a transformative force in agriculture, offering novel solutions that promise to enhance the growth and quality of a wide array of crops. Among the diverse range of nanomaterials, zinc oxide (ZnO), titanium dioxide (TiO_2), multi-walled carbon tanotubes (MWCNTs), iron oxide (FeO), and hydroxyfullerenes stand out for their remarkable ability to improve the cultivation of crops such as peanuts, soybeans, mung beans, wheat, onions, spinach, tomatoes, potatoes, and mustard. This enhancement is not just limited to yield but also extends to the overall quality of the produce, an advancement supported by the research of (Dubey and Mailapalli, 2016; Shalaby *et al.*, 2016; Shojaei *et al.*, 2019). MWCNTs, in particular, have garnered attention for their role in stimulating the growth of significant agricultural staples, including barley, soybeans, and corn. The unique properties of MWCNTs allow them to penetrate the tough outer layers of seeds and aggregate within the endosperm, the nutrient-rich tissue that nourishes the developing plant. This interaction between MWCNTs and the seed endosperm has been meticulously documented through advanced imaging techniques such as Raman spectroscopy and transmission electron microscopy (TEM), as highlighted in the work of Lahiani *et al.* (2013) (Figure 1.2).

The ability of MWCNTs and other nanomaterials to promote growth and enhance crop quality is attributed to a variety of mechanisms. These include improved water and nutrient uptake, enhanced photosynthetic activity, and increased resistance to pathogens and environmental stresses. By directly interacting with the biological processes of plants, nanomaterials can effectively address some of the most pressing challenges faced by modern agriculture, including the need for higher productivity and sustainability in the face of a changing climate and growing global population. Moreover, the application of these nanomaterials in agriculture opens up new pathways for the targeted delivery of nutrients and agrochemicals, reducing the need for excessive fertilizers and pesticides, and thereby mitigating environmental impacts. This approach not only benefits the agricultural sector by improving crop yields and quality but also contributes to the preservation of natural resources and the promotion of eco-friendly farming practices. As research in this field continues to advance, the integration of nanotechnology into agricultural practices promises to usher in a new era of farming. By harnessing the potential of nanomaterials, scientists and farmers alike are exploring innovative strategies to enhance crop productivity, ensure food security, and address the global challenges of sustainable agriculture.

Silica stands out as an essential component of plant nutrition, playing a critical role in fortifying plants against a plethora of biotic and abiotic stresses. The absence of adequate silica levels can leave plants significantly more vulnerable to these stressors, underscoring its importance in maintaining plant health and resilience. The pioneering work of Rafi, Epstein and Falk (1997)

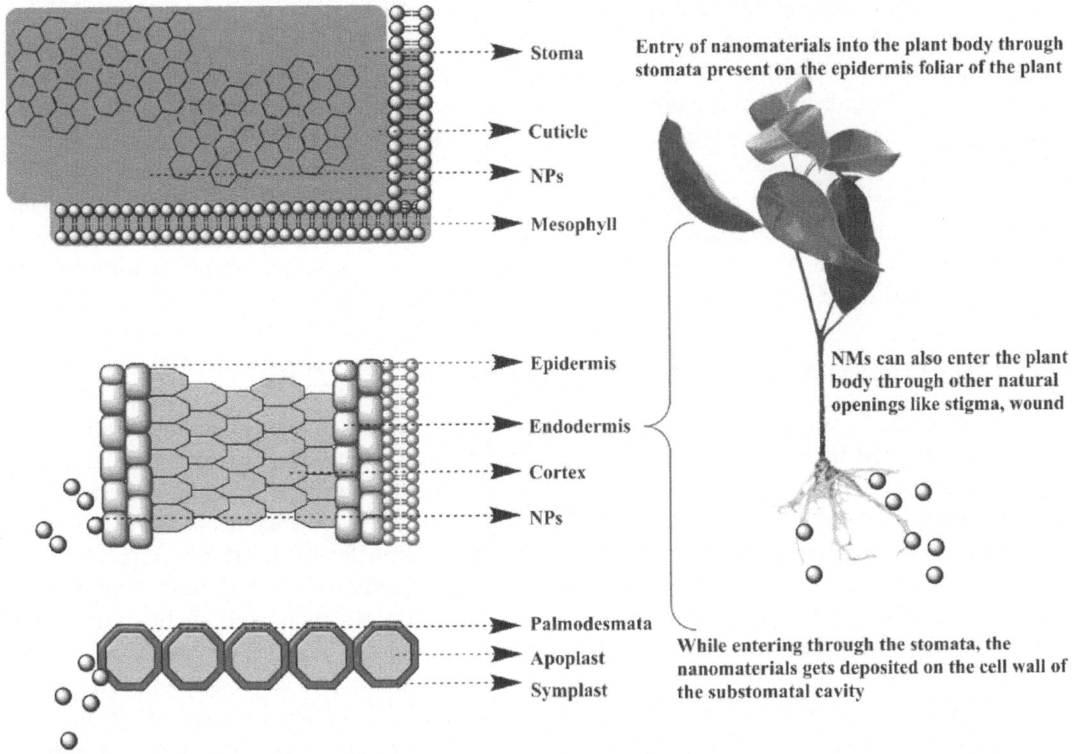

FIGURE 1.2 This figure provides a comprehensive visual representation of the multifaceted role that nanomaterials play in agriculture, specifically focusing on three critical areas: seed germination, crop growth, and the enrichment of crop quality. The figure begins by illustrating how nanomaterials can enhance seed germination, detailing the processes by which these materials improve seed vitality, water uptake, and resistance to environmental stressors. Moving forward, it showcases the impact of nanomaterials on crop growth, highlighting their role in promoting healthier and more robust plant development, efficient nutrient utilization, and improved photosynthetic activity. The latter part of the figure is dedicated to demonstrating how nanomaterials contribute to the quality enrichment of crops. It explains how these materials can increase the nutritional value of the produce; improve taste, texture, and shelf-life; and assist in reducing the accumulation of harmful substances in crops. This figure serves as an informative guide, shedding light on the innovative ways nanotechnology is being employed to revolutionize agricultural practices, leading to more sustainable and productive farming methods.

highlighted the detrimental effects of silica deficiency on plant robustness, illustrating the vital role silica plays in the overall well-being of plants. Further expanding on the benefits of silica, research conducted by Ahmad *et al.* (2007) demonstrated the material's efficacy in alleviating drought stress and bolstering biomass in plants under such conditions. This finding showcases silica's ability to enhance plant resilience against environmental challenges, thereby contributing to improved plant health and productivity.

TiO_2 has been identified as another nanomaterial with beneficial effects on plant growth. Notably, TiO_2 has been found to stimulate seed germination and increase seedling vigor, acting as a catalyst for enhanced plant development from the earliest stages of growth. This effect points to the potential of nanomaterials like TiO_2 in supporting plant growth through critical developmental phases. However, the interaction between nanomaterials and plant growth is complex and can yield varying outcomes based on several factors, including the concentration and size of the nanomaterials applied. The nuanced nature of these interactions underscores the need for careful consideration and optimization of nanomaterial application in agricultural settings. Moreover, recent studies, including those by Boykov, Shuford and Zhang (2019), have shed light on the molecular dynamics influenced by nanomaterials, revealing their ability to alter miRNA levels within plants. This adjustment in miRNA levels can trigger growth-promoting pathways, offering a molecular insight into how nanomaterials can enhance plant growth and development. These scientific advancements reveal the immense potential of nanomaterials, such as silica and TiO_2, in revolutionizing agricultural practices. By mitigating stress effects, enhancing seed germination and vigor, and activating growth-promoting pathways, nanomaterials hold the promise of significantly improving crop yields and product quality. This burgeoning field of nanotechnology in agriculture represents a pivotal shift toward more sustainable and productive farming practices, marking a significant milestone in the ongoing quest to meet global food demands while preserving environmental health.

1.5 MECHANISM AND ADVANTAGE OF NANOFERTILIZERS

The fusion of nanoscience and nanotechnology with agricultural practices is heralding a new era in farming, marked by enhanced efficiency and sustainability. This innovative approach allows for the optimized use of fertilizers and other agricultural inputs, significantly improving crop production and management while concurrently fostering soil fertility enhancements. As Mittal et al. (2020) articulate, this integration represents a monumental leap forward, offering a path toward more precise and eco-friendly agricultural practices. Despite the widespread use of traditional fertilizers, concerns have been raised regarding their long-term effects on soil quality and human health. Dimkpa and Bindraban (2017) have highlighted the potential negative consequences of conventional chemical fertilizers, including degradation of soil quality and adverse health impacts. In contrast, nanofertilizers emerge as a groundbreaking alternative, boasting a myriad of benefits over their traditional counterparts. These include novel mechanisms of action, heightened efficiency, diminished nutrient loss, and a reduced footprint on the environment. The production of nanofertilizers typically involves the encapsulation of nutrients within nanomaterials, a technique detailed by Yifen Shang *et al.* (2019). The nanoscopic scale of these fertilizers plays a critical role in their effectiveness, as their minuscule size and increased surface area significantly enhance plant absorption efficiency. Moreover, their ability to directly enter plant cells circumvents the energy-demanding processes usually required for nutrient uptake, as discussed (Brackhage *et al.*, 2013; DeRosa *et al.*, 2010). This direct pathway not only boosts the efficiency of nutrient use but also mitigates the common issues associated with traditional fertilizers, such as nitrogen loss through leaching, emissions, and the monopolization of nutrients by soil microorganisms over time.

Controlled-release nanofertilizers further elevate this efficiency by tailoring nutrient release to the needs of the plants, thereby reducing the harmful impacts of chemical fertilizers' overuse. This approach, as Suman, Jain and Varma (2010) point out, not only enhances fertilizer use efficiency (FUE) but also plays a vital role in combating soil degradation. Nanofertilizers are divided into two primary categories: macronutrient and micronutrient types, each designed to meet the specific nutritional requirements of plants. This categorization ensures that plants receive a balanced diet of essential nutrients, promoting healthier growth and development. In summary, the adoption of nanotechnology in agriculture through nanofertilizers offers a promising solution to many of the challenges faced by modern farming. By increasing nutrient use efficiency, reducing environmental impact, and improving soil health, nanofertilizers stand at the forefront of agricultural innovation, paving the way for a more sustainable and productive future in crop production and management.

1.5.1 Macronutrient Nanofertilizers

Macronutrient nanofertilizers represent a significant breakthrough in agricultural science, offering a sophisticated means to supply essential nutrients to crops. These fertilizers incorporate nanosized particles of vital elements such as nitrogen (N), phosphorus (P), potassium (K), calcium (Ca), magnesium (Mg), and sulfur (S). These elements are indispensable for plant growth and are required in substantial quantities to meet the nutritional demands of various crops. The urgency for enhanced crop nutrition stems from the global challenge of feeding an ever-growing population. As noted by Alexandratos and Bruinsma (2012), the demand for food is escalating in parallel with the world's population, necessitating increased agricultural productivity. In this context, macronutrient nanofertilizers emerge as a pivotal innovation, capable of significantly amplifying the growth and productivity of crops. One of the most compelling advantages of these nanofertilizers is their dramatically improved efficiency in nutrient delivery and utilization by plants. This efficiency translates into stronger plant growth, higher yields, and improved crop quality. Furthermore, the precise delivery of nutrients at the nanoscale reduces the wastage associated with conventional fertilization methods, where nutrients often fail to reach their target or are lost to leaching and runoff. As a result, the application of macronutrient nanofertilizers can lead to a reduction in the amount of fertilizer needed, thereby lowering production costs and minimizing the environmental impact of farming practices.

Moreover, the economic benefits of using macronutrient nanofertilizers are profound. By enhancing nutrient use efficiency and crop productivity, these nanofertilizers offer a cost-effective alternative to traditional chemical fertilizers. This economic advantage is particularly crucial for small-scale and resource-poor farmers, who can achieve higher productivity and sustainability without the burden of increased costs. In summary, macronutrient nanofertilizers are at the forefront of agricultural innovation, responding to the critical need for increased food production in a sustainable and economically viable manner. Through the delivery of essential nutrients at the nanoscale, these fertilizers promise a new era of farming that is both productive and sustainable, addressing the dual challenges of food security and environmental protection.

1.5.2 Role of Micronutrients in Nanofertilizers

Micronutrient nanofertilizers encapsulate the next frontier in precision agriculture, offering a refined approach to delivering essential nutrients that, despite their requirement in minimal quantities, play monumental roles in the physiological and metabolic processes of plants. These nutrients, which include chloride (Cl), iron (Fe), boron (B), manganese (Mn), zinc (Zn), copper (Cu), molybdenum (Mo), and nickel (Ni), are pivotal for the healthy growth and development of crops, influencing everything from enzyme function to photosynthesis. The efficacy of micronutrients in supporting plant growth extends beyond their mere presence, hinging on their bioavailability and assimilation into plant biological processes. Traditional methods of micronutrient application often face challenges related to bioavailability, with factors such as soil pH, texture, and organic matter content significantly impacting nutrient accessibility to plants, as detailed by Fageria (2009). Micronutrient nanofertilizers, through their innovative formulations, aim to surmount these barriers, ensuring that these crucial nutrients are more effectively delivered to the plants that need them. Applying these nutrients in composite formulations, particularly at the nanoscale, enhances their effectiveness, providing targeted nutrition while minimizing the environmental footprint. Such precision in nutrient delivery not only fosters optimal plant health and productivity but also aligns with sustainable farming practices by reducing nutrient runoff and accumulation in the environment.

Various application methods, including the Hoagland solution, pioneered by Hoagland and Arnon (1950), and foliar sprays, are employed to administer micronutrient nanofertilizers. The choice of method depends on several factors, including the crop species, specific nutrient requirements, and the farming context. Foliar sprays, for instance, offer a direct route for nutrient absorption through the leaves, bypassing soil-based challenges and providing a rapid response to nutrient deficiencies. In essence, micronutrient nanofertilizers represent a critical advancement in agricultural technologies, tailored to meet the nuanced nutritional needs of crops. By ensuring the precise delivery of micronutrients, these nanofertilizers not only enhance plant health and yield but also contribute to the overarching goals of environmental sustainability and food security. As agriculture continues to evolve, the role of micronutrient nanofertilizers in enabling efficient, responsible, and productive farming practices is becoming increasingly indispensable.

1.6 NANOPESTICIDES

The advent of nanopesticides heralds a new era in the realm of agricultural pest management, addressing the critical need for more sustainable and environmentally friendly plant protection strategies. As the agricultural sector grapples with the dual challenges of increasing crop productivity and minimizing economic losses due to pests and diseases, the limitations of traditional chemical pesticides have become increasingly apparent. These conventional methods, while effective in the short term, pose significant risks to environmental, human, and animal health, as outlined by Jampílek and Králová (2017). The adverse effects include toxicity to non-target organisms, reduction of soil biodiversity, and the generation of harmful byproducts. In response to these challenges, the focus of agricultural research has shifted toward the development of novel plant protection formulations that balance efficacy with minimal environmental impact. A groundbreaking advancement in this domain is the encapsulation of active pesticide ingredients in nanoformulations. This approach leverages the unique properties of nanomaterials to revolutionize the plant protection sector, offering a more controlled and precise delivery of pesticides. Nanopesticides, encompassing nanoherbicides, nanoinsecticides, and nanofungicides, stand at the forefront of sustainable agricultural development. These formulations, as described by Kah *et al.* (2013) enable the controlled release of active ingredients, targeting specific areas such as root zones or internal plant tissues. This targeted approach not only enhances the effectiveness of pest control measures but also significantly reduces the potential harm to non-target organisms and the environment.

The properties of nanopesticides, highlighted by Haq and Ijaz (2019) and Khan and Rizvi (2014), include improved stiffness, permeability, thermal stability, solubility, crystallinity, and biodegradability. These characteristics are vital for achieving a sustainable agro-environmental system, offering a promising alternative to conventional chemical pesticides. Nano-pesticides offer enhanced efficacy through improved bioavailability, reduced environmental impact by minimizing off-target effects and controlled release, decreased resistance in pest populations due to precise delivery, and increased crop productivity by enabling more efficient pest control and higher yields. As the agricultural sector continues to evolve, the integration of nanotechnology in pest management practices represents a significant stride toward more sustainable and productive farming. Nanopesticides not only offer a solution to the limitations of conventional pesticides but also align with the broader goals of environmental stewardship and human health protection. This innovative approach promises to transform agricultural practices, paving the way for a future where crop protection is both effective and harmonious with nature.

1.6.1 Effect of Nanoherbicides and Their Limitation

Nanoherbicides emerge as a pioneering solution in the quest for sustainable agriculture, specifically addressing the perennial challenge of weed management. Weeds not only compete with crops for essential nutrients but also significantly reduce agricultural productivity by encroaching on the space meant for crops. Traditional herbicides, while effective in controlling weeds, often come with a suite of drawbacks, including potential damage to crops, declining soil fertility, and adverse effects on human health. The indiscriminate application of these chemicals has prompted a reevaluation of weed control strategies, leading to the exploration of more environmentally benign alternatives. Enter nanoherbicides, a novel class of weed control agents designed to mitigate the limitations of conventional herbicides. As highlighted by Pérez-de-Luque (2017), nanoherbicides offer a targeted approach to weed management, focusing on the root receptors of weeds without depositing harmful residues in the soil. This specificity not only ensures the conservation of crop health and soil fertility but also minimizes the risk of exposure to non-target organisms, including humans. Conventional herbicides typically target the aboveground parts of weeds, leaving the subterranean sections such as tubers and rhizomes intact. These underground components can survive to propagate the next generation of weeds, as discussed by Dashora and Kanika (2018). Nanoherbicides, however, are engineered to penetrate the root system of weeds, delivering a fatal blow to these persistent underground parts. Hess (2018) emphasizes the strategic targeting of weed roots by herbicides encapsulated with nanomaterials, ensuring the comprehensive eradication of weeds.

The issue of herbicide residues in soil is a significant concern, as highlighted by Chinnamuthu and Boopathi (2009). The long-term accumulation of these chemicals can inhibit the growth of future crops and disrupt soil ecosystems, necessitating the detoxification of soils for sustainable agricultural development. Nanoherbicides, with their higher bioavailability and targeted action, present a solution to this problem, reducing the likelihood of residual buildup and facilitating easier detoxification of soils. In summary, nanoherbicides represent a significant advancement in weed management, offering a more sustainable and environmentally friendly alternative to traditional herbicides. By targeting the root receptors of weeds and minimizing non-target impacts, nanoherbicides enhance the efficacy of weed control while safeguarding soil health, crop productivity, and ecosystem integrity. As agricultural practices evolve toward greater sustainability, nanoherbicides stand as a testament to the potential of nanotechnology to address complex challenges in crop management and environmental conservation.

1.6.2 Role of Nanoinsecticides in Pest Management

Nanoinsecticides represent a groundbreaking shift in agricultural pest management, addressing some of the most pressing challenges associated with conventional pesticide use. Traditional pesticides, as noted by (Nuruzzaman *et al.*, 2016), often fall short of their intended objectives, with as much as 90% of applied chemicals failing to effectively control target insect populations or becoming dispersed in the environment. This inefficiency not only exacerbates environmental pollution but also escalates the costs associated with crop production, creating a pressing need for more effective and environmentally friendly alternatives. The advent of nano-encapsulation and nanoformulation techniques for insecticides has ushered in a new era of crop protection. These advanced methodologies facilitate the controlled release and sustained presence of active ingredients, either directly within the crops or localized to the root zones, without compromising their efficacy. Such precision ensures that the active compounds remain effective over longer periods, reducing the frequency and volume of applications needed. This controlled release mechanism also mitigates the potential for environmental contamination and minimizes the risk of harming non-target organisms.

Furthermore, traditional insecticides often suffer from limited water solubility and can inadvertently contribute to the development of resistance among target insect populations. Nanoinsecticides, through their enhanced thermal stability, crystallinity, solubility, permeability, stiffness, and biodegradability, offer vital attributes for sustainable agricultural practices. These properties, as highlighted by Dwivedi *et al.* (2016), not only improve the environmental compatibility of insecticides but also bolster their effectiveness against pests. Additionally, some research suggests that nanoformulated insecticides may even enhance plant-mediated resistance against insects. This potential for inducing a broader spectrum of plant defenses against pests further underscores the transformative impact of nanomaterials in pest-management strategies. In sum, nanoinsecticides stand at the forefront of innovative solutions to pest control, offering a host of benefits over traditional chemical pesticides. By improving the efficiency of active ingredients, reducing environmental impact, and potentially enhancing plant resistance to pests, nanoinsecticides pave the way for more sustainable and effective pest management approaches. As the agricultural sector continues to seek out advancements that balance productivity with environmental stewardship, nanoinsecticides emerge as a promising avenue for achieving these goals, heralding a new age in agricultural technology and pest control.

1.6.3 Functions of the Nanofungicides

Nanofungicides are emerging as a cutting-edge solution in the battle against fungal pathogens, which are responsible for significant losses in global agriculture. Fungal diseases are among the most prevalent threats to crops, affecting staples like rice, wheat, barley, groundnut, and cotton, and resulting in considerable yield and economic losses. As reported by (Godfray, Mason-D'Croz and Robinson, 2016), these diseases account for more than 70% of the damage inflicted on major crops, with profound implications for food security and societal well-being. The limitations of conventional fungicides, including their nonspecific action and potential to disrupt biodiversity, have prompted the search for more precise and sustainable alternatives. Nanotechnology offers a promising avenue in this regard, utilizing nanomaterials to develop fungicides that combine high efficacy with specificity toward fungal pathogens, thus mitigating the broader environmental impact.

Nanofungicides leverage the unique properties of nanoscale materials to offer broad-spectrum antifungal properties, significantly improving the management of fungal diseases. These innovations not only enhance the effectiveness of fungicidal treatments but also reduce the incidence of non-target effects, preserving the ecological balance within agricultural systems. The research by Patel *et al.* (2014) underscores the necessity for precision in fungal disease management, highlighting the potential of nanotechnology to fulfill this need. Recent studies have demonstrated the multifaceted benefits of nanofungicides beyond their primary role in combating fungal pathogens. For instance, Iftikhar *et al.* (2020), Khan, Bano and Babar (2020), and Soliman *et al.* (2020) have shown that the application of nanofungicides can enhance abiotic stress tolerance in plants, contributing to increased chlorophyll content, biomass, soluble sugar content, and seed germination rates. These findings suggest that nanofungicides not only protect crops from fungal diseases but also bolster plant resilience against environmental stressors.

Moreover, specific examples, such as the use of silver nanomaterials (AgNMs) and selenium nanomaterials (SeNMs), illustrate the potential for targeted applications of nanofungicides to improve plant growth and stress responses. Shoemaker (2020) reported increased seedling growth and leaf area in wheat treated with AgNMs, while Maruyama *et al.* (2016) observed enhanced antioxidant enzyme activity and thylakoid membrane stability in sorghum subjected to heat stress following foliar application of SeNMs. In summary, nanofungicides represent a significant advancement in agricultural practices, offering a sustainable, effective, and precise approach to managing fungal diseases. By harnessing the power of nanotechnology, these innovative treatments not only address the immediate threat of fungal pathogens but also contribute to the broader goal of enhancing plant health and resilience, paving the way for more productive and sustainable agricultural systems.

1.7 NANOMATERIALS' EFFECT IN CROP PHYSIOLOGY AND DISEASE MANAGEMENT

The integration of nanotechnology into agriculture is heralding a new age of precision farming, equipped with innovative tools designed to bolster disease resistance in crops and revamp traditional agricultural practices. The impetus for this shift is rooted in the substantial losses agriculture incurs annually due to the relentless assault of plant pathogens and pests. These adversaries not only threaten food security but also impose a heavy economic burden on farmers worldwide. Nanomaterials, particularly when formulated as nanofertilizers, have shown tremendous potential in altering crop physiology for the better. They enhance nutrient uptake efficiency, thereby directly improving plant health and vigor. This increase in nutrient assimilation supports more robust growth and may contribute to higher yields and enhanced nutritional quality of crops. Furthermore, nanotechnology facilitates the precise delivery of genetic material or proteins into plant cells. This capability is groundbreaking, offering a novel method for protecting crops against pathogens through the expression of pathogenesis-related genes. Such genetic enhancements could confer resistance to a broad spectrum of diseases, reducing the need for chemical interventions and contributing to more sustainable farming practices.

Another frontier in the nanotechnology revolution in agriculture is the development of nanosensors. These highly sensitive devices are capable of detecting plant pathogens at early stages, well before visible symptoms appear, enabling timely and targeted interventions.

Additionally, nanosensors hold promise for identifying and mitigating environmental pollutants, thus playing a crucial role in preserving the integrity of agricultural ecosystems. As outlined by Mazzaglia *et al.* (2017), the potential applications of nanotechnology in agriculture are vast and varied. From enhancing crop resilience to diseases and pests to ensuring sustainable environmental practices, nanotechnology offers a suite of tools for modernizing agriculture. This technological advancement not only aims to mitigate the challenges posed by plant pathogens and pests, but also seeks to address the broader implications of food security and environmental sustainability. As research and development in this area continue to evolve, the promise of nanotechnology in transforming agriculture becomes increasingly tangible, setting the stage for a future where farming practices are both efficient and eco-friendly.

The utilization of nanotechnology in agriculture is revolutionizing the sector, presenting innovative strategies to bolster plant defense mechanisms against a wide array of pathogens. This technological advancement is critical, as agriculture incurs significant losses each year due to the devastation caused by plant pathogens and pests. Nanomaterials, when employed as nanofertilizers, profoundly influence crop physiology by optimizing nutrient uptake and enhancing overall plant health. These nanoscale materials are also adept at facilitating the transfer of genetic material or proteins, thereby furnishing crops with a robust defense against pathogens through the activation of pathogenesis-related genes. Furthermore, nanosensors emerge as cutting-edge tools capable of pathogen detection and the elimination of environmental pollutants, adding another layer of protection to crop management practices.

In addition to these applications, nanotechnology offers novel approaches to mitigating the effects of plant pathogens, including viruses, fungi, bacteria, and insects, thereby addressing some of the primary causes of crop damage and economic losses. These modern interventions complement traditional agricultural methods, such as crop rotation, integrated pest management, and chemical pesticides, providing a more comprehensive strategy for plant disease management. Research underscores the efficacy of specific nanomaterials, such as Ag, ZnO, and alumina, in eradicating plant pathogens and stimulating plant growth, highlighting the potential of nanotechnology to significantly improve crop resilience and productivity. The challenges posed by climate change to agriculture, including various environmental stressors such as shifts in temperature, salinity, cold, and the presence of environmental pollutants, necessitate adaptive strategies to safeguard plant health. Nanotechnology emerges as a pivotal solution in this context, offering mechanisms for hormonal regulation and the activation of stress-related enzymes, thus equipping plants to withstand these adversities. Recent advancements have demonstrated the capacity of nanotechnology, particularly nanofertilizers, to enhance stress tolerance in various crop species, thereby ensuring their survival and continued productivity under challenging environmental conditions. Overall, the integration of nanotechnology into agriculture marks a significant leap forward in crop physiology and disease management. By leveraging the unique properties of nanomaterials, this innovative approach not only enhances plant health and yields but also contributes to the sustainability and resilience of agricultural systems in the face of global challenges.

1.8 FUTURE ASPECTS AND CONCLUSION

Over recent decades, nanotechnology has emerged as a transformative force in agriculture, heralding a shift toward more sustainable practices. This innovative approach seeks not only to enhance crop yields but also to minimize the environmental footprint of farming, addressing pressing challenges such as pollution reduction and the efficient use of resources. By integrating nanotechnology into agricultural systems, we are on the cusp of revolutionizing conventional farming, making strides toward a future of sustainable agriculture. The applications of nanotechnology in this field are vast and varied, encompassing rapid disease detection, nutrient enrichment, disease suppression, real-time field monitoring, and the reduction of environmental waste. Engineered nanomaterials, when combined with biosensors, have paved the way for the development of advanced tools for swift pathogen identification, marking a significant leap forward in agricultural diagnostics. Moreover, the convergence of nanotechnology with biotechnology is opening new avenues for the creation of innovative disease diagnostic tools, a development poised to significantly impact plant health management.

Despite the promising advancements brought about by nanotechnology, certain areas remain underexplored. For instance, the mechanisms underlying the resistance of some plant species to nanomaterials have yet to be fully elucidated. Similarly, the dynamics of nanomaterial uptake and translocation within plant systems require further investigation to harness their full potential effectively. As the development of smart agrochemicals, or nano-agrochemicals, continues to address a broad spectrum of agricultural challenges, it becomes imperative to establish comprehensive risk assessment, management, and mitigation frameworks. These frameworks must operate on both the international and national stages to ensure the responsible and sustainable deployment of nanotechnologies in agriculture. As the market for nanomaterials expands, there is a growing need for stringent regulations to oversee their production, handling, and disposal. Such regulations are essential to guarantee eco-friendly synthesis processes and the safe application of nanomaterials, safeguarding environmental and human health. Moreover, the synthesis, administration, and dissemination of nanomaterials must adhere to eco-friendly principles, underscoring the importance of sustainable practices in their lifecycle. However, challenges such as social acceptance, environmental costs, soil degradation, disruption of soil microbiomes, and the depletion of water resources necessitate vigilant monitoring and management. The journey toward fully integrating nanotechnology into agriculture is complex and requires a balanced approach that considers both the immense benefits and the potential risks associated with this cutting-edge technology. In conclusion, nanotechnology represents a pivotal advancement in agricultural practices, offering a path toward sustainability, increased productivity, and reduced environmental impact. As we navigate the complexities of implementing nanotechnologies in agriculture, a comprehensive and cautious approach will be essential to unlocking their full potential while safeguarding our planet for future generations.

Nanotechnology's integration into agriculture marks a significant advancement in efforts to enhance crop production and farming systems amidst the pressing challenges of global population growth, plant and animal diseases, and the impacts of climate change. This technological approach has opened up new avenues for quality enhancement in agricultural outputs, demonstrating the multifaceted roles that nanomaterials can play in modern agriculture. A key area of focus within this field has been the development and application of nanopesticides and nanofertilizers. Some of these innovative products, synthesized through environmentally friendly green methods, have already made their way to the market, reflecting the growing emphasis on sustainability in agricultural practices. The green synthesis of nanomaterials, utilizing various biological organisms, represents a burgeoning research domain. This approach not only mitigates the risks of environmental pollution but also enhances the bioactivity and efficiency of nanomaterials in agricultural applications.

The interaction between nanomaterials and plant systems is a critical aspect of research, essential for understanding how these materials influence plant physiology and biochemical responses. As we look toward the future of nanotechnology in agriculture, a comprehensive understanding of how nanomaterials interact with biological systems will be paramount. This knowledge is crucial for the functional design of nanotechnologies aimed at optimizing agricultural productivity while ensuring environmental sustainability. Innovative techniques are emerging that allow for the precise location and quantification of nanomaterials within plant tissues. These methods offer valuable insights into the transformation, mobility, and safety of nanomaterials in agricultural settings. Such advancements underscore the importance of carefully designing nanomaterials with consideration for their application methods and potential impacts on the environment. However, it's essential to recognize that while nanomaterials present numerous benefits for agricultural systems, from enhancing seed germination to improving post-harvest storage, their excessive use could pose risks of environmental pollution. Thus, a cautious and informed approach to applying nanotechnology in agriculture is imperative. Ensuring that nanomaterials are used judiciously and in ways that are beneficial to plant systems without harming the broader ecosystem will be key to realizing their full potential. In sum, nanotechnology holds promise for transforming agriculture by providing solutions to some of the most significant challenges facing the sector today. However, the journey toward fully harnessing the benefits of nanomaterials in agriculture necessitates a balanced approach, one that carefully considers both the innovative potential of these materials and their environmental footprint.

REFERENCES

Ahmad, F., *et al.* (2007) 'Effect of silicon application on wheat (*Triticum aestivum* L.) growth under water deficiency stress', *Journal of Food and Agriculture*, 7, pp. 1–7. https://doi.org/10.9755/ejfa.v12i1.5170

Alexandratos, N. and Bruinsma, J. (2012) World agriculture towards 2030/2050, the 2012 revision, ESA Working Paper No. 12–03. FAO: Rome.

Bachheti, R.K., Bachheti, A. and Husen, A. (2024) *Metal and metal-oxide based nanomaterials (synthesis, agricultural, biomedical and environmental interventions).* Singapore: Springer Nature Singapore.

Banerjee, J. and Kole, C. (2016) 'Plant nanotechnology: An overview on concepts, strategies, and tools', in Kole, C., Kumar, D., Khodakovskaya, M. (eds.) *Plant nanotechnology.* Cham, Switzerland: Springer, pp. 1–14.

Bayoumy, A.M., *et al.* (2020) 'Effect of nano metal oxides on heme molecule: Molecular and biomolecular approaches', *Biointerface Research in Applied Chemistry*, 10, pp. 4837–4845. https://doi.org/10.33263/BRIAC101.837845

Boxi, S.S., Mukherjee, K. and Paria, S. (2016) 'Ag doped hollow TiO_2 nanoparticles as an effective green fungicide against *Fusarium solani* and *Venturia inaequalis* phytopathogens', *Nanotechnology*, 27, p. 085103. https://doi.org/10.1088/0957-4484/27/8/085103

Boykov, I.N., Shuford, E. and Zhang, B. (2019) 'Nanoparticle titanium dioxide affects the growth and microRNA expression of switchgrass (*Panicum virgatum*)', *Genomics*, 111, pp. 450–456. https://doi.org/10.1016/j.ygeno.2018.03.002

Brackhage, C., *et al.* (2013) 'Silicon availability affects the stoichiometry and content of calcium and micro nutrients in the leaves of common reed', *Silicon*, 5, pp. 199–204.

Chen, J., Wang, X. and Han, H. (2013) 'A new function of graphene oxide emerges: Inactivating phytopathogenic bacterium, *Xanthomonas oryzae pv.* oryzae', *Journal of Nanoparticle Research*, 15, pp. 1–14. https://doi.org/10.1007/s11051-013-1658-6

Chinnamuthu, C.R. and Boopathi, P.M. (2009) 'Nanotechnology and agroecosystem', *Madras Agricultural Journal*, 96, pp. 17–31.

Dashora, A. and Kanika, S. (2018) 'Green synthesis of nanoparticles and their applications', *Advanced Science, Engineering and Medicine*, 10(6), pp. 523–541.

DeRosa, M.C., *et al.* (2010) 'Nanotechnology in fertilizers', *Nature Nanotechnology*, 5, pp. 91–94.

Dimkpa, C.O. and Bindraban, P.S. (2017). 'Nanofertilizers: New products for the industry?', *Journal of Agricultural and Food Chemistry*, 66, pp. 6462–6473. https://doi.org/10.1021/acs.jafc.7b02150

Dubey, A. and Mailapalli, D.R. (2016) 'Nanofertilisers, nanopesticides, nanosensors of pest and nanotoxicity in agriculture', in Lichtfouse, E. (ed.) *Sustainable agriculture reviews.* Cham, Switzerland: Springer, 19, pp. 307–330.

Dwivedi, S., *et al.* (2016) 'Understanding the role of nanomaterials in agriculture', *Microbial Inoculants in Sustainable Agricultural Productivity*, 17, pp. 271–88.

Elmer, W. and White, J.C. (2018) 'The future of nanotechnology in plant pathology', *Annual Review of Phytopathology*, 56, pp. 111–133. https://doi.org/10.1146/annurev-phyto-080417-050108

Elmer, W.H. and White, J.C. (2016) 'The use of metallic oxide nanoparticles to enhance growth of tomatoes and eggplants in disease infested soil or soilless medium', *Environmental Science: Nano*, 3, pp. 1072–1079. https://doi.org/10.1039/C6EN00146G

Etxeberria, E., *et al.* (2006) 'Fluid phase endocytic uptake of artificial nano-spheres and fluorescent quantum dots by sycamore cultured cells: Evidence for the distribution of solutes to different intracellular compartments', *Plant Signaling & Behavior*, 1, pp. 196–200. https://doi.org/10.4161/psb.1.4.3142

Fageria, N.K. (2009) *Use of nutrients in crop plants.* Boca Raton, FL: CRC Press.

Godfray, H.C., Mason-D'Croz, D. and Robinson, S. (2016) 'Food system consequences of a fungal disease epidemic in a major crop', *Philosophical Transactions of the Royal Society of London. Series B, Biological Sciences*, 371, p. 20150467. https://doi.org/10.1098/rstb.2015.0467

Gupta, A., *et al.* (2016) 'Nanoemulsions: Formation, properties and applications', *Soft Matter*, 12, pp. 2826–2841.

Haq, I.U. and Ijaz, S. (2019) 'Use of metallic nanoparticles and nanoformulations as nanofungicides for sustainable disease management in plants', in Prasad, R., Kumar, V., Kumar, M., and Choudhary D. (eds.) *In nanobiotechnology in bioformulations.* Cham, Switzerland: Springer, p. 289.

He, L., *et al.* (2011) 'Antifungal activity of zinc oxide nanoparticles against Botrytis cinerea and *Penicillium expansum*', *Microbiological Research*, 166, pp. 207–215. https://doi.org/1016/j.micres.2010.03.003

Hess, F.D. (2018) 'Herbicide effects on plant structure, physiology and biochemistry', *Pesticide Interactions in Crop Production*, 22, pp. 13–34.

Hoagland, D.R. and Arnon, D.I. (1950) 'The water-culture method for growing plants without soil', *California Agricultural Experiment Station, Circular*, 347, pp. 1–32. https://doi.org/10.33263/LIANBS104.27842803

Husen, A. (2020) 'Introduction and techniques in nanomaterials formulation', in Husen A. and Jawaid M. (eds.) *Nanomaterials for agriculture and forestry applications.* Cambridge, MA: Elsevier Inc., pp. 1–14. https://doi.org/10.1016/B978-0-12-817852-2.00001-9

Husen, A. (2022) *Engineered nanomaterials for sustainable agricultural production, soil improvement and stress management.* Cambridge, MA: Elsevier Inc. https://doi.org/10.1016/C2021-0-00054-7

Husen, A. (2023a) *Secondary metabolites based green synthesis of nanomaterials and their applications.* Singapore: Springer Nature Singapore. https://doi.org/10.1007/978-981-99-0927-8

Husen, A. (2023b) *Nanomaterials from agricultural and horticultural products.* Singapore: Springer Nature Singapore. https://link.springer.com/book/9789819934348

Husen, A. (2023c) *Nanomaterials and nanocomposites exposures to plants (response, interaction, phytotoxicity and defense mechanisms).* Singapore: Springer Nature Singapore. https://doi.org/10.1007/978-981-99-2419-6

Husen, A. and Siddiqi, K.S. (2023) *Advances in smart nanomaterials and their applications.* Cambridge, MA: Elsevier. https://doi.org/10.1016/C2021-0-02202-1

Iftikhar, A., *et al.* (2020) 'Effect of gibberellic acid on growth, biomass, and antioxidant defense system of wheat (*T. aestivum* L.) under cerium oxide nanoparticle stress', *Environmental Science and Pollution Research*, 27, pp. 33809–33820. https://doi.org/10.1007/s11356-020-09661-9

Iravani, S. (2014) 'Bacteria in nanoparticle synthesis: Current status and future prospects', *International Scholarly Research Notices*, 2014, p. 59361.

Jagana, D., Hegde, Y.R. and Lella, R. (2017) 'Green nanoparticles: A novel approach for the management of banana anthracnose caused by *Colletotrichum musae*', *International Journal of Current Microbiology and Applied Sciences*, 6, pp. 1749–1756. https://doi.org/10.20546/ijcmas.2017.610.211

Jamkhande, P.G., *et al.* (2019) 'Metal nanoparticles synthesis: An overview on methods of preparation, advantages and disadvantages, and applications', *Journal of Drug Delivery Science and Technology*, 53, p. 101174.

Jampílek, J. and Králová, K. (2017) 'Nanopesticides: Preparation, targeting, and controlled release', in Grumezescu, A.M. (ed.) *New pesticides and soil sensors.* Bucharest, Romania: Academic Press, pp. 81–127. https://doi.org/10.1016/B978-0-12-804299-1.00004-7

Kah, M., *et al.* (2013) 'Nanopesticides: State of knowledge, environmental fate, and exposure modeling', *Critical Reviews in Environmental Science and Technology*, 43, pp. 1823–1867. https://doi.org/10.1080/10643389.2012.671750

Keshvadi, M., *et al.* (2019) 'Comparative study of antibacterial inhibitory effect of silver nanoparticles and garlic oil nanoemulsion with their combination', *Biointerface Research in Applied Chemistry*, 9, pp. 4560–4566. https://doi.org/10.33263/BRIAC96.560566

Khan, M.R. and Rizvi, T.F. (2014) 'Nanotechnology: Scope and application in plant disease management', *The Plant Pathology Journal*, 13, pp. 214–231.

Khan, N., Bano, A.M.D. and Babar, A. (2020) 'Impacts of plant growth promoters and plant growth regulators on rainfed agriculture', *PLoS ONE*, 15, p. e0231426. https://doi.org/10.1371/journal.pone.0231426

Khandelwal, A. and Joshi, R. (2018) 'Synthesis of nanoparticles and their application in agriculture', *Acta Scientific Agriculture*, 2, pp. 10–13.

Lahiani, M.H., *et al.* (2013) 'Impact of carbon nanotube exposure to seeds of valuable crops', *ACS Applied Materials & Interfaces*, 5, pp. 7965–7973. https://doi.org/10.1021/am402052x

Lu, C., *et al.* (2002) 'Research of the effect of nanometer materials on germination and growth enhancement of Glycine max and its mechanism', *Soybean Science*, 21, pp. 168–171.

Madhusudhan, A., *et al.* (2024) *Functional smart nanomaterials and their theranostics approaches.* Singapore: Springer Nature Singapore.

Manjaiah, K.M., *et al.* (2019) 'Clay minerals and zeolites for environmentally sustainable agriculture', in Mariano, M., Binoy, S. and Alessio, L. (eds.) *Modified clay and zeolite nanocomposite materials.* Amsterdam, The Netherlands: Elsevier, pp. 309–329.

Maruyama, C.R., *et al.* (2016) 'Nanoparticles based on chitosan as carriers for the combined herbicides imazapic and imazapyr', *Scientific Reports*, 6, p. 19768. https://doi.org/10.1038/srep19768

Mazzaglia, A., *et al.* (2017) 'Nanomaterials in plant protection', in Axelos, M.A., Voorde, M.V. (eds.) *Nanotechnology in agriculture and food science*. 1st ed. Weinheim, Germany: Wiley-VCH Verlag GmbH & Co. KGaA, pp. 115–134. https://doi.org/10.1002/9783527697724.ch7

Mittal, D., *et al.* (2020) 'Nanoparticle-based sustainable agriculture and food science: Recent advances and future outlook', *Frontiers in Nanotechnology*, 2, p. 579954. https://doi.org/10.3389/fnano.2020.579954

Moussa, S.H., *et al.* (2013) 'Botryticidal activity of nanosized silver-chitosan composite and its application for the control of graymold in strawberry', *Journal of Food Science*, 78, pp. 589–594. https://doi.org/10.1111/1750-3841.12247

Mughal, S.S. and Hassan, S.M. (2022) 'Comparative study of AgO nanoparticles synthesize via biological, chemical and physical methods: A review', *AJMSP*, 15, p. 28.

Nel, A.E., *et al.* (2009) 'Understanding biophysicochemical interactions at the nano-bio interface', *Nature Materials*, 8, pp. 543–557. https://doi.org/10.1038/nmat2442

Nuruzzaman, M., *et al.* (2016) 'Nanoencapsulation, nano-guard for pesticides: A new window for safe application', *Journal of Agricultural and Food Chemistry*, 64, pp. 1447–1483.

Park, H.-J., *et al.* (2006) 'A new composition of nanosized silica-silver for control of various plant diseases', *The Plant Pathology Journal*, 22, pp. 295–302. https://doi.org/10.5423/PPJ.2006.22.3.295

Patel, N., *et al.* (2014) 'Agronanotechnology for plant fungal disease management: A review', *International Journal of Current Microbiology and Applied Sciences*, 3, pp. 71–84.

Pérez-de-Luque, A. (2017) 'Interaction of nanomaterials with plants: What do we need for real applications in agriculture?', *Frontiers in Environmental Science*, 5, p. 12. https://doi.org/10.3389/fenvs.2017.00012

Ponmurugan, P., *et al.* (2016) 'Antifungal activity of biosynthesized copper nanoparticles evaluated against red root-rot disease in tea plants', *Journal of Experimental Nanoscience*, 11, pp. 1019–1031. https://doi.org/10.1080/17458080.2016.1184766

Rafi, M.M., Epstein, E. and Falk, R.H. (1997) 'Silicon deprivation causes physical abnormalities in wheat (*Triticum aestivum* L.)', *Journal of Plant Physiology*, 151, pp. 497–501. https://doi.org/10.1016/S0176-1617(97)80017-X

Rao, K.J. and Paria, S. (2013) 'Use of sulfur nanoparticles as a green pesticide on *Fusarium solani* and *Venturia inaequalis* phytopathogens', RSC Advances, 3, pp. 10471–10478. https://doi.org/10.1039/C3RA40500A

Rico, C.M., *et al.* (2011) 'Interaction of nanoparticles with edible plants and their possible implications in the food chain', Journal of Agricultural and Food Chemistry, 59, pp. 3485–3498. https://doi.org/10.1021/jf104517j

Roberts, A.G. and Oparka, K.J. (2003) 'Plasmodesmata and the control of symplastic transport', *Plant, Cell & Environment*, 26, pp. 103–124. https://doi.org/10.1046/j.1365-3040.2003.00950.x

Sanzari, I., Leone, A. and Ambrosone, A. (2019) 'Nanotechnology in plant science: To make a long story short', *Frontiers in Bioengineering and Biotechnology*, 7, p. 120. https://doi.org/10.3389%2Ffbioe.2019.00120

Sardella, D., Gatt, R. and Valdramidis, V.P. (2017) 'Physiological effects and mode of action of ZnO nanoparticles against postharvest fungal contaminants', *Food Research International*, 101, pp. 274–279. https://doi.org/10.1016/j.foodres.2017.08.019

Sarkar, J., *et al.* (2020) 'Green synthesized copper oxide nanoparticles ameliorate defence and antioxidant enzymes in *Lens culinaris*', *Nanomaterials*, 10, pp. 312–339. https://doi.org/10.3390/nano10020312

Sattelmacher, B. (2001) 'The apoplast and its significance for plant mineral nutrition', *New Phytologist*, 149, pp. 167–192. https://doi.org/10.1046/j.1469-8137.2001.00034.x

Schwab, F., *et al.* (2015) 'Barriers, pathways and processes for uptake, translocation and accumulation of nanomaterials in plants-critical review', *Nanotoxicology*, 10, pp. 257–278. https://doi.org/10.3109/17435390.2015.1048326

Serag, M.F., *et al.* (2011) 'Trafficking and subcellular localization of multiwalled carbon nanotubes in plant cells', *ACS Nano*, 5, pp. 493–499. https://doi.org/10.1021/nn102344t

Sharifi, S., *et al.* (2012) 'Toxicity of nanomaterials', *Chemical Society Reviews*, 41(6), pp. 2323–2343.

Shalaby, T.A., *et al.* (2016) 'Nanoparticles, soils, plants and sustainable agriculture', in Shivendu, R., Nandita, D. and Eric, L. (eds.) *Nanoscience in food and agriculture*. Cham, Switzerland: Springer, pp. 283–312.

Shoemaker, A.G. (2020) 'The effects of titanium dioxide nanoparticles on the growth and development of *Sorghum bicolor* (L.) Moenech', *Advances in Agriculture, Horticulture and Entomology*, 132, pp. 1–15.

Shojaei, T.R., *et al.* (2019) 'Applications of nanotechnology and carbon nanoparticles in agriculture', in Suraya, A. R., Raja, N. I. R. O. and Mohd, Z. H. (eds.) *Synthesis, technology and applications of carbon nanomaterials*. Amsterdam, The Netherlands: Elsevier, pp. 247–277.

Singh, A., *et al.* (2015) 'Plant-nanoparticle interaction: An approach to improve agricultural practices and plant productivity', *International Journal of Pharmaceutical Science Invention*, 4, pp. 25–40.

Solanki, P., *et al.* (2015) 'Nano-fertilizers and their smart delivery system', in Rai, M. (ed.) *Nanotechnologies in food and agriculture*. Cham: Springer, pp. 81–101. https://doi.org/10.1007/978-3-319-14024-7_4

Soliman, M., *et al.* (2020) 'Rapid green synthesis of silver nanoparticles from blue gum augment growth and performance of maize, fenugreek, and onion by modulating plants cellular antioxidant machinery and genes expression', *Acta Physiologiae Plantarum*, 42, pp. 1–16. https://doi.org/10.3390/plants10040790

Sudha, P.N., *et al.* (2018) Nanomaterials history, classification, unique properties, production and market', in *Emerging applications of nanoparticles and architecture nanostructures*. Amsterdam: Elsevier, pp. 341–384.

Suman, P.R., Jain, V.K. and Varma, A. (2010) 'Role of nanomaterials in symbiotic fungus growth enhancement', *Current Science*, 99, pp. 1189–1191.

Usman, M., *et al.* (2020) 'Nanotechnology in agriculture: Current status, challenges and future opportunities', *Science of the Total Environment*, 721, p. 137778. https://doi.org/10.1016/j.scitotenv.2020.137778

Wang, X., Liu, X. and Han, H.Y. (2013) 'Evaluation of antibacterial effects of carbon nanomaterials against copper resistant *Ralstonia solanacearum*', *Colloids Surf. B Biointerfaces*, 103, pp. 136–142. https://doi.org/10.1016/j.colsurfb.2012.09.044

Wani, A.H. and Shah, M.A. (2012) 'A unique and profound effect of MgO and ZnO nanoparticles on some plant pathogenic fungi', *Journal of Applied Pharmaceutical Science*, 2, pp. 40–44.

Wong, M.H., *et al.* (2016) 'Lipid exchange envelope penetration (LEEP) of nanoparticles for plant engineering: A universal localization mechanism', *Nano Letters*, 16, pp. 1161–1172. https://doi.org/10.1021/acs.nanolett.5b04467

Zhai, G., *et al.* (2014) 'Transport of gold nanoparticles through plasmodesmata and precipitation of gold ions in woody poplar', *Environmental Science & Technology Letters*, 1, pp. 146–151. https://doi.org/10.1021/ez400202b

2 Role of Nanosensors for Sustainable Agricultural Practices

Murugan Karuvelan, Soumya Lakshman, Usharani Boopathy, Ramachandran Chelliah, and Deog Hwan Oh

2.1 INTRODUCTION

The agricultural sector holds paramount significance and exhibits remarkable stability on a global scale, primarily due to its pivotal role in the provision and production of raw materials for the food and feed industries. Based on data provided by the Food and Agriculture Organization, agricultural activities currently utilise about 4.9 million hectares of land, representing approximately 37.6% of the terrestrial ecosystem (FAO, 2022). It is imperative to acknowledge the concealed reality that terrestrial land masses serve as the exclusive source of fundamental life necessities, rendering every parcel of land crucial in meeting needs for sustenance. The alteration of nutrient balance in agriculture is mostly influenced by economic expansion, making the preservation of soil fertility crucial in growing nations. Therefore, the advancement of agriculture is an imperative occurrence, particularly for emerging nations, in order to alleviate poverty and address food insecurity (Srivastava *et al.*, 2023). Hence, it is imperative to proactively undertake measures to ensure the sustainable advancement of agriculture. The progress of agricultural development is influenced by various factors, such as health, climate change, social inclusion, and natural resources. Therefore, in order to facilitate the recovery of agriculture, it is necessary to implement innovative approaches that specifically prioritise the attainment of higher crop yields (Periakaruppan *et al.*, 2023).

The objective of each nation is to discover novel approaches by exploring transdisciplinary and interdisciplinary technologies, such as biotechnology and nanotechnology (NT), in order to enhance agricultural practices and crop yield. Undoubtedly, the use of scientific innovations has the potential to facilitate the sustainable development of agriculture Husen, 2022, 2023; Saleh *et al.*, 2020). The integration of novel nanoformulations of agrochemicals, such as pesticides, fertilisers, fungicides, and herbicides, through the amalgamated methods of NT, shows potential in various domains of agriculture. These methods encompass the development of nanobiofertilisers, disease identification, genetic manipulation of crops, and advancements in animal husbandry. By adopting smarter, more durable, and more cost-effective techniques, the agricultural sector can benefit from these advancements (Pirzadah *et al.*, 2020).

NT has been employed for the identification of superior genes and their integration into agricultural production and management practices to enhance yield and improve resistance against diseases. Similarly, NT has also been employed in the enhancement of animal health and reproduction. In contemporary agriculture, the overutilisation of agrochemicals presents some potential concerns, such as water contamination and the presence of residues on food products, thereby posing a risk to the well-being of organisms (Husen and Iqbal, 2019). Therefore, the implementation of advanced agricultural systems incorporating nanotools may serve as an effective approach to addressing this predicament. Similarly, there is a need to enhance both the quality and quantity of crop production. The importance of nanoparticles (NPs) in agriculture is depicted in Figure 2.1.

DOI: 10.1201/9781032630366-2

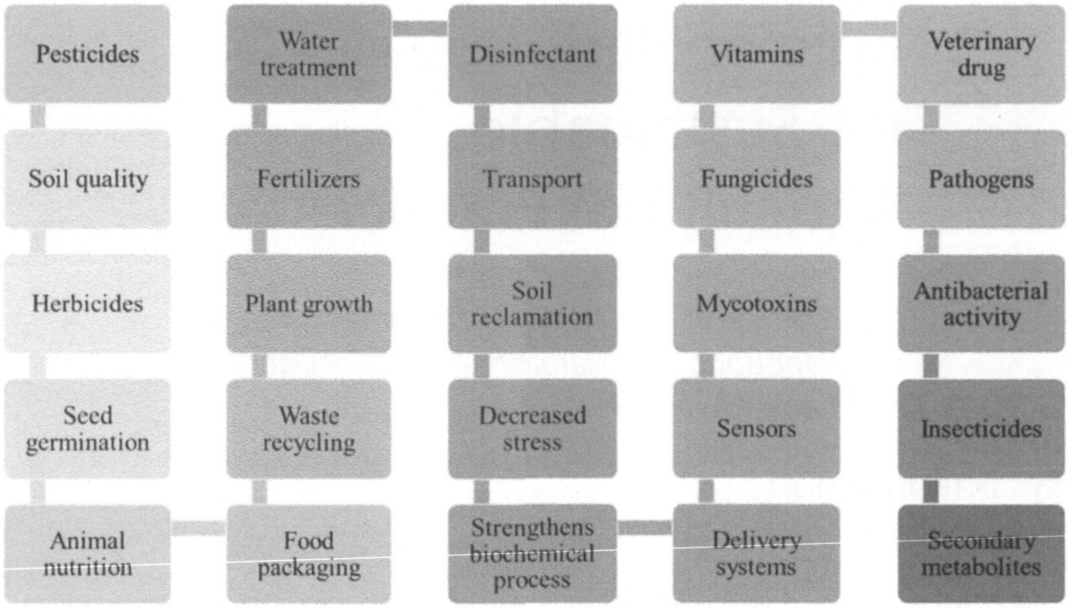

FIGURE 2.1 Overview of nanomaterials in agriculture, illustrating their applications in crop growth enhancement, pest management, and produce quality improvement.

Furthermore, the use of NT encompasses the application of nanosensors (NSs) to accurately report physiological soil conditions, facilitate targeted delivery of nanocarrier-based drugs, enhance nutrient uptake, and improve production, among other benefits. The construction of neural networks can be seen as a promising domain for leveraging the capabilities of neurotechnology. The sensitivity performances of NSs can be greatly enhanced in their applications as a result of the unique features exhibited by nanomaterials. Also, the utilisation of nanomaterials has facilitated the downsizing of these sensors, which play a crucial role in biochemical analysis.

Figure 2.1 provides a detailed and comprehensive visual representation of the multifaceted role that nanomaterials play in modern agriculture. This diagram systematically illustrates the various applications of nanomaterials across different agricultural processes and practices. It starts by highlighting the use of nanomaterials in enhancing crop growth and yield. Here, the figure shows how NPs can be used in fertilisers and soil conditioners to improve nutrient uptake efficiency and water retention in soil, leading to healthier and more robust plant growth. The figure also covers the use of NT in pest and disease management, showcasing how nano-based pesticides and fungicides can be more effective and less harmful to the environment than traditional chemical treatments. It illustrates the mechanisms by which these nanomaterials target and neutralise pests and pathogens, while reducing the risk of developing resistance. Additionally, the diagram delves into the role of nanomaterials in improving the quality and safety of agricultural produce. This includes the use of NS for monitoring environmental conditions and detecting contaminants or pathogens in crops, ensuring food safety from farm to table. Furthermore, Figure 2.1 touches upon the use of NT in agriculture-related environmental protection, such as in the management of agricultural waste and in water purification processes. Overall, Figure 2.1 serves as an educational tool, offering a holistic view of how nanomaterials are revolutionising agricultural practices. It provides insights into the ways NT can address key challenges in agriculture, leading to increased efficiency, sustainability, and safety in food production.

2.2 ADOPTION OF NANOTECHNOLOGY IN AGRO-ECOSYSTEMS

NT is widely regarded as a promising approach for addressing diverse agricultural challenges. Because of this factor, it has garnered significant attention in recent years. This has resulted in the creation of innovative and distinctive techniques to achieve high farm output, leading to enhanced agricultural production. It has offered novel agents a proficient delivery mechanism to enhance agricultural productivity. NT has enhanced agricultural production by using several delivery mechanisms, such as nanofertilisers, nanopesticides, nanoherbicides, and NSs. These technologies are employed for disease detection in crops, plant monitoring, animal husbandry, and genetic engineering, among other applications (Mukhopadhyay, 2014).

NT is utilised to enhance agricultural productivity while minimising soil degradation. Additionally, this aids in reducing nitrogen loss due to leaching and soil microbes. The use of nanocapsules is highly efficient in dispersing herbicides evenly and intentionally over plant tissues and cuticles. This is achieved through the precise and targeted delivery of various macromolecules, which are essential for boosting resistance against plant diseases, promoting plant growth, and facilitating efficient nutrient distribution. NPs can serve as targeted carriers for herbicides, pesticides, or genes, enabling them to specifically penetrate and deliver their contents to certain plant components. NT is seen as a promising remedy for addressing diverse agricultural challenges. This explains why it has been all the rage recently. It has resulted in the emergence of innovative and distinctive techniques to achieve high farm output, thereby enhancing agricultural production. It has also offered novel agents a proficient delivery mechanism to enhance agricultural productivity. NT has enhanced agricultural production through the use of several delivery agents, such as nanofertilisers, nanopesticides, nanoherbicides, and nanosensors (Kumar *et al.*, 2021a,b; 2022; Mondal *et al.*, 2022; Sharma *et al.*, 2021). These agents are employed for disease detection in crops, plant monitoring, animal husbandry, and genetic engineering, among other applications (Sekhon, 2014). Graphical depiction of the role of NSs in agroecosystem monitoring is given in Figure 2.2.

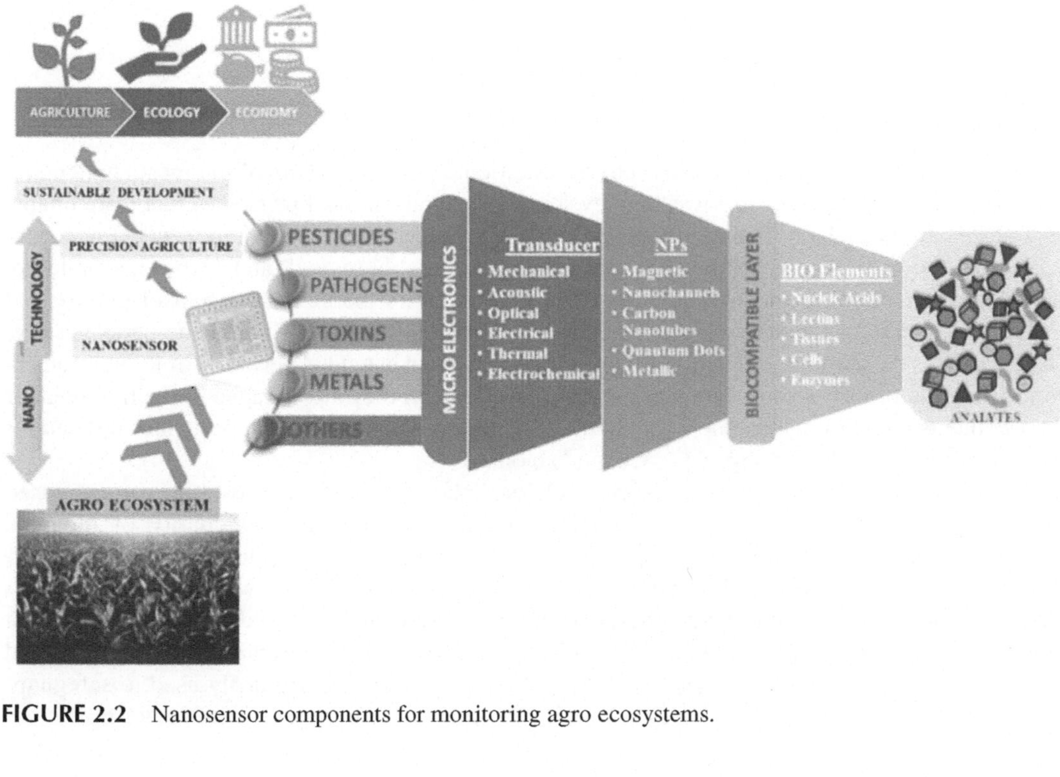

FIGURE 2.2 Nanosensor components for monitoring agro ecosystems.

The utilisation of NSs in precise agricultural practices can successfully harness agricultural natural resources, including fertilisers, chemicals, and water. Farmers may employ nanomaterials to precisely recognise disease-causing insects and various indicators of stress, such as drought. Once a crop has been detected as afflicted by a pest or bug, or if the soil is under severe drought, the NSs will lend assistance. The implementation of the automated pesticide modification application will commence. It also aids in the identification of plant diseases or soil deficiencies. Nanoencapsulated fertilisers preserve the fertilisers for future use and minimise the danger of environmental pollution (Periakaruppan *et al.*, 2023).

Figure 2.2 is an illustrative depiction that comprehensively explores the use of NT in monitoring various aspects of the agricultural ecosystem. This detailed figure presents a series of interconnected modules that demonstrate how different NT applications contribute to the efficient and effective management of agricultural resources. At the outset, the figure highlights the use of NSs in soil health monitoring. These sensors are shown detecting soil pH, moisture levels, nutrient content, and the presence of contaminants, providing critical data for optimising soil management and crop cultivation. Moving on, Figure 2.2 also shows the application of NT in monitoring crop health and growth. It illustrates how nano-based devices can detect plant diseases, stress responses, and growth patterns, enabling timely interventions for crop protection and yield optimisation. Furthermore, the figure delves into the use of nanomaterials for water quality monitoring within agricultural systems. It depicts NSs and nanofilters used for detecting and removing pollutants from irrigation water, ensuring the safety and efficiency of water resources used in farming. Additionally, Figure 2.2 addresses the role of NT in atmospheric monitoring, showing how it can help in tracking environmental factors like temperature, humidity, and the presence of harmful gases, which are crucial for maintaining a balanced agro-ecosystem. The figure explores the potential of NT in post-harvest management, demonstrating its role in preserving the quality and longevity of agricultural produce through advanced packaging and storage solutions. Overall, Figure 2.2 provides a holistic view of the numerous applications of NT in agriculture. It underscores the importance of this cutting-edge technology in enhancing the sustainability, productivity, and environmental compatibility of modern farming practices.

2.3 TANGIBLE DYNAMICS OF NANOMATERIALS

2.3.1 Nanopesticides

Nanopesticides were created as a substitute for traditional pesticides. One of the most challenging aspects for the pesticide industry was the development of conventional pesticides that were unable to produce nanoformulations containing metal NPs or polymers. The nanocomponents of nanopesticides are extremely small, yet the additions in them are highly toxic. Standard pesticides do not have these ingredients. Nanoencapsulation of pesticides offers the notable benefit of precise and progressive release of the active ingredients by manipulating the outer shell of the nanocapsules. This facilitates the administration of minute quantities over an extended period of time. This aids in minimising the runoff of superfluous pesticides. Additionally, they are engaged in the treatment of plant ailments and pest management (Sekhon, 2014). The assortment of nanopesticides utilised for plant protection includes Zn, Ag, SiO_2, Cu, and so on.

Chemical fertilisers and pesticides can enhance plant productivity, resulting in increased output. Conversely, they have the potential to impede the fertility and quality of the soil. Nanopesticides, which exhibit target specificity, mitigate the harm inflicted by non-target vegetation and the subsequent environmental dispersion of pesticides. They will enhance the soil's solubility and reduce its runoff into the environment. Nanomaterials used in the production of nanopesticides provide significant benefits such as biodegradability, thermal stability, exceptional permeability, and high tenacity (Thakur *et al.*, 2022). Pesticides are primarily used to safeguard plants from the invasion of insects, pests, fungi, and other harmful organisms. Promoting the use of environmentally friendly products is essential to protecting the environment and preventing harm to non-target animals.

Tribolium castenium, often referred to as the red flour beetle, is a pest that is particularly harmful to agricultural goods and food supplies that have been stored. As a result, the quality of these items is harmed, and the production of these products is reduced. Additionally, this insect has the ability to lower the percentage of grains that germinate. In addition to its use as a nanopesticide, copper is also employed as a nanofungicide. Copper has the potential to serve as an auxiliary instrument for the management of diseases in both organic and conventional farming practices. Copper NPs have the ability to kill fungi and insects, making them effective against pests that attack crops. Because of this, they are employed as nanopesticides, nanofungicides, nanoherbicides, and nanofertilisers. Copper oxide, copper ammonium complex, copper oxychloride, copper hydroxide, and other copper-based bioproducts are among copper-based bioproducts (Hanagasaki *et al.*, 2022; Rai and Ingle, 2012; Tefera *et al.*, 2011). Cu^{2+} ions are the active form that may be found in all goods that include copper. Copper ions, even in minute quantities, have diverse antimicrobial actions against a wide range of microorganisms, including bacteria, fungus, algae, and moulds. This is because of their contact with the nucleic acid, which disrupts enzymes, affects the transmission of energy, and also causes the integrity of the cell membrane to be disrupted (El-Saadony *et al.*, 2020)

2.3.2 Nanofertilisers

In comparison to conventional chemical fertilisers, which are an essential component of agricultural production, nanofertilisers are bioproducts that are both more efficient and less harmful to the environment. The use of nanofertilisers to stimulate plant development is a regulated process that is dependent on a wide range of conditions, including temperature, moisture, and various forms of abiotic stress, amongst others. They will be of great assistance in lowering the consumption of urea, which will result in a decrease in the amount of urea that is imported. Additionally, they will eliminate urea subsidies, which will increase the efficiency with which nitrogen is used (Pirzadah *et al.*, 2020). The preparation of nanofertilisers can be accomplished using a wide range of techniques, including physical, chemical, and biological processes, the use of derivatives derived from vegetative components of plants, or the utilisation of a modified form of conventional chemical fertilisers. There is a widespread belief that nanofertilisers represent a groundbreaking approach to the preservation of nutrients, particularly nitrogen, as well as the environment. Their purpose is to enhance the productivity of the soil as well as the fertility of the soil and the output of food productivity. Nanoscale polymers allow for the regulated and slow release of nutrients and growth stimulants, which may be accomplished in an efficient way (Srivastava *et al.*, 2023).

The results of a number of studies have demonstrated that nanofertilisers have the ability to enhance plant growth, yield, and quality. As a result, they have the endurance to improve the quality and quantity of agricultural produce that is entirely secure for eating by both humans and animals. Nanofertilisers are able to transport and supply nutrients to cells in a more efficient manner by passing through the nanoscale range between the cells, which is between 50 and 60 nanometres. As a result of their increased solubility and reactivity, they enable more precise administration and control at the intended location. And in addition to that, they have improved cuticle absorption (Javeed *et al.*, 2022). Because of their low cost and the growing knowledge among the general public about their application, zeolite-based nanofertilisers are experiencing a significant surge in demand (Preetha and Balakrishnan, 2017).

Nanofertilisers enhance the efficient and increased absorption of nutrients by plants, thanks to their large surface areas. This nanomaterial's particles can easily permeate into the plant through the applied surface, as they are smaller than the pores found in the plant roots. Reducing the particle size of the fertiliser increases the specific surface area and density of the particles, resulting in a greater surface area for the nanofertilisers. This boosts their contact with plant roots and improves the penetration and absorption of nutrients (Husen and Iqbal, 2019). Nanofertilisers augment nutrient accessibility to plants, thus enhancing total plant development by boosting dry matter production, chlorophyll levels, and photosynthesis (DeRosa *et al.*, 2010). Copper plays a crucial role in plants, since it is essential for the synthesis of chlorophyll, which is necessary for photosynthesis as well as seed germination (L. Zhang and Webster, 2009). It has a crucial role as a cofactor in metalloproteins. Copper activates various enzymes that are essential for RNA synthesis and the advancement of the photosystem (Adhikari *et al.*, 2016).

2.3.3 Nanoherbicide

Herbicides are a type of pesticide employed to inhibit the growth of undesirable plants that coexist with the primary crop. There is a chance that the presence of weeds will hinder the growth of the main crop. Herbicide application can have an impact on the growth and development of plants. The active chemicals may be administered directly into the cuticles and tissues using nanocapsules, allowing for delayed and continuous release. Herbicides exhibit greater soil stability, resulting in reduced rates of germination and dropped weights. Therefore, it is recommended to use herbicides that have inhibitory effects and can delay the process of seed germination (N. Kumar *et al.*, 2020). Conventional treatment, high levels of toxins, and a high half-life are not able to affect these herbicides, which have a high half-life.

Conversely, materials based on NT, such as nanoshells and nanopolymers, have scientific uses, such as improving the movement of soil, as demonstrated by the tobacco mild green mosaic virus (Santaella and Plancot, 2020). Copper oxide NPs are frequently employed as nanoherbicides, exhibiting effective control over weed proliferation. Nanomaterials infiltrate the plant through its roots, move through the circulatory system, and reach the photosynthetic cells, where they impede the glycolysis process. This impacts the weed by inducing famine, diminishing its growth and development, and finally leading to its demise (Itodo, Nnamonu and Wuana, 2017).

2.3.4 Nanoinsecticide

Food is a fundamental requirement for the ever-increasing population on Earth. As a consequence of this, there is a growing need for the cultivation of more food, which will result in improved protection of agricultural crops against pest attacks. Nanocapsules are the most effective means of delivering pesticides in a controlled manner among these substances. In order to dissolve the vast majority of insecticides, nanoformulations based on polymers have been utilised. Nanoinsecticides provide a number of significant benefits, including the fact that they are simple and risk-free to handle, that they are more susceptible to photodegradation, that they are less harmful to species that are not the intended targets, that they are more effective than both natural and chemical insecticides, and so on. A lot of different nanoformulations, both polymeric and non-polymeric, have been used to package a lot of different insecticides, including particle-sized ones (Chakraborty *et al.*, 2022). Owing to the greater residual rate that synthetic pesticides produce, overuse of these chemicals might potentially harm the environment. This is due to the fact that many insect pests have developed resistance to certain pesticides; hence, a new strategy for controlling pests is necessary. A few examples of nanomaterials are nanoemulsions, nanocapsules, nanofibers, nanogels, and other similar substances (Nuruzzaman *et al.*, 2016).

2.4 AID TO NANOSCALE GROWTH

We are dependent on chemical fertilisers for the purpose of increasing agricultural productivity on a worldwide scale in order to meet the growing demand for food that originates from the world's expanding population. The three most important chemical fertilisers that are in high demand are nitrogen, phosphorus, and potassium. Consequently, the production of ammonia on a global scale for a variety of agricultural and industrial applications has significantly expanded as a consequence of this. Mother Nature is now facing significant danger. Consequently, in order to guarantee food security, crop productivity, ecological security, and sustainability, contemporary agriculture is relying on an innovative strategy of nanobiotechnology for the creation of nanobiofertilisers through the use of NT. For the purpose of the formulation of nanobiofertilisers using nanoencapsulation technology, the biofertilisers that are created in an environmentally friendly manner and that contain microbes and other nutrients that promote plant development are coated with nanoscale polymers (Rana, 2021). Table 2.1 provides a list of applications for different types of NPs in the diverse agriculture industry.

Table 2.1 explores the use of nanoparticles in enhancing crop quality and yield. it explains how nanoparticles can be used to increase the shelf life of produce, improve crop resilience

TABLE 2.1
An Exhaustive Compilation of Varied Applications of Different Types of Nanoparticles in the Agricultural Sector, underscoring their Multifunctional Roles across a spectrum of Farming Activities*

Nanoparticles	Applications	Key References
AgNPs/ oxMWCNTs	Insecticide	(Hsu *et al.*, 2017)
Bionanocomposite materials	Conversion of light energy or biosensing	(Nagy *et al.*, 2014)
Cellulose nanofibers	Biomedical application-like tissue engineering	(de Lima *et al.*, 2012)
CeO_2	Stress response and tolerance of *Zeamays*	(Zhao *et al.*, 2012)
Zinc oxide	Transparent electronics, ultraviolet (UV) light emitters, piezoelectric devices, chemical sensors, spin electronics, enhancement of crop growth	(Sabir, Arshad and Chaudhari, 2014)
Silver	Control of *Colletotrichum* species	(Lamsal *et al.*, 2011)
Poly (γ-glutamic acid (γ-PGA) and chitosan (CS) polymer-based NPs	Nanofertilisers	(Pereira *et al.*, 2017)
Fe_2O_3	Insecticide in Bt-transgenic and non-transgenic cotton	(Van Nhan *et al.*, 2016)
Carbon-coated Fe Nanoparticles	Smart treatment-delivery systems in plant tissue	(González-Melendi *et al.*, 2008)
Copper (II) oxide (CuO) and multiwall carbon nanotubes (MWCNTs)	Detection of glyphosate in water	(Chang *et al.*, 2016)
Cu $(OH)_2$ nanopesticides	Enhancement of nutritional value of *Lactuca sativa*	(Zhao *et al.*, 2016)
Fe_2O_3	Nanofertiliser in *Arachis hypogaea*	(Rui *et al.*, 2016)
Nano air bubbles	Hydroponic growth of *Brassica campestris*	(Ebina *et al.*, 2013)
Optimised diuron Nanoformulation	Nanoherbicide	(Yearla and Padmasree, 2016)
Nanostructured lipid carriers	Antimicrobial activity	(Cortesi *et al.*, 2017)
Vitamin D3-loaded nanostructured lipid carriers (NLCs)	Fortifying food beverages	(Mohammadi *et al.*, 2017)
Quaternised chitosan-capped mesoporous silica NPs	Fungicidal activity against *Phomopsis asparagi*	(Cao *et al.*, 2016)

(Continued)

TABLE 2.1 *(Continued)*
An Exhaustive Compilation of Varied Applications of Different Types of Nanoparticles in the Agricultural Sector, underscoring their Multifunctional Roles across a spectrum of Farming Activities*

Nanoparticles	Applications	Key References
Thermoplastic starch/ urea medleys as a matrix with hydroxyapatite as nanocomposites	Nanofertilisers	(Giroto *et al.*, 2017)
Protein	Nanoencapsulation of hydrophobic nutraceuticals Improved functionalities in casein micelle (Meat processing)	(SEMO *et al.*, 2007)
Nanodiagnostic kit	Plant pathogen detection	(Khiyami *et al.*, 2014)
Gold NP-based electrochemical biosensor	Rapid and sensitive detection of plant pathogen DNA	(Lau *et al.*, 2017)
Nanochitosan	Supports growth of *Zea mays* and also maintains soil health	(Khati *et al.*, 2017)
NP-based sensors	Assessing food safety	(Bülbül *et al.*, 2015)
Cerium oxide and titanium Oxide	As nanofertiliser and nutrient enhancement in *Hordeum vulgare* L.	(Poščić *et al.*, 2016)
Engineered water Nanostructures	Antimicrobial platform for food safety	(Pyrgiotakis *et al.*, 2016)
Fullerene, C60 and carbon Nanotubes	Increase the water retaining capacity, biomass, and fruit yield in plants	(Husen and Siddiqi, 2014)
Magnesium hydroxide ($Mg(OH)_2$) nanocomposites	Efficient bioactive packaging of goods	(Moreira *et al.*, 2013)
Multi-walled carbon nanotubes (MWCNTs)	Contaminant carriers and translocation within *Brassica juncea*	(Chen *et al.*, 2015)
MWCNT and cotton CNF	Show ecotoxicological effects by inducing viability of *Chlorella vulgaris*	(Pereira *et al.*, 2014)
Nanobiosensors	Labelling products and automated storage	(Ali *et al.*, 2017)

Note: * This table categorizes nanoparticles such as silver, gold, copper, zinc oxide, titanium dioxide, and carbon-based nanoparticles, and delineates their specific uses in agriculture. For each type of nanoparticle, the table provides information on how they enhance plant growth, protect crops from diseases and pests, and improve nutrient utilization efficiency. the table also elaborates on the role of nanoparticles in soil amendment, demonstrating how certain nanoparticles can improve soil texture, aeration, and fertility. In the realm of pest and disease management, it illustrates the efficacy of nanoparticles as components of pesticides and fungicides, detailing their modes of action and advantages over traditional chemical agents.

to environmental stresses, and boost overall nutritional content. additionally, the table highlights the application of nanotechnology in precision agriculture, such as in the development of nano-sensors for monitoring soil and crop conditions, thus facilitating targeted and efficient use of resources. by providing a comprehensive overview, Table 2.1 serves as a vital resource for agricultural researchers, farmers, and policy makers. it offers a clear understanding of how the innovative application of nanoparticles can significantly contribute to sustainable and efficient agricultural practices, addressing current challenges and paving the way for future advancements in the sector.

2.5 THE SCIENCE BEHIND NANOBARCODES

Barcodes are playing an increasingly significant role in the management of livestock and agricultural operations in the modern world. Due to the fact that it is so effective, it has been utilised in a significant capacity as an identity tag. Both ultraviolet (UV) lights and optical microscopes are able to detect this. A significant amount of agricultural business makes considerable use of multifluorescence material that is equipped with micrometre-sized glass barcodes. In order to read NPs, which are very long-lasting, easily encodable, and smaller than microsized taggant particles, a machine makes use of these characteristics. The production of nanobarcodes (NB) is made possible by these semiautomated machines, which are extremely effective. Electroplated metallic particles, such as gold, silver, aluminium, and zinc, make up the majority of the nanorods that have been stripped. Agrofood goods are tagged in a way that is exclusive to NT. Furthermore, the use of nanobioprocessing and nanobarcoding allows for the monitoring of agricultural yield quality in addition to the detection of illnesses. For the specialised detection of reactions with particular labels, several techniques have been created.

2.5.1 Nanosensor-based Nutrient Deficiency Imaging

One of the most significant challenges that modern farmers are confronted with is the lack of nutrients that are present in the crops that they cultivate in order to get a sustainable yield from the crop over a predetermined amount of time all over the world. This is mostly the result of a decrease in the availability of fertilisers as well as the fertility of the soil. Through the utilisation of nanobiosensors, this issue may be resolved by ensuring that nutritional deficiencies are identified in a timely manner. When it comes to farming in India, this strategy is highly recommended. On the other hand, smart delivery systems assist in transporting compounds that are rich in nutrients and are preprogrammed, self-regulated, and target-specific. This helps in the exact targeting of biological carriers, which is a significant benefit (Ahmad *et al.*, 2022).

2.5.2 Nanosensors in Agriculture

The antibacterial properties of silver NPs and their ability to aid in the remediation of wastewater are the primary reasons for their development in agricultural states. According to a study, using silver NPs at a concentration of 100 ppm can prevent powdery mildew disease. The elimination of fungal populations and the promotion of conidial development in the *Cucurbitaceae* family are the means by which this is accomplished (Lamsal *et al.*, 2011). In order to determine whether or not pathogens are present in plants and postharvest food crops, silver NPs are used as both bio-NSs and electrical NSs. Biosensors that are based on silver NT can be utilised to ascertain whether or not organophosphorus pesticides are present in postharvest food crops as well as in the surrounding environment (Wu *et al.*, 2011). Several studies have reported on the sensitivity and selectivity of different NSs for certain pesticides (Table 2.2).

Furthermore, Table 2.2 discusses the technological innovations that have led to the development of these nanosensors, highlighting how they offer enhanced sensitivity and specificity compared to traditional pesticide detection methods. It also touches upon the integration of these nanosensors into smart agricultural systems, illustrating their role in ensuring food safety and environmental protection by enabling precise and timely detection of pesticide residues. Overall, Table 2.2 serves as an invaluable resource for researchers, agronomists, and environmental scientists. it provides a comprehensive overview of the state-of-the-art in nanosensor technology for pesticide detection, showcasing the advancements in this field and their significant implications for sustainable agriculture and public health.

TABLE 2.2
Detailed and Structured Presentation of the Diverse Range of Nanosensors Currently Being Utilized for the Detection of Pesticides in Agricultural Settings*.

Nanosensor	Nanomaterial Used	Pesticide Detected	Limit of Detection	Method of Nanosensor Formulation	Sensing Mechanism	Observation	Key References
Fluorescence sensor	Copper (II) oxide and multiwall carbon nanotubes (MWCNTs)	Glyphosate	0.67 ppb	CuO/MWCNT were prepared by precipitating copper nitrate by adding aqueous NaOH solution	Inhibiting the catalytic activity of CuO/MWCNTs	A selective approach or rapid screening of glyphosate	(Chang *et al.*, 2016)
Electrochemical sensor	CuO-TiO_2 hybrid nanocomposites	Methyl parathion	1.21 ppb	CuO-TiO_2 nanocomposites prepared by a facile liquid-control-precipitation method	Differential pulse voltammetry measurements assessed from a decline in current density with an increase in the methyl parathion concentration	A non-enzymatic sensor with good stability and excellent reproducibility	(Tian *et al.*, 2018)
Optical sensor	Upconverting nanoparticles (UCNPs) of the $NaYF_4$: Yb, Er type	Metribuzin	6.8 × 10– 8 M	Upconverting nanoparticles synthesised using the coprecipitation method of lanthanide metal-EDTA complexes were later used in the preparation of the sensor film by dissolving UCNPs in tetrahydrofuran along with the incorporation of an NIR dye, PVC polymer, and dioctyl phthalate	Metribuzin changes the colour of sensor film from green to blue with a significant blue shift in the absorption peak	Highly sensitive sensor with unique luminescence properties of UCNPs and great recognition abilities within a very low detection limit	(Saleh *et al.*, 2020)
Electrochemical aptasensor	Chitosan-iron oxide nanocomposite	Malathion	0.001 ng/mL	Iron oxide nanoparticles synthesised using a chemical co-precipitation method were deposited on fluorine tin oxide followed by immobilisation of aptamer onto the iron oxide dopedchitosan/FTO electrode using streptavidin	Decline in the differential pulse voltammetry peak current of the aptaelectrode with a corresponding increase in malathion concentration due to the formation of more 3D-complexes between the aptamer with malathion	A very attractive alternative to quantify and monitor malathion due to its sensitivity, stability, short analysis time, and cost-effectiveness	(Prabhakar *et al.*, 2016)

(Continued)

TABLE 2.2 *(Continued)*

Detailed and Structured Presentation of the Diverse Range of Nanosensors Currently Being Utilized for the Detection of Pesticides in Agricultural Settings*.

Nanosensor	Nanomaterial Used	Pesticide Detected	Limit of Detection	Method of Nanosensor Formulation	Sensing Mechanism	Observation	Key References
Fluorescent nanosensor	3-aminopropyl-triethoxysilane coated Yb_2O_3	Imazapyr	0.2 ppm	Hydrothermal method	Quenching of fluorescence intensity for APTES-coated Yb_2O_3 NPs with the increasing concentration of imazapyr	Among the lanthanide oxide-based nanomaterials, ytterbium (III) oxide (Yb_2O_3) NPs owe unique optical and luminescence properties with excellent efficiency for real field conditions	(Kumar *et al.*, 2020)
Surface plasmon resonance-based fiberoptic nanosensor	Tantalum (V) oxide nanoparticles	Fenitrothion	38 nM	Chemical synthesis of Ta_2O_5 nanoparticles embedded in a reduced graphene oxide matrix followed by its adhesion on a silver-coated fibre optic probe	Change in refractive index due to the interaction of fenitrothion with the silver film	The sensor is selective, repeatable, and works at ambient temperature with a response time of 23 s	(Kant, 2020)
Electrochemical luminescence sensor	Luminol-gold nanoparticles-L-cysteine-Cu(II) composites	Glyphosate	0.5 nM	Layer-by-layer assembly of graphene-gold nanoparticle composite and Lu-Au-Lcys-Cu(II) composite	Decrease in electrochemical luminescence intensity with a respective increase in the glyphosate concentration	The sensor worked on a dual inhibition strategy with excellent detection performance, high sensitivity, desirable reproducibility, stability, and accuracy	(Liu *et al.*, 2020)
Electrochemical nanosensor	CuO nanoparticles Decorated 3D graphene Nanocomposite	Malathion	0.01 nM	Copper oxide nanoparticle electro-catalysts were prepared on 3D graphene and synthesised using a hydrothermal process	Decline in peak current with increasing concentrations of malathion	Highly sensitive, reproducible, and applicable in real field conditions	(Xie *et al.*, 2018)

(Continued)

TABLE 2.2 *(Continued)*
Detailed and Structured Presentation of the Diverse Range of Nanosensors Currently Being Utilized for the Detection of Pesticides in Agricultural Settings*.

Nanosensor	Nanomaterial Used	Pesticide Detected	Limit of Detection	Method of Nanosensor Formulation	Sensing Mechanism	Observation	Key References
Optical nanosensor	Silver nanodendrites	Dimethoate	0.002 ppm	Ag nanodendrites fabricated by laser-assisted photochemical methods were immobilised on the surface of themicrosphere end-shape optical fibre	Increase in the intensity of the surface-enhanced Raman spectroscopy (SERS) signal with a proportionate increase in dimethoate concentration	A direct, rapid, real-time, and non-destructive method of detecting pesticide residue in the outdoor fields	(Pham *et al.*, 2020)
Surface plasmon resonance (SPR)-based affinity sensor	Atrazine-imprinted nanoparticles	Atrazine	0.7134 ng/mL	Atrazine-imprinted nanoparticles' synthesis using emulsion polymerisation methods followed by their attachment on the gold surface of SPR	Increase in resonance frequency in proportion to the increment in atrazine concentration	The plastic antibody-based SPR nanosensor is an attractive recognition element for the detection of atrazine with high electivity and sensitivity	(Yılmaz *et al.*, 2017)

Note: * This table meticulously categorizes various types of nanosensors based on their detection mechanisms, such as optical, electrochemical, and mass-sensitive nanosensors. Each category is further expanded to list specific examples of nanosensors, including those based on gold nanoparticles, quantum dots, carbon nanotubes, and graphene oxide. For each nanosensor type, the table provides essential details such as the principle of operation, sensitivity levels, detection limits, and the range of pesticides that can be detected. It also outlines the advantages and potential challenges associated with each type of nanosensor, offering insights into their practical applications and efficiency in real-world scenarios.

There has been the development of a triangular silver NP that possesses distinctive optical properties and heightened sensitivity to its nanoenvironment (Haes and Duyne, 2004). It has been demonstrated that chitosan-mediated biosensors have the capability to cascade the CD4-binding domain of the host virus through the disulfide bond regions. They prevent the formation of RNA and extracellular virions, which is another way in which they interact with the genetic material of the virus. They also absorb the heavy metal deposits that are found in agricultural areas, and this is dependent on the thermal and chemical stability of the materials. In a similar manner, sensors equipped with silver NPs are able to identify carcinogenic contaminants using hyperbranched polyethyleneimine. The road that will lead to the successful use of nanodevices for the delivery of particular protection in order to achieve sustainability in agriculture is still a long way off (Periakaruppan *et al.*, 2023).

2.5.3 Nanosensor-based Pathogen Detection

NSs are designed in such a manner that they may be utilised for a wide range of identifying purposes, which is one of the most essential characteristics that distinguish them from other types of identification. The influence that this has on agriculture is very beneficial. The detection of pathogens frequently makes use of NPs based on gold. The approach of seed-mediated synthesis was utilised in order to create gold nanorods, which were subsequently immobilised on the surface of the fibre core. The gold NPs were then turned on by sinking them in a solution of either the *Cymbidium* mosaic virus or the *Odontoglossom* ringspot virus antibodies. This created an optical particle plasmon resonance. In order to alter the electrode that detects *Ganoderma boninense* in electrochemical sensors, gold NPs of varying sizes were synthesised by combining gold (III) chloride trihydrate and sodium citrate dehydrate (Isha *et al.*, 2019). These NPs were then constructed in a layer-by-layer fashion. Stannic oxide and titanium dioxide were utilised as electrochemical detection components for the purpose of *Phytophthora cactorum* detection by means of an electrochemical NS. Screen-printed carbon electrodes were eventually replaced with stannic oxide and titanium dioxide NPs as a result of this factor. Amino-functionalised iron oxide/silicon dioxide metal NPs were created from metal NPs, and these NPs were covalently bonded with antibodies (Fang *et al.*, 2014). This allowed for the detection of the presence of the tomato ring spot virus, the Arabis mosaic virus, and the bean pod mottle virus (B.-T. Zhang *et al.*, 2013). The DNA of *Pseudomonas syringae* was modified by the addition of tris-2-carboxyethyl phosphine in order to facilitate the identification of the bacterium. The versatility of NSs in detecting various things is a key feature that makes them indispensable in agriculture. Nevertheless, non-target species, such as humans and animals, have been adversely impacted by the pesticides' contamination of groundwater and their buildup in food supplies, caused by the careless use of these chemicals in field situations (Figure 2.3).

Figure 2.3 is an informative and detailed illustration that vividly portrays the various impacts of pesticide exposure on human health. This comprehensive diagram delineates the different pathways through which pesticides can enter the human body, such as inhalation, ingestion, and dermal contact. It also highlights the key areas and systems within the body that are most susceptible to the adverse effects of pesticides, including the nervous system, respiratory system, endocrine system, and reproductive system. The figure goes on to explain the range of health issues that can be triggered or exacerbated by pesticide exposure. For the nervous system, it details how certain pesticides can lead to neurological disorders and cognitive impairments. Regarding the respiratory system, the diagram shows the potential for pesticides to cause respiratory problems, including asthma and bronchitis. The impact on the endocrine system is also illustrated, with an emphasis on how some pesticides can disrupt the hormone balance, potentially leading to developmental and reproductive issues. Furthermore, the figure addresses the potential carcinogenic effects of certain pesticides and their role in increasing the risk of various types of cancer. In addition to immediate health effects,

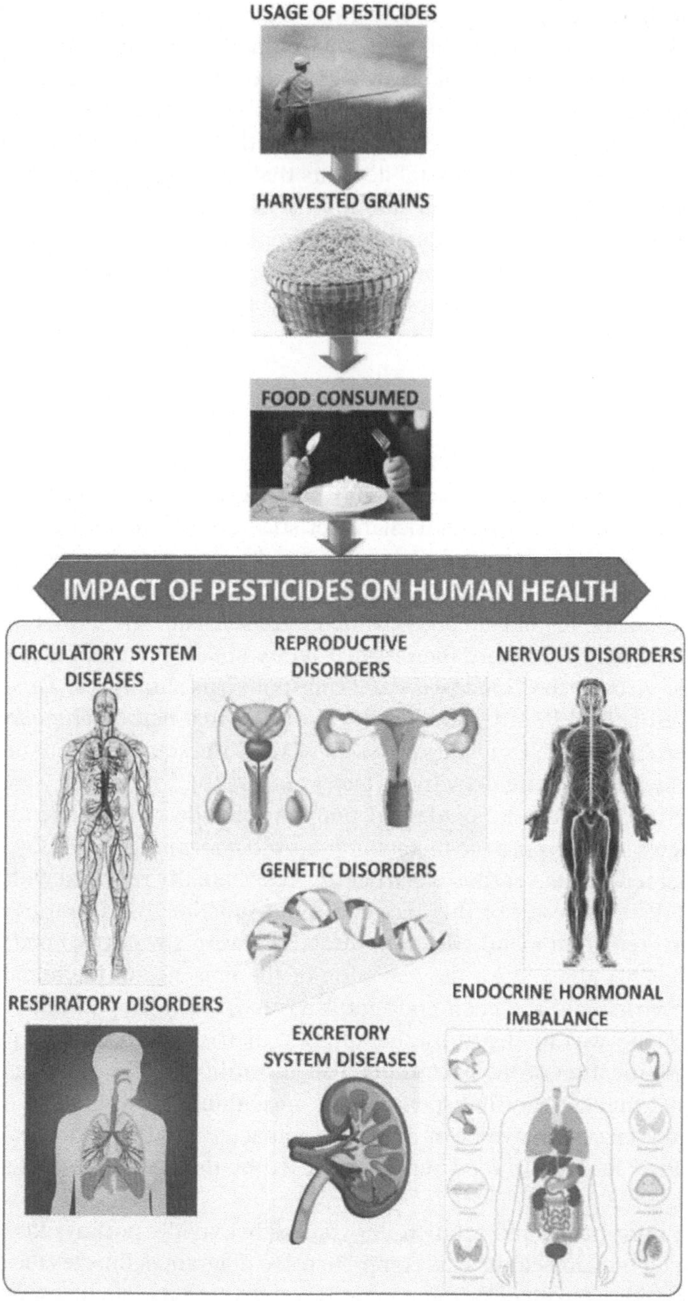

FIGURE 2.3 Impact of pesticide exposure on human health, illustrating various health risks including circulatory, reproductive, nervous, respiratory, genetic, endocrine, and excretory disorders.

Figure 2.3 also explores the long-term consequences of chronic exposure to pesticides, including the potential for bioaccumulation and the resultant health risks. The diagram is rounded out with a section on preventive measures and best practices for minimising pesticide exposure and protecting human health, particularly for those working in agriculture and individuals living in close proximity to farming areas. Overall, Figure 2.3 serves as a crucial educational tool, emphasising the importance of understanding and mitigating the risks associated with pesticide use. It provides a clear and comprehensive view of how pesticides can affect human health, highlighting the need for safer pesticide practices and stricter regulatory measures.

2.5.4 Heavy Metal Detection via Nanosensors

It is generally agreed that the presence of even minute amounts of heavy metals, such as arsenic, lead, chromium, nickel, and cadmium, is detrimental to both human health and the environment. In order to further process the nitrocellulose membrane, a colloidal gold probe was placed on a glass fibre membrane. Cadmium (II)-EDTA-BSA-antigen and IgG immunoglobulin were then dispersed throughout the membrane. Additionally, this contributes to the formation of the immunochromagraphic strip NS, which is very sensitive and has the capability of detecting cadmium levels (Xing *et al.*, 2014). The creation of cadmium tellurite that is sealed with amino acids and luminous silicon dioxide NPs was made using a modified reverse microemulsion method (Wang *et al.*, 2016). Through the process of depositing new dithiocetal-rooted stimulus molecules on a reporter dye, calorimetric NSs were produced. These NSs may be used for the detection of mercury (Jimenez-Falcao *et al.*, 2020). Surface plasmon resonance NPs are made when epicatechin and silver nitrate are mixed in different amounts to make ECAg NPs. This type of NP may be used for lead detection. The cationic 3-acetylthio-propyl-pyrazin-1 ligand was used as a calorimetric neutron source (NS) from gold NPs (II) for the purpose of palladium recognition (Ikram *et al.*, 2019). The hydrothermal synthesis of NH2-UiO-66 resulted in the production of an electrochemical sensor that is capable of detecting copper, lead, and cadmium in an efficient manner. In order to manufacture multifunctional magnetic fluorescent NPs that are capable of functioning as NSs, carboxymethyl chitosan was utilised. When it comes to packing Fe_3O_4 NPs and quantum dots, this will serve as an encapsulating agent. In this manner, it is possible to identify a wide variety of heavy metals, the presence of which holds the potential to pose a significant risk to the environment (Anwar *et al.*, 2018; Yang, Ding and Qian, 2018).

2.5.5 Nanosensor-dependent Pesticide Residue Detection

The detection of pesticides that are harmful to the environment as a whole may be accomplished with the use of a variety of different kinds of NSs. In order to detect imazapyr, a pesticide, fluorescent NPs can be used. Ytterbium (III) oxide that has been coated with 3-aminopropyl-triethoxysilane allows for the production of this through the use of the hydrothermal process (S. Kumar *et al.*, 2020). A surface-plasmon resonance-based affinity sensor for detecting atrazine is made by employing an emulsion polymerisation procedure. Atrazine-imprinted NPs are coated on the gold surface of the sensor (Yılmaz *et al.*, 2017). Fenitrothion may be detected by the chemical manufacture of tantalum pentoxide NPs in a reduced graphene oxide matrix. These NPs are attached to silver-coated fibre-optic probes, which are employed for the creation of surface-plasmon resonance-based fibreoptic NSs (Kant, 2020). For the purpose of detecting methyl parathion, electrochemical sensors are generally utilised. Copper oxide and titanium dioxide nanocomposites are used in the development of these sensors. These nanocomposites are then coated onto a glass carbon electrode by means of a straightforward liquid control precipitation procedure (Tian *et al.*, 2018). Electrochemical luminescence sensors are able to detect glyophosate in the natural environment. In order to create these, a layer-by-layer arrangement of graphene and gold NPs, as well as luminol, gold NPs, L-cysteine, and copper, is utilised (Liu *et al.*, 2020).

The development of silver nanodendrites was done in order to create optical NSs that are capable of detecting dimethoate. A laser-assisted photochemical approach is utilised in order to accomplish this. The NPs are made using a method known as coprecipitation of lanthanide metal-EDTA complexes. They are then turned into sensor film so that they can be used to make optical sensors. With the inclusion of diocytl phthalate, PVC polymer, and NIR dye, this may be used for the detection of metribuzin (Saleh *et al.*, 2020). In the presence of copper oxide-multiwalled copper nanotubes, a fluorescence sensor is utilised for the purpose of detecting glyophosate. With the addition of an aqueous solution of sodium hydroxide, copper nitrate precipitates to accomplish this (Chang *et al.*, 2016). Through the use of a chemical coprecipitation technique, NPs of iron oxide are utilised in the development of electrochemical aptasensors. This was then deposited on fluorine tin oxide, and it was immobilised in the iron oxide-doped chitosan-fluorine tin oxide electrode by means of streptavidin and an electrocatalysis of copper oxide. This resulted in the generation of electrochemical NSs, which may be used for malathion detection (Prabhakar *et al.*, 2016; Xie *et al.*, 2018).

2.5.6 Detection of Nanoparticles via Nanosensors

Excessive human activity, such as the discharge of welding fumes, diesel oil byproducts, and other similar substances, can result in the unintended production of a number of NPs in the environment. These NPs include humic acid, clay materials, and other similar substances (Picó and Andreu, 2014). These properties are being utilised in a wide variety of items that are used on a daily basis, such as pharmaceuticals, sun protection lotion, food, cosmetics, and so on, among other things, since they release nanomaterials directly into the environment (O'Brien and Cummins, 2010). In addition, the excessive usage of metal NPs has resulted in a number of problems, considering that these particles are released without any form of protection during the fabrication, shipping, and disposal processes (Hou *et al.*, 2018). Carbon-based nanomaterials are employed in the environment in a somewhat widespread manner, which results in their buildup in the environment (Farré *et al.*, 2009). Because of the larger surface area that nanomaterials possess, they are able to quickly attach and absorb onto the surface of the cell. They have the potential to cause harm to the cells in a variety of ways, including disrupting the route that transports proteins, damaging the permeability of the cell membrane, and limiting the production of essential cellular components (Kumar *et al.*, 2011). When it comes down to it, an alarmingly high quantity of manufactured NPs or their byproducts are ultimately discharged into the ecology of the soil. Some of the several kinds of manufactured NPs are elemental (such as iron, silver, and gold), fullerenes (such as carbon nanotubes and nanocones), nanocomposite materials (such as cobalt-iron-zinc oxide), quantum dots, and so on (Xie *et al.*, 2018). Other plant growth-promoting rhizobacteria, such as *Bacillus subtilis* and *Pseudomonas aeruginosa*, as well as other bacteria that are responsible for soil nitrogen fixation, are inhibited when they are exposed to varied amounts of NPs (Sharma *et al.*, 2021). As a result, there is an urgent requirement to keep a close eye on the abundance of certain micronutrients, particularly in agricultural environments. For the purpose of determining whether or not NPs are present, a number of methods may be utilised. One of these methods is the utilisation of microcavity sensors, which are in the form of whispering gallery resonators. The optical properties are disrupted as a result of the particle binding that occurs on the exterior of the microcavity. In addition to being used for the purpose of measuring particle size, the measure of change makes it possible to do real-time monitoring of binding procedures (Lu *et al.*, 2011).

2.6 FUTURE PERSPECTIVE

In order to maximise agricultural productivity, many government agencies need to adopt new technology, encourage the use of nanomaterials, and encourage urbanisation. At the same time as biotic and abiotic harmony is established in collaboration with the stability of food chains and their corresponding balance in the energy systems, sustainable agriculture must be regarded as a way that is both protective and productive. Precision agriculture is in high demand in the current world as a result of the globalisation that has occurred and the accompanying climate change that has occurred. In the field of agriculture, this kind of farming has the potential to increase crop yields while simultaneously enhancing the fertility of the soil and reducing the application of agrochemicals such as herbicides, insecticides, fertilisers, and other similar substances. It is necessary to perform constant monitoring of a variety of environmental elements in addition to their direct execution in order to practice precision farming. As a result, there is a robust demand for the implementation of cutting-edge technologies such as NT. Through the utilisation of computers, global satellites, remote sensors, and other similar technologies, this method involves direct monitoring. Farmers are able to achieve higher levels of output and make more informed decisions regarding their recovery via the use of smart sensor technology. The utilisation of smart NSs is extremely sensitive and needs to be utilised in a prudent manner in order for them to be able to function as a key instrument for the purpose of attaining sustainable agriculture in the future.

It has been demonstrated that the utilisation of NSs has resulted in an increase in the rate of agricultural productivity. These sensors will perform continuous monitoring of a variety of factors, including temperature, soil condition, moisture level, and even the microbiological composition of the soil. They are accurate, dependable, and provide previous knowledge, which even helps in minimising the forecast of crop losses in agroecosystems. These are the primary characteristics that distinguish them from other technologies. In addition, there is a requirement for multiplexed screening that is not only rapid and easy but also reliable, inexpensive, and able to identify a wide variety of plant-based nanoproducts. The creation of wide-spectrum NSs that are capable of detecting many entities will also contribute to the advancement of technology in the not-too-distant future.

2.7 CONCLUSION

Agriculture, the sole sector that supplies food to humans, should generate food from intermediate and final inputs using well-established technology. Therefore, it is of the utmost importance to make a fresh start in the agricultural sector. Despite the fact that there is a great deal of information in the field of agriculture, many nations are still struggling to ensure that their agricultural practices are sustainable. The past few years have seen an astonishing increase in the amount of research effort and innovation that has taken place in the field of NS. Many people lack understanding of the proper utilisation of nanomaterials, despite the fact that there is a great deal of information accessible about them. The amount of toxicity that many of these NPs possess is not measurable; thus, their application is restricted owing to the risk concerns that they provide as well as the influence that they have on human health.

REFERENCES

Adhikari, T., *et al.* (2016) 'Growth and enzymatic activity of maize (*Zea mays* L.) plant: Solution culture test for copper dioxide nano particles', *Journal of Plant Nutrition*, 39(1), pp. 99–115. https://doi.org/10.1080/01904167.2015.1044012

Ahmad, Z., *et al.* (2022) 'Nanotechnological interventions in agriculture', *Nanomaterials*, 12(15), p. 2667. https://doi.org/10.3390/nano12152667

Ali, J., *et al.* (2017) 'Biosensors: Their fundamentals, designs, types and most recent impactful applications: A review', *Journal of Biosensors & Bioelectronics*, 08(01). https://doi.org/10.4172/2155-6210.1000235

Anwar, A., *et al.* (2018) 'Synthesis of gold nanoparticles stabilized by a pyrazinium thioacetate ligand: A new colorimetric nanosensor for detection of heavy metal Pd(II)', *Sensors and Actuators B: Chemical*, 257, pp. 875–881. https://doi.org/10.1016/j.snb.2017.11.040

Bülbül, G., *et al.* (2015) 'Portable nanoparticle-based sensors for food safety assessment', *Sensors*, 15(12), pp. 30736–30758. https://doi.org/10.3390/s151229826

Cao, L., *et al.* (2016) 'Quaternized chitosan-capped mesoporous silica nanoparticles as nanocarriers for controlled pesticide release', *Nanomaterials*, 6(7), p. 126. https://doi.org/10.3390/nano6070126

Chakraborty, S., Singh, A. and Roychoudhury, A. (2022) 'Biogenic nanoparticles and generation of abiotic stress-resilient plants: A new approach for sustainable agriculture', *Plant Stress*, 6, p. 100117. https://doi.org/10.1016/j.stress.2022.100117

Chang, Y.-C., *et al.* (2016) 'A highly selective and sensitive nanosensor for the detection of glyphosate', *Talanta*, 161, pp. 94–98. https://doi.org/10.1016/j.talanta.2016.08.029

Chen, G., *et al.* (2015) 'Carbon nanotubes act as contaminant carriers and translocate within plants', *Scientific Reports*, 5(1), p. 15682. https://doi.org/10.1038/srep15682

Cortesi, R., *et al.* (2017) 'Nanostructured lipid carriers (NLC) for the delivery of natural molecules with antimicrobial activity: Production, characterisation and *in vitro* studies', *Journal of Microencapsulation*, 34(1), pp. 63–72. https://doi.org/10.1080/02652048.2017.1284276

de Lima, R., *et al.* (2012) 'Evaluation of the genotoxicity of cellulose nanofibers', *International Journal of Nanomedicine*, p. 3555. https://doi.org/10.2147/IJN.S30596

DeRosa, M.C., *et al.* (2010) 'Nanotechnology in fertilizers', *Nature Nanotechnology*, 5(2), pp. 91–91. https://doi.org/10.1038/nnano.2010.2

Ebina, K., *et al.* (2013) 'Oxygen and air nanobubble water solution promote the growth of plants, fishes, and mice', *PLoS One*, 8(6), p. e65339. https://doi.org/10.1371/journal.pone.0065339

El-Saadony, M.T., *et al.* (2020) 'Ecofriendly synthesis and insecticidal application of copper nanoparticles against the storage pest *Tribolium castaneum*', *Nanomaterials*, 10(3), p. 587. https://doi.org/10.3390/nano10030587

Fang, Y., Umasankar, Y. and Ramasamy, R.P. (2014) 'Electrochemical detection of p-ethylguaiacol, a fungi infected fruit volatile using metal oxide nanoparticles', *The Analyst*, 139(15), pp. 3804–3810. https://doi.org/10.1039/C4AN00384E

Farré, M., *et al.* (2009) 'Ecotoxicity and analysis of nanomaterials in the aquatic environment', *Analytical and Bioanalytical Chemistry*, 393(1), pp. 81–95. https://doi.org/10.1007/s00216-008-2458-1

Giroto, A.S., *et al.* (2017) 'Role of slow-release nanocomposite fertilizers on nitrogen and phosphate availability in soil', *Scientific Reports*, 7(1), p. 46032. https://doi.org/10.1038/srep46032

González-Melendi, P., *et al.* (2008) 'Nanoparticles as smart treatment-delivery systems in plants: Assessment of different techniques of microscopy for their visualization in plant tissues', *Annals of Botany*, 101(1), pp. 187–195. https://doi.org/10.1093/aob/mcm283

Haes, A.J. and Duyne, R.P. (2004) 'Preliminary studies and potential applications of localized surface plasmon resonance spectroscopy in medical diagnostics', *Expert Review of Molecular Diagnostics*, 4(4), pp. 527–537. https://doi.org/10.1586/14737159.4.4.527

Hanagasaki, T., *et al.* (2022) 'In vitro pesticides susceptibility of *Erwinia* sp. causing papaya (*Carica papaya*) black rot in Okinawa, Japan and captan effectiveness on papaya pot seedlings', *The Journal of General Plant Pathology*, 88, pp. 178–186. https://doi.org/10.1007/s10327-022-01056-7

Hou, J., *et al.* (2018) 'Toxic effects of different types of zinc oxide nanoparticles on algae, plants, invertebrates, vertebrates and microorganisms', *Chemosphere*, 193, pp. 852–860. https://doi.org/10.1016/j.chemosphere.2017.11.077

Hsu, C.-W., *et al.* (2017) 'Oxidized multiwalled carbon nanotubes decorated with silver nanoparticles for fluorometric detection of dimethoate', *Food Chemistry*, 224, pp. 353–358. https://doi.org/10.1016/j.foodchem.2016.12.095

Husen, A. (2022) *Engineered nanomaterials for sustainable agricultural production, soil improvement and stress management*. Cambridge, MA: Elsevier Inc. https://doi.org/10.1016/C2021-0-00054-7

Husen, A. (2023) *Nanomaterials and nanocomposites exposures to plants (response, interaction, phytotoxicity and defense mechanisms)*. Singapore: Springer Nature Singapore. https://doi.org/10.1007/978-981-99-2419-6

Husen, A. and Iqbal, M. (2019) 'Nanomaterials and plant potential: An overview', in *Nanomaterials and plant potential*. Springer International Publishing, pp. 3–29. https://doi.org/10.1007/978-3-030-05569-1_1

Husen, A. and Siddiqi, K. (2014) 'Carbon and fullerene nanomaterials in plant system', *Journal of Nanobiotechnology*, 12(1), p. 16. https://doi.org/10.1186/1477-3155-12-16

Ikram, F., *et al.* (2019) 'Epicatechin coated silver nanoparticles as highly selective nanosensor for the detection of Pb2+ in environmental samples', *Journal of Molecular Liquids*, 277, pp. 649–655. https://doi.org/10.1016/j.molliq.2018.12.146

Isha, A., *et al.* (2019) 'An NMR metabolomics approach and detection of *Ganoderma boninense* -infected oil palm leaves using MWCNT-based electrochemical sensor', *Journal of Nanomaterials*, 2019, pp. 1–12. https://doi.org/10.1155/2019/4729706

Itodo, H., Nnamonu, L. and Wuana, R. (2017) 'Green synthesis of copper chitosan nanoparticles for controlled release of pendimethalin', *Asian Journal of Chemical Sciences*, 2(3), pp. 1–10. https://doi.org/10.9734/AJOCS/2017/32937

Javeed, Z., *et al.* (2022) 'Nanofertilizers and nanopesticides: Application and impact on agriculture', in *Diverse applications of nanotechnology in the biological sciences*. Apple Academic Press, pp. 199–212. https://doi.org/10.1201/9781003277255-7

Jimenez-Falcao, S., *et al.* (2020) 'Dithioacetal-mechanized mesoporous nanosensor for Hg(II) determination', *Microporous and Mesoporous Materials*, 297, p. 110054. https://doi.org/10.1016/j.micromeso.2020.110054

Kant, R. (2020) 'Surface plasmon resonance based fiber–optic nanosensor for the pesticide fenitrothion utilizing Ta2O5 nanostructures sequestered onto a reduced graphene oxide matrix', *Microchimica Acta*, 187(1), p. 8. https://doi.org/10.1007/s00604-019-4002-8

Khati, P., *et al.* (2017) 'Nanochitosan supports growth of *Zea mays* and also maintains soil health following growth', *3 Biotech*, 7(1), p. 81. https://doi.org/10.1007/s13205-017-0668-y

Khiyami, M.A., *et al.* (2014) 'Plant pathogen nanodiagnostic techniques: Forthcoming changes?', *Biotechnology & Biotechnological Equipment*, 28(5), pp. 775–785. https://doi.org/10.1080/13102818.2014.960739

Kumar, A., *et al.* (2021b) 'Smart nanomaterial and nanocomposite with advanced agrochemical activities', *Nanoscale Research Letters*, 16(156), pp. 1–26. https://doi.org/10.1186/s11671-021-03612-0

Kumar, A., *et al.* (2021a) 'Metal-based nanoparticles, sensors and their multifaceted application in food packaging', *Journal of Nanobiotechnology*, 19(256), pp. 1–25. https://doi.org/10.1186/s12951-021-00996-0

Kumar, A., *et al.* (2022) 'Potential applications of engineered nanoparticles in plant disease management: A critical update', *Chemosphere*, 295(133798), pp. 1–21. https://doi.org/10.1016/j.chemosphere.2022.133798

Kumar, A., *et al.* (2011) 'Engineered ZnO and TiO_2 nanoparticles induce oxidative stress and DNA damage leading to reduced viability of Escherichia coli', *Free Radical Biology and Medicine*, 51(10), pp. 1872–1881. https://doi.org/10.1016/j.freeradbiomed.2011.08.025

Kumar, N., *et al.* (2020) 'Nanomaterials: Emerging trends and future prospects for economical agricultural system', in *Biogenic nano-particles and their use in Agro-ecosystems*. Singapore: Springer Singapore, pp. 281–305. https://doi.org/10.1007/978-981-15-2985-6_16

Kumar, S., *et al.* (2020) 'Assessing the potential application of bio-compatibly tuned nanosensor of Yb2O3 for selective detection of imazapyr in real samples', *Colloids and Surfaces A: Physicochemical and Engineering Aspects*, 593, p. 124612. https://doi.org/10.1016/j.colsurfa.2020.124612

Lamsal, K., *et al.* (2011a) 'Application of silver nanoparticles for the control of *Colletotrichum* species *in vitro* and pepper anthracnose disease in field', *Mycobiology*, 39(3), pp. 194–199. https://doi.org/10.5941/MYCO.2011.39.3.194

Lamsal, K., *et al.* (2011b) 'Inhibition effects of silver nanoparticles against powdery mildews on cucumber and pumpkin', *Mycobiology*, 39(1), p. 26. https://doi.org/10.4489/MYCO.2011.39.1.026

Lau, H.Y., *et al.* (2017) 'Specific and sensitive isothermal electrochemical biosensor for plant pathogen DNA detection with colloidal gold nanoparticles as probes', *Scientific Reports*, 7(1), p. 38896. https://doi.org/10.1038/srep38896

Liu, H., *et al.* (2020) 'Electrochemical luminescence sensor based on double suppression for highly sensitive detection of glyphosate', *Sensors and Actuators B: Chemical*, 304, p. 127364. https://doi.org/10.1016/j.snb.2019.127364

Lu, T., *et al.* (2011) 'High sensitivity nanoparticle detection using optical microcavities', *Proceedings of the National Academy of Sciences*, 108(15), pp. 5976–5979. https://doi.org/10.1073/pnas.1017962108

Mohammadi, M., *et al.* (2017) 'Vitamin D3-loaded nanostructured lipid carriers as a potential approach for fortifying food beverages; *In vitro* and *In vivo* evaluation', *Advanced Pharmaceutical Bulletin*, 7(1), pp. 61–71. https://doi.org/10.15171/apb.2017.008

Mondal, R., *et al.* (2022) Potential of nanobiosensor in sustainable agriculture: The state-of-art. *Heliyon*. https://doi.org/10.2139/ssrn.4134873

Moreira, F.K., *et al.* (2013) 'Nutraceutically inspired Pectin–$Mg(OH)_2$ nanocomposites for bioactive packaging applications', *Journal of Agricultural and Food Chemistry*, 61(29), pp. 7110–7119. https://doi.org/10.1021/jf402110g

Mukhopadhyay, S.S. (2014) 'Nanotechnology in agriculture: Prospects and constraints', *Nanotechnology, Science and Applications*, p. 63. https://doi.org/10.2147/NSA.S39409

Nagy, L., *et al.* (2014) 'Photosynthetic machineries in nano-systems', *Current Protein & Peptide Science*, 15(4), pp. 363–373. https://doi.org/10.2174/1389203715666140327102757

Nuruzzaman, M., *et al.* (2016) 'nanoencapsulation, nano-guard for pesticides: A new window for safe application', *Journal of Agricultural and Food Chemistry*, 64(7), pp. 1447–1483. https://doi.org/10.1021/acs.jafc.5b05214

O'Brien, N. and Cummins, E. (2010) 'Ranking initial environmental and human health risk resulting from environmentally relevant nanomaterials', *Journal of Environmental Science and Health, Part A*, 45(8), pp. 992–1007. https://doi.org/10.1080/10934521003772410

Pereira, A.E.S., *et al.* (2017) 'γ-polyglutamic acid/chitosan nanoparticles for the plant growth regulator gibberellic acid: Characterization and evaluation of biological activity', *Carbohydrate Polymers*, 157, pp. 1862–1873. https://doi.org/10.1016/j.carbpol.2016.11.073

Pereira, M.M., *et al.* (2014) 'Ecotoxicological effects of carbon nanotubes and cellulose nanofibers in *Chlorella vulgaris*', *Journal of Nanobiotechnology*, 12(1), p. 15. https://doi.org/10.1186/1477-3155-12-15

Periakaruppan, R., *et al.* (2023) 'Innovations in modern nanotechnology for the sustainable production of agriculture', *ChemEngineering*, 7(4), p. 61. https://doi.org/10.3390/chemengineering7040061

Pham, T.B., *et al.* (2020) 'Surface-enhanced Raman spectroscopy based on silver nano-dendrites on microsphere end-shape optical fibre for pesticide residue detection', *Optik*, 219, p. 165172. https://doi.org/10.1016/j.ijleo.2020.165172

Picó, Y. and Andreu, V. (2014) 'Nanosensors and other techniques for detecting nanoparticles in the environment', in *Nanosensors for chemical and biological applications*. Elsevier, pp. 295–338. https://doi.org/10.1533/9780857096722.2.295

Pirzadah, B., *et al.* (2020) Nanofertilizers: A Way Forward for Green Economy, pp. 99–112. https://doi.org/10.1007/978-3-030-39978-8_5

Pošćić, F., *et al.* (2016) 'Effects of cerium and titanium oxide nanoparticles in soil on the nutrient composition of barley (*Hordeum vulgare* L.) Kernels', *International Journal of Environmental Research and Public Health*, 13(6), p. 577. https://doi.org/10.3390/ijerph13060577

Prabhakar, N., *et al.* (2016) 'Chitosan-iron oxide nanocomposite based electrochemical aptasensor for determination of malathion', *Analytica Chimica Acta*, 939, pp. 108–116. https://doi.org/10.1016/j.aca.2016.08.015

Preetha, P.S. and Balakrishnan, N. (2017) 'A review of nano fertilizers and their use and functions in soil', *International Journal of Current Microbiology and Applied Sciences*, 6(12), pp. 3117–3133. https://doi.org/10.20546/ijcmas.2017.612.364

Pyrgiotakis, G., *et al.* (2016) 'Optimization of a nanotechnology based antimicrobial platform for food safety applications using engineered water nanostructures (EWNS)', *Scientific Reports*, 6(1), p. 21073. https://doi.org/10.1038/srep21073

Rai, M. and Ingle, A. (2012) 'Role of nanotechnology in agriculture with special reference to management of insect pests', *Applied Microbiology and Biotechnology*, 94(2), pp. 287–293. https://doi.org/10.1007/s00253-012-3969-4

Rana, T. (2021) 'Prospects and future perspectives of selenium nanoparticles: An insight of growth promoter, antioxidant and anti-bacterial potentials in productivity of poultry', *Journal of Trace Elements in Medicine and Biology*, 68, p. 126862. https://doi.org/10.1016/j.jtemb.2021.126862

Rui, M., *et al.* (2016) 'Iron oxide nanoparticles as a potential iron fertilizer for peanut (*Arachis hypogaea*)', *Frontiers in Plant Science*, 7. https://doi.org/10.3389/fpls.2016.00815

Sabir, S., Arshad, M. and Chaudhari, S.K. (2014) 'Zinc oxide nanoparticles for revolutionizing agriculture: Synthesis and applications', *The Scientific World Journal*, 2014, pp. 1–8. https://doi.org/10.1155/2014/925494

Saleh, S.M., *et al.* (2020) 'Optical sensor film for metribuzin pesticide detection', *Spectrochimica Acta Part A: Molecular and Biomolecular Spectroscopy*, 229, p. 117971. https://doi.org/10.1016/j.saa.2019.117971

Santaella, C. and Plancot, B. (2020) 'Interactions of nanoenabled agrochemicals with soil microbiome', in *Nanopesticides*. Springer International Publishing, pp. 137–163. https://doi.org/10.1007/978-3-030-44873-8_6

Sekhon, B. (2014) 'Nanotechnology in agri-food production: An overview', *Nanotechnology, Science and Applications*, p. 31. https://doi.org/10.2147/NSA.S39406

Semo, E., *et al.* (2007) 'Casein micelle as a natural nano-capsular vehicle for nutraceuticals', *Food Hydrocolloids*, 21(5–6), pp. 936–942. https://doi.org/10.1016/j.foodhyd.2006.09.006

Sharma, P., *et al.* (2021) 'A review on biosensors and nanosensors application in agroecosystems', *Nanoscale Research Letters*, 16(1), p. 136. https://doi.org/10.1186/s11671-021-03593-0

Srivastava, P., *et al.* (2023) 'Impact of nano and non-nano fertilizers on rice quality and productivity: A review', *International Journal of Environment and Climate Change*, 13(8), pp. 973–987. https://doi.org/10.9734/ijecc/2023/v13i82035

Tefera, T., *et al.* (2011) 'The metal silo: An effective grain storage technology for reducing post-harvest insect and pathogen losses in maize while improving smallholder farmers' food security in developing countries', *Crop Protection*, 30(3), pp. 240–245. https://doi.org/10.1016/j.cropro.2010.11.015

Thakur, P., *et al.* (2022) 'Nano-insecticide: Synthesis, characterization, and evaluation of insecticidal activity of ZnO NPs against *Spodoptera litura* and *Macrosiphum euphorbiae*', *Applied Nanoscience*, 12(12), pp. 3835–3850. https://doi.org/10.1007/s13204-022-02530-6

Tian, X., *et al.* (2018) 'Nonenzymatic electrochemical sensor based on CuO-TiO_2 for sensitive and selective detection of methyl parathion pesticide in ground water', *Sensors and Actuators B: Chemical*, 256, pp. 135–142. https://doi.org/10.1016/j.snb.2017.10.066

van Nhan, L., *et al.* (2016) 'The effects of Fe_2O_3 nanoparticles on physiology and insecticide activity in non-transgenic and Bt-transgenic cotton', *Frontiers in Plant Science*, 6. https://doi.org/10.3389/fpls.2015.01263

Wang, J., *et al.* (2016) 'Fabrication of an "ion-imprinting" dual-emission quantum dot nanohybrid for selective fluorescence turn-on and ratiometric detection of cadmium ions', *The Analyst*, 141(20), pp. 5886–5892. https://doi.org/10.1039/C6AN00868B

World Food and Agriculture – Statistical Yearbook 2022. (2022). FAO. https://doi.org/10.4060/cc2211en

Wu, S., *et al.* (2011) 'Controlled immobilization of acetylcholinesterase on improved hydrophobic gold nanoparticle/Prussian blue modified surface for ultra-trace organophosphate pesticide detection', *Biosensors and Bioelectronics*, 27(1), pp. 82–87. https://doi.org/10.1016/j.bios.2011.06.020

Xie, Y., *et al.* (2018) 'CuO nanoparticles decorated 3D graphene nanocomposite as non-enzymatic electrochemical sensing platform for malathion detection', *Journal of Electroanalytical Chemistry*, 812, pp. 82–89. https://doi.org/10.1016/j.jelechem.2018.01.043

Xing, C., *et al.* (2014) 'Colorimetric detection of mercury based on a strip sensor', *Anal. Methods*, 6(16), pp. 6247–6253. https://doi.org/10.1039/C3AY42002G

Yang, C.H., Ding, Y.L. and Qian, J. (2018) 'Design of magnetic-fluorescent based nanosensor for highly sensitive determination and removal of HG2+', *Ceramics International*, 44(8), pp. 9746–9752. https://doi.org/10.1016/j.ceramint.2018.02.209

Yearla, S.R. and Padmasree, K. (2016) 'Exploitation of subabul stem lignin as a matrix in controlled release agrochemical nanoformulations: A case study with herbicide diuron', *Environmental Science and Pollution Research*, 23(18), pp. 18085–18098. https://doi.org/10.1007/s11356-016-6983-8

Yılmaz, E., *et al.* (2017) 'Plastic antibody based surface plasmon resonance nanosensors for selective atrazine detection', *Materials Science and Engineering: C*, 73, pp. 603–610. https://doi.org/10.1016/j.msec.2016.12.090

Zhang, B.-T., *et al.* (2013) 'Application of carbon-based nanomaterials in sample preparation: A review', *Analytica Chimica Acta*, 784, pp. 1–17. https://doi.org/10.1016/j.aca.2013.03.054

Zhang, L. and Webster, T.J. (2009) 'Nanotechnology and nanomaterials: Promises for improved tissue regeneration', *Nano Today*, 4(1), pp. 66–80. https://doi.org/10.1016/j.nantod.2008.10.014

Zhao, L., *et al.* (2016) 'Application of metabolomics to assess the impact of Cu(OH)2 nanopesticide on the nutritional value of lettuce (*Lactuca sativa*): Enhanced Cu intake and reduced antioxidants', *NanoImpact*, 3–4, pp. 58–66. https://doi.org/10.1016/j.impact.2016.08.005

Zhao, L., *et al.* (2012) 'Stress response and tolerance of *Zea mays* to CeO_2 nanoparticles: Cross talk among H_2O_2, heat shock protein, and lipid peroxidation', *ACS Nano*, 6(11), pp. 9615–9622. https://doi.org/10.1021/nn302975u

3 Genetic Engineering and Micropropagation Techniques for Crop Plant Improvement

Farheen Bhatti, Aaliya Batool, Yousuf Shafiq, Muhammad Sheraz Hassan, Naeem Ahmad, Abida Kausar, and Noreen Zahra

3.1 INTRODUCTION

In order to cope with adverse environmental conditions, all living organisms acquire certain characteristics to enhance their survival rate (Hille *et al.*, 2018). However, these traits take many years to emerge under environmental conditions. Plants naturally develop certain adaptions like physiological adjustments and biochemical and phenological adaptations, which enhance their ability to tolerate harsh environmental conditions (Bachheti *et al.*, 2021; Hafeez *et al.*, 2024a-c; Husen, 2021a–e; Hussain *et al.*, 2023; Zahra *et al.*, 2023a; Husen and Iqbal, 2023). These processes contribute in the existence of crop plants under different stress situations (Husen, 2021b; Wu *et al.*, 2016; Zhang *et al.*, 2023). In response to different stresses, plants switch various complex signal transduction pathways that affect regulation of gene expression by activating the different signal molecules that help to tolerate the stress conditions by certain mechanisms (Liu *et al.*, 2021). To mitigate the effect of different stresses, the agricultural biotechnologists emphasize extensive research to develop new systems and strategies for the improvement of crop productivity, vigor, quality, and resistance against abiotic and biotic stress (Getnet *et al.*, 2015; Husen, 2010; Husen *et al.*, 2014; Husen *et al.*, 2016; Husen *et al.*, 2018; Husen *et al.*, 2019; Hussein *et al.*, 2017; Zahra *et al.*, 2021; Zahra *et al.*, 2023b). These challenges can be addressed by genetic modification, which is a powerful tool to create GM plants. This permits exchange of genetic material among distinct genera or species (Jha *et al.*, 2020).

Plant breeders and biotechnologists tried to commercialize crop varieties with desirable characteristics. Biotechnologists use living organisms or their parts to produce or alter the products to modify the genetic makeup of microorganisms, plants, or animals for specific purposes (Wada *et al.*, 2020). A specific area of biotechnology, agricultural biotechnology, focuses on crops and livestock development through genetic engineering. Generally, genetic transformation at the molecular level has been carried out in various plant species such as canola (*Brassica napus* L.), maize (*Zea mays* L.), cotton (*Gossypium herbaceum* L.), tobacco (*Nicotiana tabacum* L.), rice (*Oryza sativa* L.) (Li *et al.*, 2016a), tomato (*Solanum lycopersicum* L.), potato (*Solanum tuberosum* L.), alfalfa (*Medicago sativa* L.), sugarcane (*Saccharum officinarum* L.), sugar beet (*Beta vulgaris* L.), and wheat (*Triticum aestivum* L.). Many advanced approaches or modifications of the established procedures are being developed for crop improvement. New tools in genetic manipulation have been introduced day by day for crop plants' improvement and sustainable agriculture (Zhang *et al.*, 2016). These tools, such as tissue culture, micropropagation, marker-assisted selection, genetic engineering, and molecular analyses, are very useful for significant progress of agriculture.

Extensive research on gene regulation of transgenic expression of plants is accumulating in the best research laboratories worldwide. To obtain the precise genetic analysis, strong background knowledge is required of the particular traits, for example, identification of exact locations of

 DOI: 10.1201/9781032630366-3

cis-acting elements within the coding sequence that play a major role in governance of gene expression, so that chimeras' expression estimation can be possible with the least qualitative analyses. Several genetically modified crops have been released at a commercial level. The GM plants confer resistance against abiotic and biotic stress and increase food quality and production (Raymond Park *et al.*, 2011; Yu *et al.*, 2022).

Through the use of advanced genetic engineering tools, scientists may alter the genetic makeup of living organisms and have created numerous opportunities to enhance agricultural productivity (Kumar *et al.*, 2020; Sharma *et al.*, 2022). With this technique, numerous transgenic crops have been developed and marked at commercial level, such as canola, alfalfa, soybean, potato, banana, sugarcane, rice, cotton, cowpea, pigeon pea, and eggplant. Also, biotic and abiotic resistance lines have been developed through this technique. Scientists of all over the world focus on the research into weeds and insects control by using this technology (Krishna *et al.*, 2019). The success of genetic transformation depends on the compatibility of efficient transformation systems with different selectable marker genes (Dormatey *et al.*, 2021). The most commonly used selectable marker gene is neomycin phosphotransferase-II (nptII), which shows resistance to multiple antibiotics, such as kanamycin and geneticin (Kim and Kim, 2014; Kumar *et al.*, 2022).

Implementation of genetic transformation techniques has resulted in the development of different crop plants with desirable traits such as high yield, quality, and resistance to abiotic and biotic stresses and certain diseases for sustainable agriculture. Although the genetically modified plants with novel genes have been introduced with a new era of crop improvement, production, and adoption but there are some concerns of environmental challenges related to ethical issues, human health, etc. To overcome these challenges, it is important to handle them with appropriate scientific investigations and approaches (Abdul-Aziz *et al.*, 2022).

3.1.1 Methods of Genetic Engineering in Plants

There are several steps for producing genetically modified crops, with a lot of difficulties from selection of novel genes to regeneration of mutant plants (Figure 3.1).

3.1.1.1 Selection of the Target Gene and Its Isolation

Identification of the specific genes which are responsible for the required traits is a very important step, as any stress-related genes provide resistance against the stresses present in a particular plant species (Nascimento *et al.*, 2023). These specific genes are identified from the literature, and the available data about their sequence, structure, and their functions are also analyzed. After the selection of this data, then the gene is isolated, and further analysis will be conducted for its confirmation. If the data is not available at any other sites for the study of the novel gene there will be lots of difficulties for the analysis of this gene (Butler *et al.*, 2016; Janssen *et al.*, 2001).

3.1.1.2 Vector Construction

After the selection of the target gene, the next setup is to transfer the desirable gene into the host plant with the help of plasmid. Plasmid is small circular DNA present in bacteria. The vector consists of three main components: origin of replication (ORI), selectable marker site (SM), and multiple cloning site (MCS). MCS is the site where the foreign DNA is inserted (Nora *et al.*, 2019).

3.1.1.3 Cloning of the Target Gene

After PCR amplification, of millions of copies of the target gene, it is inserted to the expression cassette (a strong promoter downstream and a terminator upstream). This construct is transformed into a plasmid vector for duplication of the target gene within the bacterial competent cells. The construct with target genes (recombinant DNA) is ready to transform into the specific plant using different and indirect methods, such as using a biolistic gene gun and *Agrobacterium tumefaciens* strain, respectively (Keshavareddy *et al.*, 2018).

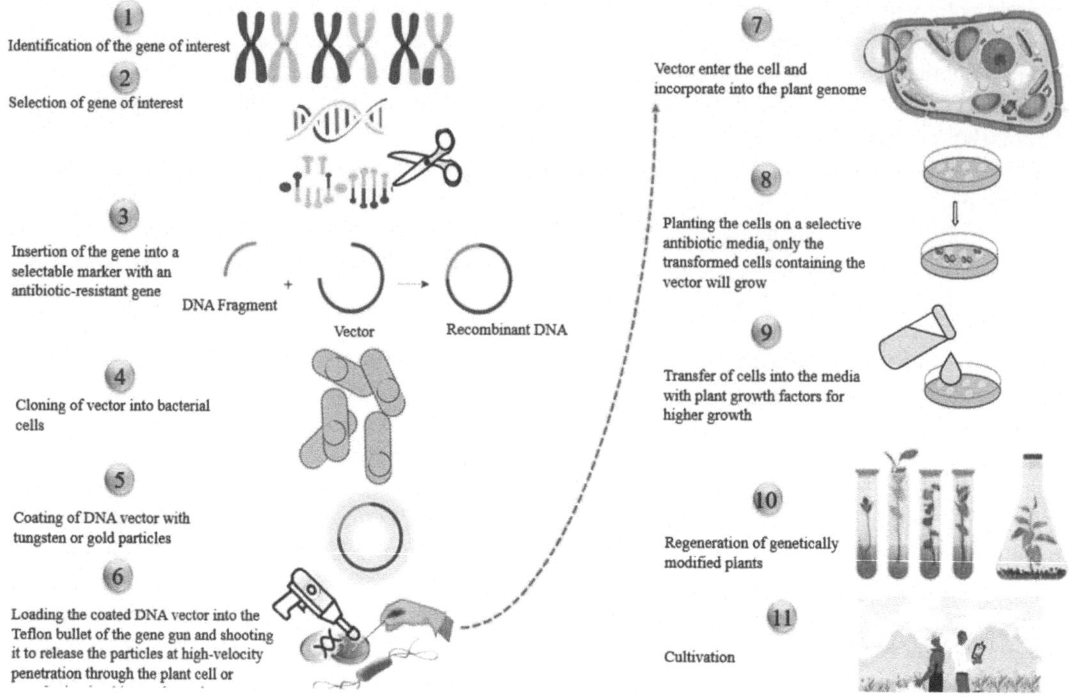

FIGURE 3.1 Mechanism of genetic engineering and micropropagation techniques for crop plant improvement.

3.1.1.4 Modified Plant Cells Selection and Plant Regeneration

The transformed cells are incubated in the media containing the antibiotic resistance selectable marker gene; only recombinant cells survive and regenerate to become the entire plants. Different molecular analyses are performed to confirm the true transgenic lines for further detection to avoid unintended changes (Davies *et al.*, 2017). The genetically altered plants are shifted to the field conditions, and biosafety assessment studies are performed to detect any negative impact on the environment or living organisms (Carroll, 2011). GM plants after biosafety regulations are released and planted in a limited area. To avoid the safety concerns of GM crops, new biotechnologies are established, for example, cisgenesis (Shukla *et al.*, 2009). In these procedures, recombinants are generated from sexually related (or identical) plant species. Besides these technologies, some other tools have been enabled for genetic engineering, such as genome editing tools (zinc finger nucleases [ZFNs], transcription activator-like effector nucleases [TALENs], and clustered regularly interspaced short palindromic repeats [CRISPR]/Cas), which are easier, more accurate, and more specific (Patel *et al.*, 2023).

3.2 GENETIC ENGINEERING TECHNOLOGIES

3.2.1 Transgenic Technology

Transgenic technology helps to overcome the hybridization barrier by utilizing a wider and more assorted range of desirable genes from genetically different species (Raymond Park *et al.*, 2011). Transgenic crops are produced via desirable gene cloning, generation of expression vectors, recipient cultivars' genetic transformation, and identification of transgenic lines in order to improve the original non-desirable traits and alter them with new useful characters (Anjanappa and Gruissem, 2021).

It may also help to knock out or mend some of the suboptimal genes to alter their non-desirable phenotypes (Hernández-Soto and Chacón-Cerdas, 2021).

Furthermore, it may help to combat biotic stress (such as parasites, nematodes, herbivory insects, bacteria, viruses, and fungi) by providing a direct approach to enhance crop resistance, with negligible effects on environment and beneficial soil microbes. Hence, these techniques may induce biotic resistance in crops via heterologous expression of exogenous genes and RNAi (Ahmar *et al.*, 2020; Yu *et al.*, 2022). For instance, a gene, namely an immune receptor gene (*AtRPL23*), was introduced in potatoes. These potatoes, when infected with *S. sclerotiorum* and *P. infestans*, exhibited smaller lesions with minimized pathogen colonization (Albert *et al.*, 2015). Similarly, the immune receptor gene (*SlVE1* and *SmELR*) from tomatoes and wild potatoes resulted in increased resistance to *Verticillium* wilt (a disease caused by two fungi, *V. dahlia* and *V. albo-atrum*) and *P. infestans* in tobacco, cotton and potatoes, correspondingly (Du *et al.*, 2015; Song *et al.*, 2018). Additionally, the heterogenous expression of genes related to chitinase, lectoferrin, lysozymes, the effectors of jasmonic acid, and salicylic acid signaling, along with other phytoalexins or antibacterial peptide genes and exopolysaccharide genes, have all been used in the transgenic advancement of different crops for bacterial resistance (Lakshman *et al.*, 2013; Rivero *et al.*, 2015)

Plants that are sessile are often exposed to continuously changing environmental conditions which may cause abiotic stress. Accordingly, this technology may also help to improve crop tolerance under abiotic stress conditions. A study reported the development of transgenic tomatoes against abiotic stresses such as salinity, heat, and drought. These tomatoes were produced by transgenes like *BADH2*, *ZAT12*, *Osmotin*, and *DREBs*. The changed expression of these genes is associated with every step of abiotic stress response, like ROS and free radicals scavenging, synthesis of compatible solutes (i.e. amino acids, polyols, sugars, and betaines), membrane and protein protection, transcriptional control, and signaling (Krishna *et al.*, 2019). Likewise, the cold shock protein genes (*cspA* and *cspB*) obtained from *E. coli* and soil bacterium *B. subtilis* were used to obtain drought tolerance in maize (Castiglioni *et al.*, 2008). Also, the overexpression of FcWRKY40 can enhance tolerance to oxidative stress in transgenic tobacco (Gong *et al.*, 2020). Another study by Gurusaravanan *et al.* (2019) used the cotton stem tip as source of explants; then, with the help of *A. tumefaciens*, the *uidA* gene was transformed in cotton by ultrasound and microinjection.

3.2.1.1 Gene Editing Technology

Gene editing tools help to bring desirable traits with DNA via insertion/deletion (indel) formation and mutation in certain gene sequences through recruitment of specific nucleases like CRISPR/CRISPR-associated protein 9 (Cas9), TALENs, ZFNs (Wada *et al.*, 2020). In addition, this process is categorized into two steps. First, a nuclease system: some extent of engineering modification cleaves the target sites of dsDNA in order to create double-stranded breaks (DSBs). Second, the cell starts two endogenous repair pathways (homologous recombination [HR] and non-homologous end junction [NHEJ] pathways) to ensure genome integrity, as the DSBs are toxic to cells in certain cases (Lieber *et al.*, 2003; Murray *et al.*, 2018).

3.2.1.2 ZFNs

For the last two decades, novel technologies have been developed to facilitate targeted variation of plant genomes. Among these technologies, ZFNs include fusion of engineered zinc-finger DNA-binding domains (having specific binding activity) and nuclease (comprising non-specific cutting activity), specifically the *Fok1* nuclease. This technique may help to produce novel genetic alterations by editing and deleting endogenous gene sequences, improving transgenic plant efficiency via targeted transgenes insertion at particular genomic loci and consequently deleting/adding transgenes from previous transgenic events (Davies *et al.*, 2017; Li *et al.*, 2016b). Some of the amino acid residues from the alpha-helix of ZFP are directly involved in target site recognition and can pair with adjacent bases (Wolfe *et al.*, 2000). Additionally, the cutting and binding domain of the Fok1 endonuclease can be detached, and the non-specific Fok1 endonuclease cutting action can be attained via removal of binding domain (Chandrasekaran and Carroll, 2016). Lastly, the target site will be cut under the Fok dimer's action, and DSBs will be produced (Wolfe *et al.*, 2000). Initially, ZFNs were effectively applied in animals as the first-generation technology related to gene editing (Kim and Kim, 2014). Afterward, they were used in rice, maize, and other crops. For instance, ZFNs help to modify maize plants in order to transmit genetic changes. Insertional disruption of *IPK1* (one target loci) in maize leads to both anticipated changes of inositol phosphate profile and herbicide tolerance in developing seeds (Shukla *et al.*, 2009). Similarly, transgenic rice plants were generated through ZFNs' intervened targeted gene disruption in the *SSIVa* gene's coding sequence, as they can effectively stimulate and cleave mutations at a *SSIVa* locus in rice (Jung *et al.*, 2018). This system may also have some inherent errors like the inter-ZFP context effect in which it is hard to attain high efficacy via joining of particular ZFPs together. Furthermore, it also includes some deficiencies like higher cytotoxicity, maximum off-target efficacy, and multiple editing unsuitability (Khan, 2019).

3.2.1.3 TALENs

The main difference between TALENs and ZFPs is in structural changes, as the TALENs' binding domain comprises transcription activator-like effectors (TALEs), and its binding capacity is generally dependent on extremely variable amino acids located on 13th and 12th position of the TALE (Boch and Bonas, 2010). In collaboration, a single base can also be precisely recognized. As reported before, Fok endonuclease is still the non-specific cutting domain which is formed by a TALENS pair binding at the target site, forming DSBs (Chandrasekaran *et al.*, 2017).

To date, TALENs have been successfully used for specific genome editing of several plant species. Ten modified crop plants via engineered nucleases are: cotton, barley, maize, tobacco, potato, tomato, wheat, and rice. It is applied for successful development of plants resistant to biotic as well as abiotic stress conditions, along with quality and yield enhancement (Khan *et al.*, 2017). For example, TALENS-based introductions of targeted mutations in an enzyme, namely acelolactate synthase (ALS), are around 30% of transformed cells of tobacco protoplasts (Zhang *et al.*, 2013). Furthermore, the particular target DNA sequence modification was reported in barley. Genes, namely green fluorescent protein (gfp)-specific TALENs in consort with a repair template, were introduced in callus cultures of barley; three out of 100 calli bombardments converted gfp to yellow fluorescent protein (*YFP*) genes successfully by using HDR (Budhagatapalli *et al.*, 2015).

Additionally, as a strong promoter insertion upstream from an anthocyanin biosynthesis controlling gene, TALENS helps to attain specific modifications via gemini-virus replicon in tomato genomes, leading to ectotropic accumulation and overexpression of this pigment in tomato tissues (Cermak *et al.*, 2015). The same technique was also found to be helpful in the production of herbicide-tolerant potatoes through the introduction of one point mutation in an ALS1 gene via gene replacement (Butler *et al.*, 2016). Similarly, another study regarding TALENs-mediated gene replacement documented herbicide resistance via double-point mutation in the ALS gene in rice (Li *et al.*, 2016c). Regarding disease resistance, TALEs are used successfully for begmovirus resistance via viral DNA targeting with engineered nucleases and particular DNA binding proteins (Cheng *et al.*, 2015).

They have some advantages over other DNA targeting systems: they are natural transcription factors (TFs), so they are very well-understood, with no target length limitations and minimum chances for off-targeting, which makes it safer, and there is rigorous technology related to gene targeting (Fu *et al.*, 2014). Conversely, one major disadvantage of TALENs is the TALEs pair use, which is mandatory for *Fok1* nuclease dimerization (Khan *et al.*, 2017).

3.2.1.4 CRISPR/Cas System

The ZFNs and TALENs have not been extensively used due to their complicated procedures and high rate of failures; however, CRISPR/Cas was more successfully utilized for several crop improvements. Genome editing via CRISPR/Cas9 is more precise, as it targets particular sites of genes associated with stress response pathways, and it may also help in gene modification to improve plants' ability to resist harsh environmental stress conditions (Jha *et al.*, 2020). It improves crop quality, which comprises the variations in nutritional components, palatability, appearance, and other preferable characteristics of several crops (Liu *et al.*, 2021). It consists of a single-guide RNA (sgRNA), along with an RNA-guided Cas endonuclease, which protects archaea and bacteria from being attacked by bacteriophages and mobile genetic elements (Hille *et al.*, 2018). Its applications are reported in different plants.

Correspondingly, the tRNA-sgRNA complexes, along with *UBIQUITIN10* promoters, developed almost 80% of the mutant phenotypes, with four allelic copies as full knockouts in *N. tobaccum*; however, the remaining 20% showed weak phenotypes, demonstrating partial allelic knockout at the T0 generation. This suggests stable and high expression of sgRNA and Cas9 leads to maximum rates of mutagenesis in T0 reproductive organs throughout plants' life (Kumar *et al.*, 2022). In addition, high-frequency, target-based mutagenesis was developed in maize via *Agrobacterium*-delivered CRISPR/Cas9. This system is comprised of *E. coli* cloning along with an *Agrobacterium* binary vector. An active transgene inheritance of CRISPR/Cas9 caused extra target-specific mutations in successive generations (Char *et al.*, 2017; Fu *et al.*, 2014).

Moreover, the structural homologs and highly conserved gene sequences among wheat sub-genomes frequently permit their concurrent targeting via CRISPR/Cas9 with a paired or single-guide RNA (sgRNA) (Smedley *et al.*, 2021). Furthermore, a tomato gene which encodes hybrid proline-rich protein 1 (*HyPRP1*), edited via CRISPR/Cas9 approach causing particular removal of its functional domains i.e. proline-rich domains (PRDs) along with eight cysteine-motifs which ultimately improves quality of fruit by enhancing texture and lessening vulnerability to environmental stressors. These edited alleles improved plant vulnerability against *F. Lycopersici* and *F. oxysporum*, which are responsible for fusarium wilt. The accurate domain editing of the *SlHyPRP1* gene through CRISPR/Cas9 produced multi-stress tolerant alleles (Tran *et al.*, 2023).

3.3 MICROPROPAGATION TECHNOLOGY

This technology is preferable to other old asexual propagation methods. It guarantees faster and huge multiplication of genetically similar copies, causing rejuvenation of cold cultivars and faster regeneration of new cultivars, which are more resistant to both biotic and abiotic stresses (El-Esawi, 2016). It is a vegetative technique which develops cell clones via several techniques comprising explant sources like shoots, roots, sterilized conditions, and plant transfer to soils. For this selected genotype, soft woody plants are utilized for vegetative growth, e.g., adventitious shoot (node formations in sugarcane) (Aslam *et al.*, 2023).

3.4 GM CROPS WITH NOVEL TRAITS FOR THE BENEFIT OF THE FARMER AND THE CONSUMER

The number of genetically altered crops with novel traits has more than doubled since 2008, following the biosafety regulations prior to commercial release (Turnbull *et al.*, 2021). The number of newly genetically altered plants under the process of commercialization is presented in "Trends in GM Approval" by ISAAA (Anonymous, 2020), expressing the occurrence of pest-resistant (PR) and herbicide-tolerant (HT) canola and soybeans with some new qualities presented in Table 3.1. The other crop plants with new characteristics exhibiting high biomass production, more yield, increased nutrient quality, resistance against diseases, and environmental stresses have been listed in ISAAA GM databases (Table 3.2). They include: the GM crop plants that are used as food, such as bananas, and show resistance to bunchy top viruses released in Uganda; golden rice enriched with β-carotene under field trials in Bangladesh; bananas with Fusarium wilt resistance and drought-tolerant wheat, disease resistance, high grain composition, and modified oil content in Australia; high-yielding

TABLE 3.1
Global Area of GM Crop (Million ha)

Rank	Country	Area (Million ha)	Biotech Crops
1	USA	72.9	Maize, soybean, cotton, canola, sugar beet, alfalfa, papaya, squash, potato
2	Brazil	49.1	Soybean, maize, cotton
3	Argentina	23.8	Soybean, maize, cotton
4	Canada	11.6	Canola, maize, soybean, sugar beet, alfalfa
5	India	10.8	Cotton
6	Paraguay	3.6	Soybean, maize, cotton
7	Pakistan	2.9	Cotton
8	China	2.8	Cotton, papaya, poplar
9	South Africa	2.7	Maize, soybean, cotton
10	Uruguay	1.3	Soybean, maize
11	Bolivia	1.2	Soybean
12	Australia	0.9	Cotton, canola
13	Philippines	0.8	Maize
14	Myanmar	0.3	Cotton
15	Spain	0.1	Maize
16	Sudan	0.1	Cotton
17	Mexico	0.1	Cotton, soybean
18	Colombia	0.1	Cotton, maize
19	Vietnam	<0.1	Maize
20	Honduras	<0.1	Maize
21	Chile	<0.1	Maize
22	Portugal	<0.1	Maize
23	Bangladesh	<0.1	Brinjal/Eggplant
24	Costa Rica	<0.1	Cotton, soybean, pineapple
25	Slovakia	<0.1	Maize
26	Czech Republic	<0.1	Maize
	Total	**185.1**	

and biomass-producing wheat in the United Kingdom; late blight-resistant potato (var. Desiree and Victoria) in Uganda; late blight- and nematode-resistant potato (var. Maris Piper with less bruising and less acrylamide) in the EU; insect pest-resistant chickpea, pigeon pea, and mustard in India; and drought-tolerant sugarcane in India and Indonesia (Ghimire *et al.*, 2023).

The global status of GM crop plants has increased more than 112-fold from 1.7 Mha (1996) to 190.4 Mha (2019) (Anonymous, 2019). From 1996 to 2019, an average of 2.7 billion ha or 6.3 billion acres was attained for commercialization. Twenty-nine countries in the world cultivated biotech crops, of which 24 developing countries grew 56% biotech crops and five industrial nations planted 46% globally. Country-wise survey reports are presented in Table 3.1 (Anonymous, 2019).

3.5 REGULATION OF GM CROPS

GM crops are strictly regulated and require biosafety evaluation for approval from relevant competent authorities prior to commercialization (Turnbull *et al.*, 2021) and must be carefully evaluated for safety to animals and humans (if consumed as feed and food) (Rozas *et al.*, 2022). International environmental protection authorities such as the World Health Organization (WHO), the Organization for Economic Cooperation and Development (OECD), and the Food and Agriculture Organization (FAO)

TABLE 3.2
Crops and Traits under Field Testing

Country	Crop	Trait	Developer
Australia	Banana	Fusarium wilt resistance	Queensland University of Technology
	Wheat	Disease resistance, drought tolerance, altered oil content, and altered grain composition	Commonwealth Scientific and Industrial Research Organization (CSIRO)
New Zealand	Ryegrass	Better nutritional quality and energy system; high metabolizable energy	AgResearch, New Zealand
United Kingdom	Wheat	Yield and biomass	Rothamstead Agricultural Research
European Union	Potato (var. Maris Piper)	Blight- and nematode-resistant, less bruising and less acrylamide	The Sainsbury Laboratory (TSL)
European Union	Camelina	Omega-3LC-PUFAS (Omega-3 long chain polyunsaturated fatty acids)	Rothamstead Research
Malawi	Bananas	Bunchy top virus	Byumbwe Research Station
Uganda	Potato (vars. Desiree and Victoria)	Late blight resistance	Kachwekano Zonal Agricultural Research and Development Institute
India	Indian mustard	Barnase-barstar system to induce heterosis	Delhi University South Campus
	Chickpea	Insect resistance with *cry7Ac*, *Cry1Aabc*	ICAR-Indian Institute of Pulses Research
	Pigeon pea	Insect resistance with *cry7Ac*, *Cry1Aabc*	ICAR-Indian Institute of Pulses Research
	Sugarcane	Drought tolerance with *DREB* gene	Sugarcane Research Institute, UP Council of Sugarcane
Philippines Bangladesh	Rice	ß-carotene	IRRI, Phil Rice, BARI

jointly collaborate and approve GM crops (Ramsay *et al.*, 2022). Genetically modified organisms (GMOs) follow uniform international labels (IPCC, 2019). It is mandatory to provide information to the consumers for this distinction (Rozas *et al.*, 2022). Mostly, the labeling process of GM food is specific in different areas of the world (Ramsay *et al.*, 2022), but the Codex Alimentarius Commission has recognized a harmonized international policy of labeling that is easy and helpful to the user (Borges *et al.*, 2018).

3.6 BIOSAFETY RISKS ASSOCIATED WITH GM CROPS

Three main health concerns are associated with genetic modification, which include allergenicity, toxicity, and gene contamination. The potential risks incorporate overexpression of recombinant proteins and pleiotropic effects (Abdul-Aziz *et al.*, 2022). Different expression patterns of novel proteins in comparison with conventional methods may produce allergic reactions, for example, genetically altered soybeans supplemented with an amino acid "methionine." Some GM crops encoded with enzymes may interrupt the natural metabolic pathways (metabolic perturbations). The other potential risk is related to inserted genes that might disturb the integrity of entire plant genomes (inactivation/modulation of endogenous genetic material leading to activation or deactivation of certain metabolic processes) (Smyth, 2022).

3.7 CONCLUSION AND FUTURE REMARKS

Genetically engineered crops have some fundamental ethical issues, but their role in sustainability of agriculture and human existence is significant, and this is the only solution for the world's problems for the next 30 years. The potential risks are not associated only with genetically engineered crops; all processes of food production have drawbacks. However, the use of genome editing tools and regulation of GM crops ensures that these crops are as safe as conventionally bred crops and can act as the drivers of sustainable food security.

The approval of GM plants for more than 24 years has been provided to alleviate global challenges, such as malnutrition, poverty, hunger, scarcity of food, plant diseases, and climate conditions. As of 2019, genetically engineered crops have reached 6.7 million acres (2.7 billion ha) since 1996. Scientists are working hard to develop new GM crops with novel traits that would be valuable to the farmer and the consumer. People are realizing and accepting the benefits of biotech crops.

REFERENCES

Abdul-Aziz, M., *et al.* (2022) 'Genetically engineered crops for sustainably enhanced food production systems', *Frontiers in Plant Science*, 13, pp. 1–24.

Ahmar, S., *et al.* (2020) 'Conventional and molecular techniques from simple breeding to speed breeding in crop plants: Recent advances and future outlook', *International Journal of Molecular Sciences*, 21(7), pp. 1–24.

Albert, I., *et al.* (2015) 'An RLP23–SOBIR1–BAK1 complex mediates NLP-triggered immunity', *Nature Plants*, 1(10), pp. 1–9.

Anjanappa, R.B. and Gruissem, W. (2021) 'Current progress and challenges in crop genetic transformation', *Journal of Plant Physiology*, 261, pp. 1–13.

Anonymous. (2019). https://www.isaaa.org/resources/publications/briefs/53/executivesummary/default.asp.

Anonymous. (2020). https://www.isaaa.org/gmapprovaldatabase/.

Aslam, A., *et al.* (2023) 'Plant tissue culture and crop improvement', in *Climate-resilient agriculture, vol 2: Agro-biotechnological advancement for crop production*. Cham, Switzerland: Springer International Publishing, pp. 841–862.

Bachheti, A., *et al.* (2021) 'Medicinal plants and their pharmaceutical properties under adverse environmental conditions', in Husen A. (ed.) *Harsh environment and plant resilience*. Cham, Switzerland: Springer International Publishing AG, pp. 457–502. https://doi.org/10.1007/978-3-030-65912-7_19

Boch, J. and Bonas, U. (2010) 'Xanthomonas *AvrBs3* family-type III effectors: Discovery and function', *Annual Review of Phytopathology*, 48, pp. 419–436.

Borges, B.J.P., *et al.* (2018) 'Genetically modified labeling policies: Moving forward or backward?, *Frontiers in Bioengineering and Biotechnology*, 6, pp. 1–10.

Budhagatapalli, N., *et al.* (2015) 'Targeted modification of gene function exploiting homology-directed repair of TALEN-mediated double-strand breaks in barley', *G3: Genes, Genomes, Genetics*, 5(9), pp. 1857–1863.

Butler, N.M., *et al.* (2016) 'Geminivirus-mediated genome editing in potato (*Solanum tuberosum* L.) using sequence-specific nucleases', *Frontiers in Plant Science*, 7, pp. 1–15.

Carroll, D. (2011) 'Genome engineering with zinc-finger nucleases', *Genetics*, 188(4), pp. 773–782.

Castiglioni, P., *et al.* (2008) 'Bacterial RNA chaperones confer abiotic stress tolerance in plants and improved grain yield in maize under water-limited conditions', *Plant Physiology*, 147(2), pp. 446–455.

Čermák, T., *et al.* (2015) 'High-frequency, precise modification of the tomato genome', *Genome Biology*, 16, pp. 1–15.

Chandrasekaran, A.P., Song, M. and Ramakrishna, S. (2017) 'Genome editing: A robust technology for human stem cells', *Cellular and Molecular Life Sciences*, 74, pp. 3335–3346.

Chandrasekaran, S. and Carroll, D. (2016) 'Origins of programmable nucleases for genome engineering', *Journal of Molecular Biology*, 428(5), pp. 963–989.

Char, S.N., *et al.* (2017) 'An agrobacterium-delivered CRISPR/Cas9 system for high-frequency targeted mutagenesis in maize', *Plant Biotechnology Journal*, 15(2), pp. 257–268.

Cheng, X., *et al.* (2015) 'Artificial TALE as a convenient protein platform for engineering broad-spectrum resistance to begomoviruses', *Viruses*, 7(8), pp. 4772–4782.

Davies, J.P., Kumar, S. and Sastry-Dent, L. (2017) 'Use of zinc-finger nucleases for crop improvement', *Progress in Molecular Biology and Translational Science*, 149, pp. 47–63.

Dormatey, R., *et al.* (2021) 'ptxD/Phi as alternative selectable marker system for genetic transformation for bio-safety concerns: A review', *Peer Journal*, 9, pp. 1–37.

Du, J., *et al.* (2015) 'Elicitin recognition confers enhanced resistance to phytophthora infestans in potato', *Nature Plants*, 1(4), pp. 1–5.

El-Esawi, M.A. (2016) 'Micropropagation technology and its applications for crop improvement', In: Anis, M., and Ahmad, N., (eds.) *Plant tissue culture: Propagation, conservation and crop improvement*, pp. 523–545. Springer Singapore.

Fu, Y., *et al.* (2014) 'Improving CRISPR-Cas nuclease specificity using truncated guide RNAs', *Nature Biotechnology*, 32(3), pp. 279–284.

Getnet, Z., *et al.* (2015) 'Growth, water status, physiological, biochemical and yield response of stay green sorghum {*Sorghum bicolor* (L.) Moench} varieties-a field trial under drought-prone area in Amhara regional state, Ethiopia', *Journal of Agronomy*, 14, pp. 188–202.

Ghimire, B.K., *et al.* (2023) 'Assessment of benefits and risk of genetically modified plants and products: Current controversies and perspective', *Sustainability*, 15(2), pp. 17–22.

Gong, Z., *et al.* (2020) 'Plant abiotic stress response and nutrient use efficiency', *Science China Life Sciences*, 63, pp. 635–674.

Gurusaravanan, P., Vinoth, S. and Jayabalan, N. (2019) 'An improved agrobacterium-mediated transformation method for cotton (*Gossypium hirsutum* L. "KC3") assisted by microinjection and sonication', *In Vitro Cellular & Developmental Biology – Plant*, 9, pp. 1–11.

Hafeez, M.B., *et al.* (2024a). 'Plant growth promoters boost the photosynthesis related mechanisms and secondary metabolism of late-sown wheat under contrasting saline regimes', *Plant Stress*, 12, p. 100480.

Hafeez, M.B., *et al.* (2024b). 'Exogenous application of plant growth regulators improves economic returns, grain yield and quality attributes of late-sown wheat under saline conditions', *International Journal of Plant Production*, 18(2), pp. 217–228.

Hafeez, M.B., *et al.* (2024c). 'Exploring the efficacy of plant growth promoters in nutrient acquisition and antioxidant defense responses of late sown wheat under saline conditions', *South African Journal of Botany*, *171*, pp. 812–824.

Hasanuzzaman, M., *et al.* (2013) 'Physiological, biochemical, and molecular mechanisms of heat stress tolerance in plants', *International Journal of Molecular Science*, 14, pp. 9643–9684.

Hernández-Soto, A. and Chacón-Cerdas, R. (2021) 'RNAi crop protection advances', *International Journal of Molecular Sciences*, 22(22), p. 12148.

Hille, F., *et al.* (2018) 'The biology of CRISPR-Cas: Backward and forward', *Cell*, 172(6), pp. 1239–1259.

Husen, A. (2010) 'Growth characteristics, physiological and metabolic responses of teak (*Tectona grandis* Linn. f.) clones differing in rejuvenation capacity subjected to drought stress', *Silvae Genetica*, 59, pp. 124–136.

Husen, A. (2021a) 'The harsh environment and resilient plants: An overview', in Husen, A. (ed.) *Harsh environment and plant resilience*. Cham, Switzerland: Springer International Publishing AG, pp. 1–23. https://doi.org/10.1007/978-3-030-65912-7_1

Husen, A. (2021b) 'Morpho-anatomical, physiological, biochemical and molecular responses of plants to air pollution', in Husen A. (ed.) *Harsh environment and plant resilience*. Cham, Switzerland: Springer International Publishing AG, pp. 203–234. https://doi.org/10.1007/978-3-030-65912-7_9

Husen, A. (2021c) 'Cross talk between autophagy and hormones for abiotic stress tolerance in plants', in Husen A. (ed.) *Plant performance under environmental stress*. Cham, Switzerland: Springer International Publishing AG, pp. 1–15. https://doi.org/10.1007/978-3-030-78521-5_1

Husen, A. (2021d) *Harsh environment and plant resilience: Molecular and functional aspects*. Springer Nature Switzerland AG. https://doi.org/10.1007/978-3-030-65912-7

Husen, A. (2021e) *Plant performance under environmental stress: Hormones, biostimulants and sustainable plant growth management)*. Cham, Switzerland: Springer Nature Switzerland AG. https://doi.org/10.1007/978-3-030-78521-5

Husen, A. and Iqbal, M. (2023) *Medicinal plants – Their response to abiotic stress*. Singapore: Springer Nature Singapore Pte Ltd. https://doi.org/10.1007/978-981-19-5611-9

Husen, A., Iqbal, M. and Aref, I. (2014) 'Growth, water status and leaf characteristics of *Brassica carinata* under drought stress and rehydration conditions', *Brazilian Journal of Botany*, 37, pp. 217–227.

Husen, A., Iqbal, M. and Aref, I. (2016) 'IAA-induced alteration in growth and photosynthesis of pea (*Pisum sativum* L.) plants grown under salt stress', *Journal of Environmental Biology*, 37, pp. 421–429.

Husen, A., et al. (2018) 'Salicylic acid alleviates salinity-caused damage to foliar functions, plant growth and antioxidant system in Ethiopian mustard (*Brassica carinata* A. Br.)', *Agriculture & Food Security*, 7(44), pp. 1–14.

Husen, A., *et al.* (2019) 'Modulation of salt-stress tolerance of Niger (*Guizotia abyssinica*), an oilseed plant, by application of salicylic acid', *Journal of Environmental Biology*, 40, pp. 94–104.

Hussain, S., *et al.* (2023) 'Deciphering the role of phytohormones and osmolytes in plant tolerance against salt stress: Implications, possible cross-talk, and prospects', *Journal of Plant Growth Regulation*, 43, pp. 38–59.

Hussein, M., *et al.* (2017) 'Salinity-induced modulation of plant growth and photosynthetic parameters in faba bean *(Vicia faba)* cultivars', *Pakistan Journal of Botany*, 49, pp. 867–877.

IPCC (2019) Climate change and land: An IPCC special report on climate change, desertification, land degradation, sustainable land management, food security, and greenhouse gas fluxes in terrestrial ecosystems. https://philpapers.org/rec/SHUCCA-2

Janssen, P.J., Audit, B. and Ouzounisa, C.A. (2001) 'Strain-specific genes of helicobacter pylori: Distribution, function and dynamics', *Nucleic Acids Research*, 29(21), pp. 4395–4404.

Jha, U.C., Bohra, A. and Nayyar, H. (2020) 'Advances in "omics" approaches to tackle drought stress in grain legumes', *Plant Breeding*, 139(1), pp. 1–27.

Jung, Y.J., *et al.* (2018) 'Application of ZFN for site directed mutagenesis of rice SSIVa gene', *Biotechnology and Bioprocess Engineering*, 23, pp. 108–115.

Keshavareddy, G., Kumar, A.R.V. and Ramu, V.S. (2018) 'Methods of plant transformation- A review', *International Journal of Current Microbiology and Applied Sciences*, 7(7), pp. 2656–2668.

Khan, S.H. (2019) 'Genome-editing technologies: Concept, pros, and cons of various genome-editing techniques and bioethical concerns for clinical application', *Molecular Therapy-Nucleic Acids*, 16, pp. 326–334.

Khan, Z., *et al.* (2017) 'Use of TALEs and TALEN technology for genetic improvement of plants', *Plant Molecular Biology Reporter*, 35, pp. 1–19.

Kim, H. and Kim, J.S. (2014) 'A guide to genome engineering with programmable nucleases', *Nature Reviews Genetics*, 15(5), pp. 321–334.

Krishna, R., *et al.* (2019) 'Transgenic tomatoes for abiotic stress tolerance: Status and way ahead', *Biotech*, 9, pp. 1–14.

Kumar, K., et al. (2020) 'Genetically modified crops: Current status and future prospects', *Planta*, 251(4), pp. 1–27.

Kumar, M., *et al.* (2022) 'Optimization of t-DNA configuration with UBIQUITIN10 promoters and tRNA-sgRNA complexes promotes highly efficient genome editing in allotetraploid tobacco', *Plant Cell Reports*, 41(1), pp. 1–20.

Lakshman, D.K., *et al.* (2013) 'Lactoferrin-derived resistance against plant pathogens in transgenic plants', *Journal of Agricultural and Food Chemistry*, 61(48), pp. 11730–11735.

Lieber, M.R., *et al.* (2003) 'Mechanism and regulation of human non-homologous DNA end-joining', *Nature Reviews Molecular Cell Biology*, 4(9), pp. 712–720.

Li, M., *et al.* (2016a) 'Reassessment of the four yield-related genes Gn1a, DEP1, GS3, and IPA1 in rice using a CRISPR/Cas9 system', *Frontiers in Plant Science*, 7, pp. 1–13.

Li, S., *et al.* (2016b) 'The OsmiR396c OsGRF4-OsGIF1 regulatory module determines grain size and yield in rice', *Plant Biotechnology Journal*, 14(11), pp. 2134–2146.

Li, T., *et al.* (2016c) 'TALEN-mediated homologous recombination produces site-directed DNA base change and herbicide-resistant rice', *Journal of Genetics and Genomics*, 43(5), pp. 297–305.

Liu, Q., *et al.* (2021) 'Application of CRISPR/Cas9 in crop quality improvement', *International Journal of Molecular Sciences*, 22(8), pp. 1–16.

Murray, D., Mirzayans, R. and McBride, W.H. (2018). 'Defenses against pro-oxidant forces—Maintenance of cellular and genomic integrity and longevity', *Radiation Research*, 190(4), pp. 331–349.

Nascimento, F.S., *et al.* (2023) 'Gene editing for plant resistance to abiotic factors: A systematic review', *Plants*, 12, pp. 1–31.

Nora, L.C., *et al.* (2019) 'The art of vector engineering: Towards the construction of next-generation genetic tools', *Microbial Biotechnology*, 12(1), pp. 125–147.

Patel, A., *et al.* (2023) 'Methods of crop improvement and applications towards fortifying food security', *Frontier in Genome Editing*, 7, pp. 1–21.

Ramsay, T., *et al.* (2022) 'Ethical aspects of GMO regulation in the EU', *EMBO Reports*, 23, pp. 1–5.

Raymond Park, J., *et al.* (2011) 'The role of transgenic crops in sustainable development', *Plant Biotechnology Journal*, 9(1), pp. 2–21.

Rivero, J., *et al.* (2015). 'Metabolic transition in mycorrhizal tomato roots', *Frontiers in Microbiology*, 6, p. 598.

Rozas, P., Kessi-Pérez, E.I. and Martínez, C. (2022) 'Genetically modified organisms: Adapting regulatory frameworks for evolving genome editing technologies', *Biological Research*, 55(1), p. 14.

Sharma, P., *et al.* (2022) 'Genetic modifications associated with sustainability aspects for sustainable developments', *Bioengineered*, 13(4), pp. 9509–9521.

Shukla, V.K., *et al.* (2009) 'Precise genome modification in the crop species *Zea mays* using zinc-finger nucleases', *Nature*, 459, pp. 437–441.

Smedley, M.A., *et al.* (2021) 'CRISPR-Cas9 based genome editing in wheat', *Current Protocols*, 1(3), pp. 1–15.

Smyth, S.J. (2022) 'Contributions of genome editing technologies towards improved nutrition, environmental sustainability and poverty reduction', *Frontiers in Genome Editing*, 4, pp. 1–9.

Song, Y., *et al.* (2018) 'Transfer of tomato immune receptor Ve1 confers Ave1-dependent verticillium resistance in tobacco and cotton', *Plant Biotechnology Journal*, 16(2), pp. 638–648.

Tran, M.T., *et al.* (2023) 'CRISPR-Cas9-based precise engineering of SlHyPRP1 protein towards multi-stress tolerance in tomato', *Frontiers in Plant Science*, 14, pp. 1–14.

Turnbull, C., Lillemo, M. and Hvoslef-Eide, T.A.K. (2021) 'Global regulation of genetically modified crops amid the gene edited crop boom – A review', *Frontiers in Plant Science*, 12, pp. 1–19.

Wada, N., *et al.* (2020) 'Precision genome editing in plants: State-of-the-art in CRISPR/Cas9-based genome engineering', *BMC Plant Biology*, 20(1), p. 12.

Wolfe, S.A., Nekludova, L. and Pabo, C.O. (2000) 'DNA recognition by Cys2His2 zinc finger proteins', *Annual Review of Biophysics and Biomolecular Structure*, 29(1), pp. 183–212.

Wu, X., *et al.* (2016) 'Advances in crop proteomics: PTMs of proteins under abiotic stress', *Proteomics*, 16(5), pp. 847–865.

Yu, H., *et al.* (2022) 'Transgenic improvement for biotic resistance of crops', *International Journal of Molecular Sciences*, 23(22), pp. 1–18.

Zahra, N., et al. (2021) 'Physiological and molecular responses to high, chilling, and freezing temperature in plant growth and production: Consequences and mitigation possibilities', in Husen, A. (ed.) *Harsh environment and plant resilience*. Cham, Switzerland: Springer. https://doi.org/10.1007/978-3-030-65912-7_10

Zahra, N., *et al.* (2023a) 'Plant photosynthetic responses under drought stress: Effects and management', *Journal of Agronomy and Crop Science*. https://doi.org/10.1111/jac.12652

Zahra, N., *et al.* (2023b) 'Influence of abiotic stresses on disease infestation in plants', *Physiological and Molecular Plant Pathology*, 127 p. 102125.

Zhang, Y., *et al.* (2013) 'Transcription activator-like effector nucleases enable efficient plant genome engineering', *Plant Physiology*, 161(1), pp. 20–27.

Zhang, Y., *et al.* (2016) 'Efficient and transgene-free genome editing in wheat through transient expression of CRISPR/Cas9 DNA or RNA', *Nature Communication*, 25, pp. 1–25.

Zhang, Y., *et al.* (2023) 'Plants' response to abiotic stress: Mechanisms and strategies', *International Journal of Molecular Science*, 24(13), pp. 1–17.

4 Engineering Soil Microbiota for Crop Plant Improvement

Megha Barot, Arpan Modi, and Azamal Husen

4.1 INTRODUCTION

Besides serving many functions like the source or sink of nutrients, preserving water quality, and supporting plants' standing, soil is also a major source of microorganisms. From the plants' perspective, microorganisms that reside there grow with the help of the plants' secretion and let the plants grow by making the soil nutrients available to the plant form the rhizosphere (Williams, Sinanaj and Hoysted, 2023). Soil biodiversity is dominated by a microbial community where most of the organisms are beneficial for plant health. In this regard, the rhizosphere may also be termed as the consortium of soil microbiomes. Microbial communities in the rhizosphere include protists, archaea, bacteria, fungi, and algae (Berg *et al.*, 2020). The secondary metabolites produced by rhizosphere microbes reduce the growth of weeds; therefore, their usage as bioherbicides is interesting to explore for the sustainable agriculture (Sindhu and Sehrawat, 2017). Next-generation sequencing methodologies, meta-'omics' tools, and high-throughput community analysis approaches have opened up new avenues for understanding the multitrophic relationships seen in the mysterious plant microbiomes (Ahmad, Nadeem and Zahir, 2019). The advance cutting-edge molecular technologies are the key to manipulate the useful microbes. Polymicrobial engineering opens the way to modify their functions to utilize them as plant growth promotors and for productivity (Prasad, Gill and Tuteja, 2018). Minimizing chemical farm input can have a greater impact on plant growth performance through planned management, augmentation, and inoculation of various microbial species (Kaul *et al.*, 2021).

In order to improve the plant's performance through a beneficial soil microbiome, one has to understand how the plant attracts target microorganisms and in return they help the plant to mitigate the adverse environmental conditions during which their growth and survival are compromised. The root architecture plays a significant role in plant-microbes interactions, particularly in rhizosphere. Plant-microbe interactions are also established during pathogenesis. However, the mechanism of pathogenesis is outside the scope of this chapter; indeed, the interactions between host-plant and beneficial microbes are elucidated. Exudates containing secondary metabolites, enzymes, sugars, fatty acid, organic acid, and signalling molecules are secreted from the root (called root exudates). They favour the growth of beneficial microbes as well as biofilm formation (Mavrodi *et al.*, 2021). This exudate has the potential to select beneficial microbes and nurture their growth under the niche. In return, the microbes provide plant growth promotion, facilitate nutrients uptake, act as chelators (siderophore), and provide resistance against biotic and abiotic stresses. Briefly, the plant secretes root exudates to which the rhizobiome can build association. They synthesize plant growth regulators, a variety of chemicals, enzymes, proteins, and antimicrobials. These include indole acetic acid (IAA), abscisic acid (ABA), gibberellic acid (GA), biological nitrogen fixation (BNF), 1-aminocyclopropane-1-carboxylate deaminase (ACCd), solubilized phosphates and potassium, 2,4-diacetylphloroglucinol (2,4-DAPG), iron-chelators (siderophores), and stress-relieving signals and chemicals (del Carmen Orozco-Mosqueda *et al.*, 2022).

With the well-established methods for genetic transformation, it is now possible to modify the genetic background (and thereby phenotypic expression) of almost all the organisms, including microbes. Genetic engineering of microbes requires the isolation of gene responsible for the desired trait, cloning of the gene using suitable vector, transfer of the gene in expression vectors, and transfer

DOI: 10.1201/9781032630366-4

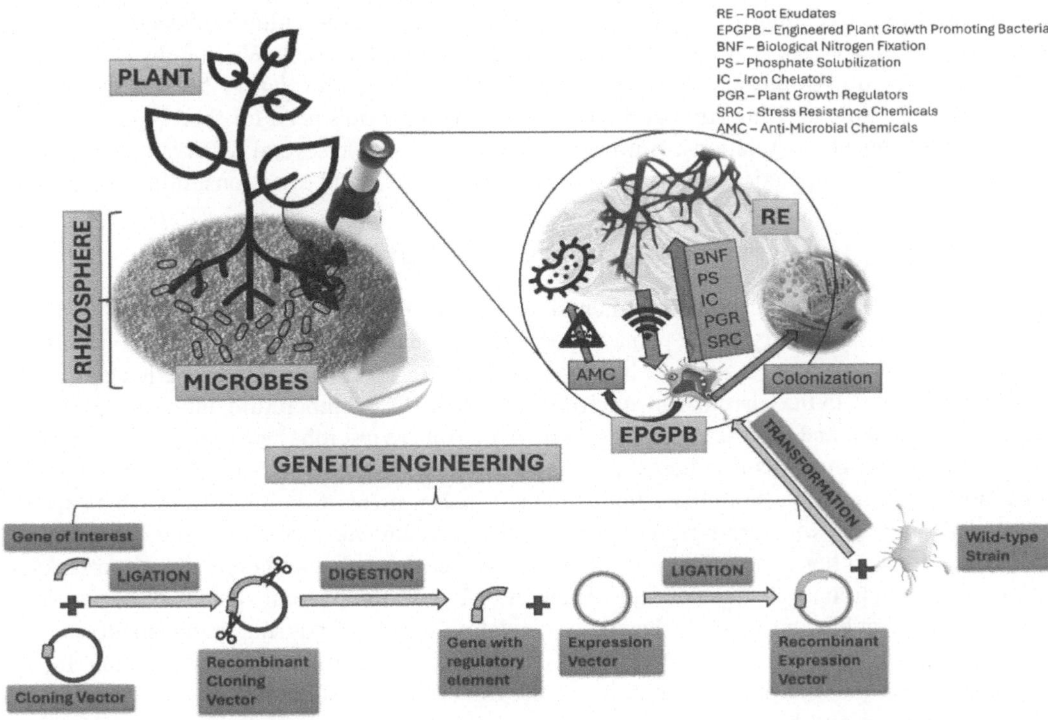

FIGURE 4.1 Interaction of engineered plant growth-promoting bacteria with the plant rhizosphere through root exudate-mediated signalling.

of expression vector in the microbes of interest. Methods for the optimal growth of the genetically modified microbes have to standardize, such as liquid culture and solid-state fermentation (Herrera-Estrella, 2020). Although the wild strain of beneficial soil microbes is capable of producing all the chemicals during symbiosis, under certain circumstances, they fail to do so. The unmet requirement of the plant can be fulfilled by genetically modifying these microbes by overexpression of desired traits (Figure 4.1). Therefore, the present chapter focuses on the impact of the beneficial soil microbial community on plant growth and development, stress management, and nutrient uptake, along with the need for their genetic manipulation as a step forward towards sustainable agriculture.

4.2 CONVENTIONAL STRATEGIES OF CROP IMPROVEMENT

There is a continuous demand for increasing crop production and productivity as the population rises. To mitigate this, breeders and scientists are incessantly engaged in the development of new varieties that impart enhanced stress resistance and crop production. Initially, the crop improvement strategies include advancement in techniques to grow field crops through agricultural engineering. The invention of the cotton gin, thresher, and harvesters made agriculture production fast and easy. These inventions helped the farmer to use land effectively to increase crop production with less labour and land. Afterwards, the chemical era began. After the discovery of the Haber process, nitrogen fertilizers raised production in less time. After the era of chemical revolution, hybrid corn development in 1930 markedly increased corn production. Since then, breeding programs have been carried out in different crop plants, and the need for better and bulk food production was fulfilled (Tahir ul Qamar *et al.*, 2020). Conventional strategies of crop improvement include pure-line selection (breeding between two pure lines with desirable characteristics), hybrid breeding (intraspecific or interspecific breeding), population breeding (selection and recombination on large populations), pedigree breeding (breeding of self-pollinating species), and tissue culture-induced somaclonal variations (Melchinger *et al.*, 2023; Rutkoski, Krause and Sorrells, 2022; Willoughby *et al.*, 2019). Conventional breeding strategies were key factors during the green revolution. Most of the hybrids were developed using breeding strategies which fulfilled the world's need for food crops. They are free from governmental regulations. However, these strategies have several disadvantages, such as requirement of large population, laborious screening procedures, and time-consuming processes.

4.3 ROLE OF RHIZOSPHERE MICROBES IN PLANT HEALTH

Plant immunity against adverse environmental conditions depends on its own genetics and supporting factors in the vicinity of the plant. Genetic factors may be enhanced through conventional breeding strategies mentioned above or biotechnological interventions such as genetic engineering and genome editing. In the surrounding environment, plants face temperature, humidity, pollution, and microbes within and outside the rhizosphere. Rhizosphere microbes surround the roots of the plant and establish mutualism (Chepsergon and Moleleki, 2023). From the plant's perspective, root exudates play a vital role in the establishment or recruitment of the rhizobiome in a species-independent manner. Root exudates contain secondary chemicals, volatile organic compounds (VOCs), sugars, and hormones. These exudates are secreted by plants under pathogen attacks as well as stress conditions such as nutrient depletion or low water. Several species of microorganisms are efficient at biocontrol, and when these exudates are released during pathogen attack, they attract such biocontrol agents. However, exact molecular mechanisms of such induction are not yet known. From the microbial community of the rhizosphere, results of such signalling attract both the beneficial and harmful organisms. In case of beneficial microorganisms, they impart very good competition to the pathogens and signalling molecules from the plant help them to dominate the community. By this means, the harmful organisms are outnumbered, and the host plant is protected (Olanrewaju and Babalola, 2022). Another way of providing help to the host plant by beneficial microbes is to produce compounds toxic to pathogens. These include antibiotics, VOCs, and siderophores (Babalola, Ayangbenro and Olanrewaju, 2019). As far as plant growth is concerned, like animals, plants also possess hormones called phytohormones. There are five classes of phytohormones *viz.*, auxin, cytokinins, gibberellins, ABA, and ethylene. IAA is a natural auxin produced by both plants and bacteria. Auxins are involved in cell division, apical dominance, seed germination, and root development. Some plant species have less ability to synthesize IAA. At the other end, more than 80% of the rhizosphere bacteria produce IAA. Plant growth promotion was observed in certain plant species using these microorganisms (Castaldi *et al.*, 2021). Auxin-producing plant growth-promoting bacteria (PGPB) also play significant roles in mitigation of stress conditions. There is a direct or indirect stress of water to the plant in abiotic conditions depending on which transport of nutrients and minerals is affected. As a result, auxin production is also compromised. It was observed that *Pseudomonas koreensis* imparted stress tolerance and growth to tomato plants by supplying auxins (Guo *et al.*, 2021). Plant growth and development require the supply of nutrients from soil. These nutrients are classified as macro- and micro-nutrients according to their required concentrations. Two major nutrients are nitrogen and phosphorous, which are often present in the form that makes them inert for the plant. Several bacterial species *viz.*, *Rhizobium*, *Azotobacter*, *Azospirillum*, *Enterobacter*, *Pseuodmonas*, and *Xanthomonas* are involved in nitrogen and phosphorous uptake from the soil (Billah *et al.*, 2019; Ke *et al.*, 2019). Lastly, during pathogen attacks, PGPB produce ACC-deaminase, siderophore, lytic enzymes, and antibiotics. Certain level of ethylene is required for the regular functioning of the plant. However, if the ethylene production is increased, for instance, during pathogen infection, chlorosis, senescence, and cell death processes start to occur. The precursor compound for the biosynthesis of ethylene is 1-aminocyclopropane-1-carboxylate (ACC) and deamination of this precursor by ACC-deaminase results in inhibition of the biosynthesis of ethylene. PGPB produce ACC-deaminase in response to the pathogen attack in order to prevent the deleterious effects of oxidative stress that occurs due to alleviated levels of ethylene in host plants (Gupta and Pandey, 2019). Siderophores are iron chelators synthesized by PGPB, which trap the iron present in the soil and make them available and unavailable for host plant and pathogen, respectively. Several siderophore-producing bacteria are reviewed in detail by Ferreira, Silva and Cunha (2019). Lipopeptide and 2,4-diacetylphlorogucinol (DAPG) are a class of antibiotics produced by PGPB under pathogen attack (Olanrewaju and Babalola, 2022). Most of the pathogenic organisms contain cell walls which are degraded by lytic enzymes. These enzymes include lipases, proteases, glucanases, and chitinases. PGPB secrete these enzymes and thereby protect the host plant during biotic stress (Olanrewaju and Babalola, 2019). Under abiotic stress conditions, direct or physiological drought and oxidative stress occur. The hormone ABA is a phytohormone synthesized during drought stress by PGPB, and it imparts drought tolerance to the plant (Ojuederie, Olanrewaju and Babalola, 2019).

4.4 TARGETED TRAITS FOR GENETIC ENGINEERING

Plant growth-promoting rhizobacteria in a natural way imparts many benefits to the host plant. However, due to several factors, they cannot be utilized with their full potential. These factors include failure to colonize, no host specificity, and undesirable modes of genetic regulation for the nutrient uptakes (Haskett, Tkacz and Poole, 2021). For these reasons, conventional modes of application of beneficial microbes (individual or consortia) need to be modified. There are genes associated with the factors mentioned. Regulating the expression of genes associated with factors and thereby regulating the trait can help in successful establishment of plant-microbe relationships and boost agricultural productivity. Two types of methods can be employed using genetic engineering approaches. First is the top-down approach in which plants are genetically engineered to favour the growth of the beneficial microbial communities. Another approach is the bottom-up approach in which soil microbes are genetically engineered to dominate the pathogenic strains, to impart host plant specificities, or to bypass stringent genetic regulation (del Carmen *et al.*, 2022). In this section, only the bottom-up approach has been discussed, as the other approach is outside the scope of this chapter.

Several agriculturally important traits are targeted during the bottom-up approach. These traits are involved in nitrogen fixation, phosphate solubilization, auxin biosynthesis, biocontrol, and host specificity. In a biological nitrogen fixation, *nif* gene cluster *viz., nifABDEFHJKLMNQRSUVWXYZ* forming nitrogenase complex is a key enzyme complex (Mahmud *et al.*, 2020). Phosphorous uptake is mostly facilitated by arbuscular mycorrhizal fungi (AMF) and phosphate solubilizing bacteria (PSB). One of the principal factors for the uptake of inorganic phosphate (Pi) is the decreased soil pH. Genes involved in glucose metabolism can be targeted to produce organic acids (and thereby to decrease the soil pH), which facilitate the phosphorous uptake by the host plant. These acids include pyruvic acid, oxalic acid, and citric acid (Abbasi, 2023). For the utilization of organic phosphate present in soil organic matter, there are genes involved in the production of enzymes such as alkaline and acid phosphatase, phytases, phosphodiesterases, and nucleases (Haskett *et al.*, 2021). Overexpression of IAA-producing genes in *E. coli* results in enhanced plant growth (Guo *et al.*, 2019). Potent and broad-spectrum antifungal compounds like pyrrolnitrin and DAPG are synthesized through the action of gene clusters *prnABCD* and *phlDACB*, respectively. Overexpression of these genes contributes significantly towards the control of fungal infestations in field crops (Haskett *et al.*, 2021). Host-specificity is a plant species-dependent process, and therefore root exudates from the host are target compounds to identify regulatory sequences in the beneficial microorganisms to be engineered. Unique secondary metabolites have been identified from the plants, and they can be used to engineer the rhizobacteria (Spöring, Finke and Hartig, 2020). To enrich the target bacterial population, genes responsible for the opine synthesis and catabolism are overexpressed to provide continuous supply of C and N for the colonization of target organisms. Traits related to rhizosphere engineering are mentioned in Table 4.1.

TABLE 4.1
Plant Growth Promoting Traits Engineered in Soil Microbial Community

Process	Bacterial Species, Engineered Trait and Plant Species	Key Findings	Key Reference
Nitrogen fixation	*Rhizobium* sp. IRBG74; *nif* cluster v3.2 and native *nif* genes; maize	Engineered species of *Pseudomonas protegens* Pf-5 fixed more N as compared to natural fixer species *K. oxytoca*	Ryu *et al.* (2020)
P. solubilization	*Pseudomonas simiae* WCS417r, *Ralstonia* sp. *strain* UNC404CL21Col, and *Pseudomonas putida* KT2440; *Phytases; Arabidopsis*	The engineered *Arabidopsis* seedling showed better plant vigor than wild-type and un-inoculated controls.	Shulse *et al.* (2019)
Auxin biosynthesis	*E. coli; aldH, aro8,* and *kdc; de novo* synthesis	–	Guo *et al.* (2019)
Biocontrol	*Pseudomonas protegens* Pf-5; *retS* (mutant); cucumber and wheat	Seedling, shoot, and root fresh weight were found to be significantly higher in *retS* mutant treated plants as compared to control inoculants	Jing *et al.* (2020)
Host-specificity	*Pseudomonas hormoni G20-18^T;* *acdS* and *acdR*; wheat seedlings	Inducible promoters were identified from the *P. hormoni* and showed strong activity of reporter genes for the proof of concept.	Sorty *et al.* (2023)

4.5 CASE STUDIES

As a conventional strategy of employment of soil microbiome, a single inoculum or consortium has a few limitations, which are targets for the modern biotechnological advancements. Genetically modified bacteria can be used as an efficient crop improvement technique in situations where the creation and release of transgenic plants presents several regulatory obstacles. Target processes for the employment of genetically engineered soil microbes are biotic and abiotic stress resistance, nutrient acquisition, and unintended genetic regulations. This section highlights current and significant studies conducted on different crop plants with engineered microbes.

Agriculturally important and natural biocontrol agents such as *Pseudomonas fluorescens* strain ATCC13525 impart many functions such as iron sequestration, phosphate solubilization, rhizosphere colonization, and nematicidal activities. Such strains are prepared for the commercial application using one of the following approaches. The first approach is liquid cell suspension, which is cost-effective but ends up with very short shelf life. The second approach is the formation of a microcapsule, which is costly. The third approach comprises freeze-dried powder, which is even more costly than microcapsule preparation. Spray and hot-air drying are the most feasible and cost-effective approaches. However, *P. fluorescens* strains are not suitable for them, as they are non-spore forming. Wu *et al.* (2021) engineered a strain of *P. fluorescens* (ATCC13525) to overexpress stress adaptation gene *rpoS* and *hfq*. The engineered bacteria were adapted to different stress conditions such as salinity, oxidative, and heat and prepared in a commercially available form afterwards. They also prepared the knock-out mutant of the same genes. Over-expressed strains (PF-pBBR2-*rpoS* and PF-pBBR2-*hfq*) showed higher survival rates than wild types, whereas mutant strains showed the lowest survival rate.

In a process of BNF in non-leguminous crops, *Azotobacter vinelandii* is a widely used bacterial strain. In the organism, BNF is negatively regulated by *nifL*. It serves as an anti-activator and controls the activity of *nifA* in response to redox, nitrogen, and carbon status. Mus *et al.* (2022) created a mutant strain of *A. vinelandii* and inoculated it to rice seedlings. They observed that release of ammonia from the organism is increased as compared to wild strain. This ammonia is available for the host plant as sole source of nitrogen. They suggested such gene editing approaches to engineer the nitrogen fixing organism to prepare promising biofertilizers. Other examples of nutrient acquisition include nitrogen fixation by engineered *Pseudomonas protegens* Pf-5 in *Arabidopsis,* alfalfa, tall fescue, and maize (Setten *et al.*, 2013); and phosphate solubilization by *Pseudomonas fluorescens* ATCC 13525 (Buch, Archana and Kumar, 2010), *P. putida*, *P. simiae*, and *Ralstionia* sp. (Shulse *et al.*, 2019).

Pseudomonas spp. WS5 is a diazotrophic endophyte capable of fixing nitrogen in the soil. However, it lacks the ability to mitigate the pathogen infestations. Compared to that, other *Pseudomonas* spp. are well-known for the antagonistic ability against pathogenic fungi. Therefore, Patel and Archana (2018) isolated gene clusters from two potential strains of *Pseudomonas* and transferred it to WS5. The engineered strain produced elevated levels of 2,4-diacetylphlorogucinol (2,4-DAPG), which is an antifungal metabolite. Trials conducted under controlled environments showed that the engineered strain was capable of producing 12–14 μg/mL of 2,4-DAPG as compared to the wild-type strain (< 2 μg/mL). Moreover, the recombinant strain showed effective control of pathogenic fungi *Magnaporthe oryzae* B157 and *Rhizoctonia solani.* Such biocontrol ability was observed until after 60 days of plant growth following infection of pathogenic fungi in rice, sorghum, and wheat.

Abiotic stress is a major limiting factor for crop loss globally. Direct physiological impact of abiotic stress is the generation of oxidative stress, which in turn enhances the ethylene biosynthesis. Excess ethylene production can give rise to senescence and cell death. Precursor molecules for the biosynthesis of ethylene are ACC, and *S*-adenosyl methionine through the action of ACC synthase and ACC oxidase. However, ACC can be degraded via the action of ACC deaminase, which is synthesized in the endophytic community. Liu *et al.* (2017) overexpressed the ACC deaminase in two of the endophytic strains of *Enterobacteriaceae* (E5 and S1). The engineered strains (E5P and S1P) promoted root and shoot biomass of rice seedlings as compared to wild-type controls under saline stress. Moreover, significantly higher activity of deaminase was found in the engineered strains as compared to controls without inoculation and inoculation with the wild-type strains.

4.6 CONCLUSION AND FUTURE PROSPECTS

As the demand for food crops increases with the growing population, employment of new strategies has become vital for agriculturists. Transgenic plant development is a versatile and reliable approach but is still not accepted by many countries. Alternative approaches within the regime of biotechnology include the development and use of genetically modified microbes. Such microbes can be easily prepared and are fast growing but are more prone to horizontal gene transfer. Therefore, engineered microbes are difficult to manage and monitor. However, the first field trials of the production of siderophore-producing genetically engineered *P. putida* opened up the gate for the utilization of engineered microbes. Afterwards, no adverse effects on the environment over three decades have been noted, which supports its safety concerns. Engineered rhizobiome can effectively enhance the nutrient uptake (nitrogen fixation and P solubilization), biotic and abiotic stress resistance, and plant growth promotion. Future strategies in this regard may include cutting-edge genome editing tools (CRISPR-Cas9) and biocontainment of the engineered microbes to make them 'regulation-friendly'.

REFERENCES

Abbasi, S. (2023) 'Plant–microbe interactions ameliorate phosphate-mediated responses in the rhizosphere: A review', *Frontiers in Plant Science*, 14, p. 1074279.

Ahmad, M., Nadeem, S.M. and Zahir, Z.A. (2019). 'Plant-microbiome interactions in agroecosystem: An application', in Kumar, V., Prasad, R., Kumar, M., and Choudhary, D. (eds.) *Microbiome in Plant Health and Disease: Challenges and Opportunities*, Singapore: Springer, pp. 251–291.

Babalola, O.O., Ayangbenro, A.S. and Olanrewaju, O.S. (2019) 'Draft genome sequences of three rhizospheric plant growth-promoting bacteria', *Microbiology Resource Announcements*, 8(26), pp. 10–1128.

Berg, G.*et al.* (2020) 'Microbiome definition re-visited: Old concepts and new challenges', *Microbiome*, 8(1), pp. 1–22.

Billah, M. *et al.* (2019) 'Phosphorus and phosphate solubilizing bacteria: Keys for sustainable agriculture', *Geomicrobiology Journal*, 36(10), pp. 904–916.

Buch, A., Archana, G. and Kumar, G.N. (2010) 'Heterologous expression of phosphoenolpyruvate carboxylase enhances the phosphate solubilizing ability of fluorescent pseudomonads by altering the glucose catabolism to improve biomass yield', *Bioresource Technology*, 101(2), pp. 679–687.

Castaldi, S. *et al.* (2021) 'Plant growth promotion function of *Bacillus* sp. strains isolated from salt-pan rhizosphere and their biocontrol potential against *Macrophomina phaseolina*', *International Journal of Molecular Sciences*, 22(7), p. 3324.

Chepsergon, J. and Moleleki, L.N. (2023) 'Rhizosphere bacterial interactions and impact on plant health', *Current Opinion in Microbiology*, 73, p. 102297.

del Carmen Orozco-Mosqueda, M. *et al.* (2022) 'Rhizobiome engineering: Unveiling complex rhizosphere interactions to enhance plant growth and health', *Microbiological Research*, 263, p. 127137.

Ferreira, M.J., Silva, H. and Cunha, A. (2019) 'Siderophore-producing rhizobacteria as a promising tool for empowering plants to cope with iron limitation in saline soils: A review', *Pedosphere*, 29(4), pp. 409–420.

Guo, D. *et al.* (2019) 'De novo biosynthesis of indole-3-acetic acid in engineered *Escherichia coli*', *Journal of Agricultural and Food Chemistry*, 67(29), pp. 8186–8190.

Guo, Q. *et al.* (2021) '*Pseudomonas koreensis* promotes tomato growth and shows potential to induce stress tolerance via auxin and polyphenol-related pathways', *Plant and Soil*, 462, pp. 141–158.

Gupta, S. and Pandey, S. (2019) 'Unravelling the biochemistry 0and genetics of ACC deaminase-an enzyme alleviating the biotic and abiotic stress in plants', *Plant Gene*, 18, p. 100175.

Haskett, T.L., Tkacz, A. and Poole, P.S. (2021) 'Engineering rhizobacteria for sustainable agriculture', *The ISME Journal*, 15(4), pp. 949–964.

Herrera-Estrella, L. (2020) 'My journey into the birth of plant transgenesis and its impact on modern plant biology', *Plant Biotechnology Journal*, 18(7), pp. 1487–1491.

Jing, X., Cui, Q., Li, X., Yin, J., Ravichandran, V., Pan, D., Fu, J., Tu, Q., Wang, H., Bian, X. and Zhang, Y. (2020) 'Engineering *Pseudomonas protegens* Pf-5 to improve its antifungal activity and nitrogen fixation', *Microbial Biotechnology*, 13(1), pp. 118–133.

Kaul, S. *et al.* (2021) 'Engineering host microbiome for crop improvement and sustainable agriculture', *Frontiers in Microbiology*, 12, p. 635917.

Ke, X. *et al.* (2019) 'Effect of inoculation with nitrogen-fixing bacterium *Pseudomonas stutzeri* A1501 on maize plant growth and the microbiome indigenous to the rhizosphere', *Systematic and Applied Microbiology*, 42(2), pp. 248–260.

Liu, Y., Cao, L., Tan, H. and Zhang, R. (2017) 'Surface display of ACC deaminase on endophytic Enterobacteriaceae strains to increase saline resistance of host rice sprouts by regulating plant ethylene synthesis', *Microbial cell factories*, 16, pp.1–9.

Mahmud K, *et al.*, (2020) 'Current progress in nitrogen fixing plants and microbiome research', *Plants*, 9(1), p. 97. https://doi.org/10.3390/plants9010097.

Mavrodi, O.V. *et al.* (2021) 'Root exudates alter the expression of diverse metabolic, transport, regulatory, and stress response genes in rhizosphere *Pseudomonas*', *Frontiers in Microbiology*, 12, p. 651282.

Melchinger, A.E. *et al.* (2023) 'Genomic prediction in hybrid breeding: I. Optimizing the training set design', *Theoretical and Applied Genetics*, 136(8), p. 176.

Mus, F. *et al.* (2022) 'Genetic determinants of ammonium excretion in *nifL* mutants of *Azotobacter vinelandii*', *Applied and Environmental Microbiology*, 88(6), p. e01876–21.

Ojuederie, O.B., Olanrewaju, O.S. and Babalola, O.O. (2019) 'Plant growth promoting rhizobacterial mitigation of drought stress in crop plants: Implications for sustainable agriculture', *Agronomy*, 9(11), p. 712.

Olanrewaju, O.S. and Babalola, O.O. (2019) 'Bacterial consortium for improved maize (*Zea mays* L.) production', *Microorganisms*, 7(11), p. 519.

Olanrewaju, O.S. and Babalola, O.O. (2022) 'The rhizosphere microbial complex in plant health: A review of interaction dynamics', *Journal of Integrative Agriculture*, 21(8), pp. 2168–2182.

Patel, J.K. and Archana, G. (2018) 'Engineered production of 2, 4-diacetylphloroglucinol in the diazotrophic endophytic bacterium *Pseudomonas* sp. WS5 and its beneficial effect in multiple plant-pathogen systems', *Applied Soil Ecology*, 124, pp. 34–44.

Prasad, R., Gill, S. S. and Tuteja, N. (eds.) (2018) *New and future developments in microbial biotechnology and bioengineering: Crop improvement through microbial biotechnology*. Elsevier.

Rutkoski, J.E., Krause, M.R. and Sorrells, M.E. (2022) 'Breeding methods: Population improvement and selection methods', in Reynolds, M.P. and Braun, H.J. (eds.) *Wheat improvement: Food security in a changing climate*. Cham, Switzerland: Springer International Publishing, pp. 83–96.

Ryu, M.H. *et al.* (2020) 'Control of nitrogen fixation in bacteria that associate with cereals', *Nature Microbiology*, 5(2), pp. 314–330.

Setten, L. *et al.* (2013) 'Engineering *Pseudomonas protegens* Pf-5 for nitrogen fixation and its application to improve plant growth under nitrogen-deficient conditions', *PLoS One*, 8(5), p. e63666.

Shulse, C.N. *et al.* (2019) 'Engineered root bacteria release plant-available phosphate from phytate', *Applied and Environmental Microbiology*, 85(18), p. e01210–19.

Sindhu, S. S. and Sehrawat A. (2017). 'Rhizosphere microorganisms: application of plant beneficial microbes in biological control of weeds', in Panpatte, D., Jhala, Y., Vyas, R. and Shelat, H. (eds.) *Microorganisms for Green Revolution: Volume 1: Microbes for Sustainable Crop Production*. Singapore: Springer, pp. 391–430.

Sorty, A.M. *et al.* (2023) 'Plant-root exudate analogues influence activity of the 1-aminocyclopropane-1-carboxylate (ACC) deaminase Gene in *Pseudomonas hormoni* G20-18T', *Microorganisms*, 11(10), p. 2504.

Spöring, M., Finke, M. and Hartig, J.S. (2020) 'Aptamers in RNA-based switches of gene expression', *Current Opinion in Biotechnology*, 63, pp. 34–40.

Tahir ul Qamar, M. *et al.* (2020) 'Effectiveness of conventional crop improvement strategies vs. omics', in Fahad, S., et al. (eds.) Environment, Climate, Plant and Vegetation Growth. Cham: Springer, pp. 253–284.

Williams, A., Sinanaj, B. and Hoysted, G.A. (2023) 'Plant-microbe interactions through a lens: Tales from the mycorrhizosphere', *Annals of Botany*, 133, p. mcad191.

Willoughby, J.R. *et al.* (2019) 'Inbreeding load and inbreeding depression estimated from lifetime reproductive success in a small, dispersal-limited population', *Heredity*, 123(2), pp. 192–201.

Wu, P. *et al.* (2021) 'Stress preadaptation and overexpression of *rpoS* and *hfq* genes increase stress resistance of *Pseudomonas fluorescens* ATCC13525', *Microbiological Research*, 250, p. 126804.

5 Molecular Breeding for Crop Plants
Improvement and Practices

Asad Azeem, Amina Hafeez, and Sami Ul-Allah

5.1 INTRODUCTION

Plant breeding has a significant influence on global food production and has become a key component in guaranteeing food security and safety (Zewdu *et al.*, 2023). But in recent times, issues with food supply and quality have surfaced worldwide as a result of the overwhelming need for food to feed the world's expanding population. Moreover, severe weather variations brought on by climate change are stressing farmers worldwide with heat waves and droughts. As a result, agricultural productivity is being severely reduced. To create plants with desired features, plant breeding can be employed. For the past 10,000 years, people have selected and bred plants with improved nutritional values through artificial plant selection (Salse *et al.*, 2023). The goal of traditional farming practices was to enhance the nutritional value of various food plants. Crop breeding underwent a revolutionary change with the discovery of Mendelian rules. Because of modern, state-of-the-art genetics, crop development has changed significantly during the past 150 years. Various methods have been employed to reduce the length of plant reproductive cycles. For the purpose of creating new plant varieties, conventional breeding methods are insufficient for enhancing the plant genome. Since the 1990s, molecular breeding has been utilized to overcome these challenges (Voss-Fels, Stahl and Hickey, 2019).

Molecular breeding (MB) exists in a number of forms, such as genome-wide selection (GWS), marker-assisted recurrent selection (MARS), marker-assisted back crossing (MABC), and marker-assisted selection (MAS). The main purpose of all these molecular techniques is genetic modification at the DNA level to enhance desirable plant characteristics. Molecular breeding, which is most commonly associated with molecular marker-assisted breeding (MAB), is the process of using molecular biotechnologies more, especially molecular markers in conjunction with genomics and linkage maps to modify and enhance plant properties based on genotypic assays (Ribaut, De Vicente and Delannay, 2010).

Through the revolutionary process of molecular breeding, plant scientists and breeders can now examine crops' DNA in great detail and learn the secrets of their genetic code. With this newfound understanding, they may choose, alter, and produce crop types with characteristics that meet the needs of time in an exact and predictable manner. In order to increase breeding programs' precision, molecular breeding has thus taken center stage in crop development initiatives. This chapter covers the molecular breeding techniques, role of different genetic markers, and high- throughput genotyping technologies.

5.2 SIGNIFICANCE OF CROP PLANT IMPROVEMENT AND MOLECULAR BREEDING

The central role of crop breeding is to improve crops' genetic makeup by generating genetic variability and selecting elite genotypes with desired traits (Aswini *et al.*, 2023). The emergence of genetic variation has been facilitated by molecular technology in two ways: First, a technique for

 DOI: 10.1201/9781032630366-5

TABLE 5.1
Role of Molecular Breeding Approaches in Crop Plant Improvements

Crop	Trait	Gene	Markers	References
Barley	Barley yellow Mosaic virus	rym4, rym9, rym11	SSR	Lema (2018)
Wheat	Processing quality	Talox-B1	SNP	Geng *et al.* (2012)
Maize	Flowering time	Dwarf8	InDel markers SNP markers:	Andersen *et al.* (2005)
Maize	Provitamin A content	PSY1	InDel markers SNP markers	Azmach *et al.* (2013)
Rice	Grain length	GS3	C/A SNP-based CAPS marker	Fan *et al.* (2009)
Rice	Spikelet fertility	S5	InDel based marker [AD1] and SNP based marker [AC2]	Ji- *et al.* (2010)
Rice	Low glutenin content	Lgc1	Two InDel based [InDel-Lgc1-1and InDel-Lgc1-2] markers	Tao *et al.* (2010)
Barley	Stripe rust	Rspx QTLs 4,5,7	SSR	Lema (2018)
Tomato	Late blight (*Phytophthora infestans*)	*Ph-3* (Chr. 9)	SNP	Jung *et al.* (2015)
Tomato	Leaf mold (*Cladosporium fulvum*)	*Cf-2*, *Cf-5* (Chr. 6)	SSR	Grushetskaya *et al.* (2007)
Tomato	Tomato mosaic virus (TMV/ToMV)	*Tm-2* (Chr. 9)	SNP	Arens *et al.* (2010)
Cowpea	Pod tenderness	pdt	SSR	Kongjaimun *et al.* (2013)
Pea	Pea weevil	SCR5a	SSR	Aryamanesh *et al.* (2014)

figuring out the relationships between sources of germplasm has been made possible by molecular marker technology. Second, the application of transgenic technology has significantly expanded the pool of potential genes for enhancing cultivated crops. The use of molecular markers in a variety of crops has enabled a significant portion of the characterization of germplasm, known as fingerprinting, the determination of the genomic regions that control the expression of the target traits, and the enhancement of crops through marker-assisted selection. Using molecular breeding, it is possible to increase the amounts of recombinant alleles for advantageous alleles and choose features that are not detectable in typical settings using other techniques. Quantitative traits are those that are controlled by multiple genes and include many important agricultural traits like quality, yield, and some types of disease resistance. Quantitative trait loci (QTLs) are the areas of genomes that contain genes linked to a specific quantitative trait (Ahmar *et al.*, 2020). The role of molecular breeding approaches in crop plant improvements is described in Table 5.1.

5.3 MOLECULAR BREEDING: TECHNIQUES, ROLE OF GENETIC MARKERS, AND HIGH-THROUGHPUT GENOTYPING TECHNOLOGIES

5.3.1 Techniques

5.3.1.1 Marker-assisted Selection (MAS)

Ever since agriculture was first practiced, people have tried to select specific traits in crops in order to improve them. Even with its present sophistication, the process still relies heavily on field observations. The current challenge is to create effective MAS strategies for plant improvement and

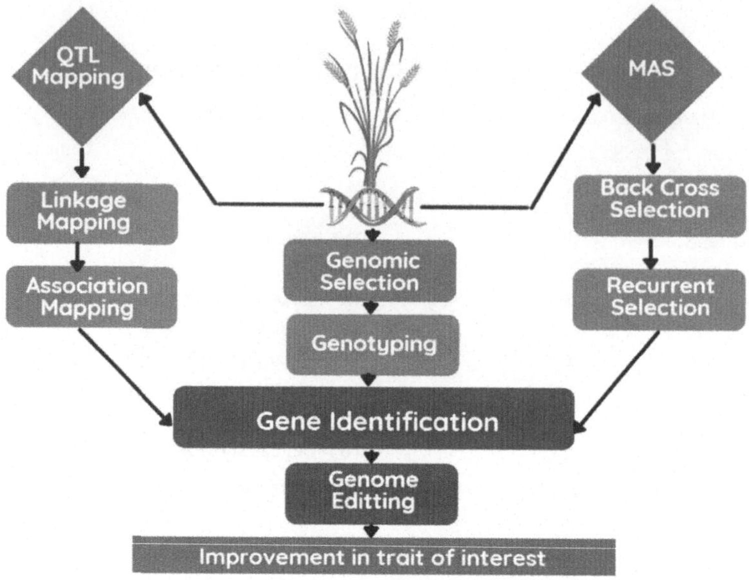

FIGURE 5.1 Molecular breeding approaches: Marker-assisted selection (MAS), QTL mapping, and genomic selection for crop improvement.

to incorporate new molecular technologies into breeding programs (Figure 5.1). With the advancement of molecular tools, MAS is now possible, regardless of whether these traits are controlled by identified genes or QTLs. The practice of selecting individuals to be parents to the next generation based on the results of DNA testing is known as MAS. The purpose of the DNA testing data and individual performance records is to increase the likelihood of identifying organisms carrying desirable and undesirable traits at an earlier developmental stage and to improve the selection's accuracy. MAS is the process of choosing particular alleles for traits that are dependent on a small number of loci. When phenotypic screening is particularly costly or time-consuming, phenotyping is only possible after flowering, the environment has a strong influence with low heritability, multiple resistance genes are pyramided, and marker profiles can be obtained from seeds, MAS is preferred over conventional breeding. For many years, transgene fixation within a cultivar has been accomplished through the use of MAS in plant breeding. Plant breeding has extensively used MAS as a crucial and practical technique to improve crop yield, quality, and tolerance to biotic or abiotic stresses (Lema, 2018).

5.3.1.2 Genomics Selection (GS)

Genomics selection (GS) is a successful breeding strategy when the goal is to optimize quantitative traits (Figure 5.1). This allows breeders to choose the best genotypes. GS plays a vital role during selection in breeding programs. It is a novel approach to plant breeding designed to address issues with multiple inheritances of complex traits. This approach uses molecular markers in both diploid and polyploid crop species (Schie, Chaudhary and Debener, 2014), but it does so in a different way. In contrast to MAS, markers in GS are not utilized to follow a trait. Any breeding population can be used for the GS of superior lines. The experimental population needs to be determined in order for GS to be successful. The population does not have to be descended from biparental crosses in order to be comparable to selection candidates in the breeding scheme to which GS will be applied (Arruda *et al.*, 2015).

GS is a subfield of quantitative genetics that predicts genetic values by phenotypic regression on indicators of genetic resemblance from DNA-based genomic sequences. GS techniques have been applied by researchers to a range of quantitative traits in plant breeding initiatives. With the use of additive genetic effects, which may suggest an individual's suitability as a parent in the following generation, GS aims to foresee responses in individuals. A thorough grasp of the connections between the genotypes being analyzed is essential to being able to apply a selection scheme such as GS at the population or breeding program level (Ficht, 2023).

5.3.1.3 Genomics-Assisted Breeding (GAB)

Plant breeding has recently benefited a great deal from the use of genomics-assisted breeding (GAB) in different crops. GAB made it possible to combine high-throughput phenotyping genomic tools with molecular markers to support breeding practices by making it easier to predict phenotype from genotype. Breeders can begin with a large population only of genotypically characterized offspring thanks to GAB, and then they can use a specific subset of these offspring for more costly phenotypic evaluation. Furthermore, genotypic assessment can be carried out in the off-season, for example, in winter nurseries, where yield trials are typically not carried out, contributing to accelerated breeding. GAB is particularly helpful for improving complex traits. GAB's ultimate objective is to facilitate crop improvement by identifying the best haplotype combinations, gene networks, and genomic regions. Thus, GAB promises to hasten the creation of novel plant varieties and advance contemporary agriculture. The latest developments in high-throughput genotyping and phenotyping technologies, marker-trait associations, natural variation exploration, and germplasm enhancement, open up new avenues for GAB in crop improvement (Leng, Lübberstedt and Xu, 2017).

5.3.2 Genetic Markers and Their Applications in Crop Improvement

5.3.2.1 Single Nucleotide Polymorphisms (SNPs)

DNA sequence variations known as single nucleotide polymorphisms (SNPs) happen when one nucleotide (A, T, G, or C) varies between individuals in the same species. The most prevalent marker system in the genomes of plants and animals is the SNP, which has recently become the new standard for molecular markers with a variety of uses. Because of their binary or co-dominant status, they can distinguish between homozygous and heterozygous alleles with ease. Furthermore, in contrast to microsatellites, their power stems from the vast number of loci that can be evaluated rather than the quantity of alleles. Once rare SNPs in a species are discovered, the genetic population discrimination power of a low-diversity species can match the number of loci in a genetically diverse species (Xu, 2010).

The successful identification and differentiation of particular genetic variations was made possible by a novel SNP-based technique that is successfully used in routine medical diagnosis because it allows for the rapid analysis of a significant number of samples. For plants with low genetic diversity that are challenging to distinguish, similar methods must be applied. Over the past ten years, SNP-based marker techniques have undergone improvements in marker density, and they have also significantly decreased the costs and time associated with SNP discoveries when compared to previous genotyping approaches. SNPs are the most prevalent type of molecular marker in a genome and can be analyzed at different genomic scales.

5.3.2.2 Simple Sequence Repeats (SSRs)

Simple sequence repeats (SSRs) are tandem repeat motifs of 1–6 nucleotides that are found at high rate in the nuclear genomes of most taxa. SSRs are also known as microsatellites. The average length of a SSR locus is 5–40 repetitions, although longer sequences of repeats are conceivable (Shehata, Al-Ghethar and Al-Homaidan, 2009). Shortly after their first description, SSRs were used in a wide range of fields because of their high variation, which makes them ideal for genome mapping in a variety of organisms. PCRs can be used to amplify SSRs for identification by employing two distinct sequences as primers that are complementary to the surrounding regions.

Microsatellites, on the other hand, are useful for a variety of purposes because of their relative abundance, codominant inheritance, reproducibility, multiallelic nature, and good genome coverage once they are created and assessed in an organism. They are utilized as genetic markers in conservation biology and plant breeding. In the fields of gene mapping, forensics, and paternity analysis, microsatellites have been shown to be incredibly useful tools for population genetics (Pickett, Miller and Ridge, 2017).

5.3.2.3 Insertion/Deletion Markers (InDels)

Insertion/deletion markers (InDels) are polymorphisms that result from nucleotide sequence insertions or deletions within the genome. The process of identifying deteriorated samples involves the use of InDels. InDels have a low mutation rate; they are also often used in kinship analysis. False exclusion is less likely in polymorphic systems with low mutation rates because they are more stable in kinship and paternity tests. SNP analysis differs from InDels in that it necessitates more involved and time-consuming steps (like the SNaPshot technique), while InDels can be used with common methods like capillary electrophoresis and PCR using fluorochrome-labeled primers. Consequently, InDels can be used in place of or in addition to STR and SNP loci, combining their respective benefits (Zhou *et al.*, 2015).

InDels allow for successful profiling in small quantities and/or deteriorated DNA samples. High-throughput genotyping systems can be used to create extended panels with high discrimination power, even though InDels have lower discrimination power than STRs. An alternative to STR and SNP in forensic identification is InDel polymorphism, particularly for samples that have been compromised. Extremely stable structures and small amplicons are the most crucial characteristics searched for in genetic markers used in forensic analyses. As encouraging next-generation markers for forensic DNA analysis, InDels' short amplicons, low mutation rates, and lack of stutter have recently made it possible for them to be used in forensic genetics labs. The 36-InDelplex panel can be used for kinship analysis and forensic identification due to its high polymorphism. The 36-InDelplex panel was shown to be reliable, sensitive, and appropriate for the examination of difficult DNA samples following a number of validation tests (Yang *et al.*, 2022).

5.3.3 High-throughput Genotyping Technologies

5.3.3.1 Genotyping by Sequencing (GBS)

The price of DNA sequencing have decreased due to advances in NGS, making genotyping by sequencing (GBS) possible for species with large genomes and high diversity (Elshire *et al.*, 2011). The Buckler lab developed GBS, a straightforward and highly multiplexed system for building decreased libraries of representation for the Illumina NGS platform. According to Beissinger, it produces a large number of SNPs that are useful for genetic analysis and genotyping. Some of this system's key features include low cost, minimal sample handling, fewer PCR and cleansing steps, no size division, no reference sequence limits, efficient bar coding, and ease of scaling up. As a novel and reasonably priced tool for genomics-assisted breeding in a variety of plant species, GBS is becoming more and more significant. Any pseudo-test-cross progeny can be genetically mapped out quickly and easily using GBS and genome-independent imputation. The GBS approach provides a significantly streamlined library production process that is easier to apply to a large number of users or lines.

Unlike other high-density genotyping techniques, which are primarily utilized for general-interest "reference" genomes, GBS's inexpensive cost makes it an effective method for identifying and genotyping SNPs in a range of crop species and populations. In numerous large crop genomes, GBS has been extensively employed to inundate mapping and breeding populations with tens of thousands to hundreds of thousands of SNP markers. It is a highly multiplexed, technically simple technology that can be used for plant genetics, breeding a variety of crops, population studies, and germplasm characterization. The basis for creating GBS libraries is the reduction of genome intricacy through restriction enzymes. This method is easy to use, quick, very specific, and incredibly repeatable, and it has the potential to access significant genomic regions that sequence capture methods are unable to reach (Chung *et al.*, 2017).

5.3.3.2 Single Nucleotide Polymorphism Arrays (SNP Arrays)

SNP arrays are automatically genotyping assays that are high throughput and reasonably priced. They have been widely utilized in crop genetic studies, such as population structure analysis, linkage map construction, genomic selection, genome-wide association studies (GWAS), and gene mapping. Up to 700K SNPs can be found in diploid rice, 487K in triploid apples, 58K in tetraploid peanuts, 820K in hexaploid wheat, 90K in octoploid strawberries, and 345K in dodecaploid sugarcane. These quantities of data points can be accommodated by the capacity of the SNP arrays that are currently available (Clevenger *et al.*, 2017; Pandey *et al.*, 2017).

For high-throughput genotyping, SNP arrays have several benefits. First, compared to data produced using NGS-based techniques, analyzing SNP array data is relatively easy, especially when taking into account the lengthy preparation of NGS libraries and the subsequent investment in bioinformatics data analysis for precise SNP calling. More specifically, Illumina can call and provide the genotypes of SNP markers, or researchers can call genotypes by using the Illumina genotype-calling pipeline in accordance with the user guidelines. However, in polyploid species, genotype calling using NGS-based data is more challenging and time-consuming. As previously stated, SNP genotype calling comprises read alignment, read trimming, SNP calling, and SNP, which calls for a foundation in bioinformatics. Secondly, SNPs from targeted genomic regions can be added to the array. Furthermore, for both the Illumina and Affymetrix platforms, there is flexibility in the number of interested SNPs that can be set on the array (Torkamaneh *et al.*, 2017).

5.3.3.3 Next-Generation Sequencing (NGS)

The creation of high-throughput sequencing applications, also known as next-generation sequencing (NGS) applications, which can generate thousands or millions of sequences at once, has been fueled by the increasing demand for inexpensive sequence data, as manifested in Figure 5.2. In order to produce several hundred million to several hundred billion DNA bases per run, NGS depends

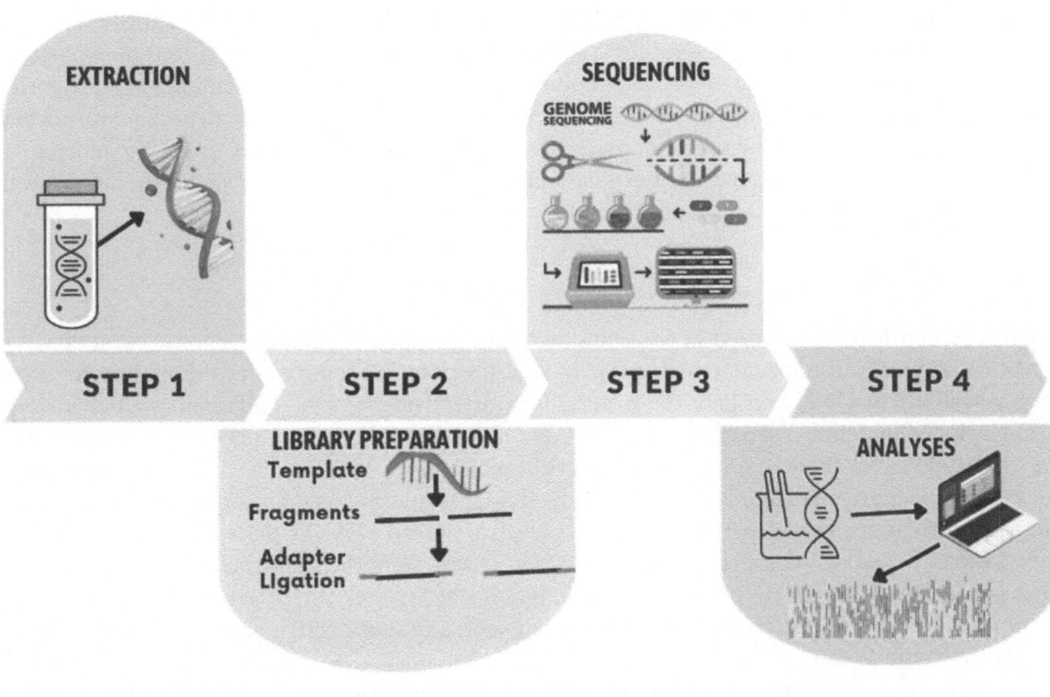

FIGURE 5.2 Process of next-generation sequencing.

on intensively parallel sequencing and imaging techniques. Up to five lac sequencing-by-synthesis operations may be carried out in parallel during ultra-high-throughput sequencing (Tanwar, Ray and Bhatt, 2023). The majority of NGS platforms can produce dependable sequences and exhibit nearly flawless coverage behavior on genomes that are neutral, GC-rich, or moderately AT-rich. The applications that data will support and the data's quality, however, differ significantly.

NGS technologies have been used recently in genome-wide sequencing and resequencing projects. In these projects, the genomes of multiple specimens are sequenced to find a large number of SNPs for investigating species diversity, producing haplotype maps, and carrying out genome-wide association studies (Henry, 2012). These applications are now feasible due to NGS technologies' accuracy, affordability, increased throughput, and ease of assaying. Another method for achieving multiplex sequencing is to pool samples into a single sequencing channel and tag randomly sheared DNA fragments from various samples with distinct, brief DNA sequences. This method (random DNA shearing followed by barcode tagging) has allowed for the rapid determination of the complete chloroplast genome sequences of several pine species as well as spruce, and it has also been used to find and map genomic SNPs in rice (Meyer *et al.*, 2016).

5.4 BEST PRACTICES IN PLANT BREEDING CONCLUSIONS AND FUTURE REMARKS

5.4.1 Germplasm Characterization and Diversity Assessment

Genetic diversity must exist in plants for successful breeding programs. Generating genetic diversity and controlling genetic variability are the two most crucial aspects of plant breeding. Although phenotypic variation relies on environmental factors and the interactions between genotype and environment, it is positively correlated with genetic diversity. Creating a genetically diverse population through parent-child crossing and inducing mutations in the breeding germplasm is a common first step in plant breeding. The difficulties in choosing plants that express recessive alleles, genes whose expression is significantly influenced by the growth environment, or genes that have negligible effects on plant phenotypes are major barriers to plant improvement. In this context, a number of biotechnological techniques have significantly increased the potential for boosting crop species' genetic diversity through the generation of transgenic plants, tissue culture-induced variability, molecularly assisted parent selection, and protoplast fusion. Progeny from each mating cycle in a breeding program that makes use of molecular markers are chosen based on the existence of a specific electrophoretic band that has been demonstrated to be closely associated with the desired trait (Choudhury, Khan and Dayanandan, 2014).

5.4.2 Marker Discovery and Validation

Visual selection is typically used in conventional plant breeding to identify genetic variation. But thanks to advances in molecular biology, it is now possible to identify it at the molecular level using variations in DNA and how those variations affect phenotypes. Numerous methods have been developed to define and amplify the DNA and to show the variation in DNA among individuals, which allows for the identification of molecular changes. After DNA is extracted from plants, particular molecules are recognizable by hybridization, polymerase chain reaction (PCR), capillary electrophoresis (CE), or polyacrylamide or gel electrophoresis (PAGE), based on their chemical compositions, sizes, and charges. DNA samples are tagged, and their genetic variation is tracked using genetic markers. Allelic forms determine biological features called genetic markers, which can be used as experimental probes to monitor a subject, a tissue, a cell, a nucleus, a chromosome, or a gene (Kindie, 2021).

The following requirements should be fulfilled by desirable genetic markers: High levels of genetic polymorphism; codominance, which allows heterozygotes and homozygotes to be easily distinguished from one another; distinct, easily identifiable allele features; uniform dispersal throughout the whole genome; neutral selection, which avoids pleiotropic effects; ease of detection, which enables process automation; low costs associated with marker development and genotyping; and high degree of replicabilities. Classical markers and DNA markers are the two kinds of genetic markers used in genetics and plant breeding. Morphological, cytological, and biochemical markers are examples of classical markers. DNA markers, including AFLP, RFLP. RAPD, SSR, and SNP, have developed into a variety of systems based on different strategies for identifying polymorphisms or approaches like DNA sequencing and polymerase chain reaction.

5.4.3 Genomic Selection and Optimization of Breeding Populations

Since 1980s, marker technology has been used by plant geneticists and breeders to identify QTLs. While MAS works well for large-effect QTLs, it has proven difficult to apply to quantitative traits that are influenced by various minor effect alleles (Boopathi, 2013). Thus, the idea of genomic selection (GS) was put forth in order to estimate all markers at the same time and thus capture all of the resulting effects. Breeding programs are still having trouble figuring out the best way to incorporate it into their operations, though to properly apply GS, conventional breeding programs must be optimized. Optimizations include changing field designs, training populations, increasing the number of lines evaluated, and utilizing the significant amount of genomic and phenotypic data collected across multiple growing seasons and environments in order to obtain heritability estimates, selection intensity, and accuracy. Breeding programs can use high-throughput phenotyping, deep learning techniques, multi-environment and multi-trait models, and genotypic and phenotypic data to optimize genetic gain and selection accuracy through GS methods.

5.4.4 Trait Introgression and Pyramiding

Comparing trait introgression with various MAS schemes to traditional breeding methods yields several advantages. One of these benefits is that fewer stages, like backcrosses, are needed for the target locus's successful introgression. Better control over the sample size and fragment length is also made possible by these MAS techniques, which can be especially helpful for effective trait incorporation.

Gene pyramiding, which is the combination of several genes governing the same trait within a single genotype, is a noteworthy application of MAS. It can be difficult to implement this strategy with traditional breeding techniques, particularly when choosing characteristics like resistance to fungus-related infections. Plants with multiple resistance genes may be difficult to identify using phenotypic data alone. On the other hand, markers provide the ability to identify genotypes that consist of gene groups at earlier stages, like in F2 populations.

5.4.5 Genomic Prediction and Phenotype Prediction Accuracy

The process of making complex trait predictions about individuals based on their observed genomic information, usually single nucleotide polymorphisms or whole genome sequence data, is known as genomic prediction. The repertoire of predictive methods has expanded significantly since then, due to a multitude of methodological and practical developments (Goddard *et al.*, 2016). The genomic prediction was estimated by the methods given below in equation (5.1).

$$y = X\beta + Zm + \varepsilon \tag{5.1}$$

where y is a vector of observed phenotypes, X represents the incidence matrix, β contains the fixed effects, Z is an incidence matrix (i.e., genotypes), m is a vector of (random) marker effects, and ε is a residual term.

In Equation (5.1), y is either continuous or categorical. In the latter case, a probit or logistic transformation is used. Equation (5.1) calls for some kind of penalization or "constraint" on the solutions because the number of markers typically surpasses the number of records, particularly in the case of sequence or high-density SNP arrays. Actually, the kind of penalty applied to marker information is what distinguishes most genomic prediction methods from one another.

Accurate phenotype prediction is a crucial element of plant breeding programs. The term "phenotype" describes an organism's observable features, including its physical attributes, growth patterns, and potential yield. For the purpose of choosing and creating new crop varieties that fulfill particular breeding goals, like disease resistance, increased yield, or better environmental adaptation, it is imperative to predict a plant's phenotype (Roth *et al.*, 2023).

5.5 CHALLENGES AND LIMITATIONS OF MOLECULAR BREEDING

Molecular marker technology has promised to revolutionize breeding programs and accelerate selection gains for over 20 years. However, for polygenic traits, MAS has not produced appreciable gains. MAS has proven to be ineffective when dealing with a large number of small-effect alleles for which no consistent, significant effects can be found, despite its success in modifying alleles with substantial effects associated with markers. The bifurcated approach of conventional MAS, in which effect estimation is required in the second step and QTL identification is the first, is the source of its shortcomings. There are various obstacles to crop enhancement with this method: (1) using biparental populations, which are frequently small and underpowered due to the lack of allelic diversity and phases in these populations; (2) the cost of creating these populations; (3) the need for validation, which requires extra work; (4) separating QTL identification from assessment results in skewed effect estimates; and (5) small-effect QTLs are frequently completely ignored because strict significance thresholds are used.

Recognizing the limitations of SSR markers is crucial to comprehending the shift to SNP markers. First of all, the genome contains insufficient SSR motifs, which limits the ability to identify gene-based markers or marinate a region with markers. Additionally, even though the high information content of SSRs stems from multiple alleles per locus, managing allele sizes in databases and consolidating SSR data from multiple platforms present challenges. Furthermore, labor-intensive procedures are needed for gel-based SSRs, and automated fragment sizing systems have restricted multiplexing capabilities. Therefore, unlike recent SNP genotyping techniques, SSR genotyping can quickly reach a point where its higher costs and low throughput become limiting factors.

The use of markers in selection entails certain costs associated with molecular methods, including development costs (finding molecular markers on the genome and proving links between markers and desirable traits) and continuous running costs. Furthermore, the potential advantages are frequently outweighed by the prohibitively high cost of importing technology from developed nations. Because functional genomic and biotechnological centers are not common in Africa, mechanical and logistical factors also come into play. Long delivery times can adversely affect the potency of markers that researchers frequently need to order well in advance, which could complicate or skew experimental results (Ebegbulem and Ozung, 2013).

5.6 FUTURE PRESPECTIVES

The fields of genetics and plant breeding have their scientific foundations firmly established thanks to the seminal contributions of Darwin and Mendel in the late 19th and early 20th centuries. Enhancing plants to meet human needs and satisfy the demands of consumers and growers has always been the main objective of plant breeding. However, the world's population is growing at an accelerated rate, and the effects of climate change on agricultural ecosystems, in addition to a host of other biological and environmental problems, have forced us to set new, urgent goals in order to protect the quality of life on Earth. In order to guarantee a sustainable food supply, we must increase food production in order to meet the growing demands of the world's population. The advancements in molecular biology in the era of genomics have created new opportunities for accelerating plant breeding processes and overcoming obstacles that impede crop productivity. Precision plant breeders now have access to invaluable tools in the form of molecular markers, especially when using MAS, which increases the efficacy of selection for a variety of desirable agronomic traits. Genes from diverse biological sources can be isolated, cloned, and inserted into plant genomes to create new plant varieties with unique characteristics that affect yield, quality, and resistance to biotic and abiotic stresses. This process has the potential to diversify the crop gene pool. Within this framework, novel genomic instruments demonstrate a critical role in bolstering traditional breeding techniques meant to accomplish sustainable food production, especially in light of evolving climate patterns and the expanding global populace.

5.7 CONCLUSION

Plant breeding has undergone a substantial change as a result of the use of markers, which has allowed cultivators to evolve new cultivars with better quality, yield, and resistance to both biotic and abiotic stresses. Using molecular markers has several benefits, such as a shorter breeding cycle, improved accuracy, more genetic diversity, cost effectiveness, and less environmental impact. However, there are some drawbacks to using markers as well, including the cost of developing new markers, limited availability of markers, low genetic diversity, the need for technical know-how, and ethical issues. Markers continue to be essential in plant breeding, which is necessary to improve crop productivity for global food security, despite these obstacles.

REFERENCES

Ahmar, S., *et al.* (2020) 'Conventional and molecular techniques from simple breeding to speed breeding in crop plants: Recent advances and future outlook', *International Journal of Molecular Sciences*, 21(7), p. 2590.

Andersen, J.R., *et al.* (2005) 'Validation of Dwarf8 polymorphisms associated with flowering time in elite European inbred lines of maize (*Zea mays* L.)', *Theoretical and Applied Genetics*, 111, pp. 206–217.

Arens, P., *et al.* (2010) 'Development and evaluation of robust molecular markers linked to disease resistance in tomato for distinctness, uniformity and stability testing', *Theoretical and Applied Genetics*, 120, pp. 655–664.

De Arruda, M.P., 2015. *Genomic selection and quantitative trait loci mapping for fusarium head blight resistance in wheat (Triticum aestivum L.).* University of Illinois at Urbana-Champaign.

Aryamanesh, N., et al. (2014) 'Identification of genome regions controlling cotyledon, pod wall/seed coat and pod wall resistance to pea weevil through QTL mapping', *Theoretical and Applied Genetics*, 127, pp. 489–497.

Aswini, M., et al. (2023) 'The role of genetics and plant breeding for crop improvement: Current progress and future prospects', *International Journal of Plant & Soil Science*, 35(20), pp. 190–202.

Azmach, G., et al. (2013) 'Marker-trait association analysis of functional gene markers for provitamin a levels across diverse tropical yellow maize inbred lines', *BMC Plant Biology*, 13(1), p. 16.

Boopathi, N.M. (2013) *Genetic mapping and marker assisted selection: Basics, practice and benefits*, pp. 343–388. New Delhi: Springer India.

Choudhury, B.I., Khan, M.L. and Dayanandan, S. (2014) 'Patterns of nucleotide diversity and phenotypes of two domestication related genes (*OsC1* and *Wx*) in indigenous rice varieties in northeast India', *BMC Genetics*, 15(1), p. 12.

Chung, Y.S., et al. (2017) 'Genotyping-by-sequencing: A promising tool for plant genetics research and breeding', *Horticulture, Environment, and Biotechnology*, 58, pp. 425–431.

Clevenger, J., *et al.* (2017) 'Genome-wide SNP genotyping resolves signatures of selection and tetrasomic recombination in peanut', *Molecular Plant, 10*(2), pp. 309–322.

Ebegbulem, V.N. and Ozung, P.O. (2013) 'Application of molecular markers in farm animal improvement: Prospects and challenges', *Online Journal of Animal and Feed Research*, 3(3), pp. 149–152.

Elshire, R.J., et al. (2011) 'A robust, simple genotyping-by-sequencing (GBS) approach for high diversity species', *PLoS One*, 6(5), p. e19379.

Fan, C., et al. (2009) 'A causal C–A mutation in the second exon of GS3 highly associated with rice grain length and validated as a functional marker', *Theoretical and Applied Genetics*, 118, pp. 465–472.

Ficht, A., et al. (2023) 'Genomic selection for agronomic traits in a winter wheat breeding program', *Theoretical and Applied Genetics*, 136(3), p. 38.

Geng, H., et al. (2012) 'Development of functional markers for a lipoxygenase Gene TaLox-B1 on chromosome 4BS in common wheat', *Crop Science*, 52(2), pp. 568–576.

Goddard, M.E., et al. (2016) 'Genetics of complex traits: prediction of phenotype, identification of causal polymorphisms and genetic architecture', *Proceedings of the Royal Society B: Biological Sciences*, 283(1835), 20160569.

Grushetskaya, Z.E., et al. (2007) 'Mapping of the cf-6 tomato leaf mould resistance locus using SSR markers', *Russian Journal of Genetics*, 43, pp. 1266–1270.

Henry, R.J. (2012) 'Next-generation sequencing for understanding and accelerating crop domestication', *Briefings in Functional Genomics*, 11(1), pp. 51–56.

Ji, Q., et al. (2010) 'Two sequence alterations, a 136 bp InDel and an A/C polymorphic site, in the S5 locus are associated with spikelet fertility of indica-japonica hybrid in rice', *Journal of Genetics and Genomics*, 37(1), pp. 57–68.

Jung, J., et al. (2015) 'Gene-based molecular marker system for multiple disease resistances in tomato against tomato yellow leaf curl virus, late blight, and verticillium wilt', *Euphytica*, 205, pp. 599–613.

Kindie, B. (2021) 'Assess molecular marker applications for genetic variety analysis in biodiversity conservation status', *Molecular Biology*, 10, p. 287.

Kongjaimun, A., et al. (2013) 'QTL mapping of pod tenderness and total soluble solid in yardlong bean [*Vigna unguiculata* (L.) Walp. subsp. *unguiculata cv.-gr. sesquipedalis*]', *Euphytica*, 189, pp. 217–223.

Lema, M. (2018) 'Marker assisted selection in comparison to conventional plant breeding', *Agricultural Research & Technology*, 14, p. 555914.

Leng, P.F., Lübberstedt, T. and Xu, M.L. (2017) 'Genomics-assisted breeding–a revolutionary strategy for crop improvement', *Journal of Integrative Agriculture*, 16(12), pp. 2674–2685.

Meyer, R.S., et al. (2016) 'Domestication history and geographical adaptation inferred from a SNP map of African rice', *Nature Genetics*, 48(9), pp. 1083–1088.

Pandey, M.K., et al. (2017) 'Development and evaluation of a high density genotyping 'Axiom_Arachis' array with 58 k SNPs for accelerating genetics and breeding in groundnut', Scientific Reports, 7(1), pp. 1–10.

Pickett, B.D., Miller, J.B. and Ridge, P.G. (2017) 'Kmer-SSR: A fast and exhaustive SSR search algorithm', *Bioinformatics*, 33(24), pp. 3922–3928.

Ribaut, J.M., De Vicente, M.C. and Delannay, X. (2010) 'Molecular breeding in developing countries: Challenges and perspectives', *Current Opinion in Plant Biology*, 13(2), pp. 213–218.

Roth, L., et al. (2023) 'Image-based phenomic prediction can provide valuable decision support in wheat breeding', *Theoretical and Applied Genetics*, 136(7), p. 162.

Salse, J., *et al.* (2023) 'Strategies for breeding crops for future environments', *Trends in Plant Science*,29 (3), pp. 303–318.

Schie, S., Chaudhary, R. and Debener, T. (2014) 'Analysis of a complex polyploid plant genome using molecular markers: Strong evidence for segmental allooctoploidy in garden dahlias', *The Plant Genome*, 7(3), plantgenome2014-2001.0002.

Shehata, A.I., Al-Ghethar, H.A. and Al-Homaidan, A.A. (2009) 'Application of simple sequence repeat (SSR) markers for molecular diversity and heterozygosity analysis in maize inbred lines', *Saudi Journal of Biological Sciences*, 16(2), pp. 57–62.

Tanwar, J., Ray, P. and Bhatt, A. (2023) 'Genotyping-by-sequencing (GBS): An important tool for crop improvement', *International Year of Millets*, 2023, p. 36.

Tao, C.H.E.N., et al. (2010) 'Development of simple functional markers for low glutelin content gene 1 (Lgcl) in rice (*Oryza sativa*)', *Rice Science*, 17(3), pp. 173–178.

Torkamaneh, D., et al. (2017) 'Fast-GBS: A new pipeline for the efficient and highly accurate calling of SNPs from genotyping-by-sequencing data', *BMC Bioinformatics*, 18, pp. 1–7.

Voss-Fels, K.P., Stahl, A. and Hickey, L.T. (2019) 'Q&A: Modern crop breeding for future food security', *BMC Biology*, 17(1), pp. 1–7.

Xu, Y. (2010) *Molecular plant breeding*. Cabi Digital Library. pp. 1–86

Zewdu, G.A., et al. (2023) 'The role of modern plant breeding to control plant disease: A review', *Agricultural Journal*, 17, pp. 19–27.

Zhou, G., et al. (2015) 'Development of genome-wide InDel markers and their integration with SSR, DArT and SNP markers in single barley map', *BMC Genomics*, 16(1), pp. 1–8.

6 Molecular, Physiological, and Biochemical Aspects of Crop Plants with PGPR under Harsh Environment

Avinash S. H., Manasa S. G, Yalavarthi Nagaraju, P. Gangadhara Rao, and Sayanti Mandal

6.1 INTRODUCTION

Plants, like sessile creatures, are continually subjected to harmful ecological irregularities like abiotic stressors. Anthropogenic influences and rising temperatures continue to intensify the negative impact of these pressures on agricultural output, posing a danger to global food security. In order to produce appropriate responses to a series of issues and ensure growth and development, plants have adapted a multi-layered, intricate, and highly regulated immune system and defenses. This system requires receiving various chemical signals and integrating this input. Acquired memory can be either temporary or permanent as a result of these stress-induced response mechanisms, which also cause phenotypic alterations and gene expression reprogramming. Being able to remember past stress experiences helps plants adapt and defend themselves, since they are constantly subjected to a variety of adverse environmental conditions (Husen and Iqbal 2023; Bachheti *et al.*, 2021; Getnet *et al.*, 2015; Husen, 2010; Husen, 2021a–e; Husen, Iqbal and Aref, 2014; Husen, Iqbal and Aref, 2016; Husen *et al.*, 2018, 2019; Hussein *et al.*, 2017). Over the last decade, scientists have made significant progress in explaining and defining stress memory processes in plants. Defense priming is one of these processes that makes the plant more sensitive to abiotic stress and equips it for the future (Arbona *et al.*, 2017).

Stress memory, also known as priming or preconditioning of plant defenses and adaptive systems, is a phenomenon in which plants become more resilient to recurring shocks and exhibit quicker and more effective defense responses. Numerous instances of stress memory in higher plants in reaction to environmental cues, including cold, salt, and drought, have been demonstrated in a variety of species and extensively examined. Sustained changes in the amounts of important signaling transcription factors, enzymes, and/or proteins are one explanation behind this, and they shed light on how exposure to different stimuli modifies and sustains the metabolism of plants. A further likely explanation is that chromatin modifications, such as DNA methylation, adjustment of histone tail, and interrupted RNA polymerase II (Pol II) contribute to the coordinated shifts in the gene expression patterns that support memory responses. Furthermore, it is currently unclear how precisely DNA changes relate to transcriptional responses in the face of abiotic stress (Avramova, 2015). Furthermore, metabolic abnormalities are created as part of the defense phenomenology during stress encounters, and some of these metabolic reactions may last until restoration, when the plant's physiology reverts to normal. Stress memory under abiotic stress has been demonstrated to be significantly impacted by changes in metabolite levels, such as the buildup of signaling chemicals. Therefore, one of the mechanisms that characterize stress memory in plants, according to current theories, is the redesign of biochemical pathways and circuits (Hilker and Schmülling, 2019).

DOI: 10.1201/9781032630366-6

6.2 DETECTION OF STRESS BY PLANTS AND THEIR RESPONSES

Plants have developed sophisticated defense systems that enable them to withstand abiotic pressures through a combination of integrated molecular and cellular responses, physical adaptation, and defense mechanisms against environmental disturbances. The recognition of stress signals and their transmission to trigger the appropriate adaptive molecular responses to guarantee survival is the first and most important stage in abiotic stress defense systems.

6.2.1 Perception of Abiotic Stress

Receptor-like kinases and histidine kinases (HKs) are examples of receptors/sensors that are involved in the perception of stress signals (RLKs). Plants have been shown to possess a large number of leucine-rich repeat (LRR) receptors that are linked to abiotic stress. These receptors fall into one of two categories: proteins or receptor-like kinases (RLKs or receptor-like proteins [RLPs]). An increasing body of research indicates that RLKs are the primary regulators of environmental stress control when it comes to the perception of abiotic stress. For instance, it has been reported that proline-rich-extensin-like rlk4 (PERK4), receptor-like protein kinase 1 (RPK1), calcium/calmodium-regulated cysteine-rich rlk (CRK36), and guard cell hydrogen peroxide-resistant1 (GHR1) regulate the moisture stress signaling in *Arabidopsis* and are involved in sensing cold and drought stress (Berr *et al.*, 2012). Moreover, a multitude of RLKs control diverse processes, including symbiosis, cellular differentiation, and the growth of roots and shoots. Signal transduction is the process by which a signal is transduced into various downstream signaling networks after being perceived as stress by receptor proteins on cell surfaces. Perception triggers a general signal transduction route that produces secondary messengers, which in turn activate a phosphorylation cascade that targets proteins involved in the control of stress defense genes. The cytosolic calcium (Ca^{2+}) increase, reactive oxygen species (ROS), and activation of the mitogen-activated protein kinase (MAPK) cascade are among the early responder signals that have been identified (Gamir, Sánchez-Bel and Flors, 2014).

6.2.2 Calcium Signaling

The specificity of the plant's cellular responses to stress is largely dependent on calcium signaling, and each stimulus detection is followed by a sharp rise in the intracellular concentration of the relevant ions. Normal physiological settings result in Ca^{2+} gradients of various magnitudes because resting cytosolic Ca^{2+} ($[Ca^{2+}]$cyt) is maintained by active transportation at nanomolar concentration into the calcium reserves or beyond the cells into the apoplast. As a response mechanism, changes in the $[Ca^{2+}]$ cyt inflow and efflux patterns are induced upon varied stimuli perception and are regulated by numerous channels and pumps. The amplitude and frequency of the signal are reliant on the stimulus type, resulting in a signature when $[Ca^{2+}]$ cyt variations are triggered in a repeated way. Every calcium signature specifies the sort of defense response by encoding information particular to a stimulus, such as the tissue type, subcellular location, size, and frequency. Calcium signal sensors that are able to detect changes in levels and transmit the information contained in the signatures to trigger the appropriate signaling cascades are essential for the decoding of Ca^{2+} signals. When calcium attaches to these sensors, a conformational change occurs that activates signaling pathways and transforms the initial stimulus perception into specific biological reactions, adding another degree of specificity to the sensor function (Fernandes *et al.*, 2008).

6.2.3 ROS Response

ROS, an essential secondary messenger and early response mechanism in plants, are produced in response to the sense of abiotic stress. ROS are effective signaling transducers, since they are present at moderate levels in plants under normal development circumstances due to the activity of antioxidants and enzymes that maintain the levels in balance. On the other hand, abiotic stressors cause an oxidative burst—an excessive generation of ROS—which can have detrimental effects such cell death. ROS, which comprise hydroxyl radical (·OH), hydrogen peroxide (H_2O_2), superoxide anion (O^{2-}), and singlet oxygen ($^1O_2$), are atmospheric oxygen intermediates of significant biological importance in plants. These intermediates are produced by specific mechanisms. Even while the generation of these chemical species varies throughout cellular compartments, the resulting ROS signal is still regarded as a signature of abiotic stress response, allowing plants to adapt to a variety of stress exposures. Since the production of ROS is ubiquitous in plant stress encounters and a combination of stressors is likely to produce varying ROS levels, various sensors may be employed to interpret these fingerprints and provide a signal unique to each stress. Known as respiratory burst oxidase homologs in plants, nicotinamide adenine dinucleotide phosphate (NADPH) oxidases (NOXs) catalyze the production of superoxides (RBOHs). When under stress, receptor kinases phosphorylate, producing increased Ca^{2+} that triggers the respiratory burst oxidase homolog D (RbohD) (Evans *et al.*, 2016). Plant cell walls become depolarized as a result of the excessive ROS generation that is induced by this. In a phenomenon referred to as a "ROS wave," ROS permeated the whole plant and simultaneously sparked cell-to-cell communication that activated one or more systemic signals. The fine-tuning of signaling is the consequence of a phenomena known as reciprocal interaction, in which calcium and ROS boost each other's induction during stress encounters (Devireddy *et al.*, 2021).

6.2.4 Activation of Mitogen-Activated Protein Kinase Pathway

Apart from the fast systemic signaling triggered by Ca^{2+} and secondary messengers of ROS, kinase cascades of MAPKs are also essential for plant signaling of various environmental stimuli. The MAPK cascade is the outcome of many phosphorylation processes that, in response to various stressors, activate relevant genes. This system regulates the expression of pertinent genes, various cellular functions, and proteins involved in the processes of development and adaptation by transferring the produced stress signals from the receptors to certain effectors. MAP kinase kinase kinase (MAPKKK), mitogen-activated kinase kinase (MAPKK), and MAP kinase (MAPK) are the three kinds of kinases involved in signal transduction by MAPK cascades (MAPKKK). First, extracellular stimuli activate the downstream MAPKKK, which is attached to certain receptors. This, in turn, phosphorylates the two serine or threonine residues in the activation loop of a downstream MAPKK, activating it. Later, MAPKK acts as a particular kinase, phosphorylating a MAPK on either of its tyrosine or threonine residues that are found in the activation loop. The latter ultimately causes additional protein kinases, enzymes, or transcription factors involved in plant stress-driven signaling pathways to become active, in addition to a variety of effector proteins found in the cytoplasm or nucleus. Moreover, MAPK cascades that are triggered by abiotic stressors, including salt, cold, and drought, exhibit interactions with phytohormones and secondary messengers like ROS, Ca^{2+}, and nitric oxide (NO). First, extracellular stimuli activate the downstream MAPKKK, which is attached to certain receptors. This, in turn, phosphorylates the two serine or threonine residues in the activation loop of a downstream MAPKK, activating it. Later, MAPKK acts as a particular kinase, phosphorylating a MAPK on either of its tyrosine or threonine residues that are found in the activation loop. The latter ultimately causes enzymes, additional protein kinases, or transcription factors involved in plant stress-driven signaling pathways to become active, in addition to a variety of effector proteins found in the cytoplasm or nucleus. Activating MAPK-driven phosphorylation cascades and secondary messengers is how phytohormones interact with one another. The control of gene expression brought about by these transduction cascades has a direct impact on the production of various phytohormones. Thus, complex crosstalk between perception, signal transduction, and phytohormone production results from the integration of many inputs and the response of several MAPK cascades to phytohormones such auxins, jasmonic acid, salicylic acid, ethylene (ET), and abscisic acid (ABA). Typically, every signaling molecule engages in unique signaling pathways, which leads to the development of a cross-talking network that synchronizes reactions to various abiotic stressors (Jagodzik *et al.*, 2018).

6.2.5 Phytohormone Production

Depending on the kind of stress experienced, plants can employ phytohormones to increase the initial stress signals during stress exposure. These phytohormone-driven signals can start a completely new signaling event with distinct components, or they can start a new signaling event with comparable components to the original signal. ABA is the phytohormone most frequently found to be involved in plant defense against abiotic stressors (Knight and Knight, 2001). Rapid generation of ROS and other early plant stress signaling events have been connected to phytohormone accumulation, highlighting the significance of both early and traditional plant responses in the control of phytohormones, which is contingent upon the kind of stress. For example, ROS generation via NADPH oxidase is required for the accumulation of ABA in conditions of severe salt and water scarcity. The buildup of ROS brought on by ABA has the ability to penetrate guard cells, open Ca^{2+} channels, raise $[Ca^{2+}]$cyt, and ultimately cause stomatal closure. Open stomata 1 are activated when ABA is initially detected by the cells through the ABA receptors RCAR/PYR1/PYL (regulatory components of ABA-receptor/pyrabactin resistant protein/PYR-like proteins) (OST1). As a member of the SNF1-related protein kinase 2 (SnRK2) family, OST1 regulates the formation of ROS by phosphorylating NADPH oxidase, which in turn causes ABA-induced stomatal closure. Additionally, ABA-mediated stress responses that occur either upstream or downstream of ROS generation have been linked to MAPK cascades. Defense genes, proteins, and metabolites are the markers of several cellular/molecular defense mechanisms that are activated or initiated as a result of these signaling events.

6.2.6 Cellular and Molecular Events

Networks of cellular and molecular systems regulate plant defense responses, which let the plant adapt to abiotic challenges by adjusting variations in growth and development. The understanding of these plant defenses against environmental disturbances, which often involve modifications to the transcriptome, proteome, and metabolome of the plant, has advanced significantly. Adaptive reactions are triggered after signal detection and transmission, which leads to the transcriptional level regulation of genes linked to stress by transcription factors (TFs). By specifically binding to the cis- and trans-elements in the promoters of target genes, a single TF may control the expression of several genes; this sort of transcriptional regulation is known as regulon. In this case, a single signal and regulatory protein governs several genes, and multiple regulons can be induced by a single environmental input. Abiotic stress in plants is regulated by ABA, a primary phytohormone that controls a complex gene regulatory system that allows plants to withstand environmental disturbances. Plants have been shown to possess a number of regulons that become active in response to abiotic stress. Myeloblastosis oncogene (MYB) and the ABA-responsive element-binding protein/ABA-binding factor (AREB/ABF) regulons function in ABA-dependent gene activation pathways, while C-repeat binding factor (CBF)/dehydration-responsive element-binding protein 1 (DREB1), DREB2, and the zinc-finger homeodomain (HD) regulons function in ABA-independent gene expression. The ABA-dependent and ABA-independent routes have been shown to converge at several locations that correspond to transcriptional enhancers and repressors. These locations have the potential to initiate synergistic interactions between salinity, drought, and cold stress by either directly or indirectly interacting with the DREB and AREB. The many TFs implicated in stress tolerance often operate in isolation from one another.

The many regulons control the expression of genes that induce abiotic stress, such as PP2Cs, cell cycle regulator genes (ICK1), and late embryogenesis abundant (LEA) class genes (RD29B, RAB18) (ABI1 and ABI2). Particular focus has been placed on RD22 and RD26 as possible targets to enhance abiotic stress tolerance. Together with their products, these stress-regulated genes play critical roles in abiotic stress responses and tolerance. For example, these genes translate into functional proteins that control a variety of abiotic stress defense mechanisms, including stomatal

TABLE 6.1
Mechanisms Induced in Crops under Different Abiotic Stress

Crop	Abiotic Stress	Mechanism	References
Arabidopsis thaliana	Salt and drought stress	Histone acetylation	Zheng *et al.* (2016)
Arabidopsis thaliana	Salinity stress	Hypomethylation	Sako *et al.* (2016)
Arabidopsis thaliana	Cold stress	Hypermethylation	Kenchanmane Raju *et al.* (2018)
Zea mays	Heat	Histone acetylation	Wang *et al.* (2015)
Zea mays	Cold	Hypomethylation	Hu *et al.* (2012)
Oryza sativa	Salt stress	Demethylation	Zhu *et al.* (2015)
Hordeum vulgare	Drought stress	Hypermethylation	Surdonja *et al.* (2017)
Hydrilla verticillata	Metal (copper) stress	Hypermethylation	Shi *et al.* (2017)

closure, membrane protection, osmolyte accumulation, and ROS scavenging. Mechanisms that are induced in crops under different abiotic stress are elucidated in Table 6.1.

6.3 RHIZOSPHERE MICROBIAL COMMUNITY

Plants have been linked to a complex microbial ecosystem, comprising beneficial microbes, since their earliest evolutionary stages. The ability of early terrestrial plants to adapt to their surroundings and overcome obstacles like diseases, new and/or stressful circumstances, and nutrient availability was facilitated by the presence of helpful microbes. Plant roots and microorganisms have varying degrees of intimacy in soil, with the plant having a higher impact on the microbial population closer to the root surface. The word "rhizosphere," which was first used by Hiltner to characterize the microbes of soil inside as well as around roots, refers to a volume of soil that is specifically impacted by plant roots or connected with the roots. Endophytes are microorganisms that live inside roots, whereas rhizosphere and rhizoplane are defined as the areas where microorganisms reside on the surface of roots. The microbial community linked to the roots (rhizobiome) is unquestionably the most populous and complex of all the microbial communities connected with high plants out of all those inhabiting the various plant areas. The process known as rhizodeposition, which includes a variety of organic acids, sugars, amino acids, vitamins, nucleosides, and other tiny compounds that function as potent chemo-attractants of the soil microbiota, is what causes the rhizosphere effect.

Rhizodeposition also describes the discharge of a class of specialized cells known as root cap border cells into the rhizosphere. Because it typically persists after decomposing from the roots, this type of cell is thought to be essential for its impact on the rhizosphere. Specifically, the majority of the passive flow of low-molecular-weight metabolites, such as sugars, amino acids, and organic acids, occurs at the root tip due to the absence of cell differentiation, which promotes metabolite diffusion to the soil. Changes in the concentration of secreted metabolites are sensed by the root tip, which then converts them into signals to adjust root development. Plants have the ability to increase the concentrations of numerous common metabolites locally through root exudate flux. These metabolites can act as sensors and integrators of the nutritional status of the plant as well as the availability of nutrients for the microbial community in the rhizosphere.

6.4 PLANT-GROWTH-PROMOTING RHIZOBACTERIA (PGPR)

Approximately 2.5–5% of all rhizosphere bacteria have PGPR. High rhizosphere competency, enhanced plant growth capacities, a broad spectrum of activity, environmental safety, compatibility with other rhizobacteria, and resistance to heat, UV light, and oxidants are all desirable qualities in a PGPR. PGPR may be divided into two groups: intracellular PGPR (iPGPR, or symbiotic

TABLE 6.2
The Effect of Microbes on Plant Responses Mitigating Salt Stress

PGPR	Crop Species	PGPR-Induced Plant Salt Stress Responses	References
Enterobacter cloacae PM23 *Serratia liquefaciens* KM4	*Zea mays*	↑ Proline, glycine betaine, free amino acids, and soluble sugars	El-Esawi *et al.* (2018a)
Streptomyces albidoflavus OsiLf-2	*Oryza sativa*	↑ Proline, total soluble sugars	Niu *et al.* (2022)
Enterobacter 64S1 and *Pseudomonas* 42P4	*Solanum lycopersicum*	↑ Proline	Pérez-Rodriguez *et al.* (2022)
Bacillus sp. Wp-6	*Triticum aestivum*	↑ Proline, soluble proteins, and soluble sugars	Zhao *et al.* (2022)
Bacillus firmus SWS	*Glycine max*	↑ Proline	El-Esawi *et al.* (2018)
Klebsiella sp. IG 3	*Avena sativa*	↑ Total soluble sugars; ↓ Proline	Sapre, Gontia-Mishra and Tiwari (2018)
Bacillus megaterium ZS-3	*Arabidopsis thaliana*	↑ K^+/Na^+ ratio, *NHX1*, and *AVP1*; ↓ *HKT1*	Shi *et al.* (2017)
Pseudomonas knackmussii	*Arabidopsis thaliana*	↑ SOS1	Rabhi *et al.* (2018)
Bacillus atrophaeus ZYH01	*Zea mays*	↑ K^+/Na^+ ratio, *NHX*, and *HKT*	Hou *et al.* (2022)
Kosakonia radicincitans KR-17	*Raphanus sativus* L.	↑ N, P, K^+, $Ca2^+$, Mg, Zn, Fe, Cu, and Na^+	Shahid *et al.* (2022)
Pseudomonas sp. M30–35	*Chenopodium quinoa*	Maintaining P content and homeostasis	Cai *et al.* (2021)

bacteria), which grow inside plants and directly exchange metabolites with them, and extracellular PGPR (ePGPR), or free-living rhizobacteria (Lephatsi *et al.*, 2021). The basis for this categorization is the degree of relationship and interaction with plants. The term "iPGPR" refers to a broad category of soil bacterial species, including *Allorhizobium, Azorhizobium*, *Bradyrhizobium, Mesorhizobium*, and *Rhizobium* of the Rhizobiaceae family, which are often found inside the unique nodular structures of root cells. Conversely, ePGPR, which includes genera like *Agrobacterium, Arthrobacter, Azotobacter, Azospirillum, Bacillus, Burkholderia, Caulobacter, Chromobacterium, Erwinia, Flavobacterium, Micrococcous, Pseudomonas*, and *Serratia*, are found in the rhizosphere, rhizoplane, or in the spaces between the root cortex cells. Three macro-effects of PGPR are observed in the host plant as a result of the beneficial action it exerts: (1) enhanced intake of nutrients; (2) heightened tolerance to biotic stressors or infections by phytopathogens, which include viruses, fungi, bacteria, nematodes, insects, etc. (Venturi and Keel., 2016); and (3) increased tolerance to abiotic stresses (Table 6.2).

6.5 MICROBIAL BIOSTIMULANTS AND THEIR EFFECT ON ENHANCEMENT OF PLANT RESPONSES

The control of plant growth and development and the mitigation of harmful impacts of abiotic stressors are important elements that affect how productively farmed plants are produced. It is commonly recognized that abiotic stressors have a detrimental impact on plant growth and development and cause crop losses on a worldwide scale. Nevertheless, little is now known about the cellular and molecular processes that help to mitigate these harmful consequences. Furthermore, the integration of biostimulants – which alter the physiological and biochemical processes of plants – into production systems is growing in order to improve plant productivity and stress tolerance. By definition, biostimulants, regardless of their nutritional content or (individual) ingredients, are a variety of compounds or microbes that are designed to stimulate the plant's natural processes to boost plant production through various modes of action (Espinas, Saze and Saijo, 2016; Mandal *et al.*, 2023).

Furthermore, biostimulants improve the metabolism of the plant, increase stress tolerance, facilitate translocation and assimilation, as well as utilization of nutrients, improve physiochemical properties of soil, and promote the growth of complementary soil microorganisms. They also contribute to plant growth and development throughout the life cycle. Plant biostimulants come in a variety of formulations with different ingredients; however, they may be broadly divided into four classes according to their source and composition. Amino acid-containing products (AACP), humic substances (HS), hormone-containing products (HCP), and microbes that are involved in plant development are some of these categories (Adedayo and Babalola, 2023; Fasusi, Babalola and Adejumo, 2023). Besides that, a significant amount of study has been conducted on a range of biostimulant categories that support plant development, including as rhizobacteria, silicon, humic, and fulvic acids, protein hydrolysates, seaweed extracts (SWEs), and plant growth-promoting rhizobacteria (PGPR). While several possible key mechanisms of action generated by biostimulants in connection to the reduction of abiotic stress have been established, the clear underlying mechanisms of action accountable for these effects remain mostly unclear. Among these systems are metal chelation, stomatal control, ion homeostasis, membrane integrity, osmoprotection, ROS scavenging, and nutrition availability.

6.5.1 Influence of PGPR-based Biostimulants on Defense Priming

PGPR formulations are being employed as biostimulants more often because they have the potential to improve plant growth, health, and yield sustainability. These rhizosphere-dwelling soil bacteria work in harmony with the plant host to promote plant development. The production of growth-stimulating phytohormones, the solubilization of mineral phosphates, nitrogen fixation, and other processes are some of the ways that this chemical communication is converted into physiological advantages. While the chemistry of the rhizosphere is still largely unknown and the establishment of mutualistic interactions between plants and the rhizomicrobiome is still poorly understood, new research has shown that different PGPR species can improve agronomic yields through direct or indirect mechanisms, such as nitrogen fixation, the production of phytohormones that stimulate growth, and the solubilization of mineral phosphates. The latest evidence suggests that PGPR-derived MAMPs may potentially be involved in the stimulation of plant growth. It has been shown that *Azospirillum brasilense* Sp245 lipopolysaccharides (LPS) increase growth (fresh weight and root length) in *Arabidopsis thaliana*. Many methods by which these organisms produce stress tolerance have been elucidated, and the application of PGPR to create abiotic stress tolerance in plants is increasingly being researched as an appealing way to manage stress.

PGPR employs many tactics to assist plants in adjusting to novel surroundings in the face of abiotic challenges (Figure 6.1). These strategies include: (1) releasing ACC deaminase (Glick, 2014), which decreases ET levels in plants; (2) secreting osmolytes, such as choline, proline, and trehalose, which are osmoprotectants; (3) releasing volatiles from bacteria, such as 2R and 3R butanediol, which cause stomatal closure, to promote stress tolerance; (4) secreting phytohormones, such as IAA, gibberellins, and cytokinins, which stimulate the lateral root formation and root hair production, increasing moisture and nutrient uptake; (5) altering the elasticity and stability of root cell membranes; and (6) secreting exopolysaccharide, which increases soil aggression and maintains high water potential across plant roots. In addition to these established pathways, PGPR can initiate physiological processes that are controlled by an intricate web of signaling events, leading to stress tolerance and priming (Adedayo and Babalola, 2023).

6.5.2 Increased Stress Tolerance by PGPR-Induced Priming

Through the process of priming, or PGPR-induced preconditioning of the plant immunity and defenses, PGPR-plant interactions result in increased resistance to abiotic stressors (Mellidou and

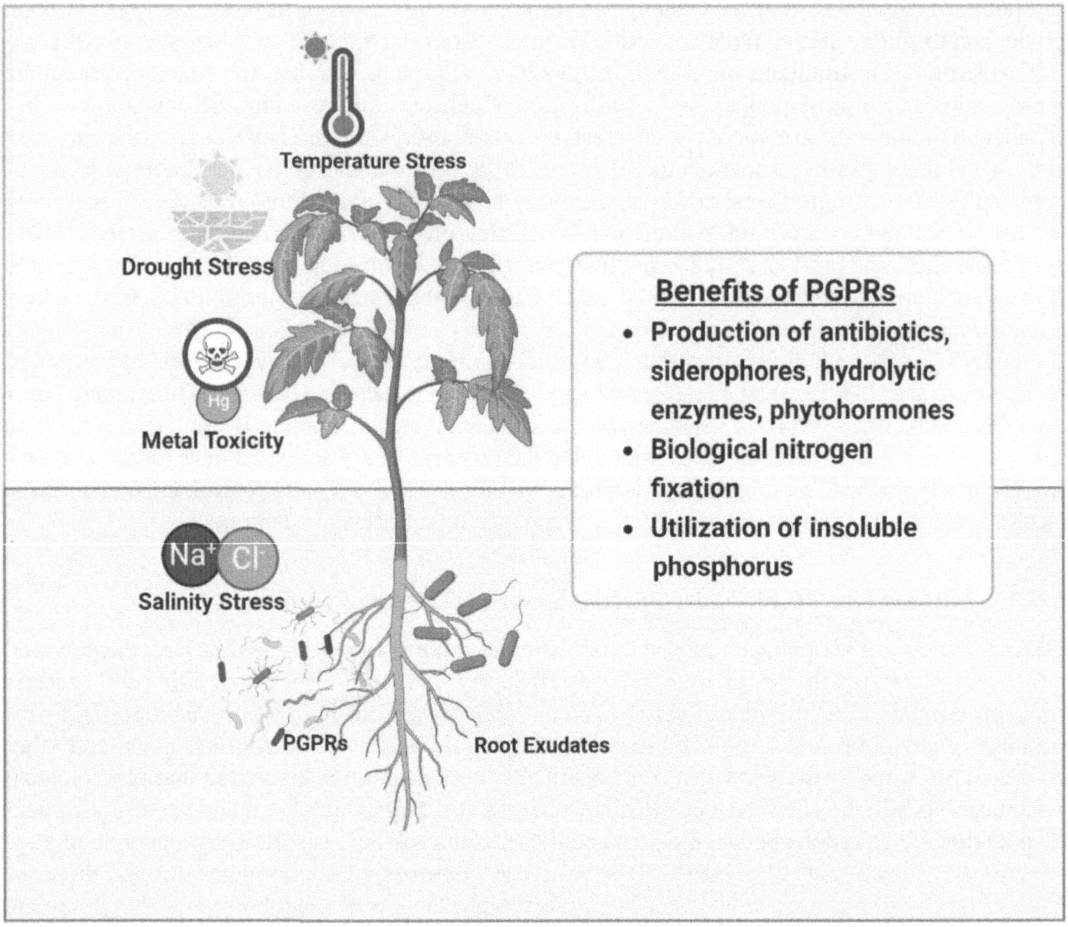

FIGURE 6.1 Schematic representation of several PGPRs influencing plant development and stress tolerance. This figure was created with BioRender.com.

Karamanoli, 2022). In comparison to unprimed plants, this condition of the plant leads to improved stress tolerance because it reacts to stress more quickly and/or robustly. The "primed state," a level of readiness attained, has been connected to effective defense response activation and increased stress resistance (Bradáčová *et al.*, 2016). Numerous methods, including systemic acquired resistance (SAR) and induced systemic resistance (ISR), can be used to define the improved resistance. SAR is a plant defense mechanism that is triggered in the plant's distal sections after a localized infection. It gives the plant resilience to other stressors and primes it to fight itself in the event of an assault. On the other side, ISR is mediated by microorganisms that mediate plant development, including PGPR, which colonizes the root system of the plant to cause resistance (Barnawal, Singh and Singh, 2019).

While ISR uses the JA and ET response pathways to produce resistance, SAR induction is dependent on SA. Although SAR and ISR have been extensively documented to emerge as important modalities of priming, there is still a gray area through which these processes of priming operate (Van Oosten *et al.*, 2017). While the exact mechanisms underlying priming remain unclear, a number of theories have been put forth. These include the build-up of dormant proteins that contribute to signal amplification, like MAPKs, the activation of transcription factors which boost the transcription of genes related to defense after stress is perceived, and epigenetic modifications involving

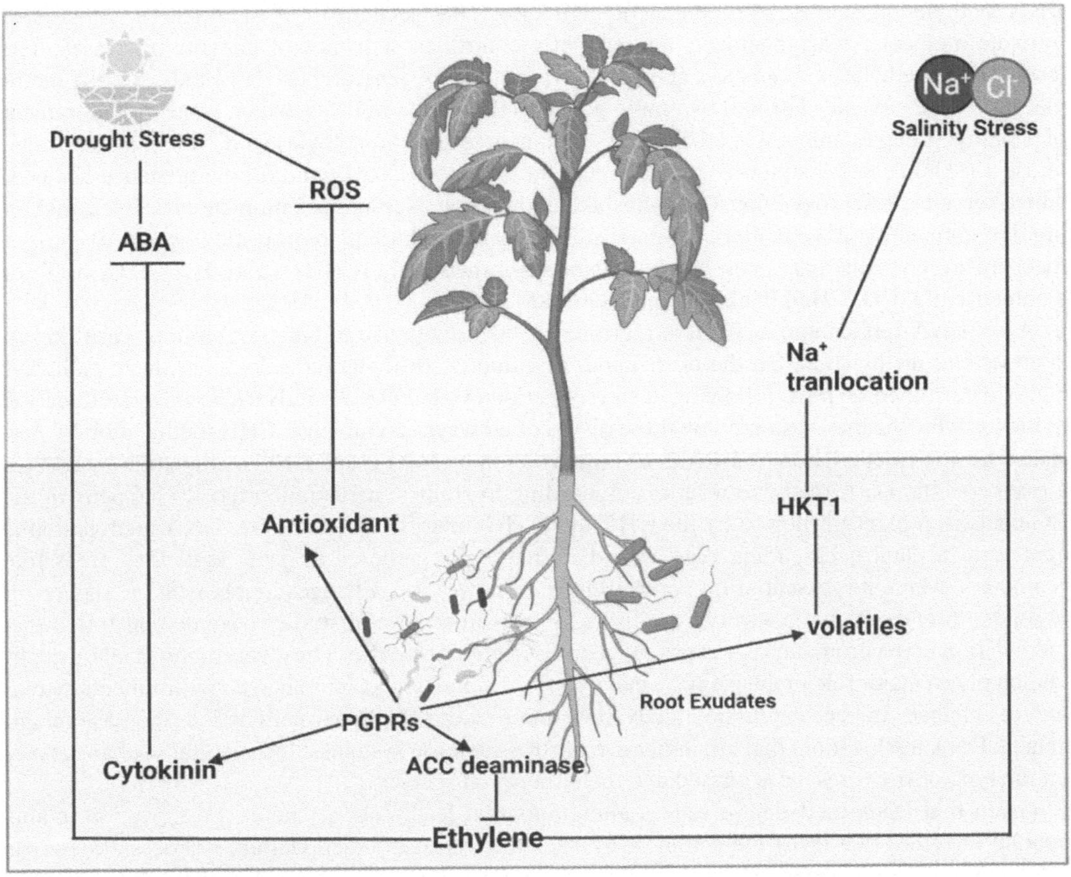

FIGURE 6.2 Summary of the PGPR-induced systemic tolerance to salt and drought stress. Compounds released by PGPR during abiotic stress, including volatiles, ACC deaminase, antioxidants, and cytokinins, are shown by inhibition arrows. Reactive oxygen species (ROS) degradation and abscisic acid (ABA) accumulation is caused by cytokinins and antioxidants, respectively. Under high salt stress, the volatiles released by PGPR upregulate high-affinity K^+ transporter 1 (HKT1) expression in shoot tissue, resulting in Na^+ recirculation throughout the plant. (Figure is created with BioRender.com.)

changes to DNA, histones, or chromatin. Since plant priming improves defense responses without compromising a plant's overall fitness and results in stress resistance or tolerance that microorganisms cannot overcome, it has been viewed as a viable technique for stress management. This resistance provides long-term protection (Conrath *et al.*, 2015).

Although there is still much to learn about the chemistry of the rhizosphere and how plant-rhizomicrobiome interactions arise, new research indicates that different PGPR species can prime plants for stronger defense responses against abiotic challenges (Figure 6.2). The defense priming caused by rhizobacteria is linked to molecular pathways that indicate a reprogramming of the plant's regulatory systems and cellular metabolism. According to current knowledge, primed plants reprogram metabolic pathways prior to environmental disturbances by changing the way different molecules like sugars, amino acids, and tricarboxylic acids are biosynthesized.

6.5.3 DNA Methylation and Plant Responses

The transfer of a methyl group from S-adenosyl methionine (SAM) to the fifth carbon of the pyrimidine ring of the cytosine nucleotide without changing the underlying DNA sequence is known as

DNA methylation, and it is an epigenetic signature. The mechanism of modification consists of intricate molecular rearrangements and a tightly controlled sequence of enzyme processes. The term "DNA methylases" refers to a class of enzymes that are generally used to catalyze DNA methylation. These enzymes fall into two main groups: (1) *de novo* MTases, which catalyze the transfer of a methyl group to unmethylated cytosines and include chromomethylase 3 (CMT3), methyltransferase 1 (MET1), and domain-rearranged methyltransferase (DRM); and (2) maintenance MTases, which serve to preserve previously established methylation. Numerous biological processes, including development, stress adaption, genomic stability, and evolution, are impacted by this epigenetic mechanism. Cytosine bases can be methylated asymmetrically (CHH; where H = A, C, or T) or symmetrically (CG, CHG) (Gehring and Henikoff, 2007).

After DNA replication, a MET is recruited to hemimethylated daughter strands, resulting in symmetrical methylation. On the other hand, asymmetric methylation lacks inheritance pathways and is determined de novo following each replication cycle. DRM2 catalyzes de novo methylation in plants, which is then sustained by three distinct pathways: asymmetric CHH methylation by persistent de novo methylation by DRM2, CG methylation by MET1, and CHG methylation by CMT3, a plant-specific DNA methyltransferase. According to studies, methylation mostly happens in the CG context in plants, followed by the CHG and CHH contexts. Furthermore, DNA methylation is also seen in plant gene-coding regions and is frequently gathered at gene regulatory areas like promoters, where it is essential for controlling gene expression. Changes in methylation may result in unique transcriptional control of the linked genes, since several studies have demonstrated that methylation in the promoter region promotes transcriptional silence. Thus, variations in DNA methylation play a major role in the plant's capacity to withstand and adapt to harsh environments. Stress induces changes in gene expression (activation/suppression) and hypermethylation (increase in epigenetic DNA methylation) that are indicative of stress defense systems. However, these changes rely on the type of stress response caused and differ between species.

Given that DNA methylation causes alterations that lead to gene silence or activation, plants contain enzymes that resist the action of MTases to remove the methylation, a process known as DNA demethylation. Transcription initiation sites are demethylated by DNA demethylases in order to regulate gene production. A base excision repair (BER) mechanism is used by these enzymes, which comprise demeter (DME), demeter-like (DML) proteins (DML2 and DML3), and regulator of silencing 1 (ROS1), to initiate the demethylation process. It seems that demethylation, which causes gene expression to be stimulated, is a fast process that is common in plants. Stress can cause changes in DNA methylation, as evidenced by a number of cases (Abid *et al.*; 2017).

6.5.4 Histone Modifications and Small RNAs

In addition to alterations in DNA molecules brought about by methylation or other processes, chemical modification of histones is another significant component of the epigenome. Histone proteins are essential for the organization and packaging of DNA molecules into nucleosomes, the fundamental structural components of chromatin. This latter consists of around 147 base pairs of DNA surrounded by an octamer of histones, including two copies of each base pair, H2A, H2B, H3, and H4. It is this core particle that controls access to genomic DNA. The N-terminal regions of histones, commonly called the tails because they extend from the broader structure, are the most frequently altered regions in terms of reversible chemical post-transcriptional modifications (PTMs) that change chromatin structure and function (Kumar *et al.*, 2017). Histone PTMs that are catalyzed by different enzymes and involved in either upregulating or downregulating gene expression include acetylation, ubiquitination, phosphorylation, de-acetylation, carbonylation, and glycosylation. Thus, the accessibility of genes for transcriptional machinery is changed by these alterations to histones. Histone acetylation, methylation, de-acetylation, and demethylation – all of which are catalyzed by histone acetyltransferases (HATs), histone methyltransferases (HMTs), histone deacetylases (HDACs), and histone demethylases (HDMs) – have demonstrated a tight relationship between plant responses to exposure to abiotic stress in the context of plant-environment interactions (Lämke and Bäurle, 2017).

Reports state that trimethylation of H3K4 is associated with transcription activation, whereas dimethylation of H3K9 and H3K27 represses gene transcription. For example, in *Arabidopsis thaliana*, a reference plant frequently used for epigenetic research on plant responses to abiotic stimuli, persistent cold stress resulted in an increase in H3K9 and H3K27 dimethylation and a decrease in H3K4 trimethylation (Jiang *et al.*, 2014). Moreover, the flowering repressor locus C (FLC) is the site of histone acetylation. The winter-annual forms of *A. thaliana* that show delays in blooming during the first season are able to flower thanks to the steady suppression of FLC caused by these histone changes. Sokol *et al.* (2007) showed that acetylation of histone H4 lys14 and enhanced phosphorylation and phosphoacetylation of histone H3 Ser-10 occurred under high salinity and cold stress conditions using western blotting to examine the reaction of nucleosomes from tobacco BY-2 and the *Arabidopsis* T87 cell line (Pagter *et al.*, 2017). Similar to DNA methylation, a number of studies have reported on histone changes as potential indicators of priming and stress memory. According to a recent study by Liu *et al.* (2010), *Arabidopsis* plants under extreme heat stress bloomed more quickly, a trait that was also observed in the unstressed offspring. The mechanism behind this phenomenon is yet unknown. Furthermore, they showed that heat-induced heat shock transcription factor (HSFA2) activated the H3K27me3 demethylase, in contrast to early flowering 6 (REF6), which de-repressed HSFA2.

The suppressor of gene silencing 3 (SGS3)-interacting protein 1 (SGIP1), which is activated by the loop formed by REF6 and HSFA2, leads to transgenerational degradation and the subsequent creation of transacting siRNA (tasiRNA) inhibition. Early blooming is induced by the REF6-HSFA2 loop, although immunity is reduced. This feedback loop is sustained by the transgenerational "ON" state of HSFA2, which functions as a long-term epigenetic memory of heat. Apart from histone changes, RNA polymerase II stalling has also been proposed as a drought stress-induced memory mark in *A. thaliana*. This might result in an active chromatin content that readies genes related to development and stimulus responses for proper expression. Additionally, it has been discovered that target gene repressions by small non-coding RNAs are activated during plant abiotic stress and transcriptional gene silencing (TGS) by heterochromatic small interfering RNAs (hc-siRNAs), which are 24 nucleotides long and act through RNA directed DNA methylation (RdDM), an epigenetic mechanism of gene regulation in plants. RdDM in plants is limited to small RNA-mediated chromatin modifications due to distinct transcriptional machinery centered around two RNA polymerase II-related enzymes that are unique to plants, namely Pol IV and Pol V. The steps involved in the canonical view of RdDM are as follows: Initially, Pol IV transcripts are duplicated into lengthy double stranded RNAs (dsRNAs). Dicer-like 3 (DCL3) then converts them into siRNAs and transports them to the cytoplasm. Ultimately, these siRNAs are re-imported into the nucleus after being loaded onto Argonaute (AGO4) on one strand. There, the siRNA targets transcripts from Pol V through sequence complementarity. Lastly, in all kinds of sequence contexts, this targeting activates DNA methyltransferase activity to facilitate de novo methylation of cytosines (Zhao *et al.*, 2022).

Although many instances of environmentally sensitive siRNAs serving as important stress and priming mechanisms have been documented, it is still unclear how epigenetic changes and acclimatization are related. Gagné-Bourque *et al.* (2015) have reported evidence of epigenetic modifications induced by microbial biostimulants. In particular, they discovered that *Bacillus subtilis* B26 boosted the global DNA methylation in *Brachypodium distachyon* by six times under typical plant development circumstances. The reported hypermethylation was associated with an increase in the number of methyltransferases involved in maintaining and controlling DNA methylation (Schwachtje *et al.*, 2019). Additionally, in contrast to the native plants, the PGB-induced hypermethylation levels remained unchanged after five and eight days of drought treatment. Very recently, De Palma *et al.* (2019) reported on the potential of the fungus *Trichoderma harzianum*, which is employed in biostimulant formulations, to induce DNA methylation and other epigenetic alterations. This finding shed light on the interaction between microbes and plants in tomato roots. The findings demonstrated a drop in DNA methylation levels at various time intervals, which was associated with transcriptional control of genes relevant to defense. Plant defense mechanisms are mediated by the observed expression of epiregulators through their interaction with *Trichoderma* and/or other effects generated by *Trichoderma*. The understanding of the epigenetic modifications caused by microbial biostimulants has expanded recently, but most of these studies have primarily examined the epigenetic mechanisms related to microbial biostimulants under biotic stress; as a result, the investigation of epigenetic mechanisms induced by abiotic stress remains largely unexplored. Apart from the epigenetic changes resulting from intricate, multi-layered defense mechanisms triggered by abiotic stress, metabolomic adaptation also involves adjustments to primary and secondary metabolites. The interplay between metabolism and epigenetics is therefore a crucial component of cellular stress response to abiotic conditions.

6.6 CONCLUSIONS

By using the natural priming process of plants, contemporary agriculture is striving for higher crop yields and crop quality. In this sense, the scientific community and the agricultural community are gradually coming to the opinion that it makes sense to use biostimulants to increase abiotic stress tolerance. One practical solution for enhancing plant growth and stress tolerance might be the use of biostimulants. Numerous research has shown the potential advantages of utilizing biostimulants in conjunction with various microbes, including PGPR; nevertheless, their exact modes of action and underlying molecular processes with regard to their priming effects and growth stimulation are still unknown. The increasing use of biostimulants has brought attention to the positive effects of these formulations, which include modifying plant physiological and biochemical processes to increase stress tolerance and productivity. The use of metabolomics and epigenetic techniques for the investigation of defense priming and plant responses to abiotic stresses is also briefly covered in this chapter.

REFERENCES

Abid, G., *et al.* (2017) 'Analysis of DNA methylation patterns associated with drought stress response in faba bean (*Vicia faba* L.) using methylation-sensitive amplification polymorphism (MSAP)', *Environmental and Experimental Botany*, 142, pp. 34–44.

Adedayo, A.A. and Babalola, O.O. (2023) The potential of biostimulants on soil microbial community: A review. *Frontiers in Industrial Microbiology*, 1, p. 1308641. https://doi.org/10.3389/finmi.2023.1308641

Arbona, V., *et al.* (2017) 'Physiological, metabolic, and molecular responses of plants to abiotic stress', *Stress Signaling in Plants: Genomics and Proteomics Perspective*, 2, pp. 1–35.

Avramova, Z. (2015) 'Transcriptional 'memory' of a stress: Transient chromatin and memory (epigenetic) marks at stress-response genes', *The Plant Journal*, 83(1), pp. 149–159.

Bachheti, A., *et al.* (2021) 'Medicinal plants and their pharmaceutical properties under adverse environmental conditions', in Husen A. (ed.) *Harsh environment and plant resilience*. Cham: Springer International Publishing AG, pp. 457–502. https://doi.org/10.1007/978-3-030-65912-7_19

Barnawal, D., Singh, R. and Singh, R.P. (2019) 'Role of plant growth promoting rhizobacteria in drought tolerance: Regulating growth hormones and osmolytes', in *PGPR amelioration in sustainable agriculture*. Kidlington, United Kingdom: Woodhead Publishing, pp. 107–128.

Berr, A., *et al.* (2012) 'Chromatin modification and remodelling: A regulatory landscape for the control of Arabidopsis defence responses upon pathogen attack', *Cellular Microbiology*, 14(6), pp. 829–839.

Bradáčová, K., *et al.* (2016) 'Micronutrients (Zn/Mn), seaweed extracts, and plant growth-promoting bacteria as cold-stress protectants in maize', *Chemical and Biological Technologies in Agriculture*, 3(1), pp. 1–10.

Cai, D., *et al.* (2021) 'Improved salt tolerance of Chenopodium quinoa Willd. contributed by *Pseudomonas sp.* strain M30-35', *PeerJ*, 9, p. e10702.

Conrath, U., *et al.* (2015) 'Priming for enhanced defense', *Annual Review of Phytopathology*, 53, pp. 97–119.

De Palma, M., *et al.* (2019) 'Transcriptome reprogramming, epigenetic modifications and alternative splicing orchestrate the tomato root response to the beneficial fungus *Trichoderma harzianum*', *Horticulture Research*, 6(1), pp. 1–15.

Devireddy, A.R., *et al.* (2021) 'Integration of reactive oxygen species and hormone signaling during abiotic stress', *The Plant Journal*, 105(2), pp. 459–476.

El-Esawi, M.A., *et al.* (2018a) 'Bacillus firmus (SW5) augments salt tolerance in soybean (*Glycine max L.*) by modulating root system architecture, antioxidant defense systems and stress-responsive genes expression', *Plant Physiology and Biochemistry*, 132, pp. 375–384.

El-Esawi, M.A., *et al.* (2018b) '*Serratia liquefaciens* KM4 improves salt stress tolerance in maize by regulating redox potential, ion homeostasis, leaf gas exchange and stress-related gene expression', *International Journal of Molecular Sciences*, 19(11), p. 3310.

Espinas, N.A., Saze, H. and Saijo, Y. (2016) 'Epigenetic control of defense signaling and priming in plants', *Frontiers in Plant Science*, 7, p. 1201.

Evans, M.J., *et al.* (2016) 'A ROS-assisted calcium wave dependent on the AtRBOHD NADPH oxidase and TPC1 cation channel propagates the systemic response to salt stress', *Plant Physiology*, 171(3), pp. 1771–1784.

Fasusi, O.A., Babalola, O.O. and Adejumo, T.O. (2023) 'Harnessing of plant growth-promoting rhizobacteria and arbuscular mycorrhizal fungi in agroecosystem sustainability', *CABI Agriculture and Bioscience*, 4, p. 26. https://doi.org/10.1186/s43170-023-00168-0

Fernandes, J., *et al.* (2008) 'Distinctive transcriptome responses to adverse environmental conditions in *Zea mays* L', *Plant Biotechnology Journal*, 6(8), pp. 782–798.

Gagné-Bourque, F., *et al.* (2015) 'Accelerated growth rate and increased drought stress resilience of the model grass *Brachypodium distachyon* colonized by bacillus subtilis B26', *PLoS One*, 10(6), p. e0130456.

Gamir, J., Sánchez-Bel, P. and Flors, V. (2014) 'Molecular and physiological stages of priming: How plants prepare for environmental challenges', *Plant Cell Reports*, 33, pp. 1935–1949.

Gehring, M. and Henikoff, S. (2007) 'DNA methylation dynamics in plant genomes', *Biochimica Et Biophysica Acta (BBA)-Gene Structure and Expression*, 1769(5–6), pp. 276–286.

Getnet, Z., *et al.* (2015) 'Growth, water status, physiological, biochemical and yield response of stay green sorghum {*Sorghum bicolor* (L.) Moench} varieties-a field trial under drought-prone area in Amhara regional state, Ethiopia', *Journal of Agronomy*, 14, pp. 188–202.

Glick, B.R. (2014) 'Bacteria with ACC deaminase can promote plant growth and help to feed the world', *Microbiological Research*, 169(1), pp. 30–39.

Hilker, M. and Schmülling, T. (2019) 'Stress priming, memory, and signalling in plants', *Plant, Cell & Environment*, 42(3), pp. 753–761.

Hou, Y., *et al.* (2022) '*Bacillus atrophaeus* WZYH01 and *Planococcus* soli WZYH02 improve salt tolerance of maize (*Zea mays* L.) in saline soil', *Frontiers in Plant Science*, 13, p. 891372.

Hu, Y., *et al.* (2012) 'Cold stress selectively unsilences tandem repeats in heterochromatin associated with accumulation of H3K9ac', *Plant, Cell & Environment*, 35(12), pp. 2130–2142.

Husen, A. (2010) 'Growth characteristics, physiological and metabolic responses of teak (*Tectona grandis* Linn. f.) Clones differing in rejuvenation capacity subjected to drought stress', *Silvae Genetica*, 59, pp. 124–136.

Husen, A. (2021a) 'The harsh environment and resilient plants: An overview', in Husen A. (ed.) *Harsh environment and plant resilience*. Cham, Switzerland: Springer International Publishing AG, pp. 1–23. https://doi.org/10.1007/978-3-030-65912-7_1

Husen, A. (2021b) 'Morpho-anatomical, physiological, biochemical and molecular responses of plants to air pollution', in Husen A. (ed.) *Harsh environment and plant resilience*. Cham, Switzerland: Springer International Publishing AG, pp. 203–234. https://doi.org/10.1007/978-3-030-65912-7_9

Husen, A. (2021c) 'Cross talk between autophagy and hormones for abiotic stress tolerance in plants', in Husen A (ed.) *Plant performance under environmental stress*. Cham, Switzerland: Springer International Publishing AG, pp. 1–15. https://doi.org/10.1007/978-3-030-78521-5_1

Husen, A. (2021d) *Harsh environment and plant resilience: Molecular and functional aspects*. Cham, Switzerland: Springer Nature Switzerland AG. https://doi.org/10.1007/978-3-030-65912-7

Husen, A. (2021e) *Plant performance under environmental stress: Hormones, biostimulants and sustainable plant growth management*. Cham, Switzerland: Springer Nature Switzerland AG. https://doi.org/10.1007/978-3-030-78521-5

Husen, A. and Iqbal, M. (2023) *Medicinal plants - Their response to abiotic stress*. Singapore: Springer Nature Singapore Pte Ltd. https://doi.org/10.1007/978-981-19-5611-9

Husen, A., Iqbal, M. and Aref, I. (2014) 'Growth, water status and leaf characteristics of *Brassica carinata* under drought stress and rehydration conditions', *Brazilian Journal of Botany*, 37, pp. 217–227.

Husen, A., Iqbal, M. and Aref, I.M. (2016) 'IAA-induced alteration in growth and photosynthesis of pea (*Pisum sativum* L.) plants grown under salt stress', *Journal of Environmental Biology*, 37, pp. 421–429.

Husen, A., *et al.* (2018) 'Salicylic acid alleviates salinity-caused damage to foliar functions, plant growth and antioxidant system in Ethiopian mustard (*Brassica carinata* A. Br.)', *Agriculture & Food Security*, 7(44), pp. 1–14.

Husen, A., *et al.* (2019) 'Modulation of salt-stress tolerance of niger (*Guizotia abyssinica*), an oilseed plant, by application of salicylic acid', *Journal of Environmental Biology*, 40, pp. 94–104.

Hussein, M., *et al.* (2017) 'Salinity-induced modulation of plant growth and photosynthetic parameters in faba bean *(Vicia faba)* cultivars', *Pakistan Journal of Botany*, 49, pp. 867–877.

Jagodzik, P., *et al.* (2018) 'Mitogen-activated protein kinase cascades in plant hormone signaling', *Frontiers in Plant Science*, 9, p. 1387.

Jiang, C., *et al.* (2014) 'Environmentally responsive genome-wide accumulation of de novo Arabidopsis thaliana mutations and epimutations', *Genome Research*, 24(11), pp. 1821–1829.

Kenchanmane Raju, S.K., *et al.* (2018) 'Epigenomic plasticity of Arabidopsis msh1 mutants under prolonged cold stress', *Plant Direct*, 2(8), p. e00079.

Knight, H. and Knight, M.R. (2001) 'Abiotic stress signalling pathways: Specificity and cross-talk', *Trends in Plant Science*, 6(6), pp. 262–267.

Kumar, S., *et al.* (2017) 'Physiological, biochemical, epigenetic and molecular analyses of wheat (*Triticum aestivum*) genotypes with contrasting salt tolerance', *Frontiers in Plant Science*, 8, p. 1151.

Lämke, J. and Bäurle, I. (2017) 'Epigenetic and chromatin-based mechanisms in environmental stress adaptation and stress memory in plants', *Genome Biology*, 18(1), pp. 1–11.

Lephatsi, M.M., *et al.* (2021) 'Plant responses to abiotic stresses and rhizobacterial biostimulants: Metabolomics and epigenetics perspectives', *Metabolites*, 11(7), p. 457.

Liu, C., et al. (2010) 'Histone methylation in higher plants', *Annual Review of Plant Biology*, 61, pp. 395–420.

Mandal, S., *et al.* (2023) 'Biostimulants and environmental stress mitigation in crops: A novel and emerging approach for agricultural sustainability under climate change', *Environmental Research*, 233, p. 116357. https://doi.org/10.1016/j.envres.2023.116357

Mellidou, I. and Karamanoli, K. (2022) 'Unlocking PGPR-mediated abiotic stress tolerance: What lies beneath', *Frontiers in Sustainable Food Systems*, 6.

Niu, S., *et al.* (2022) 'The osmolyte-producing endophyte *Streptomyces* albido flavus OsiLf-2 induces drought and salt tolerance in rice via a multi-level mechanism', *The Crop Journal*, 10(2), pp. 375–386.

Pagter, M., *et al.* (2017) 'Rapid transcriptional and metabolic regulation of the deacclimation process in cold acclimated Arabidopsis thaliana', *BMC Genomics*, 18(1), pp. 1–17.

Pérez-Rodriguez, M.M., *et al.* (2022) 'Halotolerant native bacteria *Enterobacter 64S1* and *Pseudomonas 42P4* alleviate saline stress in tomato plants', *Physiologia Plantarum*, 174(4), p. e13742.

Rabhi, N.E.H., *et al.* (2018) '*Pseudomonas knackmussii* MLR6, a rhizospheric strain isolated from halophyte, enhances salt tolerance in *Arabidopsis thaliana*', *Journal of Applied Microbiology*, 125(6), pp. 1836–1851.

Sako, K., *et al.* (2016) 'Ky-2, a histone deacetylase inhibitor, enhances high-salinity stress tolerance in *Arabidopsis thaliana*', *Plant and Cell Physiology*, 57(4), pp. 776–783.

Sapre, S., Gontia-Mishra, I. and Tiwari, S. (2018) '*Klebsiella sp.* confers enhanced tolerance to salinity and plant growth promotion in oat seedlings (*Avena sativa*)', *Microbiological Research*, 206, pp. 25–32.

Schwachtje, J., *et al.* (2019) 'Induced, imprinted, and primed responses to changing environments: Does metabolism store and process information?', *Frontiers in Plant Science*, 10, p. 106.

Shahid, M., *et al.* (2022) 'PGPR *Kosakonia Radicincitans* KR-17 increases the salt tolerance of radish by regulating ion-homeostasis, photosynthetic molecules, redox potential, and stressor metabolites', *Frontiers in Plant Science*, 13, p. 919696.

Shi, D., *et al.* (2017) '*Hydrilla* verticillata employs two different ways to affect DNA methylation under excess copper stress', *Aquatic Toxicology*, 193, pp. 97–104.

Sokol, A., *et al.* (2007) 'Up-regulation of stress-inducible genes in tobacco and Arabidopsis cells in response to abiotic stresses and ABA treatment correlates with dynamic changes in histone H3 and H4 modifications', *Planta*, 227, pp. 245–254.

Surdonja, K., *et al.* (2017) 'Increase of DNA methylation at the *HvCKX2. 1* promoter by terminal drought stress in barley', *Epigenomes*, 1(2), p. 9.

Van Oosten, M. J., *et al.* (2017) 'The role of biostimulants and bioeffectors as alleviators of abiotic stress in crop plants', *Chemical and Biological Technologies in Agriculture*, 4(1), p. 12.

Venturi, V. and Keel, C. (2016) 'Signaling in the rhizosphere', *Trends in Plant Science*, 21(3), pp. 187–198.

Wang, P., *et al.* (2015) 'Epigenetic changes are associated with programmed cell death induced by heat stress in seedling leaves of *Zea mays*', *Plant and Cell Physiology*, 56(5), pp. 965–976.

Zhao, Y., *et al.* (2022) 'Physiological, proteomic, and metabolomic analysis provide insights into *Bacillus sp.*-mediated salt tolerance in wheat', *Plant Cell Reports*, 41, pp. 95–118.

Zheng, Y., *et al.* (2016) 'Histone deacetylase HDA9 negatively regulates salt and drought stress responsiveness in Arabidopsis', *Journal of Experimental Botany*, 67(6), pp. 1703–1713.

Zhu, N., *et al.* (2015) 'The R2R3-type MYB gene OsMYB91 has a function in coordinating plant growth and salt stress tolerance in rice', *Plant Science*, 236, pp. 146–156.

7 Biochemical and Physiological Concepts for Plant Improvement

Sana Khalid, Tasveer Zahra Tariq, Behzad Murtaza, Muhammad Younis, Hafiz Muhammad Nadeem, Abdullah A. Al-Kahtani, Nabeel Khan Niazi, and Muhammad Shahid

7.1 INTRODUCTION

Crop production represents a quantitative measure of the yield. Over the last two or three decades, the production of crops has been significantly enhanced, mainly because of an increase in the harvest index, which is presently near its theoretical maximum level. Similarly, plant production can be enhanced by improving the molecular, biochemical, and physiological plant attributes (Wang and Demirer, 2023; Shahid et al., 2017c). This method is considered as a possible second green revolution. Improving any of the biochemical, physiological, and molecular attributes of plants has been reported to be linked directly or indirectly with higher agricultural production (Natasha *et al.*, 2021a). In fact, the total crop production is the result of either exclusive or combined effects of different biochemical and physiological attributes of plants (Shahid *et al.*, 2023b). Therefore, a quantitative understanding of different mechanisms and processes governing agricultural production is of great importance, especially at the cellular, genetic, molecular, and physiological levels.

The important molecular, biochemical, and physiological attributes of plants include photosynthesis, respiration, mineral nutrition, transpiration, seed germination, stomata functioning, dormancy, plant hormone functioning, tropisms, and photomorphogenesis (Figure 7.1). In fact, these and several other physiological, biochemical, and molecular processes operating within plants control the overall and entire growth and development of plants (Shahid, Pinelli and Dumat, 2012). Therefore, any modification (either positive or negative) to these physiological and molecular processes/attributes in plants can significantly affect crop production. Hence, it is critical to comprehend the mechanisms and processes occurring in plants at the genetic, cellular, molecular, and physiological levels.

Similar to all other living things, plants are subjected to various environmental stressors that are capable of severely affecting crop production and food security (Murtaza *et al.*, 2023b; Yusuf *et al.*, 2024). Generally, the environmental stressors are categorized into abiotic and biotic stresses. Abiotic stresses mainly include radiation stress, salt stress, flood stress, drought stress, temperature extreme stress, heavy metal stress, and pollution (Ahmad *et al.*, 2023b; Li *et al.*, 2023a). Owing to their sessile nature, plants cannot avoid or escape from these environmental stresses. Hence, in order to tolerate and counteract these different types of stresses, plants have evolved different kinds of molecular, biochemical, and physiological processes and traits (Bachheti *et al.*, 2021; Getnet *et al.*, 2015; Husen, 2010; Husen, 2021a, 2021b, 2021c, 2021d, 2021e; Husen, Iqbal and Aref, 2014, 2016; 2017, 2019; Hussein *et al.*, 2017; Rafiq *et al.*, 2017). Plant exposure to different environmental stresses activates certain molecular and physiological attributes and processes, thereby stimulating the production of different types of tolerance, adoptive strategies, and mechanisms (Shahid *et al.*, 2017a).

 DOI: 10.1201/9781032630366-7

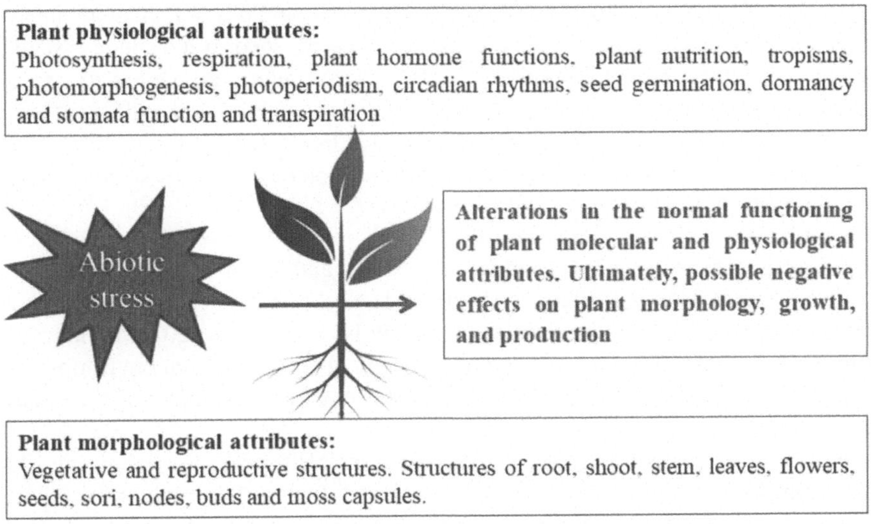

FIGURE 7.1 Plant molecular, biochemical, and physiological attributes that are affected by environmental stresses.

Reactive oxygen species (ROS) are oxygen-containing radicals that are intermediate oxidation-state species produced due to incomplete reduction of molecular oxygen. These species are capable of existing independently with one or more unpaired electrons. The radicals possess high reactivity due to the occurrence of unpaired electrons. The species are produced during normal biological functioning and under environmental stresses (Alharbi *et al.*, 2023; Shahid *et al.*, 2014b). The ROS include hydrogen peroxide (H_2O_2), singlet oxygen ($1O_2$), superoxide ($O_2.^-$), and hydroxyl radicals (·OH).

Plants are constantly and routinely exposed to these radicals. There is a widespread recognition in the literature that ROS can trigger oxidative stress, which harms key components of the cell (including proteins, DNA, and lipids) and hinders plant growth (Taheri, 2022). Overall, ROS can attack and damage almost all the macromolecules present in the cell, thereby leading to the dysfunctioning important metabolic systems, even cell death. Plants have evolved a strong antioxidant defense system to tolerate oxidative damage (Ahmad *et al.*, 2023b; Natasha *et al.*, 2018). This defense system includes non-enzymatic antioxidants such as glutathione and ascorbic acid, as well as different types of enzymes such as catalase (CAT), guaiacol peroxidases (GPOX), peroxidases (POD), superoxide dismutase (SOD), ascorbate peroxidase (APX), and glutathione reductase (GR) (Natasha *et al.*, 2019; Pi *et al.*, 2023). Collectively, these antioxidants scavenge ROS, averting oxidative stress and preserving cellular homeostasis. In addition to these antioxidants, plants have a well-established system to identify and react against environmental stresses. In fact, plants contain complex signaling pathways comprising different types of substances, such as transcription factors, hormones, and signaling proteins (Shalaby, 2024; Yusuf *et al.*, 2024). These substances are responsible for transmitting the messages across plant parts. These pathways generally activate stress-responsive genes, leading to the synthesis of stress-protective proteins and the modulation of cellular processes (Li *et al.*, 2020).

7.2 PLANT PHYSIOLOGY, GENETICS, AND PRODUCTION

Through improvements in plant molecular, biochemical, and physiological attributes and processes, recent scientific and research advancements have made it possible to increase crop productivity. Theoretically, crop production can be increased by optimizing environmental factors and growth conditions, which will enhance plant molecular, biochemical, and physiological attributes and processes. In fact, numerous techniques have recently been established to introduce essential traits in crops for improvement in production (Hu *et al.*, 2023; Khalid *et al.*, 2017). For example, He *et al.* (2013) recently revealed that the overexpression of *GhCIPK6* is correlated with significant increase in tolerance in transgenic *Arabidopsis* to environmental stresses such as drought and salt. They recommend that *GhCIPK6* be considered a potential candidate gene to tolerate environmental stresses by genetic modifications in cotton and other crops.

Moreover, different miRNAs, such as miR398 and miR398b, have been reported to regulate the expression of copper superoxide dismutase (CuSOD1 and CuSOD1), which can scavenge the reactive radicals such as ROS (Chauhan *et al.*, 2023). Hence, miRNAs help tolerate oxidative damage and safeguard photosynthetic plant systems. Recently, Yuan *et al.* (2023) reviewed the role of numerous miRNAs in plant development and tolerance to different types of environmental stresses. They revealed that the expression of miR156 may increase or decrease under different environmental stresses.

Photosynthesis is imperative for plant growth and a critical process for sustaining life on earth. Since photosynthesis is a highly stress-responsive plant function, the environmental stresses can negatively impact this crucial plant process, thereby threatening the food security. Recent advancements in genetic engineering have made it possible to cause an increase in the photosynthetic rate (Li *et al.*, 2023c). Several studies have reported an enhancement in the photosynthetic rate due to genetic modifications. For example, Fan *et al.* (2023) reported that miR171f mediates a key function in *Bermuda Grass* for the transcriptional regulation of LHC1. In this way, the oxidation-reduction properties of photosystem II acceptors are affected. This regulation causes an improvement in the efficacy of photosynthetic electron transport in both PSI and PSII (Fan *et al.*, 2023). Breeding high-yielding cultivars will therefore benefit from the identification of genomic regions linked to high photosynthetic capacity.

A technique known as "ideotype breeding" involves modeling traits beforehand and then introducing them into a model or crop species to determine how they affect yield (Ahn *et al.*, 2023; Zheng *et al.*, 2022). For ideotype breeding to be effectively implemented, understanding of the relationship between genotype and phenotype is necessary. Ideotype breeding is becoming more and more popular as a supplement to traditional breeding methods due to advances in our understanding of the genetic basis of yield-related traits, improved transformation efficiency, and high-throughput genotyping of regenerants (Ahn *et al.*, 2023; Ravasi *et al.*, 2020; Zheng *et al.*, 2022). A crop ideotype is a combination of multiple physiological and morphological attributes that improves yield or performance over existing cultivars. For the purpose of creating ideotypes and breeding high-yielding cultivars, it is now critical to identify crop ideotypic features (Ahn *et al.*, 2023).

7.3 CROP PRODUCTION UNDER STRESS CONDITIONS

Currently, there are about 8.1 billion people on the planet. According to UN estimates, world population could rise to > 9.5 billion by 2050. It means an increased production of agricultural products will be required in future to sustain the ever-increasing world population. One of the biggest obstacles to achieve high agriculture products is environmental change and stress. Regretfully, stressful conditions prevail on a large area of land. Abiotic stresses that harm plant growth include high or low water availability, extremes in temperature, heavy metals, UV radiation, and salinity (Ghasemi *et al.*, 2023; Shahid *et al.*, 2023c). These stresses also result in significant losses in agricultural productivity across the globe (Ahmad *et al.*, 2023b).

The stresses of an ever-increasing human population combined with climate change are likely to make these threats even more serious. Abiotic stress can lower crop yields by an average of more than 60% (Pi *et al.*, 2023; Shahid *et al.*, 2023a). Environmental stresses can reduce up to 50% of crop yield worldwide. The plant experiences stress when the stress threshold is crossed, which stimulates the physiological, biochemical, morphological, and molecular mechanisms (Xu *et al.*, 2023).

Plant homeostasis can be restored, and a new physiological state can emerge when these mechanisms are activated (Mousavi, Pourakbar and Siavash Moghaddam, 2022; Natasha *et al.*, 2020b).

Plants live in constantly changing environments that are usually unsuitable for their growth and development. Abiotic stresses are the major cause of these unfavorable conditions for plant growth and development. Furthermore, it is predicted that abiotic stresses will become more severe and frequent in the near future as a result of industrialization, climate change, and population growth. Thus, frequent flooding and rising temperatures may lead to declining crop yields in several regions of the world (Mesa *et al.*, 2022).

Moreover, human actions may raise the levels of pollutants in the air, water, and soil that plants have to deal with (Murtaza *et al.*, 2023a). It is estimated that abiotic stresses during the growing season can affect plant and crop growth by more than 90%. Plants respond appropriately when they receive a variety of stresses because they activate the stress signals (Natasha *et al.*, 2022b, 2022d; Shah *et al.*, 2022; Shahid *et al.*, 2020). Generally, multiple stresses act concurrently, including the combined effects of heat, salt, water, and heavy metals (Gururani, 2023; Iftikhar *et al.*, 2021). Plant growth, metabolism, and productivity are lowered as a result of these alterations interfering with the source-sink interaction and regular plant metabolism. Additionally, these stresses can alter the expression of numerous plant genes, which greatly affect crop production at a global stage (Alharbi *et al.*, 2023; Hu *et al.*, 2023; Wang *et al.*, 2023).

7.3.1 Heavy Metals Stress

As a result of fast and unchecked industrialization and urbanization, environmental contamination by heavy metals such as cadmium, arsenic, lead, mercury, and zinc is getting worse (Shen *et al.*, 2024; Shi *et al.*, 2023). Because heavy metals are widely used in the industry, high concentrations of heavy metals are found in soils and other environmental compartments (Li *et al.*, 2023b; Natasha *et al.*, 2022d). For example, 743.77 tons of Cd were released into the environment in 2009 (Zhao *et al.*, 2022). Similarly, about 800,000 tons of Pb and 30,000 tons of Cr have been discharged to the ecosystem around the world during the last five decades (Zhao *et al.*, 2022). Heavy metals are considered the most hazardous to all types of living organisms, including humans and crops, owing to their high environmental prevalence. For example, As, Cd, and Pb are highly toxic heavy metals and remain effective in the biological systems for > 50, 18–35, and 150 years (Moeen-ud-din, Yang and Wang, 2023).

As a result of increased release of heavy metals into the environment, numerous sites have become highly polluted in different regions of the world. For example, > 16% of all the soil samples analyzed in China exceeded the environmental threshold levels for different types of pollutants. The most prominent heavy metal soil contamination was 1.5%, 1.6%, 7%, 1.1%, 4.8%, 2.7%, and 2.1%, respectively, for Pb, Hg, Cd, Cr, Ni, As, and Cu (Shi *et al.*, 2023). Similarly, Jahandari and Abbasnejad (2024) evaluated the soil pollution levels, sources, distribution, environmental risks, and potential health hazards of seven heavy metals (Cd, As,Cu, Ni, Cr, Zn, and Pb) in Iran, based on 31 publications. They reported that mean Ni and Cd levels surpassed their associated permissible limits in soils. Heavy metals are highly persistent in the environment, and they can build up in the soil over time. Once high levels of heavy metals are maintained in soil, living organisms, including plants, humans, and animals, are regularly exposed to these heavy metals (Alothman and Shahid, 2022; Natasha *et al.*, 2022c). In this way, the high levels of heavy metals in soil not only deteriorate its physical and chemical characteristics but also threaten biodiversity and significantly affect crop productivity (Shi *et al.*, 2023).

Plants easily absorb the bioavailable forms of heavy metals from contaminated soils (Natasha *et al.*, 2021b). Plant uptake of heavy metals from soil is affected by different soil and plant factors (Khalid *et al.*, 2023). Plants uptake heavy metals from soil via symplastic or apoplastic pathways. Due to their role in various biological metabolisms, some heavy metals, such as Cu, Co, Ni, Mn, and Zn, are considered as essential micronutrients (Moeen-ud-din *et al.*, 2023). These heavy metals impart numerous key roles in different metabolic mechanisms, thereby having positive effects on crop growth and production. When heavy metals are present across the soil-root interface at supra-optimal levels, they impede metabolic processes that negatively impact their morphology, physiology, biochemistry, and growth (Anwar *et al.*, 2021; Noor *et al.*, 2022). Owing to enhanced production of charged radicals like ROS, crops cultivated on heavy metal-contaminated areas typically experience severe phytotoxicity (Feng *et al.*, 2023; Shahid *et al.*, 2020). These ROS cause the cell's redox state to become unbalanced, along with lipid membrane degradation and disruption of chlorophyll biosynthesis (Rafiq *et al.*, 2017) (Table 7.1).

When heavy metals interact with electron transport activities, especially in the membranes of mitochondria and chloroplasts, ROS are produced (Feng *et al.*, 2023). High ROS levels in plants can upset the cellular redox status, causing oxidative harm to exposed cells that can lead to DNA strand cleavage, biological macromolecule deterioration, membrane disintegration, lipid peroxidation, and ion leakage (Ali, Tyagi and Bae, 2023; Shahid *et al.*, 2023c). On the other hand, the harmful effects of ROS induced by heavy metals on plant macromolecules differ and are contingent upon the length of exposure, stage of plant growth, heavy metal level examined, degree of plant stress, and specific organs examined (Rafiq *et al.*, 2018).

Plants have evolved defense mechanisms to convert ROS into less toxic products, preventing injuries caused by heavy metals (Ali *et al.*, 2023). These defenses include preventing metals from entering plants, promoting metal excretion from roots, preventing toxic metal buildup in delicate tissue, chelating metals with organic ligands, preventing metal binding to cell walls, and sequestering metals in vacuoles (Shabbir *et al.*, 2020). These defense systems aid plants in maintaining their cellular redox status and lessen the harm mediated by oxidative damage. Most of these defense mechanisms rely on antioxidants such as CAT, SOD, APX, POD, GPOX, and GR (Shahid *et al.*, 2021). It is generally perceived that the elevated levels of these antioxidants are associated with stress tolerance against ROS induced by heavy metals.

7.3.2 Salt Stress

In regions where water scarcity and inadequate drainage practices on irrigated farms result in markedly reduced crop yields, salinity of the soil poses a serious threat to agriculture (Gururani, 2023; Iftikhar *et al.*, 2021). According to estimates, salinity affects about half of all irrigated land. The Food and Agricultural Organization (FAO, 2016) reports that salt conditions already affect over 6% of all land worldwide and 19.5% of all irrigated land. Over 900 million hectares of land worldwide are affected by high salt levels. Owing to human activities and climate change, the situation is getting worse (Machado and Serralheiro, 2017; Yusuf *et al.*, 2024). Salinity stress can mediate yield reduction from 5% to 50% or even higher, depending on soil salinity level, sensitivity of crops, and irrigation water management (Ahmad *et al.*, 2023a). This causes a $27.3 billion annual loss (Zhou *et al.*, 2023). It is reported that the salt-affected area in India is > 6.7 million ha (Kumar and Sharma, 2020). Studies have revealed that about 10% extra land is getting being annually, and about 50% of the arable land in India would be salt-affected by 2050 (Kumar and Sharma, 2020). Overall, reports have predicted by 2050, the area in the world affected by salinity will grow to be two to three times larger than it is now (Shao *et al.*, 2023).

Soil salinity can be caused by both natural and human factors. Extremely salinized soil can be caused by either natural (weathering, rain that contains salt from the sea) or man-induced (land clearing, irrigation) processes (Li *et al.*, 2023a; Shalaby, 2024). Human activities are responsible for 76.6 Mha of the 932.2 Mha of salt-affected soils worldwide (Kumar and Sharma, 2020). Higher concentrations of soluble or exchangeable sodium were found in the salt-affected areas, possibly as a result of inadequate cation leaching that forms the base. When soil water has high concentrations of sodium (Na^+) and chloride (Cl^-) ions, plants suffer from salt stress (Dzinyela *et al.*, 2023). In practice, soil can be categorized as saline when its electrical conductivity (EC) is greater than 4 dS m^{-1} and its exchangeable Na^+ content is 15%.

TABLE 7.1
Heavy Metal-Induced Molecular, Biochemical, and Physiological Changes in Plants

Plant	Salt Level	Physiological Changes	Reference
Brassica oleracea and *Zea mays*	Heavy metals	Results revealed an increase in H_2O_2 contents by 61% and 25%, respectively, in the roots of maize and cabbage irrigated with untreated wastewater containing high levels of heavy metals. There was also an enhancement in lipid peroxidation and reduction in pigment contents due to heavy metal toxicity.	Khalid *et al.*, 2022
Cicer arietinum	Heavy metals	Heavy metal stress caused changes in physiological (water-use efficiency, transpiration rate, photosynthesis rate, and stomatal conductance) and biochemical (antioxidant enzymes) parameters.	Kumar *et al.*, 2022
Brassica oleracea	Arsenic	Arsenic treatments caused an increase in H_2O_2 levels by 32% in roots. Arsenic stress induced lipid peroxidation both in plant roots and shoots. There was also an increase in the activities of POD and CAT. However, APX and SOD activities slightly decreased under As stress.	Natasha *et al.*, 2022a
Trifolium alexandrinum	Manganese	Manganese exposure caused membrane damage and oxidative stress by enhancing the levels of superoxide radical, H_2O_2, MDA, methylglyoxal generation, and increased activity of lipoxygenase.	Hafeez *et al.*, 2022
Spinacia oleracea	Lead oxide nanoparticles	Lead exposure induced decrease in dry weight (67%) and pigment contents (38%). Lead also mediated activation in the activities of CAT and POD.	Natasha *et al.*, 2020a
Abelmoschus esculentus	Cadmium	Exposure to Cd alone decreased the leaf area (22%) dissolved sugars (41% and 45%), and enhanced MDA (66% and 76%), H_2O_2 (67% and 53%), proline (52% and 41%), and dissolved proteins (15% and 22%) in leaves and roots, respectively	Mousavi *et al.*, 2022
Vicia faba	Cadmium	Cadmium toxicity enhanced the levels of H_2O_2, which caused lipid peroxidation and decreased pigment contents.	Shahid *et al.*, 2019
Kandelia candel and *Bruguiera gymnorrhiza*	Lead, cadmium, mercury	The activities of SOD and POD in leaves altered under heavy metal stress, while CAT activity increased. The activities of SOD, CAT, and POD in roots of heavy metal-stressed plants all increased and then decreased.	Zhang *et al.*, 2007
Pisum sativum	Cadmium	Cadmium toxicity increased ROS levels in roots and leaves of pea plants. Cadmium-mediated high levels of ROS reduced pigment contents and enhanced lipid peroxidation	Shamshad *et al.*, 2018
Medicago sativa	Heavy metals	The growth, seedling biomass, and chlorophyll contents of heavy metal-treated plants significantly decreased. There was a decrease in CAT activity and an increase in MDA level.	Duan *et al.*, 2018
Spinacia oleracea	Arsenic	Arsenic toxicity caused lipid peroxidation and decreased chlorophyll contents.	Shahid *et al.*, 2017b
Capsicum annuum	Heavy metals	Heavy metal toxicity significantly decreased plant growth by impeding leaf photosynthesis and nutrient uptake.	Altaf *et al.*, 2023
Vicia faba	Lead	Lead exposure enhanced the activities of SOD, GPX, APX, GR, and CAT in the leaves of *V. faba*.	Shahid *et al.*, 2014a

Soil salinity can negatively affect crop growth and productivity. Salinity can draw water from the crops via osmosis process, thereby declining water uptake. High Na levels in soil can alter and deteriorate soil structure. Moreover, there can be particular ion toxicity with specific crop plants. Overall, the detrimental effects of Na^+ on cellular metabolism are the main drawbacks of salt stress on cells. Salt ions, especially Na^+, profoundly inhibit enzyme activity by penetrating the cell through unidentified cationic channels (Song *et al.*, 2022). The complete disruption of metabolic pathways results from this process. Elevated levels of Na^+ and Cl^- impede the assimilation of vital nutrients during salinization (Ahmad *et al.*, 2023a; Song *et al.*, 2022). Tian *et al.* (2022) reported that 75 mM NaCl decreased nitrogen uptake by *Sophora japonica*.

Salinity stress can have numerous toxic effects on plants, thereby provoking reduction in plant growth and production. For example, decreased root and shoot growths were recorded in *Sophora japonica* under NaCl stress (Tian *et al.*, 2022). Similarly, Hou *et al.* (2023) showed that salt stress negatively influenced water relations and decreased hydraulic conductance, stomatal conductance, stomatal size, and carbon isotope discrimination in cotton plants. However, the application of certain soil amendments such as biochars can alleviate salt stress in plants (Hou *et al.*, 2023). One of the major issues of salt stress in plants is water deficit or osmotic stress (Ahmad *et al.*, 2023a; Song *et al.*, 2022). Plants with water deficit have lower cell turgor, spoiled water relations, and less efficient use of water (Wang *et al.*, 2023). Additionally, it causes disruption to photosynthetic pigments and causes stomatal closure (Arif *et al.*, 2020). When exposed to salt stress, the activity of CO_2-fixing enzymes declines and the amount of CO_2 available for fixation is reduced due to the closure of stomata induced by salt (Zhou *et al.*, 2023).

Due to osmotic stress, salt buildup slows down plant growth and reduces the ability of plants to absorb water and nutrients. Plants under salt stress exhibit inhibition of photosynthesis, reduced biomass, cell division, growth, and stomatal closure. Osmotic and oxidative stressors brought on by salt induce a range of cellular, molecular, metabolic, and physiological reactions (Dzinyela *et al.*, 2023; Zhou *et al.*, 2023). However, under mild salt and drought stress, *Zea mays* plants showed enhanced water use efficiency by reducing stomatal conductance via enhanced hydraulic resistance and osmotic adjustment (Liao *et al.*, 2022). In fact, *Zea mays* plants showed reduced net photosynthesis rate by enhancing stomatal limitations rather than decreasing photosynthetic potential (Liao *et al.*, 2022) (Table 7.2).

Different species have varying levels of resistance to salt stress. Cereal crops, like barley, can withstand moderate-to-high salinity levels. Crops that are moderately resistant to salt include bread wheat, maize, and sorghum, while wheat is less salinity-resistant (Yepes *et al.*, 2018). Vegetables that are sensitive to salt, like onions (*Allium cepa*), beans (*Phaseolus vulgaris*), and carrots (*Daucus carota*), can only withstand very low salinities. Major cultivars of main crop plants such as wheat, rice, maize, tomato, and soybean are sensitive to high salinity (Tao *et al.*, 2024). Several studies have also demonstrated that in response to salt stress, plants release root exudates that differ in composition (Kumar, Haldar and Saikia, 2023a). These specific root exudates help plants cope with salinity in the soil by drawing rhizobacteria that lower stress (Kumar *et al.*, 2023a).

Soil salinity can provoke several toxic effects inside the plants. One of the negative effects under salt stress is the overproduction of ROS. The ROS are produced rapidly in plants under salt stress owing to metabolic disturbances (Zhou *et al.*, 2023). As reported earlier, these free radicals can mediate oxidative harm to macromolecules like lipids, nucleic acids, proteins, and carbohydrates because of their high reactivity (Rafiq *et al.*, 2017). This can lead to redox imbalance and oxidative damage to crops (Figure 7.2). Plants grown in salinity generate ROS in a variety of organelles (Li *et al.*, 2023a). When exposed to salt stress, the leakage of electrons from electric transfer chains produces $O_2^{\cdot -}$, which Mn-SOD can then convert to H_2O_2. This is how mitochondrial ROS are produced. Due to increased photorespiration that produces a high level of H_2O_2, peroxisomes act as a site of ROS under salinity stress (Dzinyela *et al.*, 2023; Yusuf *et al.*, 2024).

At the cellular, metabolic, molecular, and physiological levels, salt stress in crops initiates a variety of adaptive responses (Dzinyela *et al.*, 2023; Hou *et al.*, 2023). Direct alterations to cellular

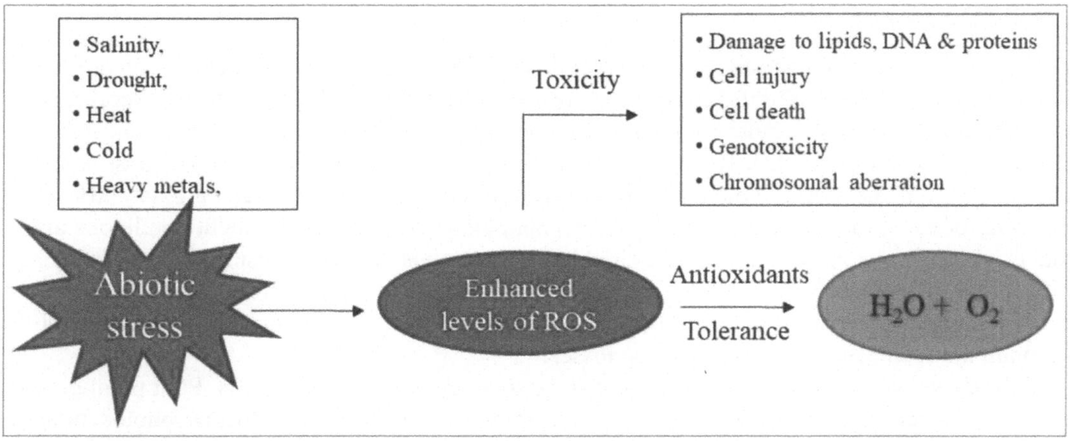

FIGURE 7.2 Possible toxic effects and tolerance mechanisms under abiotic stresses to plants.

processes are necessary for salt stress adaptation due to the changed expression of genes (Dzinyela *et al.*, 2023). Changes brought on by salt stress can cause specific metabolites to accumulate and alter the quantity of specific cellular proteins (Yepes *et al.*, 2018). However, these changes greatly vary with the type of plant species. For example, the expression of miR398 was enhanced in *Arabidopsis*, but increased in *Populus tremula* under salt stress (Jia *et al.*, 2009; Yuan *et al.*, 2023). Moreover, plants have the ability to adjust to salt stress through the regulation of phytohormones, including auxins, gibberellins (GA), ethylene (ET), salicylic acid (SA), jasmonic acid (JA), abscisic acid (ABA), brassinosteroids (BR), and cytokinins (CK) (Dzinyela *et al.*, 2023; Hou *et al.*, 2023). Plants generally tolerate salinity stress via three mechanisms, such as ion exclusion, osmotic tolerance, and tissue tolerance.

TABLE 7.2
Molecular, Biochemical and Physiological Changes in Induced in Plants under Salt Stress

Plant	Salt Level	Physiological Changes	Reference
Pisum sativum	1% sodium hypochlorite solution	Salinity and drought stress induced significant changes in growth, photosynthetic attributes, proline contents, and activity of various antioxidant enzymes.	Yusuf *et al.*, 2024
Lactuca sativa	0, 5%, and 10% moringa leaf extract	NaCl treatments enhanced levels of antioxidants (CAT, APX), relative water content, nutrient level s(N, P, and K), and pigment content.	Shalaby, 2024
Solanum lycopersicum	Not mentioned	Dwarves of SlGT-26-RNAi plants are sensitive to salinity stress in terms of relative water content, MDA, and H_2O_2 contents.	Li *et al.*, 2023a
Triticum aestivum	Not mentioned	Results showed that wild-type plants have lower proline contents and higher MDA content than TaHSP17.4- and TaHOP-overexpressing plants under salt stress.	Wang *et al.*, 2023
Brassica rapa, Salicornia fruticosa, Atriplex halimus, Cakile maritima	0, 100, 200, and 300 mM NaCl	Exposure to NaCl caused negative effects on leaf succulence, water potential, osmotic potential, hydraulic conductivity, mineral contents, protein contents, CAT, and unsaturated fatty acids.	Yepes *et al.*, 2018

7.3.3 Drought Stress

Drought stress represents limited water availability required for crop development. Climate change has greatly affected the rainfall pattern in many areas around the globe. Therefore, several regions are experiencing droughts more frequently than before owing to decreased precipitation (Munné-Bosch and Villadangos, 2023; Riyazuddin *et al.*, 2023). Drought is considered the most common abiotic stress in the world, which can significantly lower grain yield. It can seriously affect the potential of earth to meet the growing global population's food needs. Drought conditions affect plants when there is insufficient water available for their roots or when transpiration losses are significant (Gururani, 2023; Riyazuddin *et al.*, 2023). The degree of damage resulting from a drought is often unpredictable due to a multitude of factors, such as the patterns of rainfall, the soil's ability to retain moisture, and evapotranspiration losses.

Cellular dehydration and a lack of water are associated with drought stress. Poor plant growth and development are caused by a decrease in turgor pressure, stomatal closure, photosynthesis, plant hormonal level, and water potential (Ahmad *et al.*, 2023a; Ahmad *et al.*, 2023b; Munné-Bosch and Villadangos, 2023). For example, Shinohara and Leskovar (2014) reported that drought stress significantly inhibited root growth, transpiration, photosynthetic rate, and stomatal conductance of *Cynara cardunculus*. Reduced plant growth is the result of low-water stress, which affects physiological and biochemical processes such as photosynthesis, respiration, ion acquisition and chlorophyll synthesis, and the metabolism of carbohydrates and other nutrients (Liang *et al.*, 2023).

The degree of drought stress determines how a drought affects yield. As severe drought kills plants, yield loss without irrigation could be 100%. When plants are impacted by drought, their developmental stage also plays a significant role in yield loss (Sandoval *et al.*, 2024). However, recently Liao *et al.* (2022) reported enhanced water use efficiency in *Zea mays* plants due to reduced stomatal conductance via enhancing hydraulic resistance and osmotic adjustment under mild drought salt stress. This indicates that the levels of drought stress, as well as its combination with other stresses, can have different effects on plant growth. However, the mechanisms of stress tolerance may vary for different combinations and levels of stresses as well as plant types. Similarly, the duration of drought stress also governs its potential toxicity to plants. For example, under short-duration drought stress, the water-use efficiency of plants is enhanced at the leaf level, due to primarily lowered transpiration rate and reduced stomatal conductance (Hou *et al.*, 2023). On the other hand, during long-duration drought stress, water-use efficiency of plants increased due to changed stomatal density and size (Hou *et al.*, 2023).

Due to their detrimental effects on plant physiology, growth, and reproduction, severe droughts significantly reduce crop yields (Munné-Bosch and Villadangos, 2023). Limited water availability physiologically impedes vital functions like transpiration and photosynthesis, resulting in decreased growth and productivity. Drought stress has the biochemical potential to produce an excess of ROS, which can result in oxidative stress and possible cellular damage (Wang *et al.*, 2023). In a larger context, it has been stated that 50% of crop losses are attributable to abiotic stresses, with the remaining losses coming from other types of abiotic stressors, 10% coming from drought, and 20% from extremely high temperatures (Kajla *et al.*, 2015; Nazari *et al.*, 2023).

Drought stress generally encourages leaf abscission, slows cell growth, and inhibits seed germination because of low water content, which results in poor seedling growth (Munné-Bosch and Villadangos, 2023). Drought stress also causes damage to the plasma membrane, turgor loss, stomatal closure, leaf wilting, and disruptions to important metabolic processes like photosynthesis. Drought-related reductions in photosynthetic activity and consequently in plant yield are also caused by a significant inhibition of Rubisco's efficiency (Hasnain *et al.*, 2023; Sandoval *et al.*, 2024). ROS are also produced during drought stress, and these can cause oxidative damage to organelles like the nucleus, mitochondria, and chloroplasts as well as the cell membrane (Munné-Bosch and Villadangos, 2023) (Table 7.3).

TABLE 7.3
Molecular, Biochemical and Physiological Changes in Induced in Plants under Drought Stress

Plant	Physiological Changes	Reference
Pisum sativum	Drought stress affected the activity of SOD, CAT, POD, and GR. There was an enhancement in H_2O_2 and MDA contents.	Ahmad *et al.*, 2023b
Solanum lycopersicum	Drought stress induced decreased plant growth, reduced seed germination, decrease in photosynthesis and water contents, and high oxidative stress.	Liang *et al.*, 2023
Cuminum cyminum	The contents of lycopene, beta-carotene, terpenes, phenolic compounds, anthocyanin, flavonoids, and alkaloids were significantly high under drought stress.	Ghasemi *et al.*, 2023
Vaccinium corymbosum	Exposure to 50% field capacity (FC) decreased leaf area, plant water contents, photosynthetic activity, and pigment contents.	Sandoval *et al.*, 2024
Brassica rapa	The roots and leaves of *B. rapa* caused low growth and enhanced Na uptake under drought stress.	Hasnain *et al.*, 2023
Triticum aestivum	Results showed that wild-type plants have lower proline contents and higher MDA contents than TaHSP17.4- and TaHOP-overexpressing plants under drought stress.	Wang *et al.*, 2023
Cicer arietinum	There were high transcript levels (22-times higher) due to drought stress and alterations in physiological (water-use efficiency, transpiration rate, photosynthesis rate, stomatal conductance) and biochemical (antioxidant enzymes) attributes.	Kumar *et al.*, 2022

Drought disrupts photosynthesis, growth, water, and nutrient relations, and eventually causes a significant reduction in crop yields (Riyazuddin *et al.*, 2023). Depending on the plant's growth stage and other environmental factors, different species of plants show different toxicity and tolerance effects under drought stress (Ma, Dias and Freitas, 2020). In order to withstand the severe effects of drought stress, plants exhibit specific alterations in their growth patterns and physiological processes (Riyazuddin *et al.*, 2023). Moreover, the expression of different types of genes is regulated under drought stress. For example, drought stress decreased miR169 expression in *Medicago truncatula* and *Arabidopsis*, while it was reduced in *Oryza sativa* (Sunkar, Li and Jagadeeswaran, 2012; Yuan *et al.*, 2023).

7.3.4 Heat Stress

It is reported that the global temperature will continue to increase, mostly because of enhanced CO_2 release (Mondal *et al.*, 2023). If temperature increases at the current rate, the Intergovernmental Panel on Climate Change predicts that it will likely increase by 1.5°C between 2030 and 2052. With 2.16°C above average temperatures, 2020 was the warmest year on record in Europe. Exposure to temperatures above the optimal range that is high enough to permanently harm a plant's ability to grow and develop is known as heat stress (Mesa *et al.*, 2022). Plants are extremely sensitive to heat stress, and they perish at very high temperatures. Crop growth/developments are significantly influenced by temperatures below and above their optimum, where they typically thrive better (Alharbi *et al.*, 2023; Gururani, 2023). However, the effect of heat stress may vary for different plant parts. For example, Shinohara and Leskovar (2014) reported that heat stress greatly reduced shoot but not root growth of *Cynara cardunculus*. Furthermore, exposure to high temperatures during seed filling may cause accelerated filling, which will ultimately lead to poor quality and a decrease in yield (Tushabe *et al.*, 2023).

It is reported that field crops are primarily negatively impacted by heat stress with high severity; the reproductive stages are more pronounced to heat stress compared to growth stages

(Kumar *et al.*, 2023b). Due to low resource utilization by plants as a result of critical functional process impairment, even brief exposure to heat stress can drastically lower crop productivity (Mondal *et al.*, 2023; Wang *et al.*, 2023). For example, Shinohara and Leskovar (2014) reported that heat stress significantly inhibited transpiration, photosynthetic rate, and stomatal conductance of *Cynara cardunculus*. The impact of heat stress on plant growth greatly varies with exposure temperature levels and plant species. For example, Lysenko *et al.* (2023) treated barley and maize plants to moderate (37°C), genuine (42°C), and nearly lethal (46°C) degrees of heat stress for 48 hours. Under 46°C temperature, the plant growth was severely decreased. At high heat stress, maize plants survived by increasing root water uptake. On the other hand, barley plants died because they were unable to enhance root water uptake under heat stress. This shows that the type of plant greatly governs the toxic effects of heat stress on plants. Moreover, the air humidity also mediates a key role under heat stress conditions (Lysenko *et al.*, 2023). The combined heat and drought stress can also significantly affect plant growth (Shinohara and Leskovar, 2014).

By deactivating key enzymes, heat stress can seriously harm the membrane systems of essential organs and the proteins embedded in them, and disrupt their biosynthetic pathways (Kumar *et al.*, 2023b). Additionally, heat stress impedes noxious impact on cell division and growth and can alter membrane fluidity (Ahmad *et al.*, 2023a; Hu *et al.*, 2023). Such damage has the potential to severely limit plant growth, which encourages oxidative harms. However, the balance between the generation and elimination of ROS by a number of enzymatic and non-enzymatic detoxification processes is what allows the plant to survive oxidative stress (Mondal *et al.*, 2023).

Additionally, crop design for heat tolerance is aided by the knowledge of numerous plant mechanisms at various levels, including metabolites, proteins, genes, transcripts, ions, and phenotypes (Mondal *et al.*, 2023). Elevated temperatures have a direct impact on the vegetative stages of plants and reproductive processes, resulting in a significant decrease in yields (Alharbi *et al.*, 2023). Cell damage and death are accelerated by extremely high temperatures. A plant under heat stress during its vegetative phase of growth may experience a decrease in photosynthesis and a subsequent reduction in biomass (Mesa *et al.*, 2022). Increased temperatures have a major impact on the reproductive stages of development. Temperatures above 29.5°C during the time from bolting to the end of flowering reduce the number of flowers, pods, seed weight, and overall yield of *Brassica* species (Kumar *et al.*, 2023b).

The distribution and growing seasons of agricultural crops may alter as a result of the rise in temperature (Xu *et al.*, 2023). When a plant is exposed to elevated temperature, it may suffer numerous toxic effects, such as protein denaturation and misfolding, deactivation of key enzymes, deterioration of membranes, and disruption of cellular structural components (Mohan *et al.*, 2023; Xu *et al.*, 2022). All of this damage has the potential to significantly restrict plant growth and encourage oxidative damage. However, plants respond in various ways to heat stress. For example, Xu *et al.* (2022) identified SlBBX17, which plays a key role in regulating plant growth under heat stress in *Solanum lycopersicum*. They reported that SlBBX17-OE tomato plants showed enhanced tolerance to heat stress in terms of higher antioxidant enzyme activities, better membrane stability, and lower ROS level. Moreover, overexpression of SlBBX17 modulated the expression of different genes, which govern heat stress, photosynthesis, gibberellin biosynthetic mechanisms, ROS production, and other biochemical reactions (Table 7.4).

7.3.5 Cold Stress

In addition to heat stress, plants may also experience cold stress under low-temperature conditions. Cold stress is also considered an important abiotic stress, which significantly hinders crop production. In fact, exposure to cold conditions can decrease crop postharvest life and quality (Gururani, 2023; Gusain, Joshi and Joshi, 2023). Chilling between 0°C and 15°C and freezing at 0°C are known as cold stress, and it has a detrimental effect on plant growth and agricultural productivity (Verma *et al.*, 2022). Freezing stress is significantly more detrimental to plants than chilling stress. A number of important crops are still unable to adapt to the cold (Ding *et al.*, 2022; Wang *et al.*, 2022). Important crop plants can become cold-tolerant, but the process of acclimatization is

TABLE 7.4
Molecular, Biochemical and Physiological Changes in Induced in Plants under Heat Stress

Plant	Physiological Changes	Reference
Solanum lycopersicum	High temperatures caused an alteration in various physiological parameters, especially fruit production. The antioxidant contents in leaves and fruits were increased.	Mesa *et al.*, 2022
Thymus vulgaris	The transpiration rate, photosynthetic pigments, water potential, and plant nutritional values increased under high temperature. There was also an increase in cell electrolytic leakage and ROS contents.	Alharbi *et al.*, 2023
Rhododendron plants	High temperatures affected ATP contents, electron transport rate, Rubisco activity, and the activities of photosystem I and II.	Xu *et al.*, 2023
Triticum aestivum	Results showed that wild-type plants have lower proline contents and higher MDA contents than TaHSP17.4- and TaHOP-overexpressing plants under heat stress.	Wang *et al.*, 2023
Triticum aestivum	Heat stress augmented the activities of antioxidants and lower levels of MDA and H_2O_2 in the heat-tolerant cultivar ZM7698 than wild-type plants.	Hu *et al.*, 2023

thwarted by abrupt temperature changes, which are becoming more frequent as a result of climate change. Exposure to cold stress can provoke wilting, chlorosis, poor germination, flower abscission, growth retardation, pollen sterility, or decreased fruiting (Ding *et al.*, 2021; Zhang *et al.*, 2021) (Table 7.5).

Under cold stress, the levels of ROS increased in plants, thereby initiating the deterioration of essential plant molecules (Mao *et al.*, 2023). However, the supplementation of plants with melatonin can enhance cold resistance in plants (Mao *et al.*, 2023). In general, plants have evolved different types of strategies and systems to tolerate cold stress, such as altering biochemical/morphological attributes or undergoing dormant phases. Some reports also revealed the role of Ca^{2+} concentration and Ca^{2+} channels under cold stress (Xiang *et al.*, 2024). In addition, plants also have certain biochemical tolerance strategies during cold stress. For example, recently Xiang *et al.* (2024) identified 20 NtDREB homologs in plants and revealed that overexpression of NtDREB_A2.1 can significantly

TABLE 7.5
Molecular, Biochemical, and Physiological Changes in Induced in Plants under Cold Stress

Plant	Physiological Changes	Reference
Solanum lycopersicum	Plants showed cold-mediated stress by increasing ROS levels, electrolyte leakage, and MDA content.	Ding *et al.*, 2022
Camellia sinensis	Cold stress caused a decrease in the levels of primary metabolites (ascorbic acid, amino acids, and carbohydrates). Flavonoids (myricetin, quercetin, kaempferol, and catechin) also showed high levels under cold stress.	Wang *et al.*, 2022
Solanum lycopersicum	Putrescine, an organic compound responsible for unpleasant odors, accumulated under cold stress in tomato plants.	Ding *et al.*, 2021
Solanum lycopersicum	Cold stress damaged cell membranes, enhanced ROS levels, and caused photoinhibition of PSII. The activities of antioxidant enzymes were modified under cold stress.	Zhang *et al.*, 2021
Nicotiana tabacum	Cold stress mediated damage to cell membranes, leaf cells, and stomatal structures. The levels of ROS greatly increased under cold stress, while antioxidant enzyme activity and net photosynthetic rate were greatly reduced.	Pi *et al.*, 2023
Nicotiana tabacum	Identification of 20 NtDREB homologs and enhanced levels of NtDREB_A2.1 improved plant growth under cold stress.	Xiang *et al.*, 2024

enhance plant growth under cold stress. Similarly, the expression of numerous miRNAs (such as miR172, miR171, miR169 and miR408) was enhanced in *Arabidopsis* under cold stress (Liu *et al.*, 2008; Yuan *et al.*, 2023).

7.4 ROLE OF HORMONES AND GROWTH REGULATORS IN PLANT IMPROVEMENT

Numerous plant hormones and growth regulators have been identified that can control stress conditions and improve plant growth (Khan *et al.*, 2023; Zahid *et al.*, 2023). In order to improve stress tolerance and increase plant survival, these hormones function as chemical messengers, coordinating a variety of physiological and biochemical processes. A well-established network of synergistic interactions and crosstalk between various hormones governs the hormonal regulation under environmental stresses (Khan *et al.*, 2023; Sati *et al.*, 2023; Zahid *et al.*, 2023). Plants can adjust their responses to particular stressors and enhance their survival tactics thanks to these interactions.

Abscisic acid (ABA) is a crucial phytohormone that simultaneously controls a number of physiological functions, including protein storage and stomatal opening, and functions as a signaling mediator for a number of environmental stressors, including drought, salt, and cold stress (Basu and Rabara, 2017; Park *et al.*, 2021). ABA triggers physiological events like stomatal closure, bud dormancy, and seed germination (Dar *et al.*, 2017). The transcriptional and posttranscriptional regulation of genes linked to stress responses is also affected. It is necessary to comprehend the signaling pathways in order to shed light on ABA biosynthesis and its action. Studies have identified numerous locations for the biosynthesis of ABA, plasma membrane transporter proteins, and downstream elements of ABA signaling pathways to form intricate networks (Dar *et al.*, 2017; Segarra-Medina *et al.*, 2023). Moreover, ABA coordinates hormonal crosstalk between multiple stress-signaling pathways as the second messenger, thereby leading to adaptations to changing physiological and environmental conditions (Segarra-Medina *et al.*, 2023).

In addition to being a direct regulator of plant growth and development, SA also functions as a tolerance mechanism against a variety of abiotic stresses, such as salinity and heavy metal toxicity (Samanta and Roychoudhury, 2023). By controlling SA-mediated defense genes, metabolite accumulation, redox homeostasis, and photosynthetic yield, SA is crucial in helping plants withstand abiotic stresses (Shaukat *et al.*, 2022). When plants are supplemented with SA under abiotic stresses, there is possible regulation of genes, osmolytes, proteins associated with heat shocks, pathogenesis-related proteins, antioxidants, and other metabolites. By lowering the generation of ROS, both exogenous and endogenous applications of SA improved plants' resistance to temperature, drought, salt, and metal stress (Park *et al.*, 2021). In addition to its role in stress resistance and defense-related gene induction in abiotically stressed plants, SA doses affect plants' physiological and biochemical attributes.

Plants are subject to a variety of physiological effects from the pleiotropic molecule melatonin (Ahmad *et al.*, 2023a; Menhas *et al.*, 2023; Samanta and Roychoudhury, 2023). Pleiotropy is the capacity of a single molecule to have multiple effects within an organism, thereby enabling it to be a multifunctional molecule that can affect multiple physiological processes at once (Khattak *et al.*, 2023). Although melatonin has been known for the last two decades, it is only now that its role as a plant hormone with a variety of physiological activities has come to light (Khattak *et al.*, 2023). The application of melatonin has been reported to enhance cold resistance in plants (Mao *et al.*, 2023). Since melatonin is involved in a variety of stress responses and development processes, research into how it functions in plants, particularly horticultural plants, is expanding quickly.

Numerous studies have demonstrated that phytomelatonin functions as a biostimulant in plants, increasing their resistance to a range of abiotic stressors such as alkalinity, acid rain, ozone, heavy metals, drought, osmotic disturbance, and extreme temperatures (Ahmad *et al.*, 2023a; Khattak *et al.*, 2023). Melatonin appears to play roles in the scavenging of free radicals, including ROS (Menhas *et al.*, 2023). It also influences plant metabolism, both primary and secondary, controls transcription factors and enzyme transcripts related to stress, and interacts with other hormones in response to various environmental stimuli (Samanta and Roychoudhury, 2023). Because of its pleiotropy, phytomelatonin is a desirable regulator that increases plants' resistance to abiotic stress. Studies have revealed that this molecule serves as a strong biostimulator, antioxidant, and growth regulator. It facilitates photosynthesis, delays leaf senescence, and increases the antioxidant enzyme system through direct ROS scavenging under abiotic stresses (Sharma *et al.*, 2024).

7.5 CONCLUSIONS AND FUTURE PROSPECTS

There is abundant data available regarding the role of biochemical, physiological, and molecular processes and attributes toward crop growth and production. This chapter is based on numerous research studies about possible alterations in biochemical, physiological, and molecular processes and attributes of plants under abiotic stresses. Studies have revealed that the possible occurrence of abiotic stresses has increased during the last two to three decades. This is due to increased population, industrialization urbanization, and climate change. Under severe abiotic stress levels, the normal metabolic functioning of plants is disturbed due to alterations in biochemical, physiological, and molecular processes and attributes. Consequently, plant growth is affected, and a serious decline occurs in crop production. However, plant species have evolved different types of tolerance and detoxification mechanisms to cope with different types of environmental stresses, including abiotic challenges. These mechanisms include different types of antioxidants, hormones, and growth regulators. However, the intensity of activation of these tolerance and detoxification mechanisms varies with the type of plant species. There are certain plant species which can actively tolerate extreme stress conditions due to a well-established tolerance mechanism. Moreover, with recent advancements in genetic manipulation, it is now possible to establish gene-manipulation strategies, comprehend gene function, and generate crop plants equipped with desirable attributes to sustain abiotic stresses. However, efforts are still in progress to develop plant species capable of tolerating extreme stress conditions with a capacity to produce high yield.

REFERENCES

Ahmad, N. *et al.* (2023a) 'From challenges to solutions: The impact of melatonin on abiotic stress synergies in horticultural plants via redox regulation and epigenetic signaling', *Scientia Horticulturae*, 321, p. 112369.

Ahmad, R. *et al.* (2023b) 'Exogenous melatonin confers drought stress by promoting plant growth, photosynthetic efficiency and antioxidant defense system of pea (*Pisum sativum* L.)', *Scientia Horticulturae*, 322, p. 112431.

Ahn, E. *et al.* (2023) 'Ideotype breeding and genome engineering for legume crop improvement', *Current Opinion in Biotechnology*, 82, p. 102961.

Alharbi, K. *et al.* (2023) 'Synergistic effect of β-sitosterol and biochar application for improving plant growth of *Thymus vulgaris* under heat stress', *Chemosphere*, 340, p. 139832.

Ali, S., Tyagi, A. and Bae, H. (2023) 'ROS interplay between plant growth and stress biology: Challenges and future perspectives', *Plant Physiology and Biochemistry*, 203, p. 108032.

Alothman, Z.A. and Shahid, M. (2022) 'Recent advances in removal of toxic elements from water using MOFs: A critical review', *Arabian Journal of Chemistry*, 15, p. 104319.

Altaf, M.A. *et al.* (2023) 'Melatonin enhanced the heavy metal-stress tolerance of pepper by mitigating the oxidative damage and reducing the heavy metal accumulation', *Journal of Hazardous Materials*, 454, p. 131468.

Anwar, H. *et al.* (2021) 'Risk assessment of potentially toxic metal(loid)s in *Vigna radiata* L. under wastewater and freshwater irrigation', *Chemosphere*, 265, p. 129124.

Arif, Y. *et al.* (2020) 'Salinity induced physiological and biochemical changes in plants: An omic approach towards salt stress tolerance', *Plant Physiology and Biochemistry*, 156, pp. 64–77.

Bachheti, A. *et al.* (2021) 'Medicinal plants and their pharmaceutical properties under adverse environmental conditions', in Husen, A. (ed.) *Harsh environment and plant resilience*. Cham, Switzerland: Springer International Publishing AG, pp. 457–502. https://doi.org/10.1007/978-3-030-65912-7_19

Basu, S. and Rabara, R. (2017) 'Abscisic acid – An enigma in the abiotic stress tolerance of crop plants', *Plant Gene*, 11, pp. 90–98.

Chauhan, J. *et al.* (2023) 'Plant photosynthesis under abiotic stresses: Damages, adaptive, and signaling mechanisms', *Plant Stress*, 10, p. 100296.

Dar, N.A. *et al.* (2017) 'Abscisic acid: A key regulator of abiotic stress tolerance in plants', *Plant Gene*, 11, pp. 106–111.

Ding, F. *et al.* (2021) 'Jasmonic acid-regulated putrescine biosynthesis attenuates cold-induced oxidative stress in tomato plants', *Scientia Horticulturae*, 288, p. 110373.

Ding, F. *et al.* (2022) 'A jasmonate-responsive glutathione S-transferase gene SlGSTU24 mitigates cold-induced oxidative stress in tomato plants', *Scientia Horticulturae*, 303, p. 111231.

Duan, C. *et al.* (2018) 'Reveal the response of enzyme activities to heavy metals through in situ zymography', *Ecotoxicology and Environmental Safety*, 156, pp. 106–115.

Dzinyela, R. *et al.* (2023) 'Advanced study of functional proteins involved in salt stress regulatory pathways in plants', *South African Journal of Botany*, 159, pp. 425–438.

Fan, S. *et al.* (2023) 'Salt-responsive bermudagrass microRNAs and insights into light reaction photosynthetic performance', *Frontiers in Plant Science*, 14, p. 1141295.

Feng, D. *et al.* (2023) 'Heavy metal stress in plants: Ways to alleviate with exogenous substances', *Science of The Total Environment*, 897, p. 165397.

Getnet, Z. *et al.* (2015) 'Growth, water status, physiological, biochemical and yield response of stay green sorghum {*Sorghum bicolor* (L.) Moench} varieties-a field trial under drought-prone area in Amhara regional state, Ethiopia', *Journal of Agronomy*, 14, pp. 188–202.

Ghasemi, S. *et al.* (2023) 'Drought stress-mediated alterations in secondary metabolites and biosynthetic gene expression in cumin plants: Insights from gene-specific and metabolite-level analyses', *Plant Stress*, 10, p. 100241.

Gururani, M.A. (2023) 'Photobiotechnology for abiotic stress resilient crops: Recent advances and prospects', *Heliyon*, 9, p. e20158.

Gusain, S., Joshi, S. and Joshi, R. (2023) 'Sensing, signalling, and regulatory mechanism of cold-stress tolerance in plants', *Plant Physiology and Biochemistry*, 197, p. 107646.

Hafeez, A. *et al.* (2022) 'Effects of exogenous taurine on growth, photosynthesis, oxidative stress, antioxidant enzymes and nutrient accumulation by *Trifolium alexandrinum* plants under manganese stress', *Chemosphere*, 308, p. 136523.

Hasnain, Z. *et al.* (2023) 'Elucidating role of melatonin foliar spray in ameliorating adverse effects of drought stress on growth and physio-biochemical attributes of *Brassica rapa* plants', *Scientia Horticulturae*, 321, p. 112336.

He, L. *et al.* (2013) 'Molecular cloning and functional characterization of a novel cotton CBL-interacting protein kinase gene (GhCIPK6) reveals its involvement in multiple abiotic stress tolerance in transgenic plants', *Biochemical and Biophysical Research Communications*, 435, pp. 209–215.

Hou, J. *et al.* (2023) 'Combined application of biochar and partial root-zone drying irrigation improves water relations and water use efficiency of cotton plants under salt stress', *Agricultural Water Management*, 290, p. 108584.

Hu, H. *et al.* (2023) 'Lipidomics-based insights into the physiological mechanism of wheat in response to heat stress', *Plant Physiology and Biochemistry*, 205, p. 108190.

Husen, A. (2010) 'Growth characteristics, physiological and metabolic responses of teak (*Tectona grandis* Linn. f.) clones differing in rejuvenation capacity subjected to drought stress', *Silvae Genetica*, 59, pp. 124–136.

Husen, A. (2021a) 'The harsh environment and resilient plants: An overview', in Husen, A. (ed.) *Harsh environment and plant resilience*. Cham, Switzerland: Springer International Publishing AG, pp. 1–23. https://doi.org/10.1007/978-3-030-65912-7_1

Husen, A. (2021b) 'Morpho-anatomical, physiological, biochemical and molecular responses of plants to air pollution', in Husen, A. (ed.) *Harsh environment and plant resilience*. Cham, Switzerland: Springer International Publishing AG, pp. 203–234. https://doi.org/10.1007/978-3-030-65912-7_9

Husen, A. (2021c) 'Cross talk between autophagy and hormones for abiotic stress tolerance in plants', in Husen, A. (ed.) *Plant performance under environmental stress*. Cham, Switzerland: Springer International Publishing AG, pp. 1–15. https://doi.org/10.1007/978-3-030-78521-5_1

Husen, A. (2021d) *Harsh environment and plant resilience: Molecular and functional aspects*. Cham, Switzerland: Springer Nature Switzerland AG. https://doi.org/10.1007/978-3-030-65912-7

Husen, A. (2021e) *Plant performance under environmental stress: Hormones, biostimulants and sustainable plant growth management*. Cham, Switzerland: Springer Nature Switzerland AG. https://doi.org/10.1007/978-3-030-78521-5

Husen, A. and Iqbal, M. (2023) *Medicinal plants: Their response to abiotic stress*. Singapore: Springer Nature Singapore. https://doi.org/10.1007/978-981-19-5611-9

Husen, A., Iqbal, M. and Aref, I.M. (2016) 'IAA-induced alteration in growth and photosynthesis of pea (*Pisum sativum* L.) plants grown under salt stress', *Journal of Environmental Biology*, 37, pp. 421–429.

Husen, A., Iqbal, M. and Aref, I.M. (2014) 'Growth, water status and leaf characteristics of *Brassica carinata* under drought stress and rehydration conditions', *Brazilian Journal of Botany*, 37, pp. 217–227.

Husen, A. *et al.* (2019) 'Modulation of salt-stress tolerance of Niger (*Guizotia abyssinica*), an oilseed plant, by application of salicylic acid', *Journal of Environmental Biology*, 40, pp. 94–104.

Hussein, M. *et al.* (2017) 'Salinity-induced modulation of plant growth and photosynthetic parameters in faba bean *(Vicia faba)* cultivars', *Pakistan Journal of Botany*, 49, pp. 867–877.

Iftikhar, A. *et al.* (2021) 'Salinity modulates lead (Pb) tolerance and phytoremediation potential of quinoa: A multivariate comparison of physiological and biochemical attributes', *Environmental Geochemistry and Health*, 44, pp. 257–272.

Jahandari, A. and Abbasnejad, B. (2024) 'Environmental pollution status and health risk assessment of selective heavy metal(oid)s in Iran's agricultural soils: A review', *Journal of Geochemical Exploration*, 256, p. 107330.

Jia, X. *et al.* (2009) 'Differential and dynamic regulation of miR398 in response to ABA and salt stress in *Populus tremula* and *Arabidopsis thaliana*', *Plant Molecular Biology*, 71, pp. 51–59.

Kajla, M. *et al.* (2015) 'Increase in wheat production through management of abiotic stresses: A review', *Journal of Applied and Natural Science*, 7, pp. 1070–1080.

Khalid, S. *et al.* (2023) 'Predicting chemical speciation of metals in soil using visual minteq', *Soil Ecology Letters*, 5, p. 220162.

Khalid, S. *et al.* (2022) 'Plant physiological responses after fresh and sewage water irrigation: Plant health perspectives', *Gesunde Pflanzen*, 75, pp. 1289–1296.

Khalid, S. *et al.* (2017) 'A comparison of technologies for remediation of heavy metal contaminated soils', *Journal of Geochemical Exploration*, 182, pp. 247–268.

Khan, M.I.R. *et al.* (2023) 'Defensive role of plant hormones in advancing abiotic stress-resistant rice plants', *Rice Science*, 30, pp. 15–35.

Khattak, W.A. *et al.* (2023) 'Melatonin alleviating drought stress in plants: A review', *South African Journal of Botany*, 161, pp. 192–201.

Kumar, N., Haldar, S. and Saikia, R. (2023a) 'Root exudation as a strategy for plants to deal with salt stress: An updated review', *Environmental and Experimental Botany*, 216, p. 105518.

Kumar, P. and Sharma, P.K. (2020) 'Soil salinity and food security in India', *Frontiers in Sustainable Food Systems*, 4, p. 533781.

Kumar, S. *et al.* (2023b) 'Unveiling the impact of heat stress on seed biochemical composition of major cereal crops: Implications for crop resilience and nutritional value', *Plant Stress*, 9, p. 100183.

Kumar, S. *et al.* (2022) 'Metallothionein (MT1): A molecular stress marker in chickpea enhances drought and heavy metal stress adaptive efficacy in transgenic plants', *Environmental and Experimental Botany*, 199, p. 104871.

Li, F. *et al.* (2023a) 'Down-regulation of SlGT-26 gene confers dwarf plants and enhances drought and salt stress resistance in tomato', *Plant Physiology and Biochemistry*, 203, p. 108053.

Li, R. *et al.* (2023b) 'Heavy metal contamination in Shanghai agricultural soil', *Heliyon*, 9, p. e22824.

Li, X. *et al.* (2020) 'Biotic and abiotic stress-responsive genes are stimulated to resist drought stress in purple wheat', *Journal of Integrative Agriculture*, 19, pp. 33–50.

Li, Y. *et al.* (2023c) 'Variations in phenological, physiological, plant architectural and yield-related traits, their associations with grain yield and genetic basis', *Annals of Botany*, 131, pp. 503–519.

Liang, Y. *et al.* (2023) 'Alteration in the expression of tomato sucrose transporter gene SlSUT4 modulates sucrose subcellular compartmentation and affects responses of plants to drought stress', *Environmental and Experimental Botany*, 215, p. 105506.

Liao, Q. *et al.* (2022) 'Mild water and salt stress improve water use efficiency by decreasing stomatal conductance via osmotic adjustment in field maize', *Science of The Total Environment*, 805, p. 150364.

Liu, H.-H. *et al.* (2008) 'Microarray-based analysis of stress-regulated microRNAs in Arabidopsis thaliana', *RNA*, 14, pp. 836–843.

Lysenko, E.A. *et al.* (2023) 'Lower air humidity reduced both the plant growth and activities of photosystems I and II under prolonged heat stress', *Plant Physiology and Biochemistry*, 194, pp. 246–262.

Ma, Y., Dias, M.C. and Freitas, H. (2020) 'Drought and salinity stress responses and microbe-induced tolerance in plants', *Frontiers in Plant Science*, 11, p. 591911.

Machado, R.M.A. and Serralheiro, R. P. (2017) 'Soil salinity: Effect on vegetable crop growth. Management practices to prevent and mitigate soil salinization', *Horticulturae*, 3, p. 30.

Mao, K. *et al.* (2023) 'Melatonin treatment promotes cold adaptation and spring growth of tea plants', *Industrial Crops and Products*, 200, p. 116834.

Menhas, S. *et al.* (2023) 'Pleiotropic melatonin-mediated responses on growth and cadmium phytoextraction of *Brassica napus*: A bioecological trial for enhancing phytoremediation of soil cadmium', *Journal of Hazardous Materials*, 457, p. 131862.

Mesa, T. *et al.* (2022) 'Differential physiological response to heat and cold stress of tomato plants and its implication on fruit quality', *Journal of Plant Physiology*, 268, p. 153581.

Moeen-ud-din, M., Yang, S. and Wang, J. (2023) 'Auxin homeostasis in plant responses to heavy metal stress', *Plant Physiology and Biochemistry*, 205, p. 108210.

Mohan, N. *et al.* (2023) 'Acclimation response and management strategies to combat heat stress in wheat for sustainable agriculture: A state-of-the-art review', *Plant Science*, 336, p. 111834.

Mondal, S. *et al.* (2023) 'Crucial plant processes under heat stress and tolerance through heat shock proteins', *Plant Stress*, 10, p. 100227.

Mousavi, A., Pourakbar, L. and Siavash Moghaddam, S. (2022) 'Effects of malic acid and EDTA on oxidative stress and antioxidant enzymes of okra (*Abelmoschus esculentus* L.) exposed to cadmium stress', *Ecotoxicology and Environmental Safety*, 248, p. 114320.

Munné-Bosch, S. and Villadangos, S. (2023) 'Cheap, cost-effective, and quick stress biomarkers for drought stress detection and monitoring in plants', *Trends in Plant Science*, 28, pp. 527–536.

Murtaza, B. *et al.* (2023a) 'The study of removal chromium (VI) ions from aqueous solution by bimetallic ZnO/FeO nanocomposite with Siltstone: Isotherm, kinetics and reusability', *Inorganic Chemistry Communications*, 154, p. 110891.

Murtaza, B. *et al.* (2023b) 'Chromium removal from aqueous solution using bimetallic Bi0/Cu0-based nanocomposite biochar', *Environmental Geochemistry and Health*, 45(12), pp. 1–14.

Natasha *et al.* (2020a) 'Risk assessment and biophysiochemical responses of spinach to foliar application of lead oxide nanoparticles: A multivariate analysis', *Chemosphere*, 245, p. 125605.

Natasha *et al.* (2020b) 'A critical review of mercury speciation, bioavailability, toxicity and detoxification in soil-plant environment: Ecotoxicology and health risk assessment', *Science of The Total Environment*, 711, p. 134749.

Natasha *et al.* (2022a) 'Arsenic-induced oxidative stress in *Brassica oleracea*: Multivariate and literature data analyses of physiological parameters, applied levels and plant organ type', *Environmental Geochemistry and Health*, 44, pp. 1827–1839.

Natasha *et al.* (2019) 'Biogeochemistry of antimony in soil-plant system: Ecotoxicology and human health', *Applied Geochemistry*, 106, pp. 45–59.

Natasha *et al.* (2022b) 'Unrevealing arsenic and lead toxicity and antioxidant response in spinach: A human health perspective', *Environmental Geochemistry and Health*, 44, pp. 487–496.

Natasha *et al.* (2018) 'A critical review of selenium biogeochemical behavior in soil-plant system with an inference to human health', *Environmental Pollution*, 234, pp. 915–934.

Natasha, N. *et al.* (2022c) 'Zinc in soil-plant-human system: A data-analysis review', *Science of The Total Environment*, 808, p. 152024.

Natasha, N. *et al.* (2021a) 'Influence of biochar on trace element uptake, toxicity and detoxification in plants and associated health risks: A critical review', *Critical Reviews in Environmental Science and Technology*, 52, pp. 2803–2843.

Natasha, N. *et al.* (2021b) 'Health risks of arsenic buildup in soil and food crops after wastewater irrigation', *Science of The Total Environment*, 772, p. 145266.

Natasha, N. *et al.* (2022d) 'Accumulation pattern and risk assessment of potentially toxic elements in selected wastewater-irrigated soils and plants in Vehari, Pakistan', *Environmental Research*, 214, p. 114033.

Nazari, M. *et al.* (2023) 'Deciphering the response of medicinal plants to abiotic stressors: A focus on drought and salinity', *Plant Stress*, 10, p. 100255.

Noor, I. *et al.* (2022) 'Heavy metal and metalloid toxicity in horticultural plants: Tolerance mechanism and remediation strategies', *Chemosphere*, 303, p. 135196.

Park, S.-H. *et al.* (2021) 'Characterization of salicylic acid- and abscisic acid-mediated photosynthesis, Ca^{2+} and H_2O_2 accumulation in two distinct phases of drought stress intensity in *Brassica napus*', *Environmental and Experimental Botany*, 186, p. 104434.

Pi, K. *et al.* (2023) 'Negative regulation of tobacco cold stress tolerance by NtPhyA', *Plant Physiology and Biochemistry*, 204, p. 108153.

Rafiq, M. *et al.* (2017) 'Comparative effect of calcium and EDTA on arsenic uptake and physiological attributes of *Pisum sativum*', *International Journal of Phytoremediation*, 19, pp. 662–669.

Rafiq, M. *et al.* (2018) 'A comparative study to evaluate efficiency of EDTA and calcium in alleviating arsenic toxicity to germinating and young *Vicia faba* L. seedlings', *Journal of Soils and Sediments*, 18, pp. 2271–2281.

Ravasi, R. A. *et al.* (2020) 'Ideotype definition to adapt legumes to climate change: A case study for field pea in Northern Italy', *Agricultural and Forest Meteorology*, 291, p. 108081.

Riyazuddin, R. *et al.* (2023) 'Unveiling the biosynthesis, mechanisms, and impacts of miRNAs in drought stress resilience in plants', *Plant Physiology and Biochemistry*, 202, p. 107978.

Samanta, S. and Roychoudhury, A. (2023) 'Crosstalk of melatonin with major phytohormones and growth regulators in mediating abiotic stress tolerance in plants', *South African Journal of Botany*, 163, pp. 201–216.

Sandoval, Y. *et al.* (2024) 'Melatonin improves plant water status, photosynthetic performance, and antioxidant defense system in highbush blueberry (*Vaccinium corymbosum* L.) plants subjected to drought stress', *Scientia Horticulturae*, 323, p. 112528.

Sati, H. *et al.* (2023) 'Melatonin: A potential abiotic stress regulator', *Plant Stress*, 10, p. 100293.

Segarra-Medina, C. *et al.* (2023) 'Abscisic acid promotes plant acclimation to the combination of salinity and high light stress', *Plant Physiology and Biochemistry*, 203, p. 108008.

Shabbir, Z. *et al.* (2020) 'Copper uptake, essentiality, toxicity, detoxification and risk assessment in soil-plant environment', *Chemosphere*, 259, p. 127436.

Shah, A. H. *et al.* (2022) 'Risk assessment of trace element accumulation in soil and *Brassica oleracea* after wastewater irrigation', *Environmental Geochemistry and Health*, 45, pp. 8929–8942. https://doi.org/10.1007/s10653-022-01351-4

Shahid, M. *et al.* (2023a) 'Effect of freshwater and wastewater irrigation on buildup of toxic elements in soil and maize crop', *Environmental Monitoring and Assessment*, 195, p. 438.

Shahid, M. *et al.* (2023b) 'Recent trends in environmental sustainability', *Environmental Science and Pollution Research*, 30, pp. 99198–99201. doi:10.1007/s11356-023-29348-1.

Shahid, M. *et al.* (2017a) 'Cadmium bioavailability, uptake, toxicity and detoxification in soil-plant system', *Reviews of Environmental Contamination and Toxicology*, 241, pp. 73–137.

Shahid, M. *et al.* (2023c) 'Spatial distribution and accumulation of arsenic in biological samples and associated health risks by drinking groundwater in Bahawalnagar, Pakistan', *Physics and Chemistry of the Earth, Parts A/B/C*, 130, p. 103397.

Shahid, M. *et al.* (2021) 'Ecotoxicology of heavy metal(loid)-enriched particulate matter: Foliar accumulation by plants and health impacts', *Reviews of Environmental Contamination and Toxicology*, 253, pp. 65–113.

Shahid, M. *et al.* (2020) 'Trace elements-induced phytohormesis: A critical review and mechanistic interpretation', *Critical Reviews in Environmental Science and Technology*, 50, pp. 1984–2015.

Shahid, M., Pinelli, E. and Dumat, C. (2012) 'Review of Pb availability and toxicity to plants in relation with metal speciation; role of synthetic and natural organic ligands', *Journal of Hazardous Materials*, 219–220, pp. 1–12.

Shahid, M. *et al.* (2014a) 'Effect of organic ligands on lead-induced oxidative damage and enhanced antioxidant defense in the leaves of *Vicia faba* plants', *Journal of Geochemical Exploration*, 144, pp. 282–289.

Shahid, M. *et al.* (2014b) 'Heavy-metal-induced reactive oxygen species: Phytotoxicity and physicochemical changes in plants', *Reviews of Environmental Contamination and Toxicology*, 232, pp. 1–44.

Shahid, M. *et al.* (2017b) 'Arsenic accumulation and physiological attributes of spinach in the presence of amendments: An implication to reduce health risk', *Environmental Science and Pollution Research*, 24, pp. 16097–16106.

Shahid, M. *et al.* (2019) 'Comparative effect of organic amendments on physio-biochemical traits of young and old bean leaves grown under cadmium stress: A multivariate analysis', *Environmental Science and Pollution Research*, 26, pp. 11579–11590.

Shahid, M. *et al.* (2017c) 'Chromium speciation, bioavailability, uptake, toxicity and detoxification in soil-plant system: A review', *Chemosphere*, 178, pp. 513–533.

Shalaby, O. A. (2024) 'Moringa leaf extract increases tolerance to salt stress, promotes growth, increases yield, and reduces nitrate concentration in lettuce plants', *Scientia Horticulturae*, 325, p. 112654.

Shamshad, S. *et al.* (2018) 'Effect of organic amendments on cadmium stress to pea: A multivariate comparison of germinating vs young seedlings and younger vs older leaves', *Ecotoxicology and Environmental Safety*, 151, pp. 91–97.

Shao, T. *et al.* (2023) 'Salt stress affects the biomass of industrial crop Jerusalem artichoke by affecting sugar transport and metabolism', *Heliyon*, 9, p. e14107.

Sharma, P. *et al.* (2024) 'Melatonin as plant growth regulator in sustainable agriculture', *Scientia Horticulturae*, 323, p. 112421.

Shaukat, K. *et al.* (2022) 'Chapter 2 - Role of salicylic acid–induced abiotic stress tolerance and underlying mechanisms in plants, in Aftab, T. and Naeem, M. (eds.) *Emerging plant growth regulators in agriculture*. Academic Press, pp. 73–98.

Shen, F. *et al.* (2024) 'A new method for spatial three-dimensional prediction of soil heavy metals contamination', *CATENA*, 235, p. 107658.

Shi, J. *et al.* (2023) 'Spatiotemporal variation of soil heavy metals in China: The pollution status and risk assessment', *Science of The Total Environment*, 871, p. 161768.

Shinohara, T. and Leskovar, D. I. (2014) 'Effects of ABA, antitranspirants, heat and drought stress on plant growth, physiology and water status of artichoke transplants', *Scientia Horticulturae*, 165, pp. 225–234.

Song, X. *et al.* (2022) 'Biochar/vermicompost promotes Hybrid Pennisetum plant growth and soil enzyme activity in saline soils', *Plant Physiology and Biochemistry*, 183, pp. 96–110.

Sunkar, R., Li, Y.-F. and Jagadeeswaran, G. (2012) 'Functions of microRNAs in plant stress responses', *Trends in Plant Science*, 17, pp. 196–203.

Taheri, P. (2022) 'Crosstalk of nitro-oxidative stress and iron in plant immunity', *Free Radical Biology and Medicine*, 191, pp. 137–149.

Tao, J.-J. *et al.* (2024) 'The universal and divergent roles of ethylene in rice and some other crop plants under salt stress', *Environmental and Experimental Botany*, 217, p. 105555.

Tian, J. *et al.* (2022) 'Growth and nitrogen metabolism in Sophora japonica (L.) as affected by salinity under different nitrogen forms', *Plant Science*, 322, p. 111347.

Tushabe, D. *et al.* (2023) 'Negative effects of high-temperature stress on gametophyte performance and their consequences for seed reproduction in wild plants', *Environmental and Experimental Botany*, 216, p. 105532.

Verma, N. *et al.* (2022) 'Epigenetic regulation of heat and cold stress responses in crop plants', *Plant Gene*, 29, p. 100351.

Wang, L. *et al.* (2022) 'Comparative metabolomic analysis reveals the involvement of catechins in adaptation mechanism to cold stress in tea plant (*Camellia sinensis* var. *sinensis*)', *Environmental and Experimental Botany*, 201, p. 104978.

Wang, Y.-X. *et al.* (2023) 'Heat shock protein TaHSP17.4, a TaHOP interactor in wheat, improves plant stress tolerance', *International Journal of Biological Macromolecules*, 246, p. 125694.

Wang, Y. and Demirer, G. S. (2023) 'Synthetic biology for plant genetic engineering and molecular farming', *Trends in Biotechnology*, 41, pp. 1182–1198.

Xiang, Y. *et al.* (2024) 'Molecular identification of twenty NtDREB homologs and overexpression of NtDREB_A2.1 improved plant growth in response to cold-stress and P-nutrition limitation', *Environmental and Experimental Botany*, 217, p. 105570.

Xu, X. *et al.* (2022) 'Overexpression of SlBBX17 affects plant growth and enhances heat tolerance in tomato', *International Journal of Biological Macromolecules*, 206, pp. 799–811.

Xu, Y. *et al.* (2023) 'Physiological and transcriptome analyses provide new insights into the mechanism mediating the enhanced tolerance of melatonin-treated rhododendron plants to heat stress', *Journal of Integrative Agriculture*, 22, pp. 2397–2411.

Yepes, L. *et al.* (2018) 'Analysis of physiological traits in the response of Chenopodiaceae, Amaranthaceae, and Brassicaceae plants to salinity stress, *Plant Physiology and Biochemistry*, 132, pp. 145–155.

Yuan, J. *et al.* (2023) 'The roles of miR156 in abiotic and biotic stresses in plants', *Plant Physiology and Biochemistry*, 204, p. 108150.

Yusuf, M. *et al.* (2024) 'Melatonin improved efficiency of 24-epibrassinolide to counter the collective stress of drought and salt through osmoprotectant and antioxidant system in pea plants', *Scientia Horticulturae*, 323, p. 112453.

Zahid, G. *et al.* (2023) 'An overview and recent progress of plant growth regulators (PGRs) in the mitigation of abiotic stresses in fruits: A review', *Scientia Horticulturae*, 309, p. 111621.

Zhang, F.-Q. *et al.* (2007) 'Effect of heavy metal stress on antioxidative enzymes and lipid peroxidation in leaves and roots of two mangrove plant seedlings (*Kandelia candel* and *Bruguiera gymnorrhiza*)', *Chemosphere*, 67, pp. 44–50.

Zhang, L. *et al.* (2021) 'Cold-regulated gene LeCOR413PM2 confers cold stress tolerance in tomato plants', *Gene*, 764, p. 145097.

Zhao, H. *et al.* (2022) 'Comprehensive assessment of harmful heavy metals in contaminated soil in order to score pollution level', *Scientific Reports*, 12, p. 3552.

Zheng, M. *et al.* (2022) 'Integrated strategies for increasing rapeseed yield', *Trends in Plant Science*, 27, pp. 742–745.

Zhou, H. *et al.* (2023) 'Insights into plant salt stress signaling and tolerance', *Journal of Genetics and Genomics*, 51, pp. 16–34.

8 Crop Plant Response under the Influence of Plant Hormones and Adverse Situations

Eksha Guliani, Fouad Damiri, Mithilesh Yadav, and Arpit Sand

8.1 INTRODUCTION

The plant organ's growth is completely regulated by ecological input as well as by plant hormones, also known as endogenous factors (Husen, 2021a,b; Husen and Zhang, 2023). The basis for the green revolution in the past was the mutation caused in genes by controlling hormonal factors, which eventually increased the yield of crops. This development is an important factor when considering the traits involved in agriculture (Abdelaal *et al.*, 2021). The ability of a plant to access nutrients and water is determined by the architecture and size of the root system. Fertilizers can supply only limited nutrients, but these are quite expensive and could even lead to eutrophication. A correlation has been developed between water resistance and the size of the root system. Plants possessing a larger root system have an increased capacity for nutrients, which can even lead to their survival in case of any nutrient deficiency. Plants that grow on land have a very complex 3D structure of root systems in order to optimize the shoot anchorage and exploration of soil for water and nutrients. All the post-embryonic and embryonic roots have hairs that act as powerful traits to increase their surface for water and nutrient absorption. Roots of plants are exposed to various weather conditions that undergo physical, chemical and biological processes, which cause a change in the solubility, amount and mineral distribution (Jia, Giehl and von Wirén, 2022). The plant hormone activity is related to its metabolism, transport, synthesis and signal transmission pathways. The knowledge that is gained regarding the sensing of nutrients in roots is restricted to various genotypes. In this chapter, a constant change in the environmental conditions that leads to severe exposure of crops toward various kinds of stresses is discussed. There are various hormones present in plants that have the ability to improve resistance to drought in plants. The occurrence of drought conditions leads to the decrease in quality of physiological, morphological and biochemical characteristics (O'Rourke and Graham, 2021).

8.2 PLANT HORMONES

There are several plant hormones that have the capacity to improve the resistance under adverse situation (Husen, 2021c; Husen, Iqbal and Aref, 2016, 2017; Husen *et al.*, 2018, 2019; Husen and Zhang, 2023; Iqbal *et al.*, 2022; Siddiqi and Husen, 2017, 2019, 2021). Some of specific roles of plant hormones are explained in this section.

8.2.1 Auxin

Auxin is one of the most vital plant hormones involved in division of cells, their elongation formation of roots, tropic responses and embryogenesis. These are the biotechnological targets which have the ability to modify shapes, sizes and yield of the plants, which even leads to the branching of roots. For example, auxin has the ability to regulate different members of the 1-aminocyclopropane-1-carboxylate synthase (ACS) family of genes that acts as a rate-limiting enzyme involved in the biosynthesis of ethylene.

DOI: 10.1201/9781032630366-8

8.2.2 Cytokinins

The second most important phytohormones are cytokinins (CKs). These were brought into existence in 1950 and act as a stimulator in cell division, which induces variations. They are basically the byproducts of adenine, which are isolated from purine or isoprene. Cytokinins also have the ability to stimulate a transgene expression that has significantly shown drought tolerance via delayed senescence. These even take part in germination of seeds, photomorphogenesis, apical meristem development of shoots, vascular development and floral development.

8.2.3 Gibberellins

Gibberellins are basically tetracyclic diterpenoids of carboxylic acids. The purpose of these phytohormones is to inculcate resistance against the drought stress. They even enhance plant tissue development by causing elongation of cells and increasing cell division. Transgenic tomatoes are released by the reduction caused in gibberellins, which eventually increase the tolerance level during drought. It is also one of the reasons water levels are increased under drought stress due to transpiration.

8.2.4 Jasmonic Acid

The closing of stomata causes an induction of tolerance in plants during drought conditions, which also causes a shift in the reactive oxygen species (ROS). It has been found via research studies that jasmonic acid (JA) acts as a stomata-closing regulator, and it is an enhancer of various enzymes in wheat. It includes glutathione reductase, ascorbate peroxidases (APX) and monodehydroascorbate reductase under various stress conditions. JA plays an important role in conductivity of water from soil under controlled conditions of moisture. The presence of JA in roots is essential in drought stress in order to increase the content of abscisic acid.

8.2.5 Ethylene

Another important plant hormone which is actively involved in causing enhancement in drought tolerance among plants is ethylene. A study was conducted on cotton plants in which the genes that were responsive toward ethylene and reacted toward drought and heat stress were identified. An increase in resistance was observed in transgenic plants against *Alternaria alternata* and mosaic virus because of protein-coding genes-related pathogenesis. The presence of a receptor detects ethylene, and eventually it shuts down its activity, which then leads to a reflex action for drought conditions.

8.3 NUTRIENT-DEPENDENT REGULATION OF PHYTOHORMONE BIOSYNTHESIS AND METABOLISM

Plants are exposed to various weather changes which involve physical, chemical and biological changes in the soil, resulting in the fluctuations of nutrients in plant development. Phytohormone homeostasis in the roots of plants is attained by causing a balance in the metabolism and biosynthesis. It responds highly to the internal as well as the external nutrients' cues. The internal cues basically involve the deficiency of N, S, K and P, which strongly impacts the auxin level of the roots. The biosynthesis of auxin depends upon indole-3-actaldoxime (IAOx) and indole-3-pyruvic acid (IPyA). When there is a high supply of ammonium ion (NH_4^+) at elevated levels, the level of auxin in the zone of root elongation is reduced via upregulation of the GH3 type of genes. Also, the biosynthesis of CKs is highly sensitive to presence of nitrate (NO_3^-) in growth factors (Adavi and Sathee, 2021).

Abscisic acid (ABA) is stimulated by elevated levels of NO_3^- in the apical meristem of roots. It happens because of the conjugation among esters of glucose that releases bioactive ABA via β-glucosidasel (BG1). Wheat is the plant species that regulates N via the induction of NO_3^-. Also, ethylene is formed as an intermediary in signaling nitrate ions. The deficiency of N reduces the extent of JA and JA-isoleucine (JA-Ile) in the roots of *Arabidopsis* for which the mechanism is yet not known. The accumulation of salicylic acid (SA) is elevated in nitrogen-deficient roots. The presence of ammonium ions (NH_4^+) decreases when nitrate ions (NO_3^-) enhance the signaling of auxin in elongation zone. Auxin is even activated as a result of the ethylene signaling. Therefore, nitrate ions (NO_3^-) act as a receptor in these cases (Jia *et al.*, 2022).

8.4 NANOPARTICLES REGULATE PHYTOHORMONE BIOSYNTHESIS AND SIGNALING IN PLANTS

The abiotic and biotic stress factors have become important constraints that have an adverse effect on the development and metabolism of plant growth. Those regulators which emphasize plant growth are called plant growth regulators. These are basically very small endogenous molecules like CK, auxin, ethylene, salicylic acid, gibberellins and JA (Werner *et al.*, 2010). Nanoparticles have the efficiency to regulate the response of a plant, known as signaling crosstalk, which proceeds via antagonistic and synergistic actions. A lot of studies have been done to understand the concept of stress signals affecting plant growth. Nanoparticles in plants possess the ability to control phytohormones at different levels, for example, by causing regulation in the biosynthesis of particular phytohormones. This could be done by preforming modifications in the pool, which are available in order to regulate the process of signaling (Alcázar, Bueno and Tiburcio, 2020).

The responses occurring in the presence of nanoparticles depend on certain factors like plant species, type of nanoparticles and conditions for exposure. A lot of studies have been carried out demonstrating the interactions among those hormones which are responsible for plant growth and nanoparticles. But some studies must be conducted regarding the molecular mechanism relating to phytohormones' crosstalk and nanoparticles. For example, auxin plays a vital role in stimulating the plant growth process and even the stress regulation of nanoparticles. One of the main auxins is indole-3-acetic acid (IAA), which is biosynthesized by converting tryptophan into IAA. This process involves two chemical reactions with the YUC (YUCCA) family of flavin monooxygenases and with the tryptophan aminotransferase of *Arabiopsis*-1 (TAA1) family. The transportation of auxin is regulated via auxin transporter carrier 1 (AUX1) influx carriers and the pinformed (PIN) family, known as efflux carriers. The exposure of nanoparticles has the tendency to regulate plant growth along with their responses in auxin transportation and biosynthesis (Tripathi, Singh and Pandey-Rai, 2022).

8.5 HORMONAL IMPACT ON PHOTOSYNTHESIS AND PHOTOPROTECTION UNDER STRESS

Plants have been categorized as sessile organisms that have to adapt various biotic and abiotic stresses throughout their life cycle, as shown in Figure 8.1. There are certain unfavorable ecological conditions which include excessive heat and light, extreme temperature range, presence of pollutants, salinity under abiotic factors and excessive energy of excitation of in chloroplasts. The tolerance power of plants against environmental stress varies from species to species along with the mechanisms of photoprotection, but there is a general hierarchy of various responses happening in plants which are modulated by phytohormones. The impact of hormones on photoprotection and photosynthesis under the abiotic stresses is vital in order to develop an understanding toward the basic set of biological processes that is followed (Miller and Nielsen, 2021). The development of new crop plants is possible via this phenomenon with a great robustness against photosynthetic machinery.

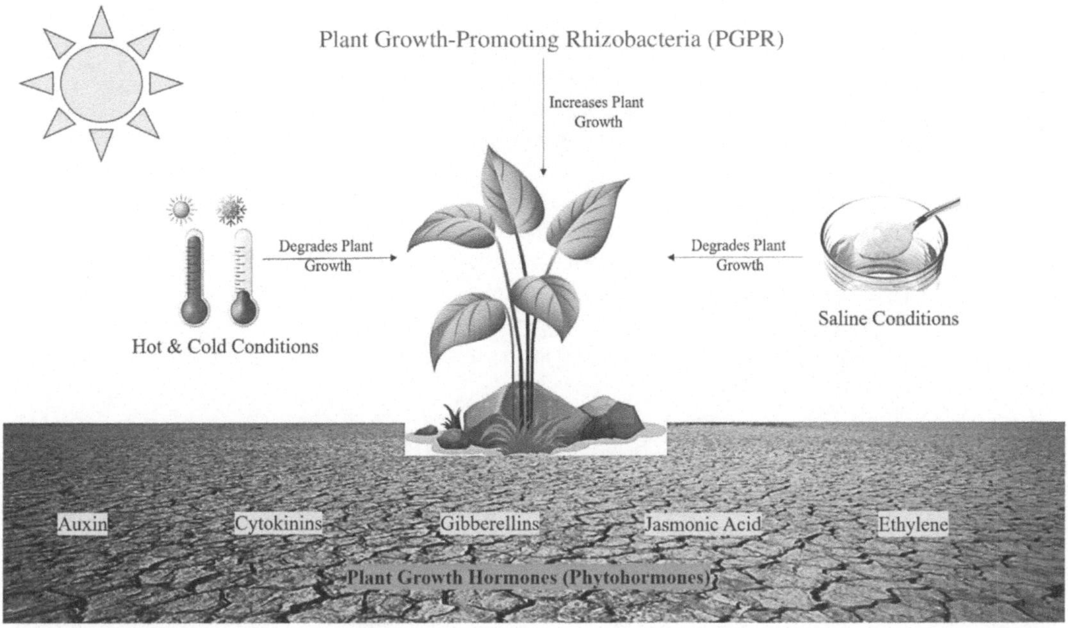

FIGURE 8.1 Environmental factors causing stress to plant growth which are overcome by the presence of plant growth hormones.

This is useful in order to predict the economic yield, which declines significantly because of sudden changes in climate that eventually become a threat to food security around the globe.

Light is one of the most variable environmental factors, and it changes rapidly. Excessive light can result in stress responses along with some interactions among the ROS and phytohormones in plants. These stress responses play an important role in plant acclimation. When energy is supplied to plants' chloroplasts in an excessive amount, it could be dissipated in the form of thermal energy, which serves as a defense to protect the plant against any kind of photodamage. ROS can result in various harmful effects like carbohydrate oxidation and peroxidation of lipids, which results in the death of the plant cells. Also, a balance is maintained among the detoxification and production of ROS by the use of non-enzymatic and enzymatic antioxidants (Müller and Munné-Bosch, 2021).

8.6 THE ROLE OF HALOPHILIC PLANT-ASSOCIATED BACTERIA IN PLANT GROWTH UNDER SALINE CONDITIONS

The salinity of soil possesses a problem for cultivation of crops, and it is one of the major issues which is faced by the agricultural industry throughout the season. The change in climatic conditions and irrigation practices has led to the problem. Halophyte microorganisms basically comprise certain microbes which promote the growth of plants (Singh *et al.*, 2020). The growth happens outside the root system, which are called ectosymbionts, and within the plant system, which are called endosymbionts. The mentioned symbionts promote plant growth even under stress conditions. These are called plant growth promoting rhizobacteria (PGPR). There are some strains which assist in the survival of halophyte under saline conditions, for which the tolerance against salt is inferred as ST-PGPR. These are basically salt-loving genera like *Halobacillus* and *Halomonas*. Some microbes from the genus, like *Kushneria*, have the ability to show the same kind of tolerance against salt. There are many glycophytes belonging to the grass family, like wheat, barley and rice, which possess salt tolerance. It has been demonstrated via experimentation that there is a 62–78%

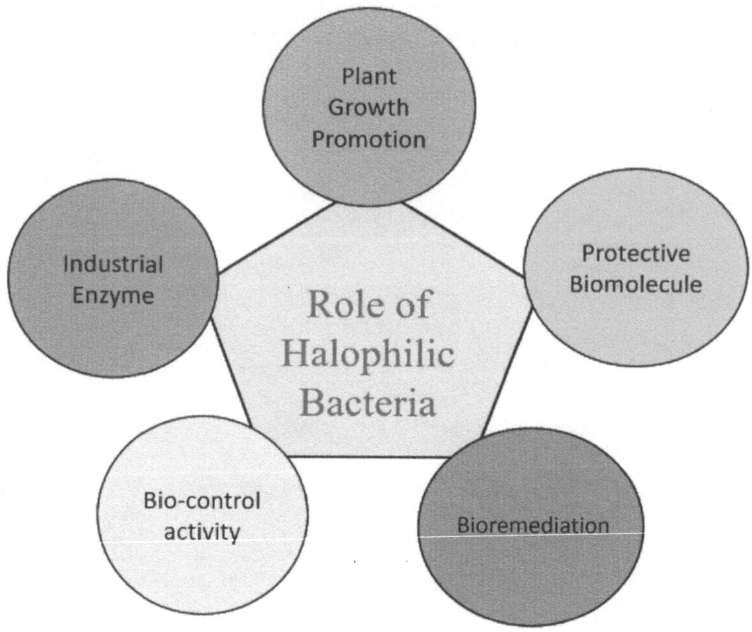

FIGURE 8.2 Role of halophilic bacteria in industry.

increase in length as well as root mass when comparisons were made with the plants grown in saline soil (Wang *et al.*, 2021).

There are different mechanisms through which endophytes increase the growth of plants that eventually include the acquisition of nutrients causing changes in the gene expression (O'Rourke, McCabe and Graham, 2020). For example ACC, which is present in various endophytes, has the ability to stimulate the acquisition of nutrients. It simultaneously reduces the amount of ACC, which is converted to ethylene. Ethylene is very well known as an inhibitor which hinders the plant growth that is released as a result of the response of plants to drought, salt or any other ecological factor. For example, various bacterial endophytes like *Pantoea, Enterobacter* and *Bacillus* possess tolerance against salt in hybrid elephant grass because of the enhancement in the acquisition of nutrients or gene expression changes. The understanding, development and applicability of various halophilic bacteria as shown in the Figure 8.2 that have been isolated through halophytes could raise various challenges in saline soil. The organic treatment would involve the incorporation of inoculated plants.

8.7 RNA-MEDIATED REGULATIONS OF GENE EXPRESSION IN CROPS

Some well-known major crops like wheat and rice have intensive GC-content transcriptomes and genomes. The RNAs have a high content of GC and possess the ability to be folded into long structures, which makes wheat and rice crops very strong and resilient. The translatome analysis in such crops has shown that mRNAs having a high content of GC could be highly translated. This infers that the evolution of RNA in such crops has induced flexibility to ensure efficient and active translation biological processes. The role of RNA has unveiled many new insights in exploring regulation by gene expression in crops. The post-transcriptional gene expression undergoing regulation could be explained by translational process.

For example, the *in vivo* landscape of RNA structures in crops like wheat and rice has unveiled that an upstream of a single stranded 5'UTR has been keenly associated along a translational efficiency which was very high. This may even cause the binding of ribosomes, which enhances its initiation. Very strong periodic triplet pattern trends of the RNA structure were observed in mRNA along with a high efficiency of translation that indicated elongation of ribosome. In wheat crops, a very significant anti-correlation was seen among BPP, which was the base pair probability and the translational efficiency in 5'UTR but not in the region of 3'UTR. The RNA structure had a noticeable significance in translational regulation which facilitates the crop editing via RNA alteration to get the desired translational levels.

The recent studies have even shown that the RNA structure regulation as a gene expression has the potential to serve as a vital genetic property that eventually guides the breeding of crops. The most effective factor considered for site binding is post-transcription regulation, which has a direct impact on the synthesis of proteins. The asymmetry of subgenomes in wheat was responsible for the genes' control, which facilitated environmental adaptations. Therefore, the maximization of asymmetry of translational subgenomes is vital for environmental responses, which offer a brand-new way of breeding and editing wheat crops (Sun, Ding and Zhang, 2022).

8.8 GENE EXPRESSION IN CROP IMPROVEMENT

There has been a lot of advancement in agricultural biotechnology as it is undergoing a revolution while leading to a development in gene editing tools like CRISPR/Cas. This type of editing is based upon the combination of two components that ultimately form the ribonucleoprotein (RNP) complex. It has the ability to identify the target site and destroy it within the genome. There are certain restrictions on those crops which are genetically modified due to which agricultural biotechnology has benefited widely from the CRISPR/Cas genome editing by causing transgenic expression along with transgene removal via genetic separation and crossing of plants. This involves the formation of an RNP complex without causing any transformation of plant genes.

Nowadays, many crops are part of gene knockout via CRISPR/Cas9-enabled improvement of crop aspects like stress resistance, improved yield and quality. For example, the translation and stability of various transcripts are highly regulated by certain mRNAs, which show a complementary sequence within target transcripts of binding sites as shown in Figure 8.3. But it even results in the generation of some undesired phenotypes, as all the targets are produced because of inhibition. Therefore, an alternative approach would be much more specific if taking rice crops into consideration. Causing a single deletion of one nucleotide in an mRNA seed pairing can efficiently release one transcript's inhibition. Hence, gene editing which is applied to the post-transcriptional

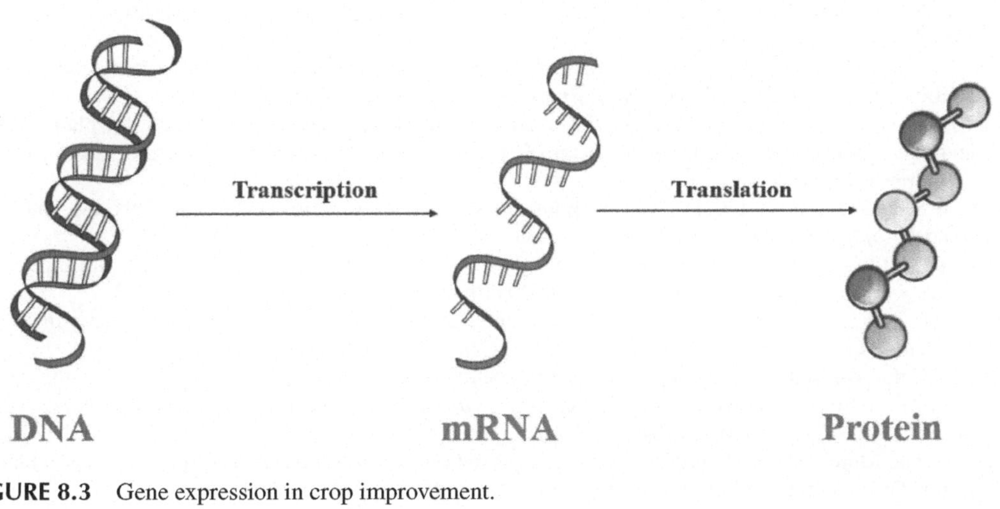

FIGURE 8.3 Gene expression in crop improvement.

regulation, including inhibition of mRNA, is the most powerful strategy in the era of crop improvement (Ferreira and Reis, 2023).

8.9 CONCLUSION

The stress caused because of saline conditions is one of the most unfortunate factors governing agricultural areas that affect plant yield and production. The molecular mechanisms involved in this sector are not well discovered. The nutrient-dependent regulation of phytohormone biosynthesis and metabolism could be inferred by the fact that the nutrient-derived signals as well as the individual nutrients have the ability to modulate metabolism, biosynthesis, transport or signaling at various levels of endogenous hormones. Therefore, it has been proven that plant microbiomes produce genes which are responsive for the induction of stress. In association with the high levels of structural complexity, the tolerance against drought in different crops by different mechanisms is poorly understood and is an area of future research.

REFERENCES

Abdelaal, K. *et al.* (2021) 'The role of plant growth-promoting bacteria in alleviating the adverse effects of drought on plants', *Biology*, 10(6), p. 520. https://doi.org/10.3390/biology10060520

Adavi, S.B. and Sathee, L. (2021) 'Elevated CO_2 alters tissue balance of nitrogen metabolism and downregulates nitrogen assimilation and signalling gene expression in wheat seedlings receiving high nitrate supply', *Protoplasma*, 258(1), pp. 219–233. https://doi.org/10.1007/s00709-020-01564-3

Alcázar, R., Bueno, M. and Tiburcio, A.F. (2020) 'Polyamines: Small amines with large effects on plant abiotic stress tolerance', *Cells*, 9(11), p. 2373. https://doi.org/10.3390/cells9112373

Ferreira, S.S. and Reis, R.S. (2023) 'Using CRISPR/Cas to enhance gene expression for crop trait improvement by editing miRNA targets', *Journal of Experimental Botany*, 74(7), pp. 2208–2212. https://doi.org/10.1093/jxb/erad003

Husen, A. and Zhang, W. (2023) *Hormonal cross-talk, plant defense and development.* Cambridge, MA: Elsevier Inc.

Husen, A. (2021) 'Cross talk between autophagy and hormones for abiotic stress tolerance in plants', in: Husen, A. (ed.) *Plant performance under environmental stress.* Cham, Switzerland: Springer International Publishing AG, pp. 1–15. https://doi.org/10.1007/978-3-030-78521-5_1

Husen, A. (2021c) *Plant performance under environmental stress: Hormones, biostimulants and sustainable plant growth management.* Cham, Switzerland: Springer Nature Switzerland AG. https://doi.org/10.1007/978-3-030-78521-5

Husen, A., Iqbal, M. and Aref, I.M. (2016) 'IAA-induced alteration in growth and photosynthesis of pea (*Pisum sativum* L.) plants grown under salt stress', *Journal of Environmental Biology*, 37, pp. 421–429.

Husen, A., Iqbal, M. and Aref, I.M. (2017) 'Plant growth and foliar characteristics of faba bean (*Vicia faba* L.) as affected by indole-acetic acid under water-sufficient and water-deficient conditions', *Journal of Environmental Biology*, 38, pp. 179–186.

Husen, A. *et al.* (2019) 'Modulation of salt-stress tolerance of Niger (*Guizotia abyssinica*), an oilseed plant, by application of salicylic acid', *Journal of Environmental Biology*, 40, pp. 94–104.

Husen, A. *et al.* (2018) 'Salicylic acid alleviates salinity-caused damage to foliar functions, plant growth and antioxidant system in Ethiopian mustard (*Brassica carinata* A. Br.)', *Agriculture & Food Security*, 7(44), pp. 1–14.

Siddiqi, K.S. and Husen, A. (2017) 'Plant response to strigolactones: Current developments and emerging trends', *Applied Soil Ecology*, 120, pp. 247–253.

Iqbal, S. *et al.* (2022) 'Phytohormones trigger drought tolerance in crop plants: Outlook and future perspectives', *Frontiers in Plant Science*, 12, p. 799318. https://doi.org/10.3389/fpls.2021.799318

Jia, Z., Giehl, R.F.H. and von Wirén, N. (2022) 'Nutrient–hormone relations: Driving root plasticity in plants', *Molecular Plant*, 15(1), pp. 86–103. https://doi.org/10.1016/j.molp.2021.12.004

Miller, A.K. and Nielsen, B.L. (2021) 'Analysis of gene expression changes in plants grown in salty soil in response to inoculation with halophilic bacteria', *International Journal of Molecular Sciences*, 22(7), pp. 3611. https://doi.org/10.3390/ijms22073611

Müller, M. and Munné-Bosch, S. (2021) 'Hormonal impact on photosynthesis and photoprotection in plants', *Plant Physiology*, 185(4), pp. 1500–1522. https://doi.org/10.1093/plphys/kiaa119

O'Rourke, J.A. and Graham, M.A. (2021) 'Gene expression responses to sequential nutrient deficiency stresses in soybean', *International Journal of Molecular Sciences*, 22(3), pp. 1–19. https://doi.org/10.3390/ijms22031252

O'Rourke, J.A., McCabe, C.E. and Graham, M.A. (2020) 'Dynamic gene expression changes in response to micronutrient, macronutrient, and multiple stress exposures in soybean', *Functional and Integrative Genomics*, 20(3), pp. 321–341. https://doi.org/10.1007/s10142-019-00709-9

Siddiqi, K.S. and Husen, A. (2019) 'Plant response to jasmonates: Current developments and their role in changing environment', *Bulletin of the National Research Centre*, 43(153), pp. 1–11.

Siddiqi, K.S. and Husen, A. (2021) 'Significance of brassinosteroids and their derivatives in the development and protection of plants under abiotic stress', *Biologia*, 76, pp. 2837–2857. https://doi.org/10.1007/s11756-021-00853-3

Singh, D.P. *et al.* (2020) 'Stage-dependent concomitant microbial fortification improves soil nutrient status, plant growth, antioxidative defense system and gene expression in rice', *Microbiological Research*, 239, pp. 126538. https://doi.org/10.1016/j.micres.2020.126538

Sun, W., Ding, L. and Zhang, H. (2022) 'The potential role of RNA structure in crop molecular breeding', *Frontiers in Plant Science*, 13, pp. 868771. https://doi.org/10.3389/fpls.2022.868771

Tripathi, D., Singh, M. and Pandey-Rai, S. (2022) 'Crosstalk of nanoparticles and phytohormones regulate plant growth and metabolism under abiotic and biotic stress', *Plant Stress*, *6*, pp. 100107. https://doi.org/10.1016/j.stress.2022.100107

Wang, Y. *et al.* (2021) 'Nanobubbles promote nutrient utilization and plant growth in rice by upregulating nutrient uptake genes and stimulating growth hormone production', *Science of the Total Environment*, 800, pp. 149627. https://doi.org/10.1016/j.scitotenv.2021.149627

Werner, T. *et al.* (2010) 'Root-specific reduction of cytokinin causes enhanced root growth, drought tolerance, and leaf mineral enrichment in Arabidopsis and tobacco', *Plant Cell*, 22(12), pp. 3905–3920. https://doi.org/10.1105/tpc.109.072694

9 Proteomics and Genomics Investigation of Micronutrients and Macronutrients Deficiency and Toxicity in Plant System

Faisal Rasheed, Ritu Chaudhary, and Altaf Ahmad

9.1 INTRODUCTION

Plant growth can be determined by the availability of their food. It utilizes inorganic nutrients in order to grow either in soil conditions or a tissue culture medium (Kerry *et al.*, 2018). Different plant species require some specific nutrients at optimum levels and minimum required levels. During the mid-19th century, Baron Justus von Liebig proved that nutrients are consequential for viability of plants and stated that nutrients are essential for plant existence; if deprived of any essential nutrients, then plants may die (Liebig, 1840; Tucker, 1999). Plants require 23 essential nutrients used for different metabolic activities (Kerry al., 2018). These nutrients can play an important role, which can be well-understood in different plant systems (Qu et al., 2020). Based on their requirements, these minerals are grouped under two broad categories: non-minerals and minerals (Al-Obaidi, 2020). Carbon (C), hydrogen (H), and oxygen (O) are non-minerals, as plants obtain them from air and water. Plants can absorb the remaining 20 elements from the soil in different forms, so these are called mineral elements (Brown et al., 2022). Moreover, these minerals are further subdivided into macro- and micronutrients. Based on the plant requirements, these macronutrients can be grouped as primary macronutrients, which include nitrogen (N), phosphorus (P), and potassium (K), whereas calcium (Ca), sulfur (S), and magnesium (Mg) are secondary or tertiary macronutrients. The micronutrients also considered as minor or trace elements are molybdenum (Mo), nickel (Ni), chlorine (Cl), boron (B), copper (Cu), iron (Fe), manganese (Mn), and zinc (Zn) are required in minute quantities (Toor et al., 2021). "Micronutrient" is a term that is always recommended by the American Society of Agronomy and the Soil Science Society of America (Ebong, Offiong and Ekpo, 2014). The beneficial elements include cobalt (Co), sodium (Na), vanadium (Va), selenium (Se), silicon (Si), and aluminum (Al) (Carpanez et al., 2022). Moreover, nutrients, particularly micronutrients, perform explicit functions in plant cells. When the internal concentration increases the threshold limit, they execute toxic effects by impairment at physiological, biochemical, and molecular levels in plants. Some micronutrients are heavy metals like Mn, Mo, Co, Zn, Fe, Cu, and Ni which are frequently disturbed in various metabolic activities due to their toxicological symptoms seen in the occurrence of supra-optimal concentrations in plants (Appenroth, 2010; Clemens, 2001) (Table 9.1, Figure 9.1).

During recent decades, researchers have made a use of transcriptome analysis to find out the expression pattern of genes during heavy metal stress conditions in plants (Jamla et al., 2021). The plant response can be better explained at a gene expression level under heavy metal stress that can be well-understood at the mRNA level (Kosakivska et al., 2021). Moreover, transcriptional analysis has certain limitations, including the differential correlations between variable expression of mRNAs with that of corresponding proteins (Kosakivska et al., 2021). Furthermore, the reciprocity between mRNA and protein has also been perceived; protein expression can be controlled at both translational and post-translational levels (Koussounadis et al., 2015). Thus, achieving information

 DOI: 10.1201/9781032630366-9

TABLE 9.1
The Roles of Various Mineral Nutrients in Plant Cells

Mineral Nutrients	Absorbed as	Role in Metabolism as a Constituent of Organic Compounds
Nitrogen	NO^{2-}, NO^{3-}, or NH^{4+}	Amino acids, proteins, purines and pyrimidine, chlorophyll, and many coenzymes
Sulfur	SO_4^{-2}	Cysteine and methionine (amino acids), Fe-S cluster, and vitamins (thiamine and biotin) as enzyme activators
Phosphorus	H_2PO^{4-}, HPO_4^{-2}	Nucleic acids, phospholipids, ADP, ATP, NAD, NADP, and phosphate esters of sugars
Potassium	K^+	Maintains anion-cation balance and cell turgidity, protein synthesis, stomatal conductance, enzyme activation
Magnesium	Mg^{+2}	Enzymes involved in carbohydrate metabolism and synthesis of DNA, RNA, chlorophyll molecule, and Mg-pectate in the middle lamellae
Iron	Fe^{2+}	Cytochromes, peroxidases, catalases, metalloflavo-proteins, and calcium hydrolysis of ATP and phospholipids
Manganese	Mn^{2+}	Nitrite reductase and hydroxylamine reductase
Molybdenum	MoO_2^{2+}	Nitrate reductase and nitrogenase (in symbiotic association)
Zinc	Zn^{+2}	Alcohol dehydrogenase, glutamic dehydrogenase, lactic dehydrogenase, and alkaline phosphatase
Calcium	Ca^{2+}	As calcium pectate in cell wall, cofactor in enzymes
Copper	Cu^{2+}	Cytochrome oxidase, ascorbic acid oxidase, polyphenol oxidase, and plastocyanin
Chlorine	Cl^-	Water splitting in photosynthesis, ionic balance in cells
Boron	BO_3^{3-} or $B_4O_2^{3-}$	Membrane functioning, pollen germination, cell elongation, carbohydrate translocation, and uptake and utilization of Ca^{+2}
Nickel	Ni^{+2}	Component of urease enzyme, nitrogen metabolism and fixation

at translational and post-translational levels can create deeper perceptions into the responses and functional interactions of proteins than can genome-based predictions. Proteomics is the systematic study of the protein expression, which is a molecular tool for explaining complete proteomes at the organelle, cell, or tissue levels, as well as comparing proteins under various kind of stressful conditions, such as heavy metals (Komatsu and Hossain, 2017; Nouri and Komatsu, 2013). Proteomics is a potential tool used to probe plant responses during stresses and alteration in biochemical processes. Still, some more elaboration is needed to expose plant responses during metal stresses through a specific or group of proteins responsible in a stress tolerance mechanism (Parvaiz et al., 2016).

Genomics is study of genomes which comprises intensive attainments to know DNA and RNA sequences using high throughput sequencing strategies, creating fine-scale maps and microchip arrays, collecting genome variations within a population, and ascertaining the transcriptional control of genes by employing digital technology and computationally intensive analysis to know the structure, function, and evolution of diverse organisms (Varma and Shrivastava, 2008). Genomics can also be described as branch of genetics which deals with the study of genes, chromosomes, and their functions, along with epistasis, heterosis, pleiotropy, and interactions among loci and alleles within the genome (Balammal and Reddy, 2012; Hartenian and Doench, 2015). The term "genomics" was introduced by Tom Roderick in 1986 during the meeting about the mapping of the human genome. Roderick worked as scientist in a laboratory in Jackson, Maine (Ibrahim et al., 2015). The concept was originally introduced by Fred Sanger, who first completed the sequencing of genomes in a virus, namely phiX174 (Anderson et al., 1981). Moreover, he developed the practice of sequencing and genome mapping along with the development of bioinformatics and data storage in the 1970s and 1980s. Presently, the genomics approach is widely used to speed up the work of conventional

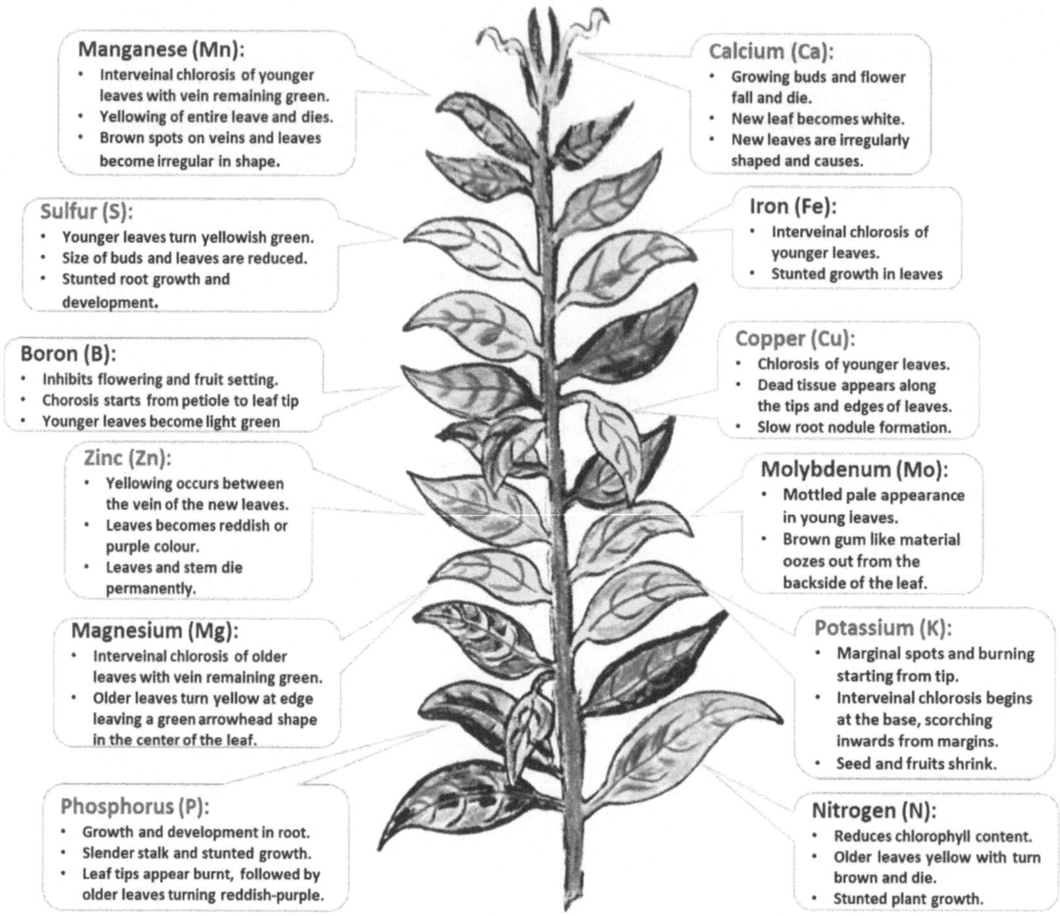

FIGURE 9.1 Roles of various mineral nutrients in plant cells.

molecular geneticists within the modern scientific field. Initially, the different organisms like plants, animals, and microbes were separated, but with this advanced technology of genomics, we can find out their resemblance based on their genome sequencing by various researchers (Varma and Shrivastava, 2008). Genomics also becomes a useful tool when dealing with complex traits as they are controlled by multigenes with their environmental influence (Liu et al., 2017a).

9.2 PROTEOMIC ALTERATION IN PLANT SYSTEMS UNDER NUTRITIONAL STRESS

Proteomics is well-defined as "omics," an advanced technique in the field of genomics that deals with the study of a cascade of proteins encoded by their genomes in organisms (Liu *et al.*, 2013). It can depict the complete protein structures and their alteration at organelle, cellular, tissue, or organ levels during various kinds of stresses, including heavy metals (Singh et al., 2016). Nowadays, the study of proteomics is expanding rapidly in the field of heavy-metals stress biology. Most of the studies of protein behavior differentially expressed under heavy metal stressed in plants depend on classical gel-based and gel-free proteomics approaches (Komatsu et al., 2017). The gel-based technique involved 2D gel electrophoresis added with Edman sequencing, matrix-assisted laser desorption/ionization time of flight (MALDI-TOF), surface-enhanced laser desorption/ionization time of flight (SELDI-TOF), mass spectrometry (MS), and electrospray ionization (ESI)-MS analysis. In contrast, the gel-free approach usually includes multidimensional protein identification technology (MudPIT), isobaric tags for relative and absolute quantification (iTRAQ), isotope-coded affinity tag (ICAT) and tandem mass tags (TMTs), etc. (Ahsan, Renaut and Komatsu, 2009; Hu, Rampitsch and Bykova, 2015). Moreover, the desired mineral elements and trace nutrients play a vital role in regulating the growth and developmental processes in plants; the increased concentration of the required amount becomes toxic to the plant systems. The formation of reactive oxygen species (ROS) causes oxidative damage, showing a direct or indirect injury at different organs in plants due to high concentrations of metal (Minic, 2015). The production of large numbers of electrons during aerobic respiration produces ROS containing free radicals of oxygen, such as hydrogen peroxide and superoxide. Hydroxyl radicals are inexorable byproducts during cellular activities. These ROS damage the structure of proteins, nucleic acids, and lipids due to denaturation of protein, DNA mutation, and lipid peroxidation, which ultimately induces apoptosis (Marrocco, Altieri and Peluso, 2017). To combat with these ROS, plants induce the production of antioxidants a-tocopherol, ascorbate, and glutathione (Rasheed *et al.*, 2022b) and enzyme-based antioxidants include catalase, superoxide dismutase, monodehydroascorbate, ascorbate peroxidase, dehydroascorbate reductase, glutathione reductase, and others (Rasheed *et al.*, 2022b). During normal conditions, the ROS production and destruction are regulated in cell metabolism. During heavy metal stress situations, production of these free radicals becomes enhanced in excess amounts to induce oxidative stress under stressful physiological conditions (Rasheed *et al.*, 2022b). So, the nutrients in excess or deficient conditions induce stress responsible for affecting plants' health. The proteomics is a powerful tool to find plant responses under stress conditions, as well as the modulation in plant metabolism at molecular level, despite different responses (Prusty *et al.*, 2022). Here are some differential protein responses during nutrient deficiency and in excess amounts in different plant systems.

Nitrogen (N) is a key element in soil fertility which directly or indirectly controls various metabolic activities that ultimately control plant productivity and their yield (Rasheed *et al.*, 2022b). During nitrogen scarcity conditions, the first symptoms show in younger and then older plant parts, such as stiffness of leaves, which become hard and erect. Researchers have found a significant role of N in elucidating the role at molecular levels in the plant cells to acquire knowledge in modulation of protein in various plant species, such as maize, wheat, and rice during N-deficient conditions by using 2-DE, MALDI-TOF-MS, and iTRAQ proteomics analysis (Chandna and Ahmad, 2015; Ding *et al.*, 2010; Nazir *et al.*, 2016). The adverse effects that cause up-and-down regulation in various N-metabolism proteins (ribulose-1,5-bisphosphate carboxylase, 6 phoshogluconate dehydrogenase, ATP synthase gamma chain, malate dehydrogenase, 5 deoxyuridine 5'-triphosphate -nucleotide hydrolase, ATPase alpha subunit, cytochrome P450, glutamate synthase, glutathione S-transferase, adenosine kinase, 4-coumaroyl-CoA synthase 3, and phenylalanine ammonia-lyase) are greatly affected, showing the dependence on nitrogen. Furthermore, it was reported by Sarasketa *et al.* (2016) that N is a potent regulator of photosynthesis capacity by enhancing chlorophyll production and positively modulating the antioxidant enzyme defense system of an organic and inorganic nature. During a trivial N-deficient condition, a non-significant alteration was observed in a model plant (*Arabidopsis thaliana*) and a differential regulation of glucosinolate metabolism and high expression of TGG1 and TGG2 myrosinases (Marino et al., 2016). The study showed the significant role of N and its potentiality by using NH_4^+ as a N source to activate various metabolisms in plants.

K is the second major macronutrient required for leaf growth, balancing anion-cation contents in plant cells (Torregrosa, Husekova and Sychrova, 2016). It is involved in photosynthesis, osmotic balance, signal transduction, and various enzyme activations. The K-deficiency can alter metabolic mechanisms at various levels and thus hamper the plant productivity and quality of products that may cause oxidative stress in plants (Takehisa *et al.*, 2015). Zhang *et al.* (2016c) reported that 258 peptides were identified in xylem sap with decreased expression due to low content of K in cotton plants by proteomics approaches. Furthermore the stress signaling compounds and recognizing proteins showed reduced expression. Sometimes the protein expression is totally absent during K-stressed conditions. Different plant proteins, mainly enzymes such as tubulin, glutathione S-transferase, plant peroxidase, NADPH-cytochrome P450 reductase, phenylalanine ammonia-lyase, and ATPase family AAA domain proteins, are altered in structure and changed their expression during K metabolism in *Nicotiana tabacum* and *Hordeum spontaneum* (Zang *et al.* 2016c; Zeng et al., 2015). Studies showed that the potassium deficiency has negative effects on plants, like growth inhibition, nutrition imbalances, and decrease in host-specific pathogenic resistance.

Phosphorus is an essential constituent for cell membranes, nucleic acids, phosphorylation, and photosynthesis processes, so it is consider as a major component for plants (Nadira *et al.*, 2016). During the various chemical reactions in photosynthesis process, the energy transfer across the membrane structure of thylakoid becomes highly affected, inactivating various important enzymes of the Calvin cycle and reducing electron transfer systems in photosynthesis under P-deficient conditions (Muneer and Jeong, 2015a, 2015b; Zhang *et al.*, 2014, 2016b). It was revealed by an experiment conducted by a two-dimensional difference in gel electrophoresis (2-D DIGE) on *A. thaliana* over expression line to know the function of low quantity of phosphate in plants. The outcome reveals that 88 differential abundant proteins were spotted, and among them 59 proteins were identified between the shoot and root of 35S: WRKY6–9 and WRKY6–1 variety plants. Li *et al.* (2017a) showed the altered structure of some important enzymes like ATP synthase, gln synthetase, nitrilase, 14–3–3 proteins, carbonic anhydrases 2, and tryptophan synthase under P-deficiency stress condition.

Calcium is a key element required for the formation of cell walls and a cofactor involved in activating various enzymes for regulating water movement and osmotic balance in plant cells (Ranty *et al.*, 2016). It was studied by Oh, Nanjo and Komatsu (2014) and Yina *et al.* (2015) in the alteration of proteomic expression on Ca content in *Glycine max*, *Hordeum vulgare* (Ciesla *et al.*, 2016), and *Arabidopsis annexin* (Wang *et al.*, 2015) as they relate to water, salinity, drought, and heat stress, respectively. Using proteomics approaches, the Ca-dependent proteins, such as protein kinase, annexin 1 (AtANN1), ACC oxidase 2, maturation polypeptide, and annexin diagnosed what was involved in signal transduction, transport, and energy transfer.

The exogenous application of C on *Cucumis sativus* observed the over-expression of enzymes of glycolysis, the citric acid cycle, fermentative metabolism, nitrogen metabolism, and ROS protection during reduced supply of oxygen in plant tissue (He *et al.*, 2012).

Magnesium is important in living plant cells that indulge in various important pathways, including photosynthesis, respiration, protein synthesis, and nucleic acid formation. It works in chlorophyll biosynthesis and plays an important role during heavy metal toxicity in *Citrus sinensis* and *A. thaliana* in plants (Fukao *et al.*, 2016; Peng *et al.*, 2015). Excess Mg stress can be determined by the application of 2-DE for better understanding of the molecular mechanisms and the importance of Mg in plants. Some proteins of *C. sinensis* were found to be important in carbohydrate, protein, nucleic acid, cell wall, cytoskeleton, and lipid metabolism under Mg deficiency conditions (Peng *et al.*, 2015). In *A. thaliana*, studies observed the alteration of specific proteomes to Mg stress in root samples by conducting iTRAQ-based quantitative techniques through GFP-expressing protoplasts. During Mg deficiency, the protoplast of the epidermis was collected from pGL2-GFPer and pMGP-GFPer marker lines, and a downregulation of proteins like Peroxin 11c, copper/zinc superoxide dismutase 1, cytosolic invertase 1, and HOPW1-1-interacting 1 was found (Fukao *et al.*, 2016). Proteome was shown to be fruitful by long term Mg-deficiency in leaves rather than roots, so the acclimatization between these tissues differed when exposed to long-term Mg-deficiency (Guo *et al.*, 2015).

Sulfur is the fourth-most-abundant nutrient and plays an important role in the formation of various co-enzymes and vitamins. Their deficiency affects chlorophyll and protein synthesis (Rasheed *et al.*, 2022a). Some researchers found the role of S in regulation of ethylene signaling, thereby mitigating heavy metal stress (Rasheed *et al.*, 2021). The differential protein functions were observed in mitigating stress, photosynthesis, protein structure, ATP formation, and enzyme function during

S deficiency in *Brassica juncea* (Rasheed *et al.*, 2021, 2022a), *Brassica napus*, and *A. thaliana* (Liu *et al.*, 2014; Yildiz and Terzi, 2016).

Boron is an important micronutrient required by plants for their growth and development. Sometimes, the increased quantity of B becomes toxic to plant species (Guo *et al.*, 2016). B toxicity was studied by a group of researchers in China, using two citrus species, namely *Citrus sinensis* and *Citrus grandis*. The researchers found that protein expression became changed, which involved energy metabolism, antioxidant and detoxification, stress responsive, co-enzyme synthesis, protein and amino acid metabolism, signal transduction, cellular transport, cytoskeleton, nucleotide metabolism, cell cycles, and DNA processing through MS analysis (Sang *et al.*, 2015). In root samples of *Citrus* species, it was found that their protein modulation and protein-related ROS scavengers were upregulated in B toxicity to combat oxidative stress (Sang *et al.*, 2017).

Mn plays a significant role in regulating various metabolic activities in different plant species, but in stressed conditions it can cause toxicity (Celma *et al.*, 2016; You *et al.*, 2014). By using proteomics approaches, *Camellia sinesis* can become tolerant against Mn toxicity due to Cs MTP8 protein present in localized form in plasma membranes of cultivar-LJCY and it has higher Mn-tolerant capacity than cultivar-YS because reduced content of chlorophyll was observed under Mn treatment (Li *et al.*, 2017b). Moreover, the Mn deficiency was studied through LC MS/MS-based iTRAQ proteomics techniques, and unambiguous determination of 11,606 different proteins was observed in the root of *A. thaliana*, which shows the great impact of Mn on plant cell (Celma *et al.*, 2016). Furthermore the study showed that Mn tolerance in *Stylosanthes guianensis* is maintained by a cascade of internal and external Mn detoxification through malate synthesis and exudation, which can be regulated by SgMDH1 at transcription and translation levels (Chen *et al.*, 2015a).

Fe is another important micronutrient which plays a significant role in smooth performance of respiration, photosynthesis, replication, and translation (Rehman *et al.*, 2021). Its deficiency becomes a serious concern in plant tissue, which is evident by proteomic analysis in pea plants to know the long-term effects of Fe deficiency. The evidence showed that the synthesis of neutralizing proteins of mitochondrial electrons transports components, maintaining the cellular energy proliferation, to reduce the ROS level and delaying mitochondrial cell death (Meisrimler *et al.*, 2016). The experiment was performed on the roots of *Cucumis sativus* and *A. thaliana* to find out if the effect of Fe deficiency and the regulation of novel proteins is Fe-dependent (Donnini *et al.*, 2010; Li and Lan, 2017). Hopff, Wienkoopb and Luthjea (2013) compare the protein expression and suppression in control in Fe-high- and Fe-low-treated maize plants. They found that 227 proteins were identified in control plants, and they were reduced to 204 in Fe-deficient and 251 in high-Fe conditions. Maximum proteins are responsible for signaling like quinone reductase, heme, and copper-containing proteins.

Zn uptake can increase the oxidative stress, evidenced from the nanoparticles-based research on *Glycine max* where imbalances in the protein expression were related to secondary metabolism, cell organization, and hormone metabolism (Hossain *et al.*, 2016). It has been found that little work has been done on advanced proteomics on Zn while studying a *Lactuca sativa L.* plant, but interesting evidence was shown by Lucini and Bernardo (2015).

Cu is an important micronutrient for the entire living world, but sometimes it becomes toxic under excess amounts and initiates the production of ROS at the cellular level (Rout *et al.*, 2013; Zhang *et al.*, 2016a). Cu-binding proteins are expressed more during excess concentrations of Cu, which works in anti-oxidative defense, protein modification, and metabolism regulation. In *Potentilla atrosanguinea,* PaSOD was shown to have overexpression in copper-zinc (SOD) at the time of seed germination during Cu stress. Furthermore, the upregulated and downregulated proteins during oxidative stress, detoxification, germination, and intermediary metabolism were identified by using MALDI-TOF and subsequent peptide mass fingerprinting analysis (Gill *et al.*, 2012).

Ni is required by plants as a micronutrient, and its deficiency induces a negative impact on growth and metabolism. Proteomics approaches by 2-DE in Indian mustard plants obtained 61 protein spots altered in expression, out of which 37 spots were ambiguously identified by MALDI-TOF MS.

The maximum identified proteins are responsible for S metabolism against oxidative stress revealing its role during Ni imbalances in cell (Wang *et al.*, 2012).

Se is a beneficial element to certain specific groups of plants for their growth and development and is also involved in protection under various kinds of abiotic stresses. Moreover, in high concentrations, it induces toxicity with denaturation in protein structure and their functions (Smolen *et al.*, 2016). There is no direct evidence found which reveals Se-mediated tolerance mechanisms under Cd stress in *Cucumis sativus*. The work of researchers showed that cadmium and Se together induce a strong interaction among proteins' response under stress condition, which affects metabolism, photosynthesis, and storage, which can be predicted with 2-DE coupled with MS (Sun *et al.*, 2016).

Si is a differentially abundant nutrient present in soils which plays an important role in plant cells. But due to lack of research in proteomics approaches, it is still unexplored (Muneer and Jeong, 2015a, 2015b; Rios *et al.*, 2017). In the exogenous supply of Si on *Dianthus caryophyllous*, the molecular mechanism behind hyperhydricity was observed. The molecular approaches in proteomics showed that Si is actively involved in various metabolisms and proteins are involved in ribosomal binding, oxido-reduction, hormone/cell signaling, metal/ion binding, defense, and photosynthesis for enhancing the recovery of the shoots from hyperhydricity (Soundararajan *et al.*, 2017).

Al poisoning is one of the major issues that has resulted in stunted growth and development (Liu *et al.*, 2017c; Zhou *et al.*, 2017). The quantitative proteomic technique based on iTRAQ was used to identify differentially expressed proteins in Al-tolerant wheat lines ET8 during AL and AL with indole-3-acetic acid (IAA) administration. Proteins such as mitogen-activated protein kinase (MAPK) 2c and MAPK1a were found to be highly expressed in the combined stress of Al and IAA, whereas auxin response factors (ARFs) and Aux/IAA were shown to be upregulated and downregulated by Al, respectively (Liu *et al.*, 2017d). Al was found to have a sensitive effect on the fruits of *Solanum lycopersicum* at the mature green, changing, and red stages. Proteins involved in protein biosynthesis, photosynthesis, and primary carbohydrate metabolism were found to be downregulated in Al-treated plants when compared to *Lanthanum* ions (Sangireddy *et al.*, 2017). iTRAQ and 2D-liquid chromatography (LC) coupled to MS/MS (2D-LC-MS/MS) were used to identify Al-responsive proteins from Al-sensitive (BR007) and Al-tolerant (SC566) sorghum lines. Proteins implicated in the antioxidant system were found in greater abundance in the tolerant line, as evidenced by protein interaction and pathway analyses (Zhou *et al.*, 2017).

9.3 GENOMICS ALTERATION IN PLANT SYSTEMS UNDER NUTRITIONAL STRESS

Plant genomics is a discipline of plant science that seeks to decode, characterize, and investigate the genetic compositions, structures, organizations, and functions of plant genomes, as well as molecular genetic interactions/networks. It aspires to develop large-scale, high-throughput technologies and efficient tools and methodologies to elucidate the basis of genetic traits/characteristics, genetic diversities, and byproduct production to understand phenotypic development throughout plant ontogenesis with genetics by environmental interactions, and to map important loci throughout the genome (Ibrokhim, 2016). The study of the entire genome, including genic and intergenic regions, is beneficial for gaining insights into molecular reactions, which will provide particular methods for future plant development (Akpınar, Lucas and Budak, 2013). However, the use of genomics in the field of abiotic or biotic stress biology is quickly expanding (Kayum *et al.*, 2016; Rout and Sahoo, 2013). Using genomic techniques, many studies are focusing on identifying abiotic stress response genes in plants. Modern genomics techniques such as recombinant DNA construction, DNA/RNA sequencing, and various types of PCR, microarray, DNA chip, fluorescent in situ hybridization, and serial analysis of gene expression are frequently used to understand genomic studies in plants (Casassola *et al.*, 2013; Monobe and Silva, 2016). Plants can be exposed to adverse environmental conditions throughout their lives, including not only nutritional stress but also extreme temperatures, drought, high salinity, light intensity and photoperiodicity, air or water pollution, and other factors that reduce growth and negatively affect the plant (Liu *et al.*, 2017e; Sett, 2017). These pressures result in disorders such as plethora, hydropenia, and others (Dash *et al.*, 2014; Roychoudhury and Banerjee, 2015). It is widely understood that nutritional stress responses cause physiological, biochemical, and morphological changes as a result of signal transduction and gene expression control of genes encoding proteins involved in plant acclimation (Curci *et al.*, 2017). The plant can respond to stress by raising the amounts of photoprotective enzymes and stress metabolites, as well as by specific long-term adaptations such as leaf shape, size, and thickness; stomata density and distribution; and chloroplast structure and function. Depending on the plant, these sorts of adaptations can occur in a matter of days or weeks (Kayum *et al.*, 2016). As a result, nutrient-induced stress is the most dominating stress factor in the plant system. Thus, using genomics tools to explore plant responses to stressors and the complexity of biochemical processes could be a useful tool in comprehending the plant system at the genomic level. Some of the differential gene expression profiles of diverse plants are explored, using various plant portions to find genes that are differentially expressed under varied nutrient concentrations.

Deng *et al.* (2016) investigated the impact of N deficits on leaf-volatile chemicals in cucumbers and discovered that C6 and C9 aldehydes are more volatile under N-stress. Cucumber LOX genes such as CsLOX19, CsLOX20, and CsLOX22 were shown to be upregulated in an N-dependent manner during genomics research. The absorption and utilization of 230 N-associated genes in sorghum lines was hypothesized, and a significant nucleotide diversity was seen in nitrate/peptide transporter 1 (NRT1/PTR) genes during nitrogen intake and utilization (Massel *et al.*, 2016). Han *et al.* (2016) achieved the usage of N efficiency (NUE), NUE-associated genes connected to plant physiology, and genetics for increasing plant performance. The expression analysis employing both microarray and RNA-seq data revealed that the genes associated with amino acid biosynthesis (HvAlaAT1-1 [MLOC_66262]) had the highest expression quality when N was abundant.

K-related stress can affect photosynthesis, transpiration, stomatal conductance, photorespiration, and respiration, which are all reduced (Shen *et al.*, 2017; Zhao *et al.*, 2016). The impact of low and high K concentrations resulted in the expression of genes linked to pear tree fruit sugar metabolism. qRT-PCR was used to validate the RNA sequencing data, and it was discovered that during the fruiting stage, expression of K transporter (gene31518) and K channel AKT1 (gene39320) genes was greatly upregulated in response to high K (Shen *et al.*, 2017). Zhang *et al.* (2017b), Zhao *et al.* (2016), and Takehisa *et al.* (2015) also provided detailed descriptions of deficiency-related works. According to Zhang and his colleagues, the mechanism of K deficits in rice plants was well-identified using transcriptome analysis, which revealed that genes involved in food transport, protein kinases, transcription processes, and plant hormones were significantly altered in roots. A phosphorylation research using the transcription factor auxin response factor 2 (ARF2) reveals regulation of the K-transporter gene HAK5 (high affinity K? transporter 5) in *A. thaliana* under low K ion concentration (Zhao *et al.*, 2016). Several groups of highly interconnected genes (encoding protein kinases, kinase-like domains containing proteins, and nutrient transporters) were significantly upregulated in response to K shortage (Takehisa *et al.*, 2015).

Jiang *et al.* (2017) evaluated P-tolerant and P-sensitive maize inbred line plants for P-starving responses. Several genes involved in P-uptake and utilization, protein and amino acid metabolism, cell development, stress tolerance, and other processes were tested quickly, using microarray technology, and showed both up- and downregulated responses in distinct inbred lines. Another study on plasma membrane H?-ATPase and its role in A. *thaliana* was proposed in relation to low P stress. RT-PCR was used to examine the expression of PM H?-ATPase genes (AHA1-AHA11) under low P concentrations and found a substantial difference in genes AHA1, AHA2, and AHA7 (Yuan *et al.*, 2017). Wu *et al.* (2016) investigated the role of P in maize by looking at ZmARF31 genes involved in P responses and root architecture regulation.

Ca-dependent protein kinase genes (MLOC_58648a, MLOC_19618a, MLOC_71733a, AK249361a, and MLOC_4965a) from *Hordeum vulgare* were identified using advanced molecular techniques. The expression and sequencing data revealed expression irregularities under various biotic and abiotic stressors (Yang *et al.*, 2017). There is no significant work on the direct implementation of Ca impact on plants. Ca-dependent protein kinase genes, on the other hand, have been tested on peanut, melon, soybean, and tomato plants with different features of plant physiology (Hettenhausen *et al.*, 2016; Hu *et al.*, 2016; Li *et al.*, 2017c; Zhang *et al.*, 2017a).

An Mg-deficiency resemblance study was performed on *Citrus sinensis* leaf samples using Illumina sequencing, and a total of 75 upregulated and 71 downregulated miRNA were identified. Ma *et al.* (2016) summarized the augmentation of cell transport, activation of lipid metabolism, and cell wall architecture. Elevated Mg in the form of magnesium sulfate is used to examine the root transcriptome in *Arabidopsis* knockout lines. The differential expression of genes encoding Mg, S, and Ca/proton transporters is examined, and it is found that Mg transporter genes MRS2-10 and MRS2-7 are upregulated. In contrast, the expression of genes encoding the S transporters SULTR3;4, SULTR3;1, and SULTR4;1 is downregulated. On the other hand, the gene encoding the vacuolar Mg2?/H? antiporter (MHX) was upregulated, while the Ca2?/H? antiporters CAX1, CAX2, and CAX3 were downregulated (Visscher *et al.*, 2010).

The S-less induced expression of S transport systems was examined in *A. thaliana* roots, while the S-inducible expression of SULTR2;1 was unclear, implying the variety of words governing transcription. The discovery adds to the mechanism of increased uptake and internal translocation of S gene expression produced by S-responsive elements (Nakashita *et al.*, 2015).

The toxicity of B was tested by extracting roots from two *Citrus* species, *Citrus sinensis* and *Citrus grandis*. In these species, 44 upregulated and 66 downregulated genes were found, with root hair defective 3 expressed in *C. sinensis* and villin4 suppressed. Furthermore, B-excess-responsive genes linked to energy, lipid, and nucleic acid metabolism have been identified, which may help to understand the mechanisms of citrus B-toxicity and B-tolerance at the transcriptional level (Guo *et al.*, 2016). Lv *et al.* (2017) demonstrated B homeostasis in *Arabidopsis* root samples by employing the SHB1/HY1 gene and raising the degree of BOR4 expression. The study concludes that excessive B absorption promotes the exclusion of excessive B by upregulating transcription of the BOR4 gene. The comprehension of genetic variety linked with B tolerance was tested in *Oryza sativa* (Neto *et al.*, 2017), where the genes involved in biochemical binding, transport, transcriptional control, and redox homeostasis were studied. Similarly, long-term B toxicity was assessed using the MiRNA assay and stem-loop qRT-PCR, with expressions of various particular genes reported (Huang *et al.*, 2016a).

Mn causing genomic analysis was discovered in *Arabidopsis* roots to provide the notion of Mn deficiency-induced genes and their activity. A diverse set of genes participating in many pathways was also discovered, indicating important functions such as photosynthesis, Calvin cycle, glycolysis, TCA cycle, respiration, and fermentation (Celma *et al.*, 2016). Another study used AK251925.1, AK249774.1, and MLOC_77860.1 candidate genes to investigate Mn-dependent regulation of photosystem II subunits and discovered that their co-localization was significant. In addition, two other types of genes, AK357955 (protein phosphatase) and AK368229 (chlorophyll a/b binding protein), were discovered to be directly engaged in Mn-dependent regulation (Leplat *et al.*, 2016). *Citrus sinensis* and *Citrus grandis* were treated with greater concentrations of Mn to better understand the processes of Mn-tolerance. The status of expression of genes involved in numerous processes such as leaf senescence, ATPase production, cell transport, protein and nucleic acid metabolism, and cell wall building was investigated using cDNA-AFLP. Under Mn excess, the genes involved in phosphorylation and dephosphorylation have both up and down expression quality, indicating the balancing nature in between phosphorylation and dephosphorylation produced by Mn-toxicity in C. *grandis* cause oxidative damage (Zhou *et al.*, 2013).

Fe uptake and chromosomal alterations in *Lens culinaris* were investigated using Fe metabolism genes (Ferritin-1, BHLH-1, and IRT-1). The up-regulation of Ferritin-1 and IRT-1 in excess Fe circumstances indicates equilibrium between Fe absorption and translocation (Gupta, McPhee and Kumar, 2017). In the leaves of Japonica rice, genes with functions such as Fe absorption, transport, hormonal control, stress oxidant, senescence, and stress were examined and extensively described.

Microarray validation by RT-qPCR revealed gene up and down regulation, but this was depending on the dose used. Out of 2525 related genes with known functions, 1773 were found to be up-regulated, while 53 were found to be down-regulated (Finatto *et al.*, 2015). In *Citrus* and *Arabidopsis*, Fe shortage causes differential gene expression by overregulating genes involved in cell wall manufacturing, ethylene and abscisic acid signal transmission, and Fe homeostasis (Fu *et al.*, 2017; Satbhai *et al.*, 2017).

Zn-induced overexpression of the NcZNT1 gene in *A. thaliana* imparted increased tolerance to Zn exposure. However, the ZNT1 Zn transporter, AtZIP4, is highly upregulated in the brain, endodermis, and pericycle during Zn shortage (Lin *et al.*, 2016).

Cu excess is responsible for the production of rice laccase genes (OsLACs) in *Arabidopsis*, notably the OsLAC10 gene, which was elevated in high Cu concentration and validated in *Escherichia coli*. The findings show that overexpression of OsLAC10 may aid in the tolerance of Cu stress, presumably by lignification in roots, which avoids excessive Cu absorption (Liu *et al.*, 2017b). Huang *et al.* (2016b) investigated the expression pattern of the OsHMA4 gene, which is responsible for Cu accumulation in rice, and discovered that OsHMA4 was highly increased by higher Cu accumulation in roots than shoots, implying a vital function for OsHMA4 in Cu homeostasis. A down-regulation of the SPL7 transcription factor (which plays important functions in plant growth and development) was also seen in *A. thaliana* under Cu stress, supporting the involvement of SPL7 under Cu stress (Gielen *et al.*, 2016). The differential expression of genes linked to the ROS generating (Rboh, AO) and superoxide scavenging (Fe-SOD, POD, CAT) systems in grapevines under Cu stress was extensively studied, revealing the upregulation of desirable genes to overcome the oxidative stress caused by Cu (Leng *et al.*, 2015). The expression of Ni-induced antioxidant enzyme genes (Fe-SOD, POD, APX, and CAT) in glycine max was examined, and it was discovered that Fe-SOD and APX were highly upregulated in Ni-treated plants, indicating that plants produced higher ROS (Sirhindi *et al.*, 2016). Another contribution was made by Theriault, Michael and Nkongolo (2016), who studied *Betula papyrifera* and discovered numerous genes involved in Ni resistance, including glutathione S-transferase, thioredoxin family protein, putative transmembrane protein, and two NRAMP metal ion transporters.

Selenocysteinemethyl-transferase (SMT) gene expression in *Lactuca sativa* was dramatically altered when Se was administered at high concentrations. Smolen *et al.* (2016) used qRT-PCR to quantify the amount of expression in leaves and discovered that the gene is upregulated under Se stress to Se-detoxification.

Si transporter (Lsi1 and Lsi2) and SOD isoform (Mn-SOD, Cu/Zn-SOD, and Fe-SOD) genes were considerably up- and downregulated in ryegrass plants under Si stress. More importantly, transporter genes (Lsi1 and Lsi2) were downregulated in roots following Si treatment alone; but, when combined with Al, an elevated pattern was established. Differential variations in gene expression of Mn-SOD, Cu/Zn-SOD, and Fe-SOD provide insight into superoxide radical production (Pontigo *et al.*, 2017). Pavlovic *et al.* (2016) show that Si functions as a stimulator for overcoming other metal deficits by applying it to cucumber plants. The investigation demonstrated that overexpression of nicotianamine synthase (NAS) associated genes (NAS1) in leaf tissues, which are responsible for NA manufacturing and accumulation, leads to improved Fe chelation. Si-induced changes in the transcript levels of 265 genes involved in suberin and lignin synthesis were investigated in *Oryza sativa* using microarray and qRT-PCR. ABC transporters, Class III peroxidase, Phytosulfokine receptors, and other genes with specialized functions were briefly characterized and discussed, emphasizing the relevance of Si nutrition in terms of reducing oxidation levels in cells (Fleck *et al.*, 2011).

Al toxicity was observed in *Setaria viridis* by using a gene (BdMATE) from the multidrug and toxic compound extrusion (MATE) family, which supports tolerance capability with a feature of overexpression (Ribeiro *et al.*, 2017). A study of the lucerne plant demonstrated the effect of Al stress by discovering and analyzing Al-stress-responsive genes at the whole-genome level. The differentially expressed genes are involved in protein biosynthesis, the citrate cycle, membrane transport, and hormone control, according to metabolic pathway enrichment analysis (Liu *et al.*, 2017c). A genome-wide transcriptome investigation of Al-response genes in *Hydrangea macrophylla* was published using an RT-PCR technique, followed by sequencing on an Illumina HiSeq 2500 platform. Under Al stress, the plasma membrane Al uptake transporter and the NRAMP Al transporter were expressed more in roots and leaves, indicating their participation in Al tolerance by lowering hazardous Al levels (Chen *et al.*, 2015b). Other important contributions on Al stress were made in *Oryza sativa*, *Glycine max*, and *Fagopyrum esculentum* using varied amounts of Al (Alvarado *et al.*, 2017; Gao *et al.*, 2017; Zhu *et al.*, 2015).

9.4 CONCLUSION

Scientists studying plants are working hard to shed light on plants' genomes and functional proteomics under various nutritional stresses. The complete proteome or plant genome sequences have yet to be widely available, so the theory still needs to be utilized. The topic has undergone significant advancements, and the latest findings from various investigations have substantially enhanced our understanding of the issue of "proteomic and genomic responsiveness of plants to nutritive stress." The expression of multiple genes that produce molecules that are involved in signaling and/or regulatory pathways, the synthesis of structural and functional proteins, and the production of metabolites that indicate the metal tolerance to stress in plant cells are the basis for plant engineering strategies that are optimized for nutritional excess or less. The proteomics and genomics discoveries that have been presented offer a broad platform for designing upcoming studies on nutritional components that cause toxicity and how to eliminate it at the molecular level. Lastly, we want to let plant enthusiasts know that there is still a lot of work to be done to identify and better characterize the important proteins and genes that underlie a plant's ability to withstand a particular stress. This can be done by utilizing large-scale throughput technologies.

The environment has become contaminated with toxic concentrations of heavy metals, particularly Cd, As, Al, Hg, and Pb, due to rigorous weathering of rocks, soil mineralization, fast industrialization, and anthropogenic intrusions. Regrettably, the ambient soil bioavailability of these heavy metals is gradually increasing globally. Through the production of reactive oxygen species, lipid peroxidation, DNA damage, sudden changes in the cell cycle, nutritional imbalance, and toxicities, these heavy metals negatively affect cells and the components that make up cells. These harmful heavy metals have also been shown to alter the expression of chaperones and proteases, impacting protein metabolism and possibly preventing the synthesis of new proteins. These heavy metals threaten plant growth, development, and crop yields, as well as the food chains, because plants frequently absorb and store even minute but hazardous concentrations of these heavy metals in their seeds and edible sections. The scientific community has been interested in studying the crucial elements of heavy metal toxicity and cultivating resilience in crop plants because of the significance of heavy metals. For phytoremediation of heavy metal-contaminated soils, there has been a persistent interest in comprehending the mechanistic perspectives of heavy metal soil uptake, transportation translocation, and sequestration/chelation in the cell compartments of heavy metal-stressed plants as well as in identifying possible heavy metal hyper-accumulating plant species. However, more thorough research is required to find novel transporters, uncover retrograde signaling pathways and cross-talks, identify potential hyper accumulators in plants, and understand the implications of co-uptake of metals in the long run. In addition to metabolomics, metallomics, metallo-proteomics, and miRNA omics, "new age omics" techniques, such as next-generation sequencing platform-driven genomics and transcriptomics, are showing promise as platforms for interpreting the mechanistic insights of heavy metal interactions with plant cells/organelles. We have reviewed recent research on harmful non-essential heavy metals, their effects on plant development and growth, and how plants respond to heavy metals stress by adapting, avoiding, or tolerating it, particularly when utilizing omics techniques. Using omics technology, researchers are gaining a better knowledge of heavy metal stress responses at the genome, transcriptome, metabolome, and miRNAome levels.

High-throughput omics techniques are being used to study the reprogramming of transcription of genes in massive amounts in exposure to and to cope with heavy metal toxicity. Numerous important genes, including heavy metal transporters, have been found by genome-wide investigations and can be efficiently exploited to modify plants to tolerate heat stress genetically. Academics are identifying crucial heavy metal-stress regulators and their evolutionary importance using several in silico technologies and databases that have been established recently. To understand molecular insights into plant-heavy metal interactions and stress responses and generate heavy metals-tolerant plants, thorough investigations are necessary before exploring omics techniques.

REFERENCES

Ahsan, N., Renaut, J. and Komatsu, S. (2009) 'Recent developments in the application of proteomics to the analysis of plant responses to heavy metals', *Proteomics*, 9, pp. 2602–2621.

Akpınar, B.A., Lucas, S.J. and Budak, H. (2013) 'Genomics approaches for crop improvement against abiotic stress', *Scientific World Journal*, 2013, pp. 1–9.

Al-Obaidi, J.R. (2020) 'Micro-and macronutrient signalling in plant cells: A proteomic standpoint under stress conditions, In: *Plant micronutrients: Deficiency and toxicity management* (Aftab, T. Hakeem, K.R., Eds.), Cham, Switzerland: Springer International Publishing, pp. 241–255.

Alvarado, M.M. *et al.* (2017) 'Aluminum enhances growth and sugar concentration, alters macronutrient status and regulates the expression of NAC transcription factors in rice', *Frontiers in Plant Science*, 8, pp. 1–16.

Anderson, S. *et al.* (1981) 'Sequence and organization of the human mitochondrial genome', *Nature*, 290, pp. 457–465.

Appenroth, K.J. (2010) 'What are "heavy metals" in plant sciences?, *Acta Physiologiae Plantarum*, 32, pp. 615–619.

Balammal, G. and Reddy, P.J. (2012) 'Proteomics & genomics—a review', *Journal of Science*, 2, pp. 91–95.

Brown, P.H., Zhao, F.J. and Dobermann, A. (2022) 'What is a plant nutrient? Changing definitions to advance science and innovation in plant nutrition', *Plant and Soil*, 476(1–2), pp. 11–23.

Carpanez, T.G. *et al.* (2022) 'Sugarcane vinasse as organo-mineral fertilizers feedstock: Opportunities and environmental risks', *Science of The Total Environment*, 4, p. 154998.

Casassola, A. *et al.* (2013) 'Gene Expression: A review on methods for the study of defense-related Gene differential Expression in plants', *American Journal of Plant Science*, 4, pp. 64–73.

Celma, J.R. *et al.* (2016) 'Systems-wide analysis of manganese deficiency induced changes in gene activity of *Arabidopsis* roots', *Scientific Reports*, 6(1), p. 16.

Chandna, R. and Ahmad, A. (2015) 'Nitrogen stress-induced alterations in the leaf proteome of two wheat varieties grown at different nitrogen levels', *Physiology and Molecular Biology of Plants*, 21, pp. 19–33.

Chen, H. *et al.* (2015b) 'Global transcriptome analysis reveals distinct aluminum-tolerance pathways in the Al-accumulating species Hydrangea macrophylla and marker identification', *PLoS ONE*, 10, pp. 1–19.

Chen, Z. *et al.* (2015a) 'Malate synthesis and secretion mediated by a manganese-enhanced malate dehydrogenase confers superior manganese tolerance in *Stylosanthes guianensis*', *Plant Physiology*, 167, pp. 176–188.

Ciesla, A. *et al.* (2016) 'Role for barley calcium-dependent protein kinase CPK2a in the response to drought', *Frontiers in Plant Science*, 7, pp. 1–15.

Clemens, S. (2001) 'Molecular mechanisms of plant metal tolerance and homeostasis', *Planta*, 212, pp. 475–486.

Curci, P.L. *et al.* (2017) 'Transcriptomic response of durum wheat to nitrogen starvation', *Scientific Reports*, 7(1), p. 14.

Dash, P.K. *et al.* (2014) 'Genome-wide analysis of drought induced gene expression changes in flax (*Linum usitatissimum*)', *GM Crops Food*, 5, pp. 106–119.

Deng, J. *et al.* (2016) 'Leaf volatile compounds and associated gene expression during short-term nitrogen deficient treatments in *Cucumis* seedlings', *International Journal of Molecular Science*, 17, pp. 1–13.

Ding, C. *et al.* (2010) 'Proteomic analysis of low nitrogen stress responsive proteins in roots of rice', *Plant Molecular Biology Reporter*, 29, pp. 618–625.

Donnini, S. *et al.* (2010) 'Proteomic characterization of iron deficiency responses in *Cucumis sativus* L. roots', *BMC Plant Biology*, 10(1), p. 268.

Ebong, G.A., Offiong, O.E. and Ekpo, B.O. (2014) 'Speciation, distribution and bioavailability of essential elements in waste impacted soils within Niger delta region of Nigeria', *Annals. Food Science and Technology*, 15, pp. 162–171.

Finatto, T. *et al.* (2015) 'Abiotic stress and genome dynamics: Specific genes and transposable elements response to iron excess in rice', *Rice*, 8, pp. 2–18.

Fleck, A.T. *et al.* (2011) 'Silicon enhances suberization and lignification in roots of rice (*Oryza sativa*)', *Journal of Experimental Botany*, 62, pp. 2001–2011.

Fu, L. *et al.* (2017) 'Physiological and transcriptional changes of three citrus rootstock seedlings under iron deficiency', *Frontiers in Plant Science*, 8, pp. 1–13.

Fukao, Y. *et al.* (2016) 'Quantitative proteomic analysis of the response to zinc, magnesium, and calcium deficiency in specific cell types of *Arabidopsis* roots', *Proteomes*, 4, pp. 1–13.

Gao, M. *et al.* (2017) 'Evaluation of reference genes for normalization of gene expression using quantitative RT-PCR under aluminum, cadmium, and heat stresses in soybean', *PLoS ONE*, 12(1), p. 15.

Gielen, H. *et al.* (2016) 'Toxicity responses of Cu and Cd: The involvement of miRNAs and the transcription factor SPL7', *BMC Plant Biology*, 16(1), p. 16.

Gill, T. *et al.* (2012) 'Protein dynamics during seed germination under copper stress in *Arabidopsis* over-expressing Potentilla superoxide dismutase', *Journal of Plant Research*, 125, pp. 165–172.

Guo, W. *et al.* (2015) 'Magnesium stress signaling in plant: Just a beginning', *Plant Signaling & Behavior*, 10, p. e992287.

Guo, P. *et al.* (2016) 'Long-term boron excess induced alterations of gene profiles in roots of two citrus species differing in boron-tolerance revealed by cDNA-AFLP', *Frontiers in Plant Science*, 7, pp. 1–16.

Gupta, D.S., McPhee, K. and Kumar, S. (2017) 'Development of molecular markers for iron metabolism related genes in lentil and their expression analysis under excess iron stress', *Frontiers in Plant Science*, 8, pp. 1–8.

Han, M. *et al.* (2016) 'Identification of nitrogen use efficiency genes in barley: Searching for QTLs controlling complex physiological traits', *Frontiers in Plant Science*, 7, pp. 1–17.

Hartenian, E. and Doench, J. G. (2015) 'Genetic screens and functional genomics using CRISPR/Cas9 technology', *FEBS Journal*, 282, pp. 1383–1393.

He, L. *et al.* (2012) 'Proteomic analysis of the effects of exogenous calcium on hypoxia- responsive proteins in cucumber roots', *Proteome Science*, 10(1), p. 15.

Hettenhausen, C. *et al.* (2016) 'Genome-wide identification of calcium-dependent protein kinases in soybean and analyses of their transcriptional responses to insect herbivory and drought stress', *Scientific Reports*, 6(1), p. 14.

Hopff, D., Wienkoopb, S. and Luthjea, S. (2013) 'The plasma membrane proteome of maize roots grown under low and high iron conditions', *Journal of Proteomics*, 91, pp. 605–618.

Hossain, Z. *et al.* (2016) 'Insights into the proteomic response of soybean towards Al_2O_3, ZnO and Ag nanoparticles stress', *Journal of Hazardous Materials*, 304, pp. 291–305.

Hu, J., Rampitsch, C. and Bykova, N. V. (2015) 'Advances in plant proteomics toward improvement of crop productivity and stress resistance', *Frontiers in Plant Science*, 6, pp. 1–15.

Hu, Z. *et al.* (2016) 'Genome wide identification and expression analysis of calcium dependent protein kinase in tomato', *Frontiers in Plant Science*, 7, pp. 1–11.

Huang, J.H. *et al.* (2016a) 'Illumina microRNA profiles reveal the involvement of miR397a in citrus adaptation to long-term boron toxicity via modulating secondary cell-wall biosynthesis', *Scientific Reports*, 6(1), p. 14.

Huang, X.Y. *et al.* (2016b) 'A heavy metal P-type ATPase OsHMA4 prevents copper accumulation in rice grain', *Nature Communications*, 7(1), p. 13.

Ibrahim, W.I.W. *et al.* (2015) 'Biomarkers for bone tumors: Discovery from genomics and proteomics studies and their challenges', *Molecular Medicine*, 21, pp. 861–872.

Ibrokhim, I.Y. (2016) 'Genomics era for plants and crop species – Advances made and needed tasks ahead', in Abdurakhmonov, I. (ed.) *Plant Genomics*. Rijeka, Croatia: Intech, pp. 3–25.

Jamla, M. *et al.* (2021) 'Omics approaches for understanding heavy metal responses and tolerance in plants', *Current Plant Biology*, 27, p. 100213.

Jiang, H. *et al.* (2017) 'Revealing new insights into different phosphorus-starving responses between two maize (*Zea mays*) inbred lines by transcriptomic and proteomic studies', *Scientific Reports*, 7(1), p. 9.

Kayum, M.A. *et al.* (2016) 'Research on biotic and abiotic stress related genes exploration and prediction in *Brassica rapa* and *B. oleracea*: A review', *Plant Breeding and Biotechnology*, 4, pp. 135–144.

Kerry, R.G. *et al.* (2018) 'Proteomic and genomic responses of plants to nutritional stress', *BioMetals*, 31, pp. 161–187. https://doi.org/10.1007/s10534-018-0083-9.

Komatsu, S. and Hossain, Z. (2017) 'Preface-plant proteomic research', *Int J Mol Sci*, 18, pp. 1–3.

Komatsu, S. *et al.* (2017) 'Integration of gel-based and gel-free proteomic data for functional analysis of proteins through Soybean Proteome Database', *Journal of Proteomics*, 163, pp. 52–66.

Kosakivska, I.V. *et al.* (2021) 'Molecular mechanisms of plant adaptive responses to heavy metals stress', *Cell Biology International*, 45(2), pp. 258–272.

Koussounadis, A. *et al.* (2015) 'Relationship between differentially expressed mRNA and mRNA-protein correlations in a xenograft model system', *Scientific Reports*, 5(1), p. 9.

Leng, X. *et al.* (2015) 'Comparative transcriptome analysis of grapevine in response to copper stress', *Scientific Reports*, 5(1), p. 17.

Leplat, F. *et al.* (2016) 'Identification of manganese efficiency candidate genes in winter barley (*Hordeum vulgare*) using genome wide association mapping', *BMC Genomics*, 17(1), p. 15.

Li, W. and Lan, P. (2017) 'The understanding of the plant iron deficiency responses in strategy I plants and the role of ethylene in this process by omic approaches', *Frontiers in Plant Science*, 8, pp. 1–15.

Liebig, B.J. (1840) *Chemistry in its application to agriculture and physiology*, London: Taylor and Walton.

Lin, Y.F. *et al.* (2016) 'Expression of the ZNT1 zinc transporter from the metal hyperaccumulator *Noccaea caerulescens* confers enhanced zinc and cadmium tolerance and accumulation to *Arabidopsis thaliana*', *PLoS ONE*, 11, pp. 1–30.

Liu, C. *et al.* (2017a) 'Prominent topologically associated domains differentiate global chromatin packing in rice from *Arabidopsis*', *Nature Plants*, 3, pp. 742–748.

Liu, Q. *et al.* (2017b) 'Comprehensive analysis of *Rice lacase* gene (OsLAC) family and ectopic expression of OsLAC10 enhances tolerance to copper stress in *Arabidopsis*', *International Journal of Molecular Sciences*, 18, pp. 1–16.

Liu, T. *et al.* (2014) 'A combined proteomic and transcriptomic analysis on sulfur metabolism pathways of *Arabidopsis thaliana* under simulated acid rain', *PLoS ONE*, 9, pp. 1–11.

Liu, W. *et al.* (2017c) 'Transcriptome analyses reveal candidate genes potentially involved in Al stress response in Alfalfa', *Frontiers in Plant Science*, 8, pp. 1–11.

Liu, X. *et al.* (2013) 'An integrated proteomic and metabolomic study on the chronic effects of mercury in *Suaeda salsa* under an environmentally relevant salinity', *PLoS ONE*, 8, pp. 1–13.

Liu, X. *et al.* (2017d) 'MAPK-mediated auxin signal transduction pathways regulate the malic acid secretion under aluminum stress in wheat (*Triticum aestivum* L.)', *Scientific Reports*, 7(1), p. 12.

Liu, Y. *et al.* (2017e) 'Effects of PEG induced drought stress on regulation of indole alkaloid biosynthesis in *Catharanthus roseus*', *Journal of Plant Interactions*, 12, pp. 87–91.

Lucini, L. and Bernardo, L. (2015) 'Comparison of proteome response to saline and zinc stress in lettuce', *Frontiers in Plant Science*, 6, pp. 1–12.

Lv, Q. *et al.* (2017) 'SHB1/HY1 alleviates excess boron stress by increasing BOR4 expression level and maintaining boron homeostasis in *Arabidopsis* roots', *Frontiers in Plant Science*, 8, pp. 1–13.

Ma, C.L. *et al.* (2016) 'MicroRNA regulatory mechanisms on *Citrus sinensis* leaves to magnesium-deficiency', *Frontiers in Plant Science*, 7, pp. 1–16.

Marino, D. *et al.* (2016) 'Quantitative proteomics reveals the importance of nitrogen source to control glucosinolate metabolism in *Arabidopsis thaliana* and *Brassica oleracea*', *Journal of Experimental Botany*, 67, pp. 3313–3323.

Marrocco, I., Altieri, F. and Peluso, I. (2017) 'Measurement and clinical significance of biomarkers of oxidative stress in humans', *Oxidative Medicine and Cellular Longevity*, 2017, pp. 1–32.

Massel, K. *et al.* (2016) 'Whole genome sequencing reveals potential new targets for improving nitrogen uptake and utilization in *Sorghum* bicolor', *Frontiers in Plant Science*, 7, pp. 1–15.

Meisrimler, C. N. *et al.* (2016) 'Long-term iron deficiency: Tracing changes in the proteome of different pea (*Pisum sativum* L.) cultivars', *Journal of Proteomics*, 140, pp. 13–23.

Minic, Z. (2015) 'Proteomic studies of the effects of different stress conditions on central carbon metabolism in microorganisms', *Journal of Proteomics & Bioinformatics*, 8, pp. 80–90.

Monobe, M.M.S. and Silva, R.C. (2016) 'Gene expression: An overview of methods and applications for cancer research', *Veterinaria e Zootecnia*, 23, pp. 532–546.

Muneer, S. and Jeong, B.R. (2015a) 'Proteomic analysis of salt-stress responsive proteins in roots of tomato (*Lycopersicon esculentum* L.) plants towards silicon efficiency', *Plant Growth Regulation*, 77, pp. 133–146.

Muneer, S. and Jeong, B.R. (2015b) 'Proteomic analysis provides new insights in phosphorus homeostasis subjected to pi (inorganic phosphate) starvation in tomato plants (*Solanum lycopersicum* L.)', *PLoS ONE*, 10, pp. 1–18.

Nadira, U.A. *et al.* (2016) 'Identification of the differentially accumulated proteins associated with low phosphorus tolerance in a Tibetan wild barley accession', *Journal of Plant Physiology*, 198, pp. 10–22.

Nakashita, A.M. *et al.* (2015) 'Sulfur-responsive elements in the 39-nontranscribed intergenic region are essential for the induction of sulfate transporter 2; 1 gene expression in *Arabidopsis* roots under sulfur deficiency', *Plant Cell*, 27, pp. 1279–1296.

Nazir, M. *et al.* (2016) 'Nitrogen-deficiency stress induces protein expression differentially in low-N tolerant and low-N sensitive maize genotypes', *Frontiers in Plant Science*, 7, pp. 1–16.

Neto, J.B.A. *et al.* (2017) 'Genetic factors underlying boron toxicity tolerance in rice: Genome- wide association study and transcriptomic analysis', *Journal of Experimental Botany*, 68, pp. 687–700.

Nouri, M.Z. and Komatsu, S. (2013) 'Subcellular protein overexpression to develop abiotic stress tolerant plants', *Frontiers in Plant Science*, 4, pp. 1–7.

Oh, M.W., Nanjo, Y. and Komatsu, S. (2014) 'Gel free proteomic analysis of soybean root proteins affected by calcium under flooding stress', *Frontiers in Plant Science*, 5, pp. 1–15.

Parvaiz, A. *et al.* (2016) 'Role of proteomics in crop stress tolerance', *Frontiers in Plant Science*, 7, pp. 1–14.

Pavlovic, J. *et al.* (2016) 'Silicon enhances leaf remobilization of iron in cucumber under limited iron conditions', *Annals of Botany*, 118, pp. 271–280.

Peng, H.Y. *et al.* (2015) 'Proteomic analysis of *Citrus sinensis* roots and leaves in response to long-term magnesium-deficiency', *BMC Genomics*, 16(1), p. 24.

Pontigo, S. *et al.* (2017) 'Silicon-mediated alleviation of aluminum toxicity by modulation of Al/Si uptake and antioxidant performance in ryegrass plants', *Frontiers in Plant Science*, 8, pp. 1–15.

Prusty, S. *et al.* (2022) 'Proteomic and genomic studies of micronutrient deficiency and toxicity in plants', *Plants*, 11, p. 2424. https://doi.org/10.3390/plants11182424

Qu, Q. *et al.* (2020) 'Rhizosphere microbiome assembly and its impact on plant growth', *Journal of Agricultural and Food Chemistry*, 68(18), pp. 5024–5038.

Ranty, B. *et al.* (2016) 'Calcium sensors as key hubs in plant responses to biotic and abiotic stresses', *Frontiers in Plant Science*, 7, pp. 1–7.

Rasheed, F. *et al.* (2021) 'Involvement of ethylene in reversal of salt stress by salicylic acid in the presence of sulfur in mustard (*Brassica juncea* L.)', *Journal of Plant Growth Regulation*, 41, pp. 3449–3466. https://doi.org/10.1007/s00344-021-10526-9

Rasheed, F. *et al.* (2022a) 'The key roles of salicylic acid and sulfur in plant salinity stress tolerance', *Journal of Plant Growth Regulation*, 41, pp. 1891–1904. https://doi.org/10.1007/s00344-020-10257-3

Rasheed, F. *et al.* (2022b) 'Nitric oxide and salicylic acid regulate glutathione and ethylene production to enhance heat stress acclimation in wheat involving sulfur assimilation', *Plants*, 11, p. 3131. https://doi.org/10.3390/plants11223131

Rehman, A.U. *et al.* (2021) 'Toxicity of heavy metals in plants and animals and their uptake by magnetic iron oxide nanoparticles', *Journal of Molecular Liquids*, 321, p. 114455.

Ribeiro, A.P. *et al.* (2017) 'Overexpression of BdMATE gene improves aluminum tolerance in *Setaria viridis*', *Frontiers in Plant Science*, 8, pp. 1–12.

Rios, J.J. *et al.* (2017) 'Silicon-mediated improvement in plant salinity tolerance: The role of Aquaporins', *Frontiers in Plant Science*, 8, pp. 1–10.

Rout, J.R. and Sahoo, S.L. (2013) 'Antioxidant enzyme gene expression in response to copper stress in *Withania somnifera* L', *Plant Growth Regulation*, 71, pp. 95–99.

Rout, J.R. *et al.* (2013) 'Copper-stress induced alterations in protein profile and antioxidant enzymes activities in the in vitro grown *Withania somnifera* L', *Physiology and Molecular Biology of Plants*, 19, pp. 353–361.

Roychoudhury, A. and Banerjee, A. (2015) 'Transcriptome analysis of abiotic stress response in plants', *Transcriptomics*, 3, pp. 1–2.

Sang, W. *et al.* (2015) 'Two-dimensional gel electrophoresis data in support of leaf comparative proteomics of two citrus species differing in boron-tolerance', *Data Brief*, 4, pp. 44–46.

Sang, W. *et al.* (2017) 'Effects of high toxic boron concentration on protein profiles in roots of two citrus species differing in boron-tolerance revealed by a 2-DE based MS approach', *Frontiers in Plant Science*, 8, pp. 1–19.

Sangireddy, S. *et al.* (2017) 'Effects of Al^{3+} and La^{3+} trivalent metal ions on tomato fruit proteomes', *Proteomes*, 5(1), p. 7.

Sarasketa, A. *et al.* (2016) 'Nitrogen source and external medium pH interaction differentially affects root and shoot metabolism in *Arabidopsis*', *Frontiers in Plant Science*, 7, pp. 1–12.

Satbhai, S.B. *et al.* (2017) 'Natural allelic variation of *FRO2* modulates *Arabidopsis* root growth under iron deficiency', *Nature Communications*, 8, pp. 1–10.

Sett, R. (2017) 'Tolerance of plant in response to abiotic stress factor', *Recent Advances in Petrochemical Science*, 1, p. 555573.

Shen, C. *et al.* (2017) 'Transcriptome analysis of differentially expressed genes induced by low and high potassium levels provides insight into fruit sugar metabolism of pea', *Frontiers in Plant Science*, 8, pp. 1–16.

Singh, S. *et al.* (2016) 'Heavy metal tolerance in plants: Role of transcriptomics, proteomics, metabolomics and ionomics', *Frontiers in Plant Science*, 6, pp. 1–36.

Sirhindi, G. *et al.* (2016) 'Jasmonic acid modulates the physio-biochemical attributes, antioxidant enzyme activity, and gene expression in *Glycine max* under nickel toxicity', *Frontiers in Plant Science*, 7, pp. 1–12.

Smolen, S. *et al.* (2016) 'Iodine and selenium biofertification with additional application of salicylic acid affects yield, selected molecular parameters and chemical composition of lettuce plants (*Lactuca sativa* L. var. capitata)', *Frontiers in Plant Science*, 7, pp. 1–16.

Soundararajan, P. *et al.* (2017) 'Exogenous supplementation of silicon improved the recovery of hyperhydric shoots in *Dianthus caryophyllus* L. by stabilizing the physiology and protein expression', *Frontiers in Plant Science*, 8, pp. 1–17.

Sun, H. *et al.* (2016) 'Physiological and proteomic analysis of selenium-mediated tolerance to Cd stress in cucumber (*Cucumis sativus* L.)', *Ecotoxicology and Environmental Safety*, 133, pp. 114–126.

Takehisa, H. *et al.* (2015) 'Coexpression network analysis of macronutrient deficiency response genes in rice', *Rice*, 8(1), p. 7.

Theriault, G., Michael, P. and Nkongolo, K. (2016) 'Comprehensive transcriptome analysis of response to nickel stress in white birch (*Betula papyrifera*)', *PLoS ONE*, 11, pp. 1–21.

Toor, M. D. *et al.* (2021). 'Nutrients and their importance in agriculture crop production: A review', *Indian Journal of Pure & Applied Biosciences*, 9(1), pp. 1–6.

Torregrosa, V.L., Husekova, B. and Sychrova, H. (2016) 'Potassium uptake mediated by Trk1 is crucial for *Candida glabrata* growth and fitness', *PLoS ONE*, 11, pp. 1–18.

Tucker, M.R. (1999) Essential plant nutrients: their presence in North Carolina soils and role in plant nutrition, Agronomic division, NCDA & CS bulletin-Agronomic Division, USA, pp. 1–9.

Varma, A. and Shrivastava, N. (2008) 'The role of plant genomics in biotechnology,' in Doelle H.W. and DaSilva E.J. (eds.) *Biotechnology* (vol VIII). Oxford, UK: Encyclopedia of Life Support Systems (EOLSS), Eolss Publishers, pp. 1–12.

Visscher, A.M. *et al.* (2010) 'Growth performance and root transcriptome remodeling of *Arabidopsis* in response to mars-like levels of magnesium sulfate', *PLoS ONE*, 5, pp. 1–16.

Wang, Y. *et al.* (2012) 'Response to nickel in the proteome of the metal accumulator plant *Brassica juncea*', *Journal of Plant Interactions*, 7, pp. 230–237.

Wang, X. *et al.* (2015) 'Proteomic study of microsomal proteins reveals a key role for *Arabidopsis annexin* 1 in mediating heat stress-induced increase in intracellular calcium levels', *Molecular & Cellular Proteomics*, 14, pp. 686–694.

Wu, F. *et al.* (2016) 'Molecular evolution and association of natural variation in ZmARF31 with low phosphorus tolerance in maize', *Frontiers in Plant Science*, 7, pp. 1–14.

Yang, Y. *et al.* (2017) 'Genome wide survey indicates diverse physiological roles of the barley (*Hordeum vulgare* L.) calcium-dependent protein kinase genes', *Scientific Reports*, 7(1), p. 14.

Yildiz, M. and Terzi, H. (2016) 'Proteomic analysis of chromium stress and sulfur deficiency responses in leaves of two canola (*Brassica napus* L.) cultivars differing in Cr(VI) tolerance', *Ecotoxicology and Environmental Safety*, 124, pp. 255–266.

Yina, Y. *et al.* (2015) 'Comparative proteomic and physiological analyses reveal the protective effect of exogenous calcium on the germinating soybean response to salt stress', *Journal of Proteomics*, 113, pp. 110–126.

You, X. *et al.* (2014) 'Proteomic changes of citrus roots in response to long-term manganese toxicity', *Trees*, 28, pp. 1383–1399.

Yuan, W. *et al.* (2017) 'Arabidopsis plasma membrane H+-ATPase genes AHA2 and AHA7 have distinct and overlapping roles in the modulation of root tip H+ efflux in response to low-phosphorus stress', *Journal of Experimental Botany*, 68, pp. 1731–1741.

Zeng, J. *et al.* (2015) 'Identification of the proteins associated with low potassium tolerance in cultivated and Tibetan wild barley', *Journal of Proteomics*, 126, pp. 1–11.

Zhang, H. *et al.* (2016a) 'Analysis of copper binding proteins in rice radicles exposed to excess copper and hydrogen peroxide stress', *Frontiers in Plant Science*, 7, pp. 1–15.

Zhang, H. *et al.* (2017a) 'Genome-wide identification and expression analysis of calcium-dependent protein kinase and its related kinase gene families in melon (*Cucumis melo* L)', *PLoS ONE*, 12, pp. 1–23.

Zhang, K. *et al.* (2014) 'Comparative proteomic analyses provide new insights into low phosphorus stress responses in maize leaves', *PLoS ONE*, 9, pp. 1–16.

Zhang, S. *et al.* (2016b) 'iTRAQ based quantitative proteomic analysis gives insight into sexually different metabolic processes of poplars under nitrogen and phosphorus deficiencies', *Proteomics*, 16, pp. 614–628.

Zhang, X. *et al.* (2017b) 'Transcriptome analysis of rice seedling roots in response to potassium deficiency', *Scientific Reports*, 7(1), p. 11.

Zhang, Z. *et al.* (2016c) 'Proteome quantification of cotton xylem sap suggests the mechanisms of potassium deficiency-induced changes in plant resistance to environmental stresses', *Scientific Reports*, 6(1), p. 15.

Zhao, S. *et al.* (2016) 'Phosphorylation of ARF2 relieves its repression of transcription of the K + transporter gene HAK5 in response to low potassium stress', *Plant Cell*, 28, pp. 3005–3019.

Zhou, C.P. *et al.* (2013) 'Leaf cDNA-AFLP analysis of two citrus species differing in manganese tolerance in response to long-term manganese-toxicity', *BMC Genomics*, 14(1), p. 19.

Zhou, D. *et al.* (2017) 'Quantitative iTRAQ proteomics revealed possible roles for antioxidant proteins in sorghum aluminum tolerance', *Frontiers in Plant Science*, 7, pp. 1–14.

Zhu, H. *et al.* (2015) 'Genome-wide transcriptomic and phylogenetic analyses reveal distinct aluminum-tolerance mechanisms in the aluminum-accumulating species buckwheat (*Fagopyrum tataricum*)', *BMC Plant Biology*, 15, pp. 1–13.

10 Molecular and Physiological Studies of the Role of Micronutrients in Sustainable Growth and Development of Crop Plants

Aryadeep Roychoudhury, Swayambhik Mukherjee, and Anindya Ghosh

10.1 INTRODUCTION

Micronutrients, also referred to as trace elements or essential minerals, are fundamental elements required by plants in minute quantities for their optimal growth and development. Despite their minimal requirement, micronutrients are pivotal for various physiological and biochemical processes that underlie plant health, productivity, and overall sustainability (Husen, 2024). Along with their role in plant growth, they are also involved in detection and response to various kinds of stressors, contributing to innate immunity of plants (Jan *et al.*, 2022). Despite the established beneficial roles micronutrients exert in plants, excessive application of these elements can lead to a variety of problems, such as contamination of the soil due to metal accumulation, toxicity in the plants, and detrimental effects on health (Rehman *et al.*, 2019). In recent years, significant advancements in molecular and physiological research have unraveled the intricate mechanisms through which micronutrients exert their influence on plant growth and development. This chapter delves into the contemporary comprehension of the multifaceted role of micronutrients in fostering sustainable growth and development in crop plants, with a specific focus on the molecular and physiological dimensions.

10.2 MICRONUTRIENTS AND PLANT PHYSIOLOGY

Micronutrients encompass a diverse array of elements, including iron (Fe), zinc (Zn), copper (Cu), manganese (Mn), nickel (Ni), molybdenum (Mo), boron (B), and chlorine (Cl). Cobalt (Co), known to be essential for bacterial fixation of atmospheric nitrogen (N) in leguminous plants, is not, however, generally considered to be essential for all higher plants. Nevertheless, it has been shown to have a beneficial impact on crops in other plant families, such as *Gramineae* (e.g., wheat, *Triticum* spp.) (Asher, 1991), and hence is referred to as a "beneficial" element. Each of these micronutrients plays a unique and indispensable role in the physiological functioning of plants.

Iron serves as a critical component for chlorophyll synthesis, electron transport, and nitrogen fixation. Molecular studies elucidated the significance of iron-regulated transcription factors in orchestrating iron uptake and homeostasis, thereby facilitating plants to adapt to fluctuating iron availability in the soil (Curie and Briat, 2003).

Zinc, another crucial micronutrient, is indispensable for the activation of enzymes and the synthesis of auxins, which are pivotal plant hormones regulating various aspects of growth and development. Recent molecular investigations have led to the identification of zinc-regulated genes that are intricately associated with root development and stress responses, thus enhancing our understanding

DOI: 10.1201/9781032630366-10

of the molecular underpinnings of the impact of zinc on plant physiology (Broadley *et al.*, 2007). Zinc enzymes play a pivotal role in regulating transcription, processing of RNA, and translation. It has been found that the zinc-finger motif (Kawagashira *et al.*, 2001), prominent in transcription factors, was present in up to 4% of the predicted proteins in *Arabidopsis thaliana* and was crucial in protein-protein interactions. Several other enzymes, for instance, all types of DNA and RNA polymerases, as well as RNA-editing enzymes involved in mitochondria and chloroplasts, are also dependent on zinc (Kraemer and Clemens, 2005). Zinc-dependent enzymes have major roles in various processes in the chloroplast. For instance, proteolytic activities taking place in the chloroplast are zinc-dependent, like processes during the repair of photosystem II by turning over the photodamaged D1 protein (Bailey *et al.*, 2002). Several enzymes rely on zinc for their activities, for example, the enzyme carbonic anhydrase required for fixation of carbon dioxide (CO_2) in C4 plants, is dependent on zinc. Therefore, during zinc deficiency, carbonic anhydrase activity is reduced and contributes to an overall reduction in the net photosynthesis rate (Jelakovic *et al.*, 2003).

Boron, an essential micronutrient, plays an important role in plant development. When plant roots absorb boron, it is first distributed to meristems, organs for reproduction, and other developing tissues, suggesting the important role of boron in plant maturation and movement from the vegetative stage to the reproductive stage (Pereira *et al.*, 2021).

Chlorine is often considered as an essential plant micronutrient due to its participation in a wide range of physico-chemical processes (Colmenero *et al.*, 2019; Raven, 2017). Chlorine, in the form of chloride, is primarily found to accumulate in vacuoles and reach concentration up to 150 mM, serving as a permeable substance to drive the flow of water (Geilfus, 2018). Furthermore, chloride accumulation could play a role in improving nitrogen utilization by decreasing nitrate compartmentalization in the vacuole and promoting its assimilation. In a study conducted by Rosales *et al.* (2020), nitrogen utilization was found to increase by 0–22% in citrus and olive and by 60–80% in tomato and tobacco when aided by chlorine.

Nickel is required by plants for proper germination of seeds as well as for nitrogen metabolism in plants in which ureides (compounds derived from urea) are metabolically important (Gerendas *et al.*, 1999). This is because nickel is the metal component in the enzyme urease, which catalyzes the reaction, converting urea to ammonium (Havlin *et al.*, 1999).

Cobalt has been proposed as an essential micronutrient for plants. It has been found to be present in a variety of proteins and enzymes that function in plant metabolism, with possible roles in key plant metabolic pathways, such as nitrogen fixation (Hu *et al.*, 2021)

Molybdenum is present in only a few plant proteins but plays a vital role in plant growth and development. These molybdenum-containing proteins have important roles in the assimilation of nitrogen, biosynthesis of phytohormones, and metabolism of sulfur (Shwarz and Mendel, 2006).

10.3 MICRONUTRIENTS AND NUTRITIONAL QUALITY

In addition to their essential roles in plant growth and development, micronutrients significantly influence the nutritional quality of crops. The nutritional value of food crops is of paramount importance for human health, and deficiencies in certain micronutrients can lead to severe health issues in humans. Mineral deficiencies can trigger a series of interactions between macro-elements, micro-elements or beneficial elements (Figure 10.1). Micronutrient interactions significantly modulate or influence the nutrient dynamics in crop plants (summarized in Table 10.1). This section explores how various micronutrients impact the nutritional quality of crops.

Iron and Anemia: Iron deficiency is a widespread nutritional problem, leading to anemia in many parts of the world. Iron-fortified crops, such as biofortified beans, rice, and wheat, have been developed to address this issue (Bouis, 2003). These crops are genetically modified to accumulate higher levels of iron in their edible parts, thus increasing iron intake for consumers.

Zinc and Zinc-deficiency Disorders: Zinc plays a crucial role in human health, impacting various physiological processes. Zinc deficiency can lead to growth retardation, immune

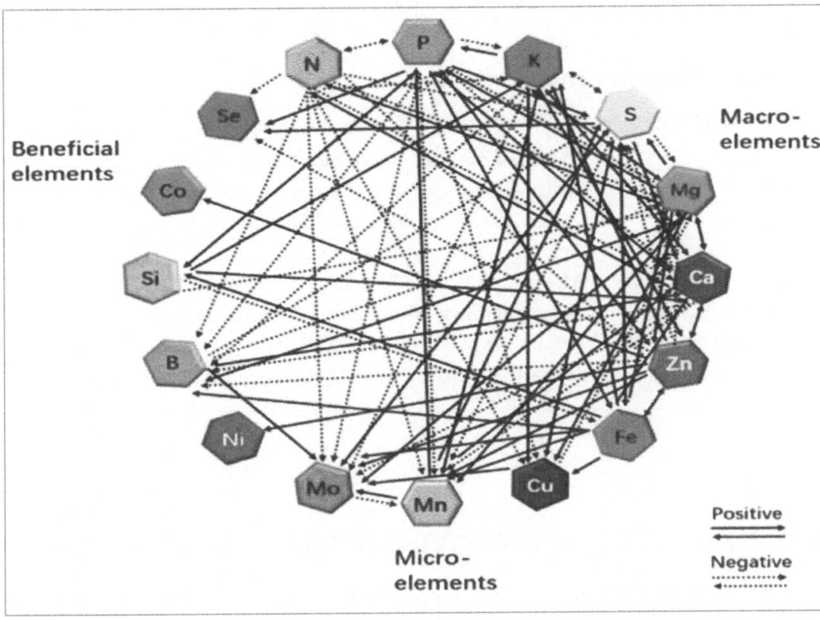

FIGURE 10.1 Interactions among macro- and micro-elements or beneficial elements in plants in response to individual mineral deficiencies. Single-element deficiency (any one of 16 elements) can lead to increased (solid lines) or decreased (dashed lines) uptake of other minerals (Fan *et al.*,2021).

system dysfunction, and other health problems (Hotz and Brown, 2004). Biofortified crops, especially zinc-enriched rice and wheat, have been developed to combat zinc deficiency (Gómez-Galera *et al.*, 2010). These crops accumulate higher levels of zinc in their grains, providing a dietary source of this essential nutrient.

Selenium and Antioxidant Capacity: Selenium is an essential micronutrient for humans, and its deficiency can result in health issues. Selenium-enriched crops, like selenium-fortified broccoli and rice, have been developed to enhance selenium intake (Rayman, 2012). Selenium has antioxidant properties and can enhance the antioxidant capacity of crops, contributing to their nutritional quality.

Copper and Metabolic Processes: Copper is vital for various metabolic processes in the human body. While copper deficiencies are relatively rare, ensuring an adequate dietary intake of copper through crops like legumes and whole grains is essential for overall health (Bost *et al.*, 2016).

Manganese and Bone Health: Manganese is essential for bone health and other physiological functions. It contributes to the synthesis of connective tissues and plays a role in antioxidant defense mechanisms. Manganese-enriched crops can help ensure an adequate intake of this micronutrient (Hirschi, 2009).

Boron and Plant Metabolism: Although required in very small amounts, boron is essential for plant metabolism. Crops that are rich in boron contribute to the overall dietary boron intake. Boron is involved in various metabolic processes, including carbohydrate metabolism and nucleic acid synthesis (Warington, 1923).

Chlorine and Osmotic Regulation: Chlorine, in the form of chloride ions, plays a role in osmotic regulation in plants. While chlorine deficiencies in humans are rare, consuming crops like lettuce and celery, which contain chloride ions, indirectly contributes to dietary chlorine intake (Gong *et al.*, 2011).

TABLE 10.1
Micronutrient Interactions in Crop Plants

Crop	Micronutrients	Interactions
Wheat	Iron (Fe)	Enhances zinc (Zn) uptake and utilization (Curie and Briat, 2003)
Rice	Zinc (Zn)	Influences phosphorus (P) uptake and translocation (Broadley *et al.*, 2007)
Maize	Manganese (Mn)	Affects iron (Fe) homeostasis. (Dong *et al.*, 2019)
Soybean	Copper (Cu)	Interacts with molybdenum (Mo) in nitrogen fixation (Jan *et al.*, 2022)
Tomato	Boron (B)	Modulates calcium (Ca) transport (Pereira *et al.*, 2021)
Spinach	Molybdenum (Mo)	Affects nitrogen (N) metabolism (Hänsch and Mendel, 2009)
Rice	Boron (B)	Affects potassium (K) uptake (Wimmer and Eichert, 2013)
Potato	Manganese (Mn)	Interacts with magnesium (Mg) in chlorophyll synthesis (Gong *et al.*, 2011)

10.4 CONSIDERATIONS FOR IMPROVING CROP NUTRITIONAL QUALITY

While the development of biofortified crops is a promising approach to enhance the nutritional quality of food, several challenges must be addressed:

Bioavailability: Simply increasing the concentration of a micronutrient in a crop does not guarantee improved human absorption and utilization. Factors like the form of the nutrient and its interactions with other compounds in the food matrix can affect bioavailability (Lombi *et al.*, 2011). Researchers need to consider these factors while developing biofortified crops.

Acceptance and Palatability: The success of biofortified crops depends on consumer acceptance. These crops should not only provide increased nutrient levels, but also maintain proper taste and appearance. Developing biofortified crops that are appealing to consumers is a critical challenge (Talsma, Melse-Boonstra and Brouwer, 2017).

Sustainable Agriculture: Sustainable cultivation practices are essential to ensure the long-term availability of biofortified crops. Ensuring that biofortification does not deplete soil nutrients or harm the environment is crucial for the sustainability of agricultural systems (Nestel *et al.*, 2006).

Regulatory and Safety Considerations: The development and release of biofortified crops involve regulatory and safety assessments. These assessments ensure that the crops are safe for consumption and the environment. Compliance with regulatory requirements is a complex process that can impact the adoption of biofortified crops (Gómez-Galera *et al.*, 2012).

10.5 SOIL FACTORS INFLUENCING MICRONUTRIENT AVAILABILITY

The availability of micronutrients in soil is a critical determinant of their uptake by plants. Several soil factors can influence micronutrient availability, including soil pH, organic matter content and microbial activity.

Soil pH: Soil pH plays a significant role in micronutrient availability. For instance, iron (Fe) and manganese (Mn) become less available to plants in alkaline soil (high pH), while zinc (Zn) availability decreases in acidic soil (low pH) (Kabata-Pendias and Mukherjee, 2007). Soil amendments such as lime or sulfur can be used to adjust soil pH and improve micronutrient availability.

Organic Matter: Organic matter in soil contributes to the chelation of micronutrients, preventing them from forming insoluble compounds (Gupta *et al.*, 2008). Soil with higher organic matter content tends to have better micronutrient availability.

Microbial Activity: Soil microorganisms can influence the availability of micronutrients through various processes, including mineralization of organic matter and nutrient cycling (Bünemann *et al.*, 2018). Microbial activities can release micronutrients from organic matter, making them more accessible to plants.

10.6 BIOTECHNOLOGICAL APPROACHES FOR MICRONUTRIENT ENRICHMENT

Biotechnological approaches offer innovative solutions for enhancing the micronutrient content of crops, ultimately improving human nutrition. Genetic engineering and breeding techniques have been employed to develop crop varieties with increased micronutrient concentrations.

Genetic Engineering: Genetic modification allows scientists to introduce or enhance specific genes involved in micronutrient uptake, transport, or storage. For example, researchers have successfully engineered rice to accumulate higher levels of iron and zinc, addressing prevalent micronutrient deficiencies in regions where rice is a dietary staple (Goto *et al.*, 1999).

Marker-assisted Selection (MAS): MAS is a breeding technique that identifies plants with desirable micronutrient traits based on genetic markers. This approach expedites the development of crop varieties with improved micronutrient content. An example is the development of zinc-enriched wheat varieties using MAS (Velu *et al.*, 2016).

10.7 MOLECULAR MECHANISMS OF MICRONUTRIENT UPTAKE

Plants have evolved a diverse array of strategies to effectively acquire and transport micronutrients from the soil to different plant tissues. Advances in molecular studies have shed light on the intricate mechanisms governing the uptake and transport of micronutrients. The iron-regulated transporter (IRT) gene family plays a pivotal role in mediating iron uptake from soil. These genes exhibit regulation in response to iron availability, enabling plants to dynamically adjust their iron uptake strategies according to the prevailing environmental conditions (Curie and Briat, 2003). Likewise, the cation diffusion facilitator (CDF) transporter family mediates the transport of divalent cations, for example, zinc, iron, and manganese, from the cytoplasm to subcellular compartments or the extracellular spaces (Gustin *et al.*, 2011). The CDF family plays a vital role by regulating metal homeostasis, thus having critical importance in metal tolerance in plant cells (Migocka *et al.*, 2014; Migocka *et al.*, 2015) Additionally, GmMTP8.1, a manganese transporter located in the endoplasmic reticulum, was found to trigger movement of manganese out of the leaf cells, along with increasing the sequestration of manganese within cellular compartments. Thus, it plays a critical role in providing manganese tolerance in plants (Li *et al.*, 2021a).

Plants have developed mechanisms for the absorption and uptake of iron through roots. When aerobic conditions are prevalent in the soil, the bioavailability of iron is low, since it is mostly present in the form of oxides and hydroxides that are insoluble. Thus, to facilitate the absorption and uptake of iron, plants have developed two strategies. The first strategy is observed mostly in dicot and non-graminaceous monocot plants. In these plants, the roots reduce the pH of the rhizosphere through the extrusion of protons, after which iron undergoes reduction from its ferric (Fe^{3+}) to ferrous (Fe^{2+}) form. This reduction is carried out by an inducible enzyme, Fe^{3+} reductase, located at the plasma membrane. Iron in its ferrous form is then transported across the membrane through the ferrous ion (Fe^{2+})-specific transporter system. The second mechanism involves the secretion of siderophores, which in Greek means "iron carriers." However, no reduction from ferric (Fe^{3+}) to ferrous (Fe^{2+}) iron takes place here. This strategy is mostly observed in the *Gramineae* family (Eskandari, 2011; Tavakoli *et al.*, 2014).

10.8 MICRONUTRIENTS AND STRESS RESPONSES

Micronutrients possess a dual role, not only influencing regular plant growth but also actively participating in stress responses. Zinc, for instance, has been shown to enhance plant tolerance to heavy metal stress. Molecular investigations have highlighted that zinc-induced expression of stress-related genes plays a pivotal role in augmenting plant stress tolerance (Palmgren *et al.*, 2008).

Likewise, boron is a vital micronutrient crucial for cell wall formation and stability. Research has demonstrated that boron deficiency leads to impaired cell elongation and root development. Recent molecular studies have led to the identification of genes responsible for boron transport and utilization, thus enhancing our comprehension of plant responses to boron deficiency stress (Wimmer and Eichert, 2013).

Manganese has been found to play an important role in plant metabolism. It functions as a catalytically active metal, as well as serves as an activator of multiple enzymes which help plants to combat stress situations. The enzyme superoxide dismutase contains manganese, which plays a vital role in providing protection to plants from the damaging effects of reactive oxygen species and free radicals, thereby functioning in oxidative stress responses (Hänsch and Mendel, 2009).

Copper plays a very important role in regulating redox metabolism in plants by functioning as both an oxidizing and reducing agent in biochemical reactions. This is due to the fact that copper exists in two oxidation states under physiological conditions, Cu^{+} and Cu^{2+}, and can interchange between the forms. However, the presence of excess copper ions can become toxic for the plants as they catalyze the production of free radicals and may lead to oxidative stress. Thus, to prevent such a situation, the majority of the copper ions are bound by scavenging proteins immediately after uptake, such as metallothioneins, which prevent them from accumulating in a form which would lead to toxicity (Hänsch and Mendel, 2009).

Another study has uncovered a molybdenum-enzyme mitochondrial amidoxime-reducing component (mARC), which covers the mammalian mitochondria (Havemeyer *et al.*, 2006). This enzyme was found to catalyze the reduction of N-hydroxylated amidines concomitantly with cytochrome b_5 and cytochrome b_5 reductase enzymes, which might play an important role in detoxification. Although the enzyme was initially identified in mammalian systems, it was found to have homologs in eubacteria and plants (Kozmin *et al.*, 2008).

10.9 MICRONUTRIENT DEFICIENCIES

Micronutrients are required by plants in minute quantities for their optimum growth and functioning. In plants, micronutrients play an important role in regulating a wide range of enzymatic reactions; hence, their deficiency leads to stress and malfunctioning of the plant activities.

Very little information on chlorine deficiencies was documented until recently, since the symptoms were often considered as physiological leaf spots. However, more recent studies have shown that chlorine deficiencies exist in Montana, where visual symptoms are observed in winter wheat and durum wheat cultivars (Engel, Bruckner and Eckhoff, 1998). Chlorine deficiency symptoms are majorly found in cereal crops and generally include chlorotic and necrotic leaf-spotting with abrupt boundaries between the living and dead tissues, leaf wilting at the margins, and high levels of branching in the root system. Diagnosis of chlorine deficiency symptoms is difficult, as they are highly cultivar-specific and can often be mistaken for leaf diseases (Mengel and Kirkby, 2001). The symptoms of cobalt deficiency include retardation of growth, leaf chlorosis and necrosis, and reduced yield of crops, resembling the deficiency of nitrogen (Liu, 1998). In legumes, cobalt deficiency showed symptoms such as reduction in plant size, smaller-sized, pale yellow leaves, smaller pods, smaller size, and reduced number of nodules when compared to non-deficient control plants. The growth of roots is also affected, as observed through an overall reduction of root length and volume in case of cobalt deficiency. It has been found that deficiency of cobalt causes reduction in the synthesis of methionine, which limits protein synthesis and contributes to the smaller-sized bacteroids (Marschner, 2011). Cobalt deficiency has also been reported to cause growth retardation in non-leguminous plants like rubber and tomato (Wilson and Nicholas, 1967)

The deficiency symptoms of nickel have not been well-documented. However, in graminaceous plants, nickel deficiency has been found to cause chlorosis, similar to iron deficiency, along with a marked enhancement of senescence with reduction in tissue iron concentrations. Patchy necrosis has also been observed in the youngest leaves of the plant (Brown, 2007).

Manganese deficiency is known to cause oxidative stress, along with reduced photosynthetic electron transport, efficiency of water use, and functioning of roots (Schmidt *et al.*, 2016). In maize (*Zea mays* L.) plants, manganese deficiency has been found to restrict nitrate (NO_3^-) uptake and transport through production of inhibiting enzymes such as nitrate reductase and glutamine synthase, disrupting nitrogen metabolism. This leads to an overall reduction in chlorophyll synthesis and solubility of proteins (Gong *et al.*, 2011).

Another important micronutrient for plants is iron, whose deficiency has been seen to cause several detrimental effects. Iron mainly exists in soil in the form of insoluble ferric hydroxide, which cannot be utilized by plants, causing significant reduction in iron bioavailability (Guerinot and Yi, 1994). Symptoms of iron deficiency mainly include yellowing of leaves with green veins, as observed in citrus (Fu *et al.*, 2017). The effect of iron deficiency was studied in crop plants (Li *et al,* 2021b) in *Areca catechu* (L.) seedlings, which showed instances of chloroplast degeneration and reduction in chlorophyll synthesis in chlorotic leaves. Since iron is required to produce chlorophyll, the most prominent symptom of iron deficiency is chlorosis. Other symptoms mainly include yellowing of the leaves, with the appearance of dark-green colored veins, a condition referred to as interveinal chlorosis. In agricultural fields where iron is below the required level, proper translocation of iron to the different parts of the plant fails to occur, which is why the symptoms of iron deficiency mostly appear first on the new leaves and growth. In cases of moderate deficiency levels, plants can recover from stress, but in instances of severe iron deficiency, the entire plant turns yellowish or white in color. In crop plants such as sorghum and corn, the plants appear to be striped due to interveinal chlorosis (Tavakoli *et al.*, 2014). Iron deficiency was also found to cause a reduction in the levels of expression of genes encoding the enzymes, such as nitrate reductase and glutamate synthase.

At elevated pH, plants suffer from copper deficiency, showing symptoms like cupping and slight chlorosis of entire leaves or between the veins in leaves that are newly formed along with necrotic spots on the margins of leaves. When copper deficiency reaches an acute level, lateral growth can be inhibited due to the death of the apical meristem. This necrosis has been found to cause excessive branching in dicot plants, while in cereal crops, it leads to a higher degree of tillering (Rehman *et al.*, 2019).

Molybdenum deficiency is often found to resemble phenotypic effects observed in cases of nitrogen deficiency, with stunted growth and leaf chlorosis being the most prominent symptoms observed in legumes. However, in plants deficient in molybdenum, leaves turn pale, accompanied by rolling, cupping, and scorching. The leaves become brittle before gradually withering off, leaving behind the midrib (Sherefu and Zewide, 2021).

Zinc is an essential micronutrient for plants, and its deficiency has severe consequences in plant growth. Causes of zinc deficiency are related to the type of soil as well as weather conditions. For instance, in calcareous soil types, the adsorption of zinc in soil by particles of limestone and clay determines the levels of zinc in the soil, and lower adsorption leads to lower levels of zinc available for plants. On the other hand, in cases of eroded soil types, the main cause of zinc deficiency is the deficiency of organic matter. Zinc levels are also found to decrease during cold and moist weather conditions. This may be due to the limitations in root growth in cooler soil or due to the reduction in microbial activity, leading to lesser amounts of zinc extracted from organic matter. It has been found that an increase in the levels of bicarbonate (HCO_3) provides a resistance to the shoot-mediated zinc uptake (Mousavi, Galavi and Rezaei, 2012).

10.10 MICRONUTRIENT TOXICITY

Micronutrient toxicity occurs when soluble nutrients are present in excess amounts in the soil beyond a tolerance threshold. Despite the vital role played by micronutrients, the effect becomes toxic when present beyond the threshold concentration and exerts negative effects. Excess amounts of micronutrients promote oxidative stress by triggering overproduction of reactive oxygen species (ROS) in the form of hydrogen peroxide and superoxide radicals, coupled with insufficient activity of the antioxidant enzymes, such as catalase and glutathione peroxidase.

Manganese toxicity has been found to be responsible for reduction in plant growth in poorly drained, acidic soil. Manganese toxicity is characterized by the formation of brown spots along with chlorosis and necrosis in leaves, which gradually lead to complete inhibition of plant growth (Li *et al.*, 2019; Millaleo *et al.*, 2010). Manganese tolerance was studied by Li *et al.* (2021), where it was observed that the presence of metal tolerance proteins (MTPs) might be involved in manganese tolerance in plants, for instance, soybean (*Glycine max*).

Boron, when present in excess concentrations, can produce toxic effects. The impacts of the toxic effects of boron were studied by Khan *et al.* (2021) in 19 *Aegilops* accessions and one bread wheat cultivar. It was found that the stress induced by boron toxicity affected plant growth, with a more pronounced effect on shoots compared to roots.

Another essential micronutrient, copper, has been shown to produce toxic effects when present in higher concentrations. Increased levels of copper in the soil interfere with the nutrient balance in the soil through binding with clay minerals, organic matter and hydrated iron, manganese, and aluminum oxides, directly affecting the productivity of the crop plants. Plants suffering from copper toxicity show stunted growth, appear bluish in color, and gradually turn brown or yellow (Rehman *et al.*, 2019). Excess copper can trigger intracellular signals, which may cause cell damage through the generation of ROS. It was observed that the presence of excess Cu^{2+} led to the accumulation of H_2O_2 in maize leaves, resulting in oxidative stress (Liu *et al.*, 2018).

Iron, when present in excess concentration, can produce toxic effects in plants. Iron toxicity is mostly found to occur in flooding soil where, due to the lower redox potential created, the iron concentration increases manifold. Iron toxicity can be identified through its symptoms of increased flaccidity of the roots accompanied by the development of black color in the roots. There is a reduction in the level of branching of the roots along with root softening. Other symptoms include leaf mottling and shoot die-back (Eskandari, 2011; Tavakoli *et al.*, 2014).

10.11 CROP YIELD AND MICRONUTRIENT MANAGEMENT

Effective management of micronutrients is of paramount importance for achieving sustainable crop yield. However, imbalances or deficiencies in micronutrient levels can exert a detrimental impact on plant growth and productivity. To combat such a scenario, the application of micronutrient-based fertilizers to the soil, seeds, or foliage on a regular basis needs to be carried out, based on regular testing of soil and plant tissues. Molecular studies have provided invaluable insights into the genetic determinants governing micronutrient uptake and utilization efficiency. For instance, a study investigating the importance of chloride-containing fertilizers on the production of citrus reported an increase in yield, vitamin C content, and improvement in the yield and flavor of juice via an enhancement in the content of nitrogen and potassium (Liu *et al.*, 2021)

Manganese, an essential micronutrient, was found to be deficient mostly in dry, sandy, and calcareous soil types, which resulted in a decrease in the yield of crops (Ijaz *et al.*, 2021). A study was done on manganese solubilization and the manganese solubilizing ability of *Bacillus* spp. strains.

The strains which were able to solubilize manganese were isolated from the maize rhizosphere and could function as potential bioinoculants to promote the growth of plants under manganese deficient conditions.

Biofortification of zinc had an important role in promoting increased yield of zinc. The study conducted by Ali *et al.* (2021) showed that when 15 varieties of wheat were treated with a bioactive nutrient-fortified fertilizer containing zinc, there was an increase in wheat yield as well as uptake of zinc.

Moreover, contemporary molecular techniques such as genetic engineering and marker-assisted selection offer promising avenues for developing crop varieties endowed with improved micronutrient uptake and utilization traits. These innovations have the potential to enhance the nutritional quality of crops and mitigate micronutrient deficiencies in human diets (Genc *et al.*, 2008).

10.12 CONCLUSION

Molecular and physiological investigations have profoundly enriched our understanding of how micronutrients contribute to sustainable crop growth and development. These studies have unraveled complex mechanisms of micronutrient uptake, transport, and utilization, thereby providing insights into how plants respond to nutrient availability and stress conditions. Micronutrients are of utmost importance for the maintenance and increase in the yield of food crops for an expanding global population. The micronutrient composition of crops directly affects the intake of these elements for human and animal nutrition, and therefore it has to be ensured that they are present in adequate quantities. As agricultural practices continue to evolve, the incorporation of this knowledge into crop management strategies holds immense potential to enhance global food security and advance the cause of sustainable agriculture.

The insights gleaned from the comprehensive exploration of the role of micronutrients in sustainable crop growth and development reveal a profound connection between plant physiology and their molecular underpinnings. The array of micronutrients, encompassing iron, zinc, copper, manganese, molybdenum, boron, and chlorine, orchestrates diverse functions vital for plant health and productivity. These functions extend beyond regular growth to encompass responses to stressors, contributing to the innate immunity of the plants.

Micronutrient deficiencies can lead to a cascade of stress-induced reactions, ultimately hampering plant activities. However, excessive application of these micronutrients can induce toxicity, raising concerns about soil contamination, plant health, and human safety. This dual role highlights the delicate balance required to harness the benefits of micronutrients while mitigating their potential risks. The molecular mechanisms underlying micronutrient uptake and transport have come to light through various studies. Notably, the IRT gene family and CDF transporter family exemplify the complexity of these processes. The orchestrated regulation of these genes showcases plant adaptability to varying nutrient availability, thereby enhancing their nutrient acquisition efficiency. Moreover, the vital role of micronutrients in stress responses accentuates their significance in plant resilience. The contribution of zinc to stress tolerance, exemplified by zinc-regulated genes, highlights the intricate molecular mechanisms that enhance plant ability to withstand environmental challenges. The identification of genes responsible for boron transport adds a layer of understanding to how plants counter boron deficiency stress. In the realm of crop yield enhancement, the strategic management of micronutrients takes the central stage. Understanding the genetic determinants governing micronutrient uptake efficiency can aid in optimizing crop management strategies and ensuring the proper utilization of these elements. Advances in genetic engineering and MAS also hold the potential to engineer crop varieties with improved micronutrient traits, addressing both yield and nutritional quality concerns. Despite these strides, challenges and future directions remain. Unraveling the intricate interplay between different micronutrients and their interactions with other nutrients presents a complex puzzle. The integration of omics technologies, such as genomics, transcriptomics, proteomics, and metabolomics, can provide a holistic perspective on micronutrient-related processes, offering a deeper understanding of their roles and interactions.

Overall, the collaborative efforts of molecular and physiological research have illuminated the multifaceted roles of micronutrients in sustaining crop growth and development. The intricate molecular mechanisms governing their functions, responses to stress, and potential for enhancing crop yield showcase their immense importance. As agricultural practices continue to evolve, integrating these insights into innovative crop management strategies holds the potential to fortify global food security and propel the cause of sustainable agriculture forward.

10.13 FUTURE DIRECTIONS AND CHALLENGES

Although a significant progress has been made to understand the factors governing the functions of micronutrients in plants, some challenges still remain to derive a comprehensive idea of the intricate molecular mechanisms that direct micronutrient functions in plants. Future research should be aimed at understanding the micronutrient homeostasis to derive a comprehensive idea regarding the uptake of micronutrients, their transport and distribution to generate sustainable strategies to overcome deficiency of micronutrients in crop plants. In order to ensure global food security, the effects of micronutrients on consumption patterns and health of the plant also need to be taken into account. Such an integrative approach is the key to develop sustainable solutions to bolster crop production while maintaining their nutritional qualities.

ACKNOWLEDGMENTS

Financial assistance from the Science and Engineering Research Board, Government of India through the grant [EMR/2016/004799] and Department of Higher Education, Science and Technology and Biotechnology, Government of West Bengal, through the grant [264(Sanc.)/ST/P/S&T/1G-80/2017] to Prof. Aryadeep Roychoudhury is gratefully acknowledged.

REFERENCES

Ali, M.A. *et al.* (2021) 'Bioactive nutrient fortified fertilizer: A novel hybrid approach for the enrichment of wheat grains with zinc', *Frontiers in Plant Science*, 12(December), pp. 1–14. https://doi.org/10.3389/fpls.2021.743378

Asher, C. J. (1991). Beneficial elements, functional nutrients, and possible new essential elements. In J. J. Mortvedt, ed., *SSSA Book Series* (2nd ed., Chap. 18). Soil Science Society of America. https://doi.org/10.2136/sssabookser4.2ed.c18

Bailey, S. *et al.* (2002) 'A critical role for the Var2 FtsH homologue of Arabidopsis thaliana in the photosystem II repair cycle *in vivo*', *Journal of Biological Chemistry*, 277(3), pp. 2006–2011. https://doi.org/10.1074/jbc.M105878200

Bost, M. et al. (2016). 'Dietary copper and human health: Current evidence and unresolved issues', *Journal of Trace Elements in Medicine and Biology*, 35, 107–115. https://doi.org/10.1016/j.jtemb.2016.02.006

Bouis, H.E. (2003) 'Micronutrient fortification of plants through plant breeding: Can it improve nutrition in man at low cost?', *Proceedings of the Nutrition Society*, 62(2), pp. 403–411. https://doi.org/10.1079/PNS2003262

Broadley, M.R. *et al.* (2007) 'Zinc in plants: Tansley review', *New Phytologist*, 173(4), pp. 677–702. https://doi.org/10.1111/j.1469-8137.2007.01996.x

Brown, P.H. (2007) 'Nickel', in Barker, A.V. and Pilbeam, D.J. (eds.) *Handbook of plant nutrition*. Boca Raton, FL: CRC, Taylor & Francis, pp. 395–410.

Bünemann, E. *et al.* (2018). Soil quality—A critical review. *Soil Biology and Biochemistry*, 120, pp. 105–125. https://doi.org/10.1016/j.soilbio.2018.01.030

Colmenero, J. *et al.* (2019). 'Chloride as a beneficial macronutrient in higher plants: New roles and regulation', *International Journal of Molecular Sciences*, 20(19), p. 4686. https://doi.org/10.3390/ijms20194686

Curie, C. and Briat, J.F. (2003)'Iron transport and signaling in plants', *Annual Review of Plant Biology*, 54, pp. 183–206. https://doi.org/10.1146/annurev.arplant.54.031902.135018

Dong, J. *et al.* (2019). 'Inositol pyrophosphate InsP8 acts as an intracellular phosphate signal in Arabidopsis', *Molecular Plant*, 12(11), pp. 1463–1473. https://doi.org/10.1016/j.molp.2019.08.002

Engel, R.E., Bruckner, P.L. and Eckhoff, J. (1998) 'Critical tissue concentration and chloride requirements for wheat', *Soil Science Society of America Journal*, 62(2), p. 401. https://doi.org/10.2136/sssaj1998.03615995006200020016x

Eskandari, H. (2011). 'The importance of iron (Fe) in plant products and mechanism of its uptake by plants', *Journal of Applied Environmental and Biological Sciences*, 1(10), pp. 448–452.

Fan, X. *et al.* (2021) 'Cross-talks between macro- and micronutrient uptake and signaling in plants', *Frontiers in Plant Science*, 12(October), pp. 1–20. https://doi.org/10.3389/fpls.2021.663477

Fu, L. *et al.* (2017) 'Physiological and transcriptional changes of three citrus rootstock seedlings under iron deficiency', *Frontiers in Plant Science*, 8(June), pp. 1–13. https://doi.org/10.3389/fpls.2017.01104

Geilfus, C.M. (2018) 'Chloride: From nutrient to toxicant', *Plant and Cell Physiology*, 59(5), pp. 877–886. https://doi.org/10.1093/pcp/pcy071

Genc, H. *et al.* (2008). 'Vegetation indices as indicators of damage by the Sunn pest (Hemiptera: Scutelleridae) to field-grown wheat', *African Journal of Biotechnology*, *7*(2).

Gerendas, J. *et al.* (1999). 'Significance of nickel for plant growth and metabolism', *Journal of Plant Nutrition and Soil Science*, 162, pp. 241–256.

Gómez-Galera, S. *et al.* (2010) 'Critical evaluation of strategies for mineral fortification of staple food crops', *Transgenic Research*, 19(2), pp. 165–180. https://doi.org/10.1007/s11248-009-9311-y

Gómez-Galera, S. *et al.* (2012) 'Constitutive expression of a barley Fe phytosiderophore transporter increases alkaline soil tolerance and results in iron partitioning between vegetative and storage tissues under stress', *Plant Physiology and Biochemistry*, 53, 46–53. https://doi.org/10.1016/j.plaphy.2012.01.009

Gong, W. *et al.* (2011) 'Long-term applications of chemical and organic fertilizers on plant-available nitrogen pools and nitrogen management index', *Biology and Fertility of Soils*, 47(7), pp. 767–775. https://doi.org/10.1007/s00374-011-0585-x

Goto, F. *et al.* (1999). 'Iron fortification of rice seed by the soybean ferritin gene', *Nature Biotechnology*, 17, pp. 282–286. https://doi.org/10.1038/7029

Guerinot, M.L. and Yi, Y. (1994) 'Iron: Nutritious, noxious, and not readily available', *Plant Physiology*, 104(3), pp. 815–820. https://doi.org/10.1104/pp.104.3.815

Gupta, U., Wu, K. and Liang, S. (2008). 'Micronutrients in soils, crops, and livestock', *Earth Science Frontiers*, 15, 110–125. https://doi.org/10.1016/S1872-5791(09)60003-8

Gustin, J.L., Zanis, M.J. and Salt, D.E. (2011) 'Structure and evolution of the plant cation diffusion facilitator family of ion transporters', *BMC Evolutionary Biology*, 11(1). https://doi.org/10.1186/1471-2148-11-76

Hänsch, R. and Mendel, R.R. (2009) 'Physiological functions of mineral micronutrients (Cu, Zn, Mn, Fe, Ni, Mo, B, Cl)', *Current Opinion in Plant Biology*, 12(3), pp. 259–266. https://doi.org/10.1016/j.pbi.2009.05.006

Havemeyer, A. *et al.* (2006) 'Identification of the missing component in the mitochondrial benzamidoxime prodrug-converting system as a novel molybdenum enzyme', *Journal of Biological Chemistry*, 281(46), pp. 34796–34802. https://doi.org/10.1074/jbc.M607697200

Havlin, J.L. *et al.* (1999) *Soil fertility and fertilizers*. 6thedn. Upper Saddle River, NJ: Prentice Hall.

Hirschi, K.D. (2009) 'Nutrient biofortification of food crops', *Annual Review of Nutrition*, 29, pp. 401–421. https://doi.org/10.1146/annurev-nutr-080508-141143

Hotz, C. and Brown, K. H. (2004). 'International Zinc Nutrition Consultative Group (IZiNCG) Technical Document No. 1. Assessment of the risk of zinc deficiency in populations and options for its control', *Food and Nutrition Bulletin*, 25(Suppl. 2), pp. S94–S203.

Hu, X. *et al.* (2021) 'Cobalt: An essential micronutrient for plant growth?', *Frontiers in Plant Science*, 12(November), pp. 1–24. https://doi.org/10.3389/fpls.2021.768523

Husen, A. (2024) *Essential minerals in plant-soil system (Coordination, signalling, and interaction under adverse situations)*. Cambridge, MA: Elsevier Inc.

Ijaz, A. *et al.* (2021) 'Insights into manganese solubilizing *Bacillus* spp. for improving plant growth and manganese uptake in maize', *Frontiers in Plant Science*, 12(November), pp. 1–18. https://doi.org/10.3389/fpls.2021.719504

Jan, A.U. *et al.* (2022) 'Zinc-induced anti-oxidative defense and osmotic adjustments to enhance drought stress tolerance in sunflower (*Helianthus annuus* L.)', *Environmental and Experimental Botany*, 193, p. 104682. https://doi.org/10.1016/j.envexpbot.2021.104682

Jelakovic, S. *et al.* (2003) 'Structure and catalytic mechanism of the cytosolic D-ribulose-5-phosphate 3-epimerase from rice', *Journal of Molecular Biology*, 326(1), pp. 127–135. https://doi.org/10.1016/S0022-2836(02)01374-8

Kabata-Pendias, A., & Mukherjee, A. B. (2007). *Trace elements from soil to human*. Berlin, Heidelberg: Springer. https://doi.org/10.1007/978-3-540-32714-1

Kawagashira, N. *et al.* (2001). 'Multiple zinc finger motifs with comparison of plant and insect', *Genome Informatics*, 12, pp. 368–369.

Khan, M.K. *et al.* (2021) 'Variability in physiological traits reveals boron toxicity tolerance in *Aegilops* species', *Frontiers in Plant Science*, 12(October). https://doi.org/10.3389/fpls.2021.736614

Kozmin, S.G. *et al.* (2008) 'YcbX and yiiM, two novel determinants for resistance of Escherichia coli to n-hydroxylated base analogues', *Molecular Microbiology*, 68(1), pp. 51–65. https://doi.org/10.1111/j.1365-2958.2008.06128.x

Kraemer, U., & Clemens, S. (2005). 'Functions and homeostasis of zinc, copper, and nickel in plants', In *Molecular biology of metal homeostasis and detoxification* (Vol. 14, pp. 216–271). Berlin, Heidelberg: Springer. https://doi.org/10.1007/4735_96

Li, J. *et al.* (2019) 'Advances in the mechanisms of plant tolerance to manganese toxicity', *International Journal of Molecular Sciences*, 20(20). https://doi.org/10.3390/ijms20205096

Li, J. *et al.* (2021a) 'Characterization of metal tolerance proteins and functional analysis of GmMTP8.1 involved in manganese tolerance in soybean', *Frontiers in Plant Science*, 12(November), pp. 1–16. https://doi.org/10.3389/fpls.2021.683813

Li, J. *et al.* (2021b). 'Iron deficiency leads to chlorosis through impacting chlorophyll synthesis and nitrogen metabolism in *Areca catechu* L.', *Frontiers in Plant Science*, 12. https://doi.org/10.3389/fpls.2021.710093

Liu, J. (1998) *Cobalt: Physiological effects and uptake mechanisms in plants.* (Doctoral dissertation). https://digital.library.adelaide.edu.au/dspace/bitstream/2440/19399/1/01front.pdf

Liu, J. *et al.* (2018) 'Copper-caused oxidative stress triggers the activation of antioxidant enzymes via ZmMPK3 in maize leaves', *PLoS ONE*, 13(9), pp. 1–12. https://doi.org/10.1371/journal.pone.0203612

Liu, X. *et al.* (2021) 'Migration of chlorine in Plant–Soil–Leaching system and its effects on the yield and fruit quality of sweet orange', *Frontiers in Plant Science*, 12(October), pp. 1–12. https://doi.org/10.3389/fpls.2021.744843

Lombi, E. *et al.* (2011) 'Megapixel imaging of (micro)nutrients in mature barley grains', *Journal of Experimental Botany*, 62(1), pp. 273–282. https://doi.org/10.1093/jxb/erq270

Marschner, P. (2011) *Marschner's mineral nutrition of higher plants.* Amsterdam: Elsevier/Academic Press.

Mengel, K., & Kirkby, E. A. (2001). *Principles of plant nutrition.* Dordrecht: Kluwer Academic Publishers. https://doi.org/10.1007/978-94-010-1009-2

Migocka, M. *et al.* (2014) 'Two metal-tolerance proteins, MTP1 and MTP4, are involved in Zn homeostasis and Cd sequestration in cucumber cells', *Journal of Experimental Botany*, 66(3), pp. 1001–1015. https://doi.org/10.1093/jxb/eru459

Migocka, M. *et al.* (2015) 'Cucumber metal tolerance protein CsMTP9 is a plasma membrane H^+-coupled antiporter involved in the Mn^{2+} and Cd^{2+} efflux from root cells', *Plant Journal*, 84(6), pp. 1045–1058. https://doi.org/10.1111/tpj.13056

Millaleo, R. *et al.* (2010) 'Manganese as essential and toxic element for plants: Transport, accumulation and resistance mechanisms', *Journal of Soil Science and Plant Nutrition*, 10(4), pp. 476–494. https://doi.org/10.4067/s0718-95162010000200008

Mousavi, S.R., Galavi, M. and Rezaei, M. (2012) 'The interaction of zinc with other elements in plants : A review', *International Journal of Agriculture and Crop Sciences*, 4(24), pp. 1881–1884.

Nelson, S.K. *et al.* (1992) 'Copper-marginal and copper-deficient diets decrease aortic prostacyclin production and copper-dependent superoxide dismutase activity, and increase aortic lipid peroxidation in rats', *Journal of Nutrition*, 122(11), pp. 2101–2108. https://doi.org/10.1093/jn/122.11.2101

Nestel, P. *et al.* (2006) 'Biofortification of staple food crops', *Journal of Nutrition*, 136(4), pp. 1064–1067. https://doi.org/10.1093/jn/136.4.1064

Palmgren, M.G. *et al.* (2008) 'Zinc biofortification of cereals: Problems and solutions', *Trends in Plant Science*, 13(9), pp. 464–473. https://doi.org/10.1016/j.tplants.2008.06.005

Pereira, G.L. *et al.* (2021) 'Boron: More than an essential element for land plants?', *Frontiers in Plant Science*, 11(January). https://doi.org/10.3389/fpls.2020.610307

Raven, J.A. (2017) 'Chloride: Essential micronutrient and multifunctional beneficial ion', *Journal of Experimental Botany*, 68(3), pp. 359–367. https://doi.org/10.1093/jxb/erw421

Rayman, M.P. (2012) 'Selenium and human health', *The Lancet*, 379(9822), pp. 1256–1268. https://doi.org/10.1016/S0140-6736(11)61452-9

Rehman, M. *et al.* (2019) 'Toxicología ambiental del cobre, avances recientes y perspectivas futuras: una revisión', *Investigación En Ciencias Ambientales y Contaminación*, 26(18), pp. 18003–18016.

Rosales, M.A. *et al.* (2020) 'Chloride improves nitrate utilization and NUE in plants', *Frontiers in Plant Science*, 11(3), pp. 1–13. https://doi.org/10.3389/fpls.2020.00442

Sherefu, A. and Zewide, I. (2021). 'Review paper on effect of micronutrients for crop production', *Nutrition and Food Processing*, 4. https://doi.org/10.31579/2637-8914/063

Schmidt, S.B., Jensen, P.E. and Husted, S. (2016) 'Manganese deficiency in plants: The impact on photosystem II', *Trends in Plant Science*, 21(7), pp. 622–632. https://doi.org/10.1016/j.tplants.2016.03.001

Schwarz, G. and Mendel, R.R. (2006)'Molybdenum cofactor biosynthesis and molybdenum enzymes', *Annual Review of Plant Biology*, 57, 623–647. https://doi.org/10.1146/annurev.arplant.57.032905.105437

Talsma, E.F., Melse-Boonstra, A. and Brouwer, I.D. (2017) 'Acceptance and adoption of biofortified crops in low- and middle-income countries: A systematic review', *Nutrition Reviews*, 75(10), pp. 798–829. https://doi.org/10.1093/nutrit/nux037

Tavakoli, M. T. *et al.* (2014). 'The importance of micronutrients in agricultural production', *Advances in Environmental Biology*, 8(10), pp. 31–35.

Velu, G. *et al.* (2016). 'Effect of drought and elevated temperature on grain zinc and iron concentrations in CIMMYT spring wheat', *Journal of Cereal Science*, 69, pp. 182–186. https://doi.org/10.1016/j.jcs.2016.03.006

Warington, K. (1923) 'The effect of boric acid and borax on the broad bean and certain other plants', *Annals of Botany*, os-37(4), pp. 629–672. https://doi.org/10.1093/oxfordjournals.aob.a089871

Wilson, B. and Nicholas, D.J.D. (1967) 'A cobalt requirement for non-nodulated legumes and for wheat', *Phytochemistry,* 6, pp. 1057–1066.

Wimmer, M. A. and Eichert, T. (2013). "Review: Mechanisms for boron deficiency-mediated changes in plant water relations', *Plant Science*, 203–204, pp. 25–32. https://doi.org/10.1016/j.plantsci.2012.12.012

11 Molecular and Physiological Concepts
Macronutrients in Crop Plant Growth and Development

Noopur Khare, Pragati Khare, and Sachidanand Singh

11.1 INTRODUCTION

Macronutrients are the lifeblood of crops, enabling them to grow, develop and ultimately sustain global food production. These vital substances – nitrogen (N), phosphorus (P), potassium (K), calcium (Ca), magnesium (Mg), and sulphur (S) – are crucial for the biochemical processes that support the health and productivity of crop plants. It is crucial for modern agriculture to comprehend the molecular and physiological principles guiding the absorption, transportation, and utilization of these macronutrients (Husen, 2024; Javed *et al.*, 2022a). The complex interaction between agricultural plants and macronutrients focuses on the molecular and physiological factors that affect the growth and development. By figuring out the complex systems that control the absorption and assimilation of macronutrients, we can improve food security, increase crop productivity, and practise sustainable agriculture (de Bang *et al.*, 2021). The physiological and molecular mechanisms control how macronutrients are taken up by plants from the soil, distributed throughout them, and used to catalyse the creation of essential biomolecules (Khan *et al.*, 2014). In this chapter, we'll also look at how to provide enough macronutrients for plant growth while minimizing the negative effects on the environment, correcting nutrient deficits, and reducing the effects of nutrient surpluses. We will emphasize the crucial role of macronutrients in crop plant health as we proceed through this investigation, clarify the molecular players and genetic regulators involved, and talk about the constantly evolving approaches and technologies that have the potential to completely transform nutrient management in agriculture. By doing this, we hope to provide readers with a solid grounding in the intricate molecular and physiological workings of macronutrients, enabling them to promote the development of effective crop production systems that are crucial for feeding a growing world population.

11.1.1 Importance of Macronutrients in Crop Plants

The cornerstones of agricultural prosperity and global food security, macronutrients are of utmost importance to crop plants. The vital elements like nitrogen (N), phosphorus (P), potassium (K), calcium (Ca), magnesium (Mg), and sulphur (S) serve as the foundation for essential components, including proteins, nucleic acids, and sugars, which are responsible for photosynthesis, cell division, and general plant development (Leroux *et al.*, 2022). Macronutrients control nutrient absorption and assimilation by functioning as co-factors for enzymes involved in a variety of metabolic processes, as seen in Figure 11.1. Additionally, they help the plant transfer nutrients, maintain structural integrity, and control turgor pressure (Zenda *et al.*, 2021). The availability and balance of macronutrients are closely connected to crop yield, quality, and resistance to environmental stressors. Beyond their importance to plants, macronutrients play a crucial role in

DOI: 10.1201/9781032630366-11

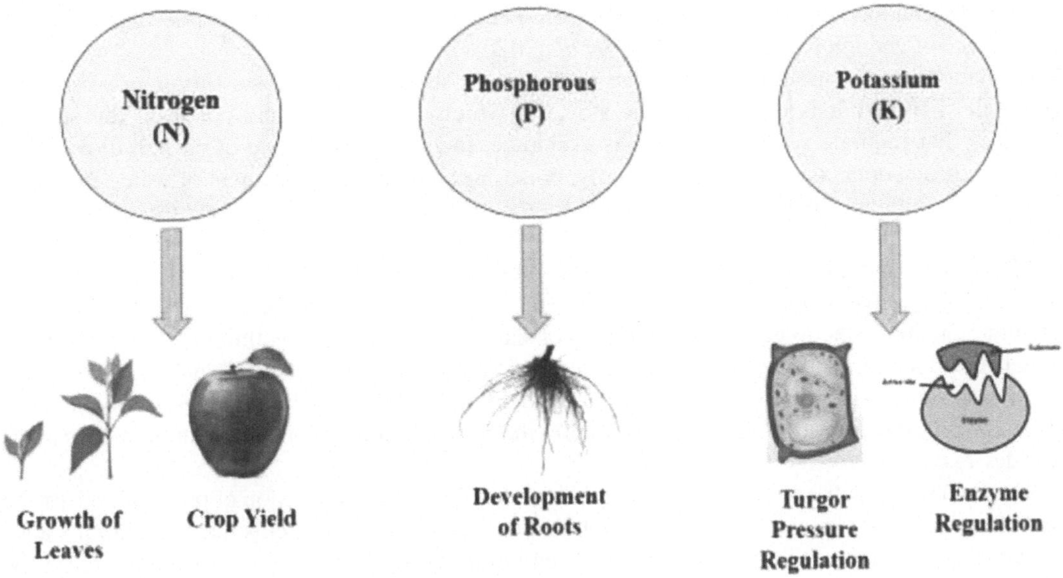

FIGURE 11.1 Importance of macronutrients in crop plants.

human nutrition, as they are mostly obtained through the crops we grow (Bulgari *et al.*, 2019). We can ensure sustainable crop production, reduce environmental impact, and responsibly and ecologically meet the world's food demand by understanding and successfully controlling macronutrients in agriculture.

11.1.2 Overview of Macronutrients

Macronutrients play critical roles in agricultural plant growth and development, as shown by an overview of these nutrients (Ahmad *et al.*, 2016). Nitrogen is crucial for photosynthesis and protein synthesis, since it is required for the production of amino acids and chlorophyll. In order for nucleic acids to develop, phosphorus, a crucial component of ATP, the energy unit of cells, must be present (Karthika *et al.*, 2018). Potassium maintains cell turgor, controls osmotic pressure, and facilitates the movement of nutrients and water. Magnesium and calcium are essential for intracellular signalling, enzyme activity, and cell wall stability (Tränkner *et al.*, 2018). Sulphur is essential for the production of proteins and amino acids, which increases plant vigour. The health, growth, and productivity of agricultural plants are determined by this broad set of macronutrients, underlining their critical role in agriculture and the world's food security. Successful crop management and sustainable agricultural practices depend on an understanding of their unique roles and optimal supply (Javed *et al.*, 2022b).

Protein synthesis and photosynthesis are both powered by nitrogen. It is a crucial part of chlorophyll, the green pigment in plants that absorbs sunlight and uses it to create energy through photosynthesis (Blankenship, 2021). Additionally, nitrogen is a crucial component of amino acids, which are the fundamental chemicals that make up proteins. Therefore, it is not overstated to claim that nitrogen is the main factor in plant growth, as it has a direct impact on how leaves, stems, and overall biomass develop (Patani *et al.*, 2023). Another important macronutrient, phosphorus, is essential for the transport of energy within plant cells. It is a crucial component in adenosine triphosphate (ATP), the molecule that stores and transmits energy in cells. In addition, phosphorus is necessary for the synthesis of nucleic acids like DNA and RNA, which are in charge of transmitting genetic information and facilitating cell division. Plants struggle to capture and develop slowly in the absence of enough phosphorus (Zhang *et al.*, 2020).

In plant cells, potassium controls the osmotic pressure, which affects turgor pressure and total cell stiffness. It is crucial to how stomata, which are microscopic pores on the surface of leaves that regulate water loss and gas exchange, including the intake of carbon dioxide for photosynthesis, open and close. Additionally, potassium aids in the movement of water and other nutrients throughout the plant, ensuring that vital substances get to where they are required (Na *et al.*, 2022).

Magnesium and calcium play a key role in preserving the structural integrity of plant cells. Pectin in cell walls is stabilized by calcium, which promotes cell adhesion and proliferation. In addition, calcium functions as a secondary messenger in a variety of cellular activities, such as signal transduction and enzyme activation. On the other hand, magnesium is necessary for photosynthesis, since it serves as the primary component of the chlorophyll molecule. Additionally, it plays a role in the activation of enzymes, notably those involved in the metabolism of nucleic acids (Pandey *et al.*, 2013).

The final macronutrient on the list is sulphur, which aids in the production of proteins and amino acids. It is essential for the creation of disulphide bonds, which give proteins their structural stability. Sulphur shortage can diminish crop vigour and inhibit protein synthesis, which can impact the general health and production of plants (Bhattacharya, 2021).

As the primary building blocks of plant nutrition, macronutrients have an effect on almost every aspect of crop growth and development. For maximizing agricultural output, guaranteeing global food security, and implementing sustainable farming practices that have the lowest negative environmental impact, a balanced and careful supply of these vital materials is crucial. The key to effective nutrient management in contemporary agriculture is understanding the distinct functions and needs of each macronutrient (Koch *et al.*, 2020).

11.1.3 Objective and Scope of This Chapter

This chapter's main goal is to provide an in-depth examination of the molecular and physiological theories that underlie the crucial role of macronutrients in the growth and development of agricultural plants. We seek to provide readers with a profound grasp of how macronutrients shape the basic foundation of agriculture and global food security through a methodical and comprehensive approach.

11.2 MACRONUTRIENT UPTAKE AND TRANSPORT

It is crucial to comprehend how plants take up and transport these vital components if one is to understand the molecular and physiological ideas of macronutrients in crop plant growth and development (Jose, 2023). Intricate processes called macronutrient absorption and transport make sure that plants get enough nitrogen, phosphorus, potassium, calcium, magnesium, and sulphur, as seen in Figure 11.2. Here, we explore the key aspects of macronutrient uptake and transport:

11.2.1 Root Uptake Mechanisms

11.2.1.1 Active Transport

Mechanisms for active transport are used for various macronutrients, including calcium and potassium. These methods entail specialized transport proteins in the membranes of the root cell, which actively pump macronutrients into the plant against gradients in their concentrations. ATP is frequently needed as energy for this process (Colmenero-Flores *et al.*, 2019).

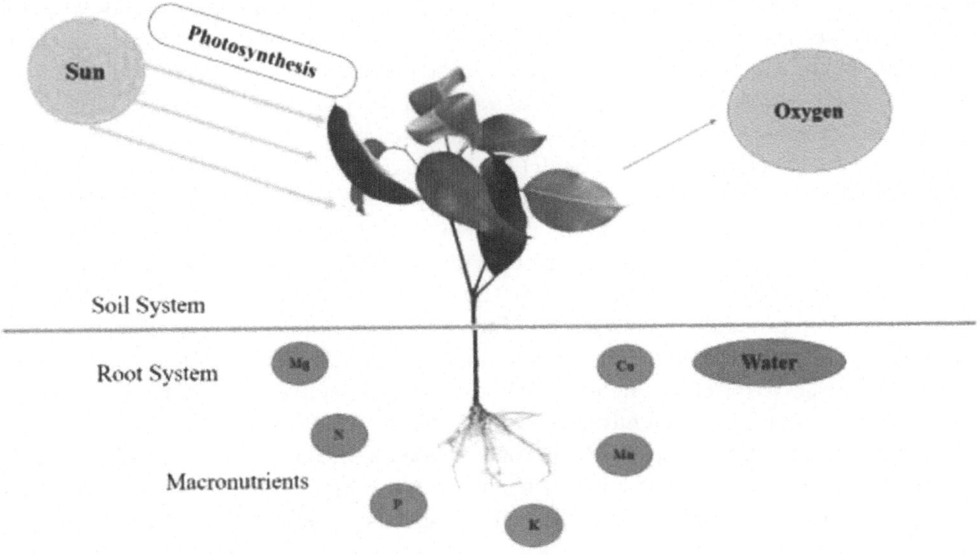

FIGURE 11.2 Macronutrient uptake and transport.

11.2.1.2 Passive Transport

Nitrate and sulphate are two macronutrients that are absorbed predominantly by diffusion, which is made possible by concentration gradients. The plant does not need to exert energy for passive uptake (Karthika *et al.*, 2018).

11.2.2 Macronutrient Mobility in the Plant

11.2.2.1 Translocation

Macronutrients must be delivered within the plant after being taken up by the root cells in order to reach the different tissues and organs where they are required. The vascular system of the plant, which is made up of the xylem and phloem, is where this movement takes place. From the roots to the shoots, the xylem predominantly transports water and dissolved minerals, including macronutrients. On the other side, the phloem is in charge of moving macronutrients and photosynthates (sugars) all over the plant (Bhatla *et al.*, 2018).

11.2.2.2 Phloem and Xylem Transport

Macronutrients are loaded into the phloem in source tissues (such as leaves) and transferred to sink tissues (such as roots and growing fruits), where they are consumed or stored. Photosynthates, in particular, are particularly rich in sugars. Distribution of macronutrients to various plant sections is made easier by phloem movement, facilitating growth and development.

While the phloem is the main pathway for the movement of macronutrients, some minerals and water-soluble nutrients may also move through the xylem. The distribution of important macronutrients, particularly calcium and magnesium, to different plant tissues is facilitated by this transport (Hawkesford *et al.*, 2012).

11.3 NITROGEN IN CROP PLANT GROWTH

Crop plant growth and development are fundamentally reliant on nitrogen, which has a significant impact on agricultural output and food security worldwide. Nitrogen is a crucial macronutrient for all of the plants' functions, making it an integral part of the complex molecular and physiological interactions that occur in crops. In particular, nitrogen is essential for the production of the green pigment chlorophyll, which powers photosynthesis, the process by which plants use light to create energy and carbohydrates. Because nitrogen is essential to chlorophyll's capacity to absorb light energy, its absence causes chlorosis, which is characterized by yellowing leaves, decreased photosynthetic activity, and restricted plant growth (Tan *et al.*, 2020).

In addition to its role in photosynthesis, nitrogen is a key ingredient in amino acids, which are the building blocks of proteins. These proteins, which control almost every aspect of plant biology, are the building blocks of life. Nitrogen is involved in a wide range of processes, from enzymes that catalyse biochemical events to structural proteins that control plant form and function. Additionally, nitrogen affects crop productivity and quality. Increased crop production, which results in higher harvests, is frequently associated with adequate nitrogen supply. Finding the ideal mix is essential, though. Rich, vegetative development at the expense of reproductive components like flowers and fruits can be the result of too much nitrogen (Nunes-Nesi *et al.*, 2010).

Efficiency in nitrogen use is a major issue in contemporary agriculture, since excessive or ineffective nitrogen usage can result in environmental problems, such nutrient runoff and groundwater pollution. To ensure that nitrogen is provided to crops optimally, crop management techniques and technology are continually improving. This maximizes productivity and sustainability (Anas *et al.*, 2020).

11.3.1 Role of Nitrogen in Crop Plants

Nitrogen has several different roles in crop plants and is essential to their development and yield. Nitrogen is a crucial macronutrient that supports a number of vital biological functions in plants. In the beginning, it is an essential part of chlorophyll, the green pigment that aids in photosynthesis. Nitrogen enables plants to generate carbohydrates, which serve as their primary energy source and the foundation for development, by absorbing and converting light energy into chemical energy (Leghari *et al.*, 2016). In addition, nitrogen is a crucial component of amino acids, the chemical substances that make up proteins. The mainstays of plant biology, proteins operate as structural building blocks for plant tissues, enzymes that catalyse biochemical events, and regulators of many cellular activities. Without enough nitrogen, plants would find it difficult to make these proteins, which would restrict their development, reduce their vigour, and lower food yields (Zheng *et al.*, 2020).

The impact of nitrogen also extends to crop quality. Increased crop production, which results in higher harvests, is frequently associated with adequate nitrogen supply. The balance is tricky, though, since too much nitrogen can encourage fruit and flower production at the expense of vegetative growth, thus lowering crop quality overall. Effective management of nitrogen is crucial in modern agriculture. A significant difficulty is striking a balance between providing enough nitrogen for agricultural growth and reducing environmental effects, such as nutrient runoff and groundwater pollution. In order to satisfy both the nutritional needs of a growing global population and the responsible stewardship of the environment, sustainable agricultural practices and technology are constantly being improved (Gawankar *et al.*, 2019).

11.3.2 Nitrogen Assimilation

Crop plants obtain and assimilate necessary nitrogen from their surroundings through a key mechanism called nitrogen assimilation. Nitrate (NO_3^-) and ammonium (NH_4^+), the two main inorganic forms of nitrogen, are the principal sources of nitrogen absorption for plants (Mokhele *et al.*, 2012).

11.3.3 Nitrate and Ammonium Uptake

Nitrate Uptake: Crop plants' roots have the ability to take up nitrate ions from the soil. Specialized transporters, such as nitrate transporters (NRTs), which are found on the root cell membranes, are involved in this process. These transporters enable the active uptake of nitrate despite its gradient of concentration while frequently requiring ATP as energy. Once inside the root cells, nitrate can undergo enzymatic processes to change into nitrite (NO_2) and then into ammonium. The production of proteins and amino acids is one of the many metabolic activities that can employ this ammonium (Yadav *et al.*, 2021).

Ammonium Uptake: Direct uptake of ammonium ions from the soil is a different method of nitrogen assimilation. This activity is carried out by ammonium transporters (AMTs) that are present on the membranes of root cells. Because ammonium uptake takes place along the concentration gradient, it doesn't require energy expenditure like nitrate uptake does. Ammonium is an efficient and direct source of nitrogen inside plants because it is easily converted into amino acids and then used for protein production (Xu *et al.*, 2012).

11.3.4 Nitrogen Fixation

It is possible for some plants, especially legumes like soybeans, peas, and clover, to transform atmospheric nitrogen gas (N_2) into a form that is useful for plants and, later, animals, thanks to nitrogen fixation, a remarkable and crucial mechanism in the nitrogen cycle, as seen in Figure 11.3. Specialized bacteria known as nitrogen-fixing bacteria, which create symbiotic partnerships with these plants, are responsible for this process (Saha *et al.*, 2017). Nitrogen-fixing bacteria, like *Rhizobia*, live in the root nodules of the host plants in these symbiotic relationships. These bacteria are endowed with the nitrogenase enzyme, which allows them to dissolve atmospheric nitrogen (N_2) into ammonia (NH_3) or ammonium ions (NH_4^+) by dissolving its robust triple bond. Plants may easily assimilate the type of nitrogen known as ammonium into their metabolic activities (Hassen *et al.*, 2023).

Through this mutualistic relationship, the nitrogen-fixing bacteria receive a source of energy in the form of carbohydrates from the leguminous plants, and the plants receive a biologically usable form of nitrogen from the bacteria. Because of this, legumes have a special capacity to grow on soils that are low in nitrogen while also adding nitrogen to the soil, which helps nearby crops (Bhat *et al.*, 2015).

Sustainable agriculture relies less on synthetic nitrogen fertilizers, which can have negative effects on the environment, thanks to nitrogen fixation. Farmers may increase soil fertility, boost crop yields, and reduce nitrogen pollution by utilizing nitrogen-fixing plants and their bacterial companions. This helps to create more sustainable and resilient agricultural systems. To fully use

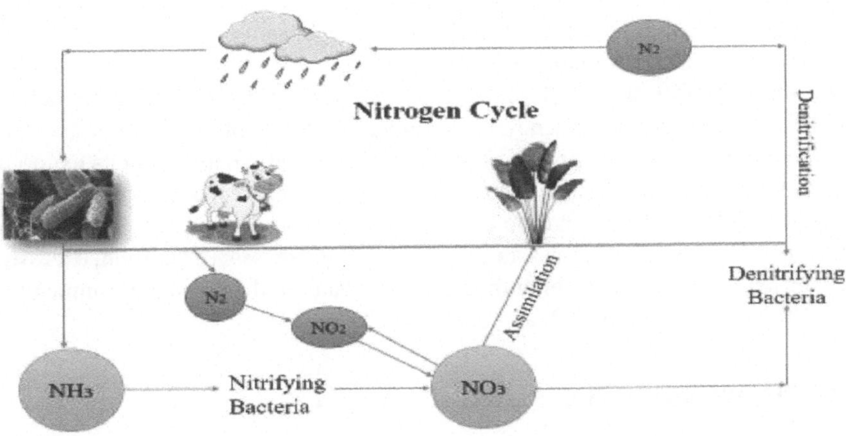

FIGURE 11.3 Nitrogen assimilation.

nitrogen fixation in agriculture and address nitrogen-related issues, it is essential to comprehend its complex molecular and ecological features (Soumare *et al.*, 2020).

11.3.5 Nitrogen Use Efficiency

Nitrogen use efficiency (NUE) is a critical concept within the molecular and physiological context of macronutrients in crop plant growth and development. It quantifies the effectiveness with which plants utilize the available nitrogen for biomass production, crop yield, and overall performance while minimizing losses to the environment. Achieving high NUE is not only economically advantageous for farmers but also environmentally responsible, as it reduces nitrogen pollution, including nitrate leaching into groundwater and emissions of nitrogenous gases into the atmosphere (Lal *et al.*, 2023).

NUE encompasses two main components:

Nitrogen Uptake Efficiency: This aspect of NUE focuses on the plant's ability to acquire nitrogen from the soil efficiently. It involves mechanisms such as root morphology, nutrient transporter proteins, and the plant's capacity to capture and utilize different nitrogen forms (nitrate and ammonium). Optimizing these processes ensures that plants can access an ample nitrogen supply, even in nitrogen-limited soils (Rasmussen *et al.*, 2015).

Nitrogen Utilization Efficiency: This component evaluates how effectively plants allocate and utilize the absorbed nitrogen for growth and yield. It encompasses the molecular and physiological processes that govern the allocation of nitrogen to essential functions like photosynthesis, protein synthesis, and structural growth. Efficient nitrogen utilization means that a higher proportion of absorbed nitrogen is converted into valuable biomass, resulting in improved crop yields (de Oliveira Silva *et al.*, 2020).

11.3.6 Symptoms of Nitrogen Deficiency and Excess

Crop plant growth and development are significantly influenced by nitrogen, a crucial macronutrient. But both nitrogen excess and deficiency can have different and frequently negative impacts on plants (Guo *et al.*, 2019).

Nitrogen Deficiency Symptoms: There are several obvious symptoms that crop plants are nitrogen-deficient. Chlorosis, when older leaves turn pale green or yellow due to insufficient chlorophyll production, is one of the most obvious signs. Reduced photosynthetic activity, stunted growth, and lower agricultural output are the outcomes of this. Because nitrogen deprivation can direct resources away from reproductive structures in favour of supporting essential metabolic processes, plants also display delayed flowering and fruiting. Overall, a lack of nitrogen can result in worse crop quality and financial losses for farmers (Huang *et al.*, 2020).

Nitrogen Excess Symptoms: On the other hand, too much nitrogen might be bad. Overly lush foliage, lengthened stems, and a tendency for lodging, where the plants bend or break under their own weight, are all characteristics of luxuriant vegetative development in plants exposed to excess nitrogen. Such quick, excessive vegetative development can rob reproductive organs of nutrients, resulting in lower yields and poorer crop quality. Furthermore, too much nitrogen might render plants more vulnerable to some illnesses and environmental stresses. In addition, excessive nitrogen leaking from agricultural areas can contaminate groundwater and worsen environmental problems, such as algae blooms in waterways (Huang *et al.*, 2020).

11.4 PHOSPHORUS IN CROP PLANT GROWTH

An important macronutrient for crop plants, phosphorus is crucial for several physiological and molecular processes that control growth and development. Phosphorus is a component of several essential biomolecules, including DNA, RNA, and ATP, at the molecular level. Plants have genetic blueprints in the form of DNA and RNA, and their cells' main source of energy is ATP. This means that phosphorus is essential for activities like cell division, the transfer of genetic information, and the transmission of energy, all of which are essential for the growth and reproduction of plants (Malhotra *et al.*, 2018).

Phosphorus plays a physiological role in the development of robust root systems, which are necessary for nutrition and water intake. It encourages early root formation and improves root branching, enabling the plant to search a broader area of soil for crucial materials. Additionally, phosphorus helps reproductive structures like flowers and seeds mature, which affects agricultural productivity and quality (Bhatla *et al.*, 2018).

11.4.1 Role of Phosphorus in Crop Plants

Phosphorus, as one of the important macronutrients for agricultural plants, has a significant impact on a variety of molecular and physiological processes that are essential for plant growth and development. ATP, DNA, and RNA are among the most important macromolecules that contain phosphorus at the molecular level (Zhu *et al.*, 2018). These molecules are essential for the transfer of energy, cellular metabolism, and genetic information in plants. The presence of phosphorus in DNA and RNA is critical for gene expression as well as the production of proteins and enzymes, which are necessary for nearly every biological process in plants, such as photosynthesis, respiration, and nutrient uptake (Uchida, 2000).

Phosphorus physiologically has a significant influence on the growth and operation of roots. It encourages the development of strong root systems, which makes it easier for the plant to absorb water and nutrients from the soil. Better access to resources and increased plant resistance to environmental challenges like drought and nutrient deficiency are both benefits of this greater root development (Liu *et al.*, 2019).

11.4.2 Phosphorus Uptake

The crucial process of phosphorus uptake by crop plants depends on both molecular and physiological mechanisms. It is also intimately connected to the expansion and maturation of the plant as a whole. Specialized transporter proteins, primarily phosphate transporters, take in inorganic phosphate (Pi) into the root cells on a molecular level. These transporters frequently need ATP as energy and assist Pi in crossing the root cell membrane against the gradient of its concentration. To suit plant needs, phosphorus intake is closely regulated (Uchida, 2000). Plants can release compounds like organic acids that solubilize bound phosphates and make them more easily absorbable when there is a phosphate deficiency in the soil. The efficiency of phosphorus uptake can also be increased via mycorrhizal interactions, which are partnerships between plants and beneficial fungi. These fungi expand the plant's root system and have the ability to obtain phosphorous in sporadically soluble forms, releasing it for usage by the plant (Bechtaoui *et al.*, 2021).

Numerous physiological aspects of plant growth depend on phosphorus. Phosphorus is essential to the metabolism of energy because it is required for ATP, the primary type of energy in plant cells. Phosphorus is also necessary for the synthesis of nucleic acids, such as DNA and RNA, which influence how genetic information is transmitted and genes are expressed. Phosphorus also makes up a significant portion of adenosine diphosphate (ADP) and adenosine monophosphate (AMP), which are essential for cellular respiration and energy transfer (Malhotra *et al.*, 2018).

11.4.3 Phosphate Transporters

Phosphate transporters are critical components of the molecular machinery that regulates the uptake and utilization of phosphorus. These specific proteins are responsible for the plant's efficient uptake of Pi from the soil and are found on the membranes of the root cell. Phosphate transporters are typically categorized into different groups based on their structural and functional characteristics. Phosphate transporters play a crucial role in the plant's ability to scavenge and obtain this essential macronutrient, even in soils with low soil phosphorus supply (Wang *et al.*, 2018). They facilitate the movement of Pi ions across the root cell membrane against its gradient of concentration, a process that typically requires ATP. The activity of these transporters is tightly regulated to ensure that the plant can maintain a balanced supply of phosphorus in accordance with its growth requirements. Understanding the function of phosphate transporters is necessary for optimizing phosphorus uptake in agricultural plants. The genetic and molecular mechanisms that control these transporters' regulation are constantly being studied by scientists in an effort to improve their efficacy (Viollet *et al.*, 2012).

11.4.4 Mycorrhizal Symbiosis

A key factor in the uptake of macronutrients and, consequently, the growth and development of agricultural plants, mycorrhizal symbiosis is an intriguing molecular and physiological concept. Mycorrhizal fungus and plants have a mutually beneficial relationship in which one partner considerably benefits the other. At the molecular level, mycorrhizal connections are marked by a remarkable exchange of resources. The plant provides nutrients to the mycorrhizal fungi, which the fungi use as fuel, with the carbohydrates it produces during photosynthesis (Ferrol *et al.*, 2019). In exchange, the mycorrhizal fungi enhance the plant's ability to absorb nutrients, namely, phosphorus and, to a lesser extent, other macronutrients like nitrogen. The plant's capacity to absorb nutrients is effectively increased by these fungi's ability to enter soil regions outside of the plant's root system (Nath *et al.*, 2018).

The growth of agricultural plants is significantly physiologically impacted by mycorrhizal symbiosis. Mycorrhizal associations increase nutrient uptake, particularly phosphorus, to assist plants growing in soils with poor nutrient availability. This is especially useful in agriculture, because it reduces the need for extensive fertilizer use and promotes environmentally friendly agricultural practices. In addition, mycorrhizal symbiosis can make a plant more resilient to environmental hazards like drought and disease attacks, which improves crop resilience and yield (Fiorilli *et al.*, 2022).

11.4.5 Phosphorus Use Efficiency

A key idea in the field of molecular and physiological theories guiding the use of macronutrients in crop plant growth and development is the concept of phosphorus use efficiency (PUE). PUE measures a crop plant's ability to use available phosphorus, a crucial macronutrient, as effectively as possible while reducing environmental losses (Hossain *et al.*, 2017). Optimizing PUE at the molecular level necessitates a thorough knowledge of the genetic and biochemical systems that control phosphorus absorption, transport, and utilization in plants. It entails optimizing phosphorus assimilation routes, improving root architecture to access phosphorus-rich soil patches, and fine-tuning phosphate transporter genes (Hasan *et al.*, 2016).

When a plant produces organic acids and exudates that improve phosphorus solubility and availability, it is able to take phosphorus from a range of soil phosphorus pools, including sparingly soluble forms. High PUE has several physiological effects that enhance crop performance (Jemo *et al.*, 2006). Effective phosphorus use promotes improved root growth, which facilitates improved nutrient and water intake. Even on phosphorus-deficient soils, the plant can access phosphorus thanks to its strong root system. High PUE also promotes healthy vegetative growth, prompt blooming, and ideal seed and fruit production, all of which lead to higher crop yields and improved crop quality (Emmanuel and Babalola, 2020).

11.4.6 Symptoms of Phosphorus Deficiency and Excess

Phosphorus, as a fundamental macronutrient in crop plant growth and development, exhibits distinct symptoms when deficient or present in excess, reflecting its intricate molecular and physiological role (Soetan *et al.*, 2010).

Phosphorus Deficiency Symptoms: When crops face phosphorus deficiency, visible signs emerge as a consequence of impaired molecular processes. The most prominent symptom is stunted growth, often accompanied by a dark green to bluish-green coloration of leaves. This darkening results from an increase in anthocyanin pigments, a response to stress. Phosphorus deficiency also leads to reduced root development, limiting the plant's ability to explore the soil for nutrients and water. Consequently, there's delayed flowering and fruiting, resulting in diminished crop yield. Additionally, phosphorus-deficient plants may exhibit purpling of leaves and a decline in overall vigour (Mostofa *et al.*, 2018).

Phosphorus Excess Symptoms: Conversely, an excess of phosphorus can also detrimentally affect crop plants. When plants are exposed to excessive phosphorus, they may display symptoms such as reduced root growth, reduced nutrient uptake efficiency, and imbalanced nutrient ratios. The excessive presence of phosphorus can lead to nutrient imbalances, interfering with the uptake of other essential macronutrients, such as iron and zinc. This can result in nutrient deficiencies even when they are present in the soil. Furthermore, the buildup of excessive phosphorus in soils can have environmental repercussions, including water pollution, as excess phosphorus runoff into water bodies can lead to eutrophication, harming aquatic ecosystems (Osman, 2013).

11.5 POTASSIUM IN CROP PLANT GROWTH

11.5.1 Role of Potassium in Crop Plants

Potassium is a crucial macronutrient with a variety of physiological and molecular functions that are vital to the growth and development of crop plants. Potassium plays a crucial role in a number of biochemical and physiological processes at the molecular level. The main regulator of osmotic potential in plant cells is potassium, which affects cell turgor pressure and water uptake. This function is essential for preserving healthy cell structures and avoiding wilting (Ahanger *et al.*, 2017). In addition, potassium is essential for the activation of the enzymes responsible for protein synthesis, respiration, and photosynthesis. It enhances the transfer of sugars and other nutrients into the plant and aids in the creation of ATP, the main source of energy for plant cells. Potassium also influences the regulation of stomatal openings in leaves, affecting gas exchange and transpiration rates. This, in turn, impacts water use efficiency and the plant's ability to withstand environmental stressors such as drought and salinity (Kapoor *et al.*, 2020).

Potassium is crucial for several physiological processes, including root growth, blooming, and fruit formation. Strong root systems let plants obtain water and nutrients when potassium levels are adequate. Due to potassium's role in blooming and fruit development, crop yields and quality are increased (Hasanuzzaman *et al.*, 2018). With its involvement ranging from physiological activities like sustaining cell turgor pressure to molecular processes like enzyme activation, potassium is an essential macronutrient for agricultural plant growth and development. In order to improve crop productivity, guarantee crop quality, and address environmental issues, all within the framework of sustainable farming practices, it is essential to optimize potassium management in agriculture (Vasseghian *et al.*, 2022).

11.5.2 Potassium Uptake

In terms of the molecular and physiological principles guiding the utilization of macronutrients in plant growth and development, the process of potassium uptake by crop plants marks a crucial turning point. Potassium uptake from the soil into the plant root cells is mediated at the molecular level by specialized transport proteins known as potassium transporters (Rawat *et al.*, 2016).

For the plant to maintain potassium homeostasis, these transporters are crucial. They ensure that the plant may obtain an appropriate amount of this crucial macronutrient by facilitating the flow of potassium ions (K+) across the root cell membranes. Notably, potassium uptake frequently happens passively, fuelled by the electrochemical gradients produced by other ion transport processes, and does not always need energy expenditure (in the form of ATP). In terms of physiology, potassium uptake is essential for the growth and development of crop plants (Vishwakarma *et al.*, 2019). Potassium ions play a role in the regulation of osmotic pressure, the activation of enzymes, and the movement of nutrients throughout the plant. Osmotic control of potassium affects cell turgor pressure, preserving normal cell structure and averting wilting. Additionally, potassium affects energy generation and plant metabolism by serving as a cofactor for enzymes involved in photosynthesis, respiration, and protein synthesis (Hänsch and Mendel, 2009).

11.6 MOLECULAR REGULATION OF MACRONUTRIENT HOMEOSTASIS

11.6.1 Transcription Factors and Signalling Pathways

The complex molecular control of macronutrient balance in plants is intricately regulated by transcription factors and signalling pathways. These molecular parts come together to form a complex network that controls how the plant reacts to variations in nutrient availability and promotes the best possible uptake, utilization, and distribution of macronutrients. The expression of genes involved in nutrient absorption and transport is tightly regulated by transcription factors, which are frequently unique to particular macronutrients like nitrogen, phosphorus, or potassium. These transcription factors are triggered when the plant's macronutrient levels are insufficient, which causes the transcription of genes encoding the nutrient transporters, enzymes, and other proteins required for effective nutrient acquisition (López-Arredondo, Sánchez-Calderón and Yong-Villalobos, 2017).

11.6.2 Nutrient-Sensing Mechanisms

Crop plants use nutrient sensing systems as their molecular sentinels to detect and react to changes in macronutrient availability, which is crucial for preserving macronutrient homeostasis. These systems are essential for the plant to be able to adjust to changing environmental conditions and maximize nutrient uptake and utilization (O'Brien *et al.*, 2016).

The activation of receptor-like kinases (RLKs) or receptor-like proteins (RLPs) in the plant's cell membranes is one of the main mechanisms for nutrient sensing. The availability of specific macronutrients, such as nitrate or phosphate, can be signalled by these receptors, which can recognize the signals specifically. These receptors start signalling cascades when nutrition levels are low, changing gene expression and triggering physiological reactions that improve nutrient uptake (Liang and Zhou, 2018). Transcription factors that respond to nutrients are essential for nutrition sensing. These transcription factors undergo post-translational modifications brought on by alterations in nutrient levels, which in turn activate them when they bind to signalling molecules. Once activated, they control the expression of genes related to nutrient transporters, nutrient metabolism enzymes, and other elements of the plant's machinery for nutrient uptake and utilization (Mitra, 2017).

Additionally, hormone signalling pathways and nutrient sensing mechanisms are intricately linked in plants. Auxins, cytokinins, and abscisic acid (ABA), among other phytohormones, can influence nutritional responses by affecting procedures such root development, nutrient transport, and stomatal regulation (Nguyen *et al.*, 2021).

11.6.3 Genetic Variation and Crop Improvement

The pursuit of harnessing the molecular control of macronutrient homeostasis for crop development is based on genetic variety, which acts as a cornerstone. The ability of crop plants to adapt to various nutritional circumstances is supported by their inherent genetic variety. In order to create crop varieties with improved nutrient usage efficiency and adaptability, breeding programmes can make use of this diversity as a significant resource (Nelson *et al.*, 2016).

Researchers and breeders use genetic variation to find and choose plants that exhibit advantageous features for acquiring and using macronutrients. This procedure frequently entails the discovery of certain genes, alleles, or molecular markers linked to enhanced nutrition absorption or utilization. These genetic components can then be added to breeding programmes using conventional breeding methods or cutting-edge biotechnological techniques like marker-assisted selection or gene editing (Elango *et al.*, 2021).

11.7 INTERACTIONS AMONG MACRONUTRIENTS

The biochemical and physiological principles that control macronutrients in crop plant growth and development are intimately linked, with these crucial components interacting in a variety of complex ways. To flourish and experience optimal growth, crop plants need a precise balance of macronutrients, including nitrogen, phosphorus, potassium, calcium, magnesium, and sulphur. These macronutrients participate in a variety of complimentary biochemical processes at the molecular level. For instance, although phosphorus is an essential element for the production of nucleic acids and ATP, nitrogen is necessary for the synthesis of amino acids and proteins. In order to sustain cell turgor pressure, potassium is necessary, while calcium is crucial for the creation of cell walls (Rosier *et al.*, 2018). Magnesium is a structural element of chlorophyll, which is essential for photosynthesis, while sulphur is required for the creation of coenzymes and amino acids that include sulphur. Macronutrient interactions have physiological effects on nutrient absorption, transport, and utilization. A macronutrient's absorption and utilization can be improved by having adequate quantities of another macronutrient. For instance, calcium can affect the uptake of other cations like potassium and magnesium, while potassium can enhance the intake of magnesium. The ability of a plant to absorb and use other macronutrients might be hampered by imbalances or deficiencies in one of them, which can result in nutrient-related ailments and decreased agricultural yields (Awuchi *et al.*, 2020).

11.8 ENVIRONMENTAL FACTORS INFLUENCING MACRONUTRIENT UPTAKE

Crop plants' biochemical and physiological mechanisms for absorbing macronutrients are closely entwined with a variety of environmental variables. These elements have a significant impact on how nutrients are acquired, affecting how plants grow and develop. The availability and uptake of macronutrients are greatly influenced by soil factors like as pH, texture, moisture, temperature, salt, and other stresses (Jose, 2023).

The ionization of macronutrients, for example, is controlled by soil pH, which affects their solubility and accessibility to plant roots. While alkaline soils can impact the availability of micronutrients, acidic soils may restrict the intake of important macronutrients like phosphate and calcium. Nutrient retention and release rates are influenced by soil texture, which is defined by its composition in sand, silt, and clay. Due to their lesser capacity to hold nutrients, sandy soils may require more frequent nutrient applications, whereas clay soils may be able to hold nutrients but release them more slowly (Dhaliwal *et al.*, 2019).

11.9 FUTURE TRENDS IN MACRONUTRIENT MANAGEMENT FOR CROP PRODUCTION

Future advances in macronutrient management for crop productivity are poised to blend molecular and physiological principles with novel technologies and sustainable practices as agriculture continues to change in response to shifting environmental and societal issues. Several key trends are likely to shape the future of macronutrient management:

Precision Agriculture: With the help of remote sensing, data analytics, and automation improvements, farmers will be able to customize the application of macronutrients to the unique requirements of different fields or even individual plants. This precise method guarantees effective nutrient utilization, decreased waste, and increased crop yields (Tian *et al.*, 2020).

Nutrient Use Efficiency: Research into crop varieties with improved nitrogen uptake and utilization features will be prompted by the increased emphasis on nutrient usage efficiency. The development of nutrient-efficient crops that can grow in nutrient-deficient or stressful conditions will rely heavily on molecular breeding and genetic engineering (Hirel *et al.*, 2011).

Nutrient Recycling and Circular Agriculture: The idea of "circular agriculture," which emphasizes nutrient recycling and minimizing nutrient losses, will become more popular. There will be less demand for external nutrient inputs thanks to cutting-edge cropping techniques like agroforestry and cover crops as well as technologies for absorbing and recycling nutrients from waste streams (Rodríguez-Espinosa *et al.*, 2023).

Biostimulants and Microbiome Management: Utilizing advantageous microbes and biostimulants to improve nutrient uptake and root health will become commonplace. Strategies for maximizing nutrient availability will be guided by an understanding of the interactions between plant roots and the soil microbiota (Sible, Seebauer and Below, 2021).

Smart Fertilizers: Nutrient management will undergo a revolution with the creation of intelligent fertilizers that release nutrients in response to plant needs or environmental circumstances. These fertilizers will lessen nutrient leaching and environmental effects because they were created utilizing nanotechnology and controlled-release technologies (Al-Juthery *et al.*, 2021).

11.10 CONCLUSION

In conclusion, the investigation of molecular and physiological ideas related to macronutrients in crop plant growth and development has shown a complex and interrelated web of processes that control nutrient absorption, utilization, and homeostasis. Each of the essential macronutrients – nitrogen, phosphorus, potassium, calcium, magnesium, and sulphur – plays a crucial function at the molecular and physiological levels. For instance, nitrogen, which is assimilated from nitrate and ammonium sources, increases protein synthesis and total plant development. Phosphate transporters and signalling pathways coordinate the efficient intake of phosphorus, which is essential for energy metabolism and nucleic acid synthesis. Potassium impacts a number of physiological processes by regulating enzyme activation and maintaining cell turgor pressure.

Even though they are needed in smaller amounts, calcium, magnesium, and sulphur are similarly important, since they take part in procedures like the production of cell walls, photosynthesis, and amino acid synthesis. Further highlighting the intricacy of plant nutrition are the complicated interactions among these macronutrients that are controlled by transcription factors, signalling pathways, and nutrient sensing mechanisms. A comprehensive strategy for managing macronutrients in agriculture is required, since environmental conditions such as soil pH, temperature, and salinity have a significant impact on nutrient availability and uptake. Precision farming, nutrient-efficient crop types, circular agriculture, and cutting-edge technologies like smart fertilizers and digital farming platforms will all be included in macronutrient management in the future.

Sustainable nutrient management techniques will be essential in this changing climate for addressing food security and maintaining environmental integrity. We pave the way for a future where agriculture not only satisfies the expanding demands of a global population, but also does it sustainably, resiliently, and effectively as we delve further into the molecular and physiological intricacies of macronutrients in crop plants. We will navigate the difficult realm of managing macronutrients in crop production through this integration of knowledge, technology, and sustainability.

REFERENCES

Ahanger, M.A. *et al.* (2017) 'Plant growth under water/salt stress: ROS production; Antioxidants and significance of added potassium under such conditions', *Physiology and Molecular Biology of Plants*, 23, pp. 731–744.

Ahmad, M. *et al.* (2016) 'Potassium-solubilizing bacteria and their application in agriculture', in *Potassium solubilizing microorganisms for sustainable agriculture*. New Delhi: Springer, pp. 293–313.

Al-Juthery, H.W., Lahmod, N.R. and Al-Taee, R.A. (2021) 'Intelligent, nano-fertilizers: A new technology for improvement nutrient use efficiency (article review)', *IOP Conference Series: Earth and Environmental Science*, 735(1), p. 012086.

Anas, M. *et al.* (2020) 'Fate of nitrogen in agriculture and environment: Agronomic, eco-physiological and molecular approaches to improve nitrogen use efficiency', *Biological Research*, 53(1), pp. 1–20.

Awuchi, C.G., Igwe, V.S. and Amagwula, I.O. (2020) 'Nutritional diseases and nutrient toxicities: A systematic review of the diets and nutrition for prevention and treatment', *International Journal of Advanced Academic Research*, 6(1), pp. 1–46.

Bechtaoui, N. *et al.* (2021) 'Phosphate-dependent regulation of growth and stresses management in plants', *Frontiers in Plant Science*, 12, p. 679916.

Bhat, T.A. *et al.* (2015) 'Nitrogen fixing biofertilizers; Mechanism and growth promotion: A review', *Journal of Pure and Applied Microbiology*, 9(2), pp. 1675–1690.

Bhatla, S.C. *et al.* (2018) 'Plant mineral nutrition', in *Plant physiology, development and metabolism*. New Delhi: Springer India, pp. 37–81.

Bhattacharya, A. (2021) 'Mineral nutrition of plants under soil water deficit condition: A review', in *Soil water deficit and physiological issues in plants*. Singapore: Springer, pp. 287–391.

Blankenship, R.E. (2021) *Molecular mechanisms of photosynthesis*. UK: John Wiley & Sons.

Bulgari, R., Franzoni, G. and Ferrante, A. (2019) 'Biostimulants application in horticultural crops under abiotic stress conditions', *Agronomy*, 9(6), p. 306.

Colmenero-Flores, J.M. *et al.* (2019) 'Chloride as a beneficial macronutrient in higher plants: New roles and regulation', *International Journal of Molecular Sciences*, 20(19), p. 4686.

de Bang, T.C. *et al.* (2021) 'The molecular–physiological functions of mineral macronutrients and their consequences for deficiency symptoms in plants', *New Phytologist*, 229(5), pp. 2446–2469.

de Oliveira Silva, A. *et al.* (2020) 'Nitrogen utilization efficiency in wheat: A global perspective', *European Journal of Agronomy*, 114, p. 126008.

Dhaliwal, S.S. *et al.* (2019) 'Dynamics and transformations of micronutrients in agricultural soils as influenced by organic matter build-up: A review', *Environmental and Sustainability Indicators*, 1, p. 100007.

Elango, D., Sandoya, G. and Chopra, S. (2021) 'Techniques and tools of modern plant breeding', in *Plant biotechnology: Experience and future prospects*. Cham: Springer, pp. 17–26.

Emmanuel, O.C. and Babalola, O.O. (2020) 'Productivity and quality of horticultural crops through co-inoculation of arbuscular mycorrhizal fungi and plant growth promoting bacteria', *Microbiological Research*, 239, p. 126569.

Ferrol, N., Azcón-Aguilar, C. and Pérez-Tienda, J. (2019) 'Arbuscular mycorrhizas as key players in sustainable plant phosphorus acquisition: An overview on the mechanisms involved', *Plant Science*, 280, pp. 441–447.

Fiorilli, V. *et al.* (2022) 'Arbuscular mycorrhizal symbiosis differentially affects the nutritional status of two durum wheat genotypes under drought conditions', *Plants*, 11(6), p. 804.

Gawankar, M.S. *et al.* (2019) 'Effect of girdling on induction of flowering and quality of fruits in horticultural crops-a review', *Advanced Agricultural Research and Technology Journal*, 3, pp. 201–15.

Guo, J. *et al.* (2019) 'Growth, photosynthesis, and nutrient uptake in wheat are affected by differences in nitrogen levels and forms and potassium supply', *Scientific Reports*, 9(1), p. 1248.

Hänsch, R. and Mendel, R.R. (2009) 'Physiological functions of mineral micronutrients (Cu, Zn, Mn, Fe, Ni, Mo, B, Cl)', *Current Opinion in Plant Biology*, 12(3), pp. 259–266.

Hasan, M.M. *et al.* (2016) 'Regulation of phosphorus uptake and utilization: Transitioning from current knowledge to practical strategies', *Cellular & Molecular Biology Letters*, 21, pp. 1–19.

Hasanuzzaman, M. *et al.* (2018) 'Potassium: A vital regulator of plant responses and tolerance to abiotic stresses', *Agronomy*, 8(3), p. 31.

Hassen, A.I., Babalola, O.O. and Carlson, R. (2023) 'Rhizobacterial-mediated interactions for enhanced symbiotic performance of the root nodule *Rhizobia* in legumes', in *Sustainable agrobiology: Design and development of microbial consortia*. Singapore: Springer Nature Singapore, pp. 41–56.

Hawkesford, M. *et al.* (2012) 'Functions of macronutrients', in *Marschner's mineral nutrition of higher plants*. Academic Press, pp. 135–189.

Hirel, B. *et al.* (2011) 'Improving nitrogen use efficiency in crops for sustainable agriculture', *Sustainability*, 3(9), pp. 1452–1485.

Hossain, M.A. *et al.* (eds.) (2017) *Plant macronutrient use efficiency: Molecular and genomic perspectives in crop plants*. Academic Press.

Huang, C.H. *et al.* (2020) 'Early diagnosis and management of nitrogen deficiency in plants utilizing Raman spectroscopy', *Frontiers in Plant Science*, 11, p. 663.

Husen, A. (2024) *Essential minerals in plant-soil system (Coordination, signalling, and interaction under adverse situations)*. Cambridge, MA: Elsevier Inc.

Javed, A. *et al.* (2022a) 'Soil fertility: Factors affecting soil fertility, and biodiversity responsible for soil fertility', *International Journal of Plant, Animal and Environmental Sciences*, 12(1), pp. 21–33.

Javed, T. *et al.* (2022b) 'Recent advances in agronomic and physio-molecular approaches for improving nitrogen use efficiency in crop plants', *Frontiers in Plant Science*, 13, p. 877544.

Jemo, M. *et al.* (2006) 'Phosphorus benefits from grain-legume crops to subsequent maize grown on acid soils of southern Cameroon', *Plant and Soil*, 284, pp. 385–397.

Jose, J.V. (2023) 'Physiological and molecular aspects of macronutrient uptake by higher plants', in *Sustainable plant nutrition*. Academic Press, pp. 1–21.

Kapoor, D. *et al.* (2020) 'The impact of drought in plant metabolism: How to exploit tolerance mechanisms to increase crop production', *Applied Sciences*, 10(16), p. 5692.

Karthika, K.S., Rashmi, I. and Parvathi, M.S. (2018) 'Biological functions, uptake and transport of essential nutrients in relation to plant growth', in *Plant nutrients and abiotic stress tolerance*. Singapore: Springer, pp. 1–49.

Khan, N.A. *et al.* (2014) 'Salinity tolerance in plants: Revisiting the role of sulfur metabolites', *Journal of Plant Biochemistry & Physiology*, 2(120), p. 2.

Koch, M. *et al.* (2020) 'The importance of nutrient management for potato production part I: Plant nutrition and yield', *Potato Research*, 63, pp. 97–119.

Lal, S.K. *et al.* (2023) 'Concurrent overexpression of rice GS1; 1 and GS2 genes to enhance the nitrogen use efficiency (NUE) in transgenic rice', *Journal of Plant Growth Regulation*, 1(1), pp. 1–22.

Leghari, S.J. *et al.* (2016) 'Role of nitrogen for plant growth and development: A review', *Advances in Environmental Biology*, 10(9), pp. 209–219.

Leroux, L. *et al.* (2022) 'Exploring the agricultural landscape diversity-food security nexus: An Analysis in two contrasted parklands of Central Senegal', *Agricultural Systems*, 196, p. 103312.

Liang, X. and Zhou, J.M. (2018) 'Receptor-like cytoplasmic kinases: central Players in plant receptor kinase–mediated signaling', *Annual Review of Plant Biology*, 69, pp. 267–299.

Liu, X. *et al.* (2019) 'Phosphate-solubilizing *Pseudomonas sp.* strain P34-L promotes wheat growth by colonizing the wheat rhizosphere and improving the wheat root system and soil phosphorus nutritional status', *Journal of Plant Growth Regulation*, 38, pp. 1314–1324.

López-Arredondo, D.L., Sánchez-Calderón, L. and Yong-Villalobos, L. (2017) 'Molecular and genetic basis of plant macronutrient use efficiency: Concepts, opportunities, and challenges', in *Plant macronutrient use efficiency*. Academic Press, pp. 1–29.

Malhotra, H. *et al.* (2018) 'Phosphorus nutrition: plant growth in response to deficiency and excess', in *Plant nutrients and abiotic stress tolerance*. Singapore: Springer, pp. 171–190.

Mitra, G. (2017) 'Essential plant nutrients and recent concepts about their uptake', in *Essential plant nutrients: Uptake, use efficiency, and management*. Singapore: Springer, pp. 3–36.

Mokhele, B. *et al.* (2012) 'Nitrogen assimilation in crop plants and its affecting factors', *Canadian Journal of Plant Science*, 92(3), pp. 399–405.

Mostofa, M.G. *et al.* (2018) 'Strigolactones in plant adaptation to abiotic stresses: An emerging avenue of plant research', *Plant, Cell & Environment*, 41(10), pp. 2227–2243.

Na, H. *et al.* (2022) 'Hydrogel-based strong and fast actuators by electroosmotic turgor pressure', *Science*, 376(6590), pp. 301–307.

Nath, M. *et al.* (2018) 'Microbe-mediated enhancement of nitrogen and phosphorus content for crop improvement', in *Crop improvement through microbial biotechnology*. Elsevier, pp. 293–304.

Nelson, C.S. *et al.* (2016) 'Cross-phenotype association tests uncover genes mediating nutrient response in drosophila', *BMC Genomics*, 17, pp. 1–14.

Nguyen, T.Q. *et al.* (2021) 'Phytohormonal roles in plant responses to heavy metal stress: Implications for using macrophytes in phytoremediation of aquatic ecosystems', *Environmental Toxicology and Chemistry*, 40(1), pp. 7–22.

Nunes-Nesi, A., Fernie, A.R. and Stitt, M. (2010) 'Metabolic and signaling aspects underpinning the regulation of plant carbon nitrogen interactions', *Molecular Plant*, 3(6), pp. 973–996.

O'Brien, J.A. *et al.* (2016) 'Nitrate transport, sensing, and responses in plants', *Molecular Plant*, 9(6), pp. 837–856.

Osman, K.T. (2013) 'Plant nutrients and soil fertility management', in *Soils: Principles, properties and management*. Bangladesh: Springer, pp. 129–159.

Pandey, P., Srivastava, R.K. and Dubey, R.S. (2013) 'Salicylic acid alleviates aluminum toxicity in rice seedlings better than magnesium and calcium by reducing aluminum uptake, suppressing oxidative damage and increasing antioxidative defense', *Ecotoxicology*, 22, pp. 656–670.

Patani, A. *et al.* (2023) 'Various agriculture crop plant-based bioactive compounds and their use in nanomaterial synthesis and applications', in *Nanomaterials from agricultural and horticultural products*. Singapore: Springer Nature Singapore, pp. 223–241.

Rasmussen, I.S., Dresbøll, D.B. and Thorup-Kristensen, K. (2015) 'Winter wheat cultivars and nitrogen (N) fertilization—Effects on root growth, N uptake efficiency and N use efficiency', *European Journal of Agronomy*, 68, pp. 38–49.

Rawat, J., Sanwal, P. and Saxena, J. (2016) 'Potassium and its role in sustainable agriculture', in *Potassium solubilizing microorganisms for sustainable agriculture*. New Delhi: Springer India, pp. 235–253.

Rodríguez-Espinosa, T. *et al.* (2023) 'Nitrogen management in farming systems under the use of agricultural wastes and circular economy', *Science of The Total Environment*, 876, p. 162666.

Rosier, A., Medeiros, F.H. and Bais, H.P. (2018) 'Defining plant growth promoting rhizobacteria molecular and biochemical networks in beneficial plant-microbe interactions', *Plant and Soil*, 428, pp. 35–55.

Saha, B. *et al.* (2017) 'Biological nitrogen fixation for sustainable agriculture', in *Agriculturally important microbes for sustainable agriculture: volume 2: Applications in crop production and protection*. Singapore: Springer, pp. 81–128.

Sible, C.N., Seebauer, J.R. and Below, F.E. (2021) 'Plant biostimulants: A categorical review, their implications for row crop production, and relation to soil health indicators', *Agronomy*, 11(7), p. 1297.

Soetan, K.O., Olaiya, C.O. and Oyewole, O.E. (2010) 'The importance of mineral elements for humans, domestic animals and plants: A review', *African Journal of Food Science*, 4(5), pp. 200–222.

Soumare, A. *et al.* (2020) 'Exploiting biological nitrogen fixation: A route towards a sustainable agriculture', *Plants*, 9(8), p. 1011.

Tan, X.L. *et al.* (2020) 'Bambara groundnut: An underutilized leguminous crop for global food security and nutrition', *Frontiers in Nutrition*, 7, p. 601496.

Tian, H. *et al.* (2020) 'Computer vision technology in agricultural automation—A review', *Information Processing in Agriculture*, 7(1), pp. 1–19.

Tränkner, M., Tavakol, E. and Jákli, B. (2018) 'Functioning of potassium and magnesium in photosynthesis, photosynthate translocation and photoprotection', *Physiologia Plantarum*, 163(3), pp. 414–431.

Uchida, R. (2000) 'Essential nutrients for plant growth: Nutrient functions and deficiency symptoms', *Plant Nutrient Management in Hawaii's Soils*, 4, pp. 31–55.

Vasseghian, Y. *et al.* (2022) 'Metal-organic framework-enabled pesticides are an emerging tool for sustainable cleaner production and environmental hazard reduction', *Journal of Cleaner Production*, 373, p. 133966.

Viollet, B. *et al.* (2012) 'Cellular and molecular mechanisms of metformin: An overview', *Clinical Science*, 122(6), pp. 253–270.

Vishwakarma, K. *et al.* (2019) 'Avenues of the membrane transport system in adaptation of plants to abiotic stresses', *Critical Reviews in Biotechnology*, 39(7), pp. 861–883.

Wang, F. *et al.* (2018 February) 'Molecular mechanisms of phosphate transport and signaling in higher plants', in *Seminars in cell & developmental biology*. Academic Press, vol. 74, pp. 114–122.

Xu, G., Fan, X. and Miller, A.J. (2012) 'Plant nitrogen assimilation and use efficiency', *Annual Review of Plant Biology*, 63, pp. 153–182.

Yadav, B. *et al.* (2021) 'Plant mineral transport systems and the potential for crop improvement', *Planta*, 253, pp. 1–30.

Zenda, T. *et al.* (2021) 'Revisiting sulphur—The once neglected nutrient: It's roles in plant growth, metabolism, stress tolerance and crop production', *Agriculture*, 11(7), p. 626.

Zhang, X., Ward, B.B. and Sigman, D.M. (2020) 'Global nitrogen cycle: Critical enzymes, organisms, and processes for nitrogen budgets and dynamics', *Chemical Reviews*, 120(12), pp. 5308–5351.

Zheng, J. *et al.* (2020) 'Mass spectrometry-based protein complex profiling in time and space', *Analytical Chemistry*, 93(1), pp. 598–619.

Zhu, J., Li, M. and Whelan, M. (2018) 'Phosphorus activators contribute to legacy phosphorus availability in agricultural soils: A review', *Science of the Total Environment*, 612, pp. 522–537.

12 Interactive Role of Mineral Nutrient and Plant Hormones in Crop Plant Growth and Development

Sandra Fabiola Velasco-Ramírez, Ana Cristina Ramírez Anguiano, Ana Paulina Velasco-Ramírez, Alejandro Velasco-Ramírez, and Milagros Melissa Flores Fonseca

12.1 INTRODUCTION

Justus von Liebig (1803–1873) is considered the founder of agricultural chemistry. One of the most relevant contributions to this field was the invention of nitrogen fertiliser, which was described by von Liebig in 1840 in the book *Organic Chemistry and its Application to Agriculture.* For this, he is considered to be the first scientist to search for and synthesise information concerning the importance of mineral elements in plant growth. He thus became the founder of mineral nutrition as a scientific discipline. The terms "essential mineral element" and "essential nutrient" were proposed by Arnon and Stout (1939); these authors conclude that in order for an element to be considered essential it must fulfil three criteria:

- The absence of the element in question must result in abnormal growth, failure of the entire life cycle, or premature death of the plant.
- The element must be specific and not replaceable by any other element.
- The element must exert its effect directly on growth or metabolism, and neither should cause indirect effects such as antagonism with other elements or toxicity problems.

Plants contain virtually all natural elements, but only 17 of them are considered essential elements for their nutrition. The components of this group play a basic role in plant metabolism and physiology as constituents of metabolites and macromolecules (e.g., nitrogen, phosphorus, boron, and sulphur), enzyme cofactors (magnesium, zinc, iron, copper, molybdenum, and nickel), electrolytes (potassium, sodium, and chlorine) or in signalling cascades as second messengers (e.g., calcium). In many cases, minerals may have several of these functions (e.g., phosphorus, as its assimilated form phosphate [Pi], is part of the nucleic acids and acts as an electrolyte); phospholipids and proteins play an essential role in energy transfer reactions and signalling cascades. On the other hand, and in accordance with this strict definition of essential elements, we can also consider that those mineral elements which block the toxic effects of other elements – as well as those which replace others in specific minor functions, such as osmotic pressure regulators – are not essential. However, they may be considered beneficial elements (Clarkson, 1996). Among the latter group are Na and Si (Elawad *et al.*, 1982; Malavolta, 2006).

The essentiality of 14 mineral elements has been established for superior plants, although knowledge of the requirements of chlorine and nickel is still restricted to a limited number of plant species. Nitrogen (N), phosphorus (P), and potassium (K) are identified as primary macronutrients due

DOI: 10.1201/9781032630366-12

to the fact that they are absorbed in high amounts by plants; their application as fertiliser is often required (Martin and Marschner, 1988). Calcium (Ca), magnesium (Mg), and sulphur (S) are called secondary macronutrients because, although they are required in relatively high amounts by plants, their intrinsic deficiencies are rare, and their application to the soil is due to the necessary correction of problems other than their possible deficiency (Grusak *et al.*, 2016). Boron (B), chlorine (CI), copper (Cu), iron (Fe), molybdenum (Mo), manganese (Mn), zinc (Zn), and nickel (Ni) are known as micronutrients because although they are essential for the plant, they are required in smaller proportions (< 50 ppm), as their functions in the plant are metabolic and not structural. Of these, the first two are non-metallic and the rest are considered as heavy metals (Clarkson, 1996; Lawlor, 2004).

Plants, lacking a selective nutrient uptake system, may contain large amounts of non-nutrient and even toxic elements, such as aluminium (in acid soils), selenium, and fluoride (in soils where industrial waste is deposited). Cobalt, in turn, is essential for nitrogen-fixing micro-organisms living in the roots of legumes (Nunes Da Silva *et al.*, 2022). Hence, the priority for plants to achieve optimal growth and development is to maintain the best possible levels of essential minerals despite their limited availability in the environment in which they are growing. Lack of these elements for plants usually occurs when there are not enough nutrients in the soil. But there is also another reason for nutrient deficiency in plants, and that is high salinisation of the soil, which prevents plants from accessing these nutrients even if they are present in large quantities (Schachtman and Shin, 2007). Also, absorption capacity is highly dependent on the condition of the plants, so that nutrient uptake and transport can be affected by problems such as damaged or undeveloped roots, insufficient or excessive irrigation, and poor drainage. Inadequate pH values are also harmful. To cope with reduced nutrient availability, plants modify their physiology and metabolism and increase nutrient acquisition (López- Bucio *et al.*, 2003). They achieve this by adapting their entire structure to stimulate the growth of the organs directly involved (Hermans *et al.*, 2006; Reymond *et al.*, 2006; Svistoonoff *et al.*, 2007). Anthocyanins, phenolic compounds of the flavonoid group, which are mainly found in fruits, flowers, and leaves (Do *et al.*, 2023), are characterised by their high reducing power and can be rapidly induced in response to stress conditions caused by abiotic or biotic factors (Petroni and Tonelli, 2011). Some examples of these factors are high light radiation, low temperatures, or deficiencies of mineral elements such as nitrogen, phosphorus, boron, and manganese as well as cases of pathogen incidence (Tripathi *et al.*, 2022). These conditions lead to an overproduction of reactive oxygen species. In this way, anthocyanins can neutralise the harmful effects of free radicals that cause cell damage and death or dissipate radiant energy, which cannot be used in the photosynthetic process (Neill and Gould, 2003). Finally, plants also make use of alternative metabolic pathways in order to optimise the amount of nutrients that are already low in the place where they are growing (Clarkson, 1996).

As discussed above, responses to variations in nutrient supply have been researched; however, our knowledge of how plants monitor internal and external mineral levels is still very poor, so knowledge of the nature of all plant nutrient sensors remains elusive. Furthermore, the control of nutrient signalling involves different metabolites, reactive oxygen species, and hormones reflecting the complexity of these networks and the responses they may elicit. This chapter will therefore focus on providing evidence for the involvement of the role of mineral nutrients and plant hormones (phytohormones) in plant growth and their development.

12.2 PHYTOHORMONES AS GROWTH REGULATORS

Plant growth regulators are defined as synthetic or natural organic compounds that influence the biological growth and development of plants, usually acting in low concentrations (Sajjad, 2017). Due to progressive global warming, desertification, salinity, flooding, etc., fewer productive areas around the world exist. This has contributed to a decrease in agricultural production. The most outstanding issues involve drought, heat and cold, salinity, toxic elements such as heavy metals, etc. (Bachheti *et al.*, 2021; Getnet *et al.*, 2015; Husen, 2010, 2021a, 2021b, 2021c, 2021d, 2021e, 2024a,

2024b; Husen, Iqbal and Aref, 2014, 2016; Husen *et al.*, 2018, 2019; Hussein *et al.*, 2017; Raza *et al.*, 2019). The major environmental challenge we face is to prevent these environmental constraints from affecting seed germination and seedling growth. We therefore find it easily understandable that various techniques have been developed to attempt to increase productivity and make crops more tolerant to stress. One of the techniques that has been applied in recent years with good results is the application of plant growth regulators (PGRs). Regulatory chemical vectors (RCVs) are synthetic compounds or compounds produced by the plant itself (phytohormones). They are found in very low concentrations and perform various physiological functions. These growth regulators can either increase or decrease growth and development by altering their normal biological processes, thus acting as inhibitors or stimulants. The application of RCVs is a feasible and practical means to minimise yield losses induced by the above factors. RCVs such as auxin, cytokinin, brassinosteroids, gibberellin, salicylic acid, abscisic acid, jasmonates, ethylene, and polyamine have been used for improving yield under saline and other adverse conditions (Quamruzzaman *et al.*, 2021). However, how it is applied for optimal performance is also discussed in the scientific community (Sajjad, 2017). Biotechnology has been able to synthetically manufacture growth regulators that can mimic the role of fitohormones in a natural way. A variety of regulators exist able to promote or inhibit plant growth. Some authors have suggested the existence of chemical compounds capable of controlling growth in a specific way; therefore, regulators have been classified into ten different types, according to the activity or stimulating capacity that each one may have on plant growth, in a single organ or procedure such as photosynthesis or fruit ripening. As a result, these regulators have made it possible to enhance the cultivation process in plant organisms, as it is one of the ideal sources in terms of achieving the objective that biotechnology has aimed for in recent years with the integration of techniques that manage to eliminate many of the most common problems that damage crops, such as microbial phytopathogens and entomopathogens (Didi *et al.*, 2022). We could also mention environmental changes and changes in the soil.

The idea of understanding the functionality of the metabolism of plant organisms, such as plant hormone control and how this can depend on nutrients and other factors, generates basic knowledge in the scientific field about plant physiology. This is required for the plants to develop different biochemical processes in a more regulated way according to a specific need. It could even improve the time they require for their development, maintaining an isolation of the organism from the biotic and abiotic conditions that are present; fully optimising its growth conditions.

12.3 INTERACTIONS BETWEEN PHYTOHORMONES AND NUTRIENTS

As explained in the introduction, nutrients are elements that the plant takes from its environment (air and soil) to create organic compounds; however, the normal functioning of a plant not only depends on nutrients but also requires certain mechanisms that allow it to regulate and/or coordinate the different activities of its cells, tissues and, organs. At the same time it must be able to sense and respond to changes in the environment (Mimmo *et al.*, 2018). Among the possible regulatory mechanisms, the best-known is the system of chemical messengers (chemical signals). This chemical communication is mainly established through plant hormones. A phytohormone or plant hormone is defined as an organic substance, different from nutrients, active at very low concentrations, sometimes produced in certain tissues and transported to another tissue, where it exerts its effects. It can also be active in the tissues where it is synthesised (Vaishnav and Chowdhury, 2023).

The presence of both nutrients and hormones is essential for normal plant development. The hormonal actions of regulating tissue growth (quantitative change) as well as differentiation events (qualitative change) exert an important influence on nutrient requirements, as well as on their uptake and movement. Plants treated with hormones and induced to develop more vegetative structures will productively undergo changes in nutritional demands in terms of required concentration, site of required supply, and type of element needed to fully express the hormonal effect (Vaishnav and

TABLE 12.1
Essential Nutrients and Their Function in the Plant

Nutrient	Function in the Plant
	Macronutrients
Nitrogen	Constituent of all proteins and chlorophyll It is found in coenzymes and nucleic acids
Phosphorus	Important in energy transmission as part of adenosine triphosphate. They are made of many proteins, coenzymes, nucleic acids, and metabolic substrates.
Potassium	Works in regulatory mechanisms such as photosynthesis, the translocation of carbohydrates, and protein synthesis.
Calcium	Component of the cell wall It plays a role in the structure and permeability of cell membranes.
Magnesium	Constituent of chlorophyll and enzyme activator
Sulphur	Important constituent of vegetable proteins
	Micronutrients
Boron	Important in the transfer of sugars and carbohydrate metabolism
Iron	Participates in the synthesis of chlorophyll and enzymes for electron transfer
Manganese	Controls several oxidation-reduction systems in the formation of 0_2 during photosynthesis
Copper	Catalyst for respiration, enzymatic constituent
Zinc	Participates in enzymatic systems that regulate various metabolic activities
Molybdenum	In nitrogenase, it is necessary for the fixation of nitrogen and its transformation into amino acids and proteins
Cobalt	Essential for symbiotic nitrogen fixation
Chlorine	Activates the production of 0_2 in photosynthesis
Nickel	Activator of enzymatic processes and urea metabolism

Chowdhury, 2023). Hormones are formed from various chemical elements, mainly C, H, O, and N; the first three are taken from the air and from reserves in the tissues, while the last is taken from the air and soil via fertilisers. To give an example, the main hormones that have been characterised as having a high nitrogen content are auxins, cytokinins, and polyamines, all of which are growth stimulants; this partly explains why an N deficiency is reflected in poor plant growth.

Since the use of phytohormones to manipulate crop growth and productivity has become widespread, it is important to know whether such applications have an influence on the uptake, translocation and concentration of nutrients in the plants where they are applied (Rubio *et al.*, 2009) (Table 12.1).

12.3.1 Cytokinins (CKs)

Cytokinins (CKs) were originally discovered in a search for proliferation-promoting factors in plant cell culture in conjunction with a second phytohormone, auxin. CKs are essential hormones in the activation of several processes linked to plant growth and development, and they are related to the action of various different genes. They are derivatives of the base adenine which at its N6 position shows several substitutions, adenine alone having no hormonal effect. Among the most relevant physiological effects of CKs are: they promote cell division; cause sprout initiation, organogenesis, and androgenesis; delay or retard senescence; activate dormant lateral buds; and intensify the expression of "demand" in the transport of processed sap at the phloem level (Schaller, Bishopp and Kieber, 2015) (Figure 12.1).

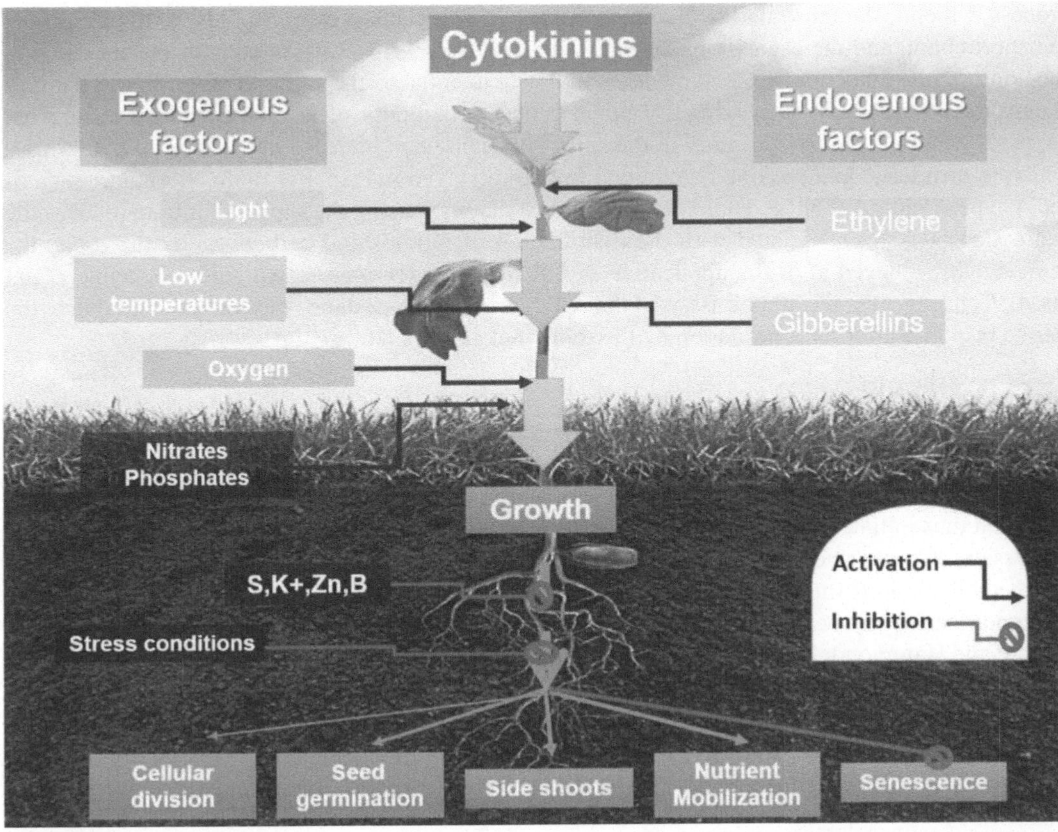

FIGURE 12.1 Factors that can activate/induce (arrows in black) or decrease/inhibit (arrows in red) the activation of the Cytokinins (CK) as well as physiological processes in which they participate (arrows in grey).

12.3.1.1 CK and Its Influence on Nutrient Homeostasis (Nitrogen, Phosphorus, Sulphur, and Iron)

12.3.1.1.1 Nitrate (NO_3^-)

Inorganic nitrogen greatly affects plant growth and development. Plants are constantly sensing this nutrient and modulate their metabolic activities and program their development to efficiently adapt to the nutritional environment. CK is the phytohormone in charge of communicating the status of nitrogen-derived nutrients from the root to the xylem vessels (Sakakibara, 2006). Recent studies reveal a molecular mechanism of nitrogen-dependent CK biosynthesis and its regulation by available nitrogen sources (Miyawaki *et al.*, 2004; Takei *et al.*, 2004). For example, it has been observed that very low levels of nitrogen in the soil (NO_3 or NH_4) result in a reduction in the level of CKs. In this regard, Rubio Wilhelmi *et al.* (2011, 2012) evaluated the effect of CKs on oxidative stress in tobacco plants subjected to N deficiency, both wild type (WT) and transgenic plants overexpressing the isopentenyltransferase (ipt) gene, the enzyme that encodes the key step in CK synthesis, which grew under N-deficient conditions (7 and 1 mM NO_3^-). The results of these researchers showed that WT plants subjected to N deficiency reduced biomass and relative growth rate (RGR) by increasing oxidative damage and reducing leaf concentrations of the different N forms. However, the transgenic plants, despite showing a significant decrease in leaf N forms, did not show alterations in oxidative metabolism, thus maintaining leaf biomass and RGR under N deficiency. This study also concluded that the expression of the IPT gene in tobacco plants subjected to N deficiency leads to a maintenance of photosynthesis in such a way that the CKs could act to protect this physiological process, which is demonstrated by the maintenance of sm-Rubisco transcript levels, the electronic transport chain, and the capacity of the chloroplast to use P-trioses. In relation to the source-sink relationship, the induction of CK synthesis could help to enhance the strength of the sink. That represents the young leaves, improving leaf biomass under conditions of severe N deficiency. But what happens if there is an increase in N in the soil? This causes an increase in the level of cytokinins, which in turn leads to a positive regulation of genes involved in response to this hormone, such as the A-type ARRs (A-ARRs). Therefore, it is known that increasing the supply of nitrate induces the expression of genes associated with the assimilation of nitrates and carbonates by the roots, the same action achieved after the application of CK to plants (Brenner *et al.*, 2005; Scheible *et al.*, 2004). Consequently, it could be asserted that CK acts as a long-distance messenger to modulate the nitrate response after its assimilation by the roots (Sakakibara, 2006).

12.3.1.1.2 Phosphate [Pi]

The need for P in plants generates the accumulation of sugars and starch in the tissues. In cases of significant deficiencies, some species express a purple colour in the foliage. In addition to this, a sucrose load is generated in the phloem, causing carbon to be fixed. The carbon is relocated to the roots and thus initiates a cascade of signalling of sugars and secondary metabolites that modify the expression of genes and transporters. These allow optimising the acquisition of P through different methods such as the secretion of acids (including acid phosphatases and alkaline), as well as the modification of some structures, which will help the plant to generate a nutritional balance (White and Hammond, 2008). CK negatively regulates Pi starvation responses (PSRs). Many of the morphological responses to the need for Pi, such as those already mentioned above, are similar to those obtained by the lack of N, and both are reminiscent of the effects of alterations in auxin or cytokinin levels. There are studies that show that if the cytokinin content is decreased, it is because there is a shortage of both Pi and N (Ei-D *et al.*, 1979; Horgan and Wareing, 1980; Wagner and Beck, 1993). Consequently, exogenously administering CK leads to the inhibition of root growth due to the absence of nutrients. Contrary to what is observed with the application of Pi or NO_3^-, the supply of these nutrients leads to the development of lateral roots. On the other hand, CK inhibits anthocyanin accumulation induced by Pi decrement, while PSR gene expression is not significantly affected in shoots, suggesting a more complex interaction between CK and Pi starvation signalling (Franco-Zorrilla *et al.*, 2002; Martín *et al.*, 2000). Contrary to the effect of CK on PSR, Pi limitation repressed the accumulation of CK and the expression of the CRE1 gene (a tissue-specific transcription enhancer) (Ei-D *et al.*, 1979; Franco-Zorrilla *et al.*, 2002). During senescence, cytokinins (CKs) levels naturally decrease, leading to the progression of leaf aging and the associated catabolic processes, such as chlorophyll degradation, nutrient remobilisation, and cellular breakdown. However, by externally spraying CKs or genetically modifying plants to overexpress the **ipt gene**, which encodes the **enzyme isopentenyltransferase (IPT)** from *Agrobacterium tumefaciens*, it is possible to delay these senescence-related processes (Lim *et al.*, 2007; McCabe *et al.*, 2001). In other topics regarding the participation of Pi, despite the notable effect that CKs have to delay senescence in plants, the mechanism of action is still not well-understood. It is proposed that during senescence, CKs are perceived by receptors associated with histidine kinases and that the signal generated is translated through a cascade of phosphorylations that finally stimulate target genes responsible for the program of delaying foliar senescence, where three modules are mainly involved. These modules are: receptor histidine kinases (HKs), protein phosphotransferases (HPs), and response regulators (RRs) (Lim *et al.*, 2007). In this regard, in *Arabidopsis spp.* one of the three cytokinin receptors, histidine kinase 3 (AHK3), has a primary function in controlling leaf longevity mediated by specific phosphorylation of the response regulator ARR2 (Kim *et al.*, 2006).

12.3.1.1.3 Sulphur (S)

Knowledge of CK-mediated sulphur assimilation is more limited than for nitrogen and phosphorus. There is evidence that cytokinin positively regulates the expression of the -S-sensitive seed storage protein, [3-conglycinin [subunit 3, in *Arabidopsis* roots (Ohkama *et al.*, 2002). These results apparently suggested that the cytokinin and current signalling cascades are integrated into the regulation of nutrient responses. However, no evidence has been proven, showing a direct connection between cytokinin-dependent signalling and nutrient uptake. Maruyama-Nakashita *et al.* (2004) examined the effects of plant hormones on SULTR1;2 mRNA expression to define their regulatory roles in sulphate uptake in *Arabidopsis* roots. The results demonstrated CK-dependent repression of sulphate influx in *Arabidopsis* roots and indicated a direct involvement of the CRE1/WOL/AHK4-derived signal in the negative regulation of the high-affinity sulphate transporters SULTR1;1 and SULTR1;2. Their results suggested that -S-dependent induction and CK-dependent repression reciprocally operate the sulphate uptake machinery for optimisation of sulphur assimilation in plants (Maruyama-Nakashita *et al.*, 2004).

12.3.1.1.4 Iron

Recently, Séguéla *et al.* (2008) have shown that CK represses the expression of genes in *Arabidopsis* grown in low-iron conditions, because when these hormones were supplied, a decrease was found in the expression of the genes that encode both the iron transporter (IRT1) and the ferric reductase (FRO2) in addition to the FIT gene, which is also closely linked to the decrease in its activities. This same study demonstrated that CK-mediated repression in relation to the response of genes to iron deficiency depends on the receptors of CK, CRE1, and AHK3. However, this is independent of iron supply, indicating that CK signalling modulates the expression of these genes but does not regulate their specific expression in response to low iron levels under growth conditions. Therefore, CKs can control iron-responsive gene expression, and consequently iron homeostasis, at least in part, by regulating the growth rate in meristems (Lai *et al.*, 2007; Séguéla *et al.*, 2008).

12.3.2 Abscisic Acid (ABA)

The phytohormone abscisic acid (ABA) was identified in the 1960s after studies were performed on abscission (programmed loss of a plant organ such as a leaf, flower, or fruit) from fruits and bud dormancy. The group, led by F. Addicott (Addicott and Cams, 1983), isolated compounds that caused abscission of cotton fruits and in 1963 identified one of them, abscisin II, as an ABA. Shortly afterwards, another research group led by Phillips and Wareing (1958) isolated a substance from leaves of *Acer pseudoplatanus* that promoted bud dormancy. This, called dormin, was also identified as an ABA. Since then, ABA has been implicated in multiple physiological processes such as growth regulation, seed dormancy, germination, senescence, cell division, control of stomata opening, and responses to environmental stresses such as drought, salinity, low temperature, pathogen attacks, and ultraviolet radiation (Addicott and Cams, 1983; Assmann and Shimazaki, 1999; Leung and Giraudat, 1998) (Figure 12.2).

12.3.2.1 ABA and Its Influence on Nutrient Homeostasis (Nitrogen, Phosphorus, Sulphur, and Potassium)

12.3.2.1.1 Nitrate (NO_3^-)

The mutual relationship between ABA and the ability to maintain a stable internal condition among nitrates was discovered by analysing the results of affecting ABA signalling on the response of plants to nitrate replenishment and, on the other hand, by studying the control of nitrates in the biosynthesis of ABA (Zhang, Rong and Pilbeam, 2007). In the first case, the inhibition of the lateral root is induced by high levels of nitrate, where ABA plays a crucial role in mediating the inhibition of this structure due to high levels of nitrate, since it is found that this inhibition is significantly lower in known ABA-deficient mutants (abal-1, aba2-3, aba2-4, and aba3-2) and in some ABA-sensitive mutants (abi4-1, abi4-2, and abi5-1) (Signora *et al.*, 2001). While ABA concentrations in the split

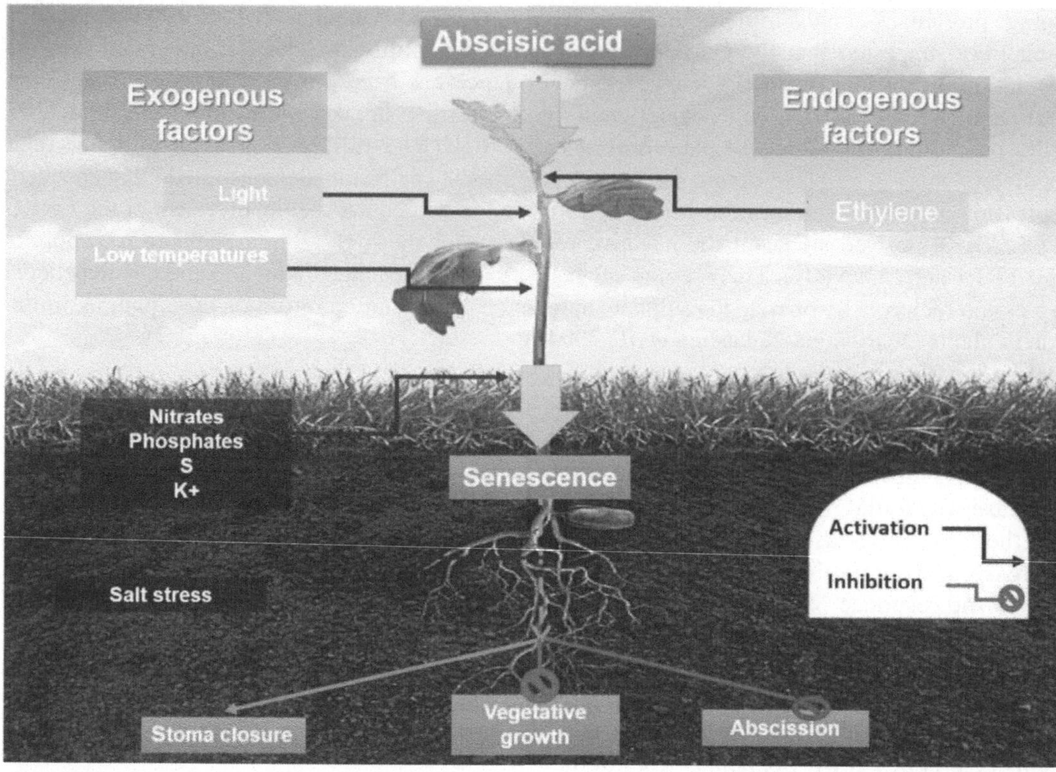

FIGURE 12.2 Factors that can activate/induce (arrows in black) or decrease/inhibit (arrows in red) the activation of the Abscisic acid (ABA) as well as physiological processes in which they participate (arrows in grey).

root of barley plants grown under steady-state conditions demonstrated that it is not related to the rate of nitrate supply, there were rapid but transient changes in ABA concentrations after short-term changes. All of the above support the idea that ABA acts directly as a messenger that controls plant responses to high nitrate concentrations (Brewitz, Larsson and Larsson, 1995), although it could also reflect a mechanism of adaptation to a sudden change in nitrate availability. On the other hand, if roots are found to be receiving a high supply of nitrate, this will lead to plant-wide changes in ABA levels. Exogenous supply of ABA has been shown to mimic the inhibitory effect induced by high nitrate levels. Elevated levels of NO_3^- stimulate the accumulation of ABA in the endodermis and in the quiescent centre of the root apical meristem (a pool of stem cells that rarely divide within the root apical meristem [RAM]), where they contribute to the maintenance of the meristem. This response results from the bioactive release of ABA from conjugation with glucose esters through β-glucosidase 1 (BG1) that is transcriptionally induced by NO_3^- (Ondzighi-Assoume, Chakraborty and Harris, 2016). This process seems to be conserved in all plant species.

12.3.2.1.2 Phosphate [Pi]

ABA also appears to mediate PSRs, as demonstrated by ABA treatment in Pi-deprived *Arabidopsis* plants, which represses transcription of several members of the At4 family (a group of riboregulators conserved in different plant species) (Shin *et al.*, 2006). In addition, *Arabidopsis* mutants show slight alterations with respect to ABA biosynthesis (abi2-1 and aba1). Another gene that is minimised in its expression is PSR, which ultimately leads to an accumulation of anthocyanins when grown with a low Pi supply (Ciereszko and Kleczkowski, 2002; Trull *et al.*, 1997). However, defects in ABA signalling in these mutants did not affect other responses such as low activity of phosphate-using enzymes (Trull *et al.*, 1997). Taken as a whole, this evidence shows that ABA differentially regulates in a discrete way some Pi signalling pathways, but not all of them.

12.3.2.1.3 Sulphur (S)

Evidence exists that ABA impacts the ability to maintain stable internal sulphur conditions by increasing the levels of glutathione (GSH), which from a chemical point of view is the most prominent low-molecular-weight compound in plants, owning a sulphhydryl group (-SH); it plays a critical role in maintaining the redox state of plants. GSH is one of the main antioxidant metabolites used by plant species to resist biotic and abiotic stresses, leading to the creation of reactive oxygen species. Therefore, ABA-mediated control explains their joint function in detoxifying electrophilic xenobiotics in biological systems and thus providing protection to plants (Jiang and Zhang, 2001; Shao *et al.*, 2008).

12.3.2.1.4 Potassium (K^+)

ABA also induces hardiness in adverse environmental conditions such as saline, water-poor soils, or low temperatures. ABA stimulates the movement of potassium ions in the roots, thus promoting the penetration of water and its retention in these tissues. This process is also carried out by regulating ion channels in the plasma membrane of root cells. Furthermore, ABA reduces the resistance of water passing through the apoplast and membrane, thereby increasing water flux in the root tissues. This process also contributes to the regulation of water channels or aquaporins, although the relationship with the action of ABA has not been elucidated. As in seeds, some leaf-type genes, including the rab (ABA-responsive) and dhn (dehydration) genes, can also be expressed in vegetative tissues under aquatic or osmotic stress or exogenous ABA treatment. This suggests that hormones may also provide some tolerance to lack of water in these tissues (Busk and Pagès, 1998).

12.3.3 ETHYLENE

Ethylene is the only gaseous plant hormone, simple and small, present in angiosperms and gymnosperms but also in bacteria and fungi as well as mosses, liverworts, ferns, and other organisms. Being a gas, it can move rapidly through tissues, not so much by transport as by diffusion. Its effect also starts with very small amounts, which already provoke responses. Physiological effects of this phytohormone are: cell expansion, epinasty, breaking of dormancy in seeds and buds, induction of flowering, fruit ripening, and acceleration of senescence and leaf and flower drop (Gamalero *et al.*, 2023) (Figure 12.3).

12.3.3.1 Ethylene and Its Influence on Nutrient Homeostasis (Nitrogen, Phosphorus, Iron, and Potassium)

12.3.3.1.1 Nitrate (NO_3^-)

According to Zheng *et al.*, elevated ethylene production was detected when *Arabidopsis* seedlings were subjected to an NO_3^- shift (from high to low NO_3^-), which may suggest that ethylene acts as an intermediate in NO_3^- signalling, especially in response to recovery from NO_3^- deficiency (Zheng *et al.*, 2013).

12.3.3.1.2 Phosphate (Pi)

Ethylene induces primary root elongation in Pi-deficient plants, but if they are grown under appropriate Pi conditions, the effect is the opposite. If during plant growth, they are treated with an ethylene synthesis inhibitor, a decrease of root elongation may be observed (Ma *et al.*, 2003).

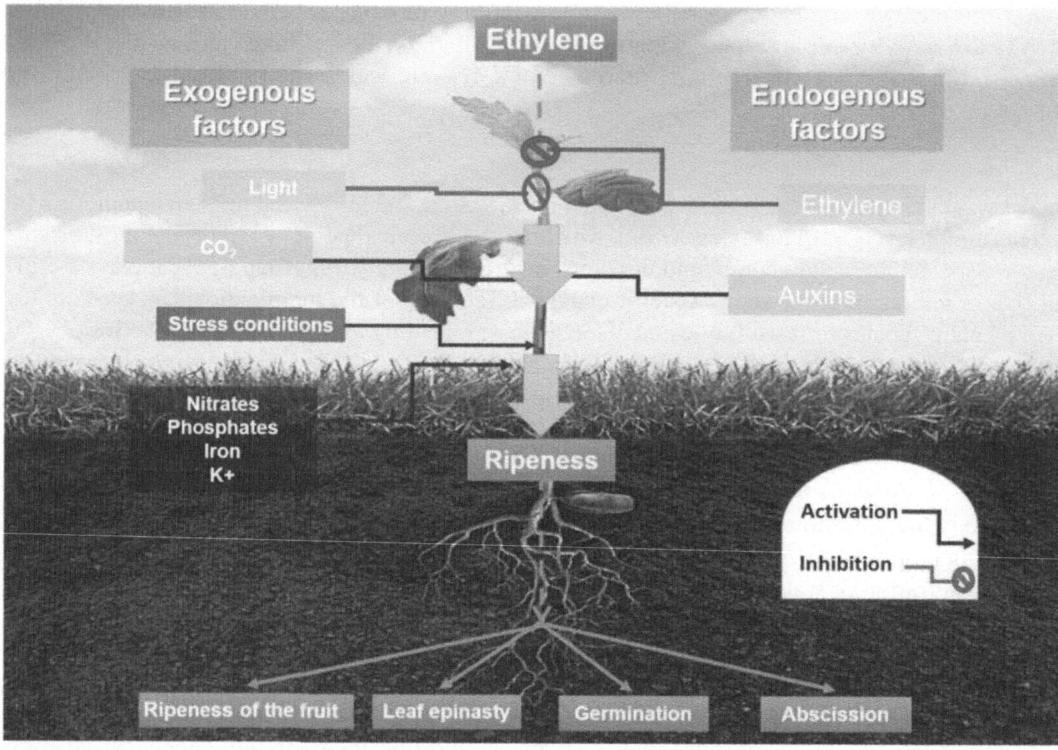

FIGURE 12.3 Factors that can activate/induce (arrows in black) or decrease/inhibit (arrows in red) the activation of the Ethylene as well as physiological processes in which they participate (arrows in grey).

12.3.3.1.3 Iron

Iron in plants presents a regulation that goes hand in hand with ethylene, since genes that respond to this mineral (members of the FER/FIT family) have been described which promote gene expression in response to low supply of iron. This response was also favoured by the ethylene conditions; however, to reach definitive knowledge, said expression must be analysed with mutants of said members. In another point, it has been shown that the addition of the ethylene precursor, 1-aminocyclopropane-1-carboxylic acid (ACC) to *Arabidopsis*, tomato, and cucumber plants improves the expression of genes that respond to low iron and mediate iron absorption and assimilation (e.g., genes encoding iron transporters, iron reductases, or H+-ATPases). Furthermore, an increase in the activity of the ferric reductase enzyme has been observed in the roots of plants grown in low-iron conditions treated with ACC. The application of ethylene synthesis inhibitors repressed the expression of this set of genes. However, to fully demonstrate this regulatory mechanism, said expression must be analysed in plants with altered FER/FIT function (Lucena *et al.*, 2006).

12.3.3.1.4 Potassium (K^+)

There is likely evidence that ethylene is involved in the control of potassium deficiency responses. In fact, low potassium levels induced the expression of several genes that control ethylene biogenesis, which correlates with an increase in ethylene biosynthesis (Shin and Schachtman, 2004). This increase in ethylene levels could explain the reduced number of lateral roots found in potassium-deprived *Arabidopsis* plants. Increased ethylene has been reported following boron and magnesium deficiency (Camacho-Cristóbal *et al.*, 2015; Li *et al.*, 2009; Li, Wu and Wang, 2017; Schachtman and Shin, 2007).

12.3.3.1.5 Other Minerals

The increase in ethylene has been reported after the deficiency of boron and magnesium (Camacho-Cristóbal *et al.*, 2015; Li *et al.*, 2009; Li *et al.*, 2017; Schachtman and Shin, 2007).

12.3.4 Auxins

Auxins are a group of natural plant hormones that regulate various aspects of plant growth and development. The predominant form in plants is indole acetic acid (IAA), which is highly active in bioassays and commonly present in nanomolar concentrations. Other natural forms of auxins are 4-chloro-indolacetic acid (4-Cl-IAA), phenylacetic acid (PAA), indole butyric acid (IBA), and indole propionic acid (IPA). Physiological effects of these hormones on plants include: growth and root formation, tropism regulation, apical dominance, organ abscission, flower and fruit development, and vascular differentiation (Schaller *et al.*, 2015) (Figure 12.4).

12.3.4.1 Auxins and Their Influence on Nutrient Homeostasis (Nitrogen, Phosphorus, Sulphur, and Potassium)

12.3.4.1.1 Nitrate (NO_3^-)

Local auxin biosynthesis depends mainly on the indole-3-pyruvate (IPyA) and indole-3-aldehyde oxime (IAOx) pathways. The IPyA pathway converts tryptophan to IPyA in two steps through the tryptophan aminotransferase (TAA1/TAR) family and converts IPyA to indole-3-acetic acid (IAA) through the YUCCA family flavin monooxygenase to produce auxin (Zhao, 2018). Nitrogen

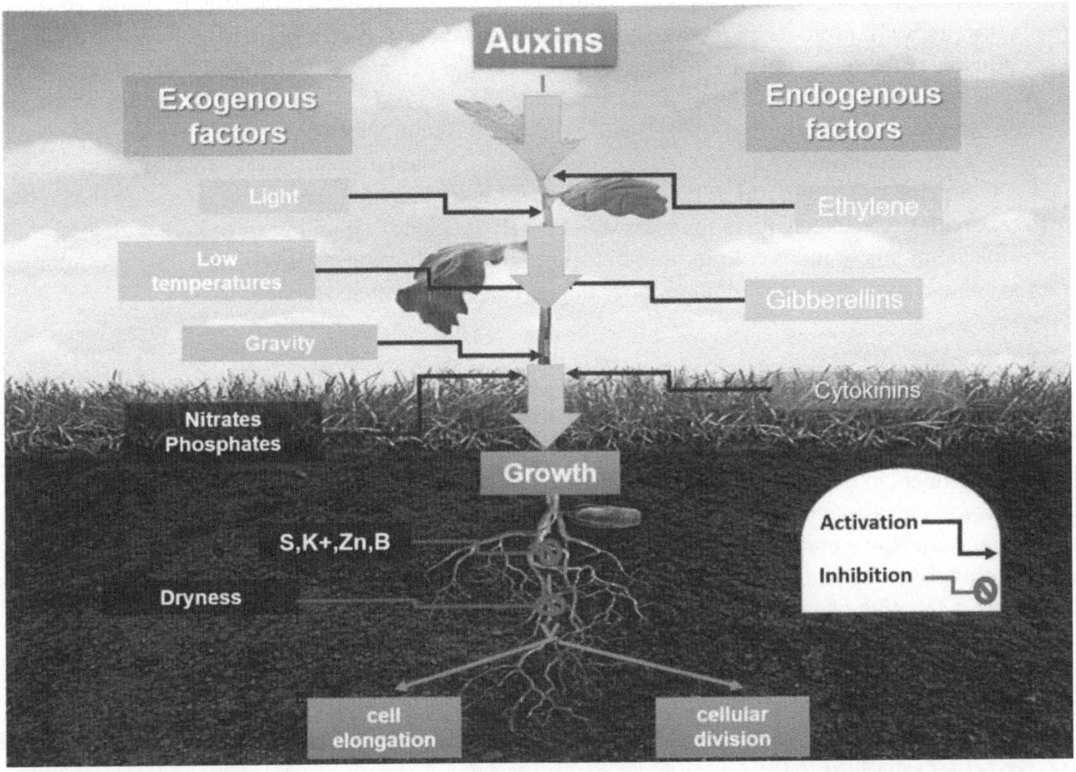

FIGURE 12.4 Factors that can activate/induce (arrows in black) or decrease/inhibit (arrows in red) the activation of the Auxins as well as physiological processes in which they participate (arrows in grey).

deficiency targets these two stages by upregulating TAA1/TAR2 and YUC3/5/7/8 to increase *de novo* biosynthesis in the mature zone as well as in the root apical meristem (Jia *et al.*, 2021; Ma *et al.*, 2014). This nitrogen deficiency-dependent local activation of the IPyA pathway appears to be conserved in all plant species grown under low-nitrate conditions, with roots accumulating at higher levels than plants grown under high-nitrate conditions. In comparison, auxin content was lower in nitrate-deficient shoots than in plants grown in nitrate-rich media (Walch-Liu *et al.*, 2006). This alters the pattern of auxin accumulation, which may indicate a role for these hormones in suppressing the effects of high nitrate supply on root development, such as preventing lateral root proliferation. However, this hypothesis requires further investigation because auxin application did not reduce the inhibition of lateral root development in plants grown under high nitrate conditions, suggesting that these hormones do not directly control nitrate signalling (Zhang *et al.*, 2007).

12.3.4.1.2 Phosphate [Pi]

Pi deficiency also increases auxin content. For example, Pi limitation has been observed to increase the effect of auxin to promote the proliferation of lateral roots and inhibit the growth of primary roots in *Arabidopsis* plants (López-Bucio *et al.*, 2002). This involves the regulation of tryptophan aminotransferase (TAA1) in the root apical meristem (Bhosale *et al.*, 2018). However, analysis of *Arabidopsis* mutants with altered auxin signalling or polar transport, such as axr1-3, axr2-1, and axr4-1 or lpr1, revealed that these plants still exhibit starvation-induced root responses. Root hair formation suggests that the regulatory mechanisms driving root changes in response to low Pi may be independent of auxin transport/signalling (López-Bucio *et al.*, 2003, 2005). Furthermore, a Pi deficiency response mutant (pdr2) has impaired local Pi sensing, thereby exhibiting hypersensitive responses to low Pi, such as increased accumulation of antianthocyanins, induction of PSR gene expression, and increased inhibition of primary roots despite elongation. The growth rate is higher, but under low Pi conditions, auxin treatment cannot rescue or mimic its root defects, demonstrating that auxin does not promote Pi induction (Ticconi *et al.*, 2004). Auxin homeostasis during Pi starvation involves the induction of DAO1, which leads to the oxidation of IAA and balances the level of physiological IAA activity (Bhosale *et al.*, 2018; Mellor *et al.*, 2016). The phrase indicates that certain studies, specifically by López-Bucio et al. (2002) and Nacry et al. (2005), have suggested that the effects of auxin on root responses under low inorganic phosphorus (Pi) conditions are independent of primary phosphorus deficiency signals. This implies that auxin may be acting through a different pathway or mechanism than the standard Pi deficiency signaling in roots. It suggests that auxin plays an autonomous role in adapting roots to low phosphorus levels, possibly modulating root growth or architecture without directly depending on Pi deficiency signals.

12.3.4.1.3 Sulphur (S)

S deficiency appears to downregulate the IPyA and IAOx pathways, thereby reducing auxin accumulation; this was observed in the expression of YUC2/3/9 and CYP79B2 genes in roots, which were significantly downregulated by S deficiency (Zhao *et al.*, 2014). Transcriptional activation of several members of the auxin/indole-3-acetic acid (Aux/IAA) gene family was observed in sulphur deficiency experiments (Nikiforova *et al.*, 2003). Since Aux/IAA proteins are inhibitors of auxin signalling (Guilfoyle and Hagen, 2007), it is reasonable to assume that sulphur deficiency causes auxin inhibition. On the other hand, auxin treatment repressed the expression of a low-sulphur gene encoding a putative glucosinolate enzyme that can release thiol groups from glucosinolates, indicating the bidirectional nature of the sulphur-auxin interaction. The fact that low sulphur induces SULTR1; 2 and APR2 independently of auxin suggests that these hormones control only a part of the plant response to sulphur starvation (Dan *et al.*, 2006).

12.3.4.1.4 Potassium (K^+)

Potassium deficiency appears to downregulate the IPyA and IAOx pathways, thereby reducing auxin accumulation, as the expression of YUC8, CYP79B2, and CYP79B3 in roots was significantly reduced by potassium deficiency (Liu *et al.*, 2023). In contrast, the role of auxin in the control of potassium homeostasis is supported by the fact that auxin treatment of maize plants increases the expression of the gene encoding the potassium transporter ZMK1 (Philippar *et al.*, 1999). This transcriptional induction is consistent with the increase in potassium transport observed in structures such as coleoptiles and protoplasts of maize after auxin application.

12.3.4.1.5 Other Minerals

Some nutrients such as Zn and B are closely related to auxins, since their deficiency results in a lower number of auxins in the tissue, which reduces cell division and elongation processes. In several crops, internode shortening is characteristic, partly due to low auxin synthesis (Hagen *et al.*, 2010).

12.3.5 Jasmonic Acid (JA)

Jasmonic acid (JA)-related molecules and their derivatives are called jasmonates (JAs). They are compounds of lipid origin with a molecular structure similar to that of prostaglandins in animals. They act as signal molecules in plant responses to various stress situations (wounds, injuries, pathogen and pest attacks, exposure to drought and ozone) and are involved in a variety of growth and developmental processes (Creelman and Mullet, 1997; Ghasemi Pirbalouti *et al.*, 2014). During the 1980s, JA and some of its derivatives were initially discovered as growth inhibitors and senescence promoters in several plant species, both in monocots and dicots. In cereals, for example, JA inhibits the growth of both roots and coleoptiles. From then on, chemical analyses in plants have determined the existence of several stereoisomers of JA (Creelman and Mullet, 1997) (Figure 12.5).

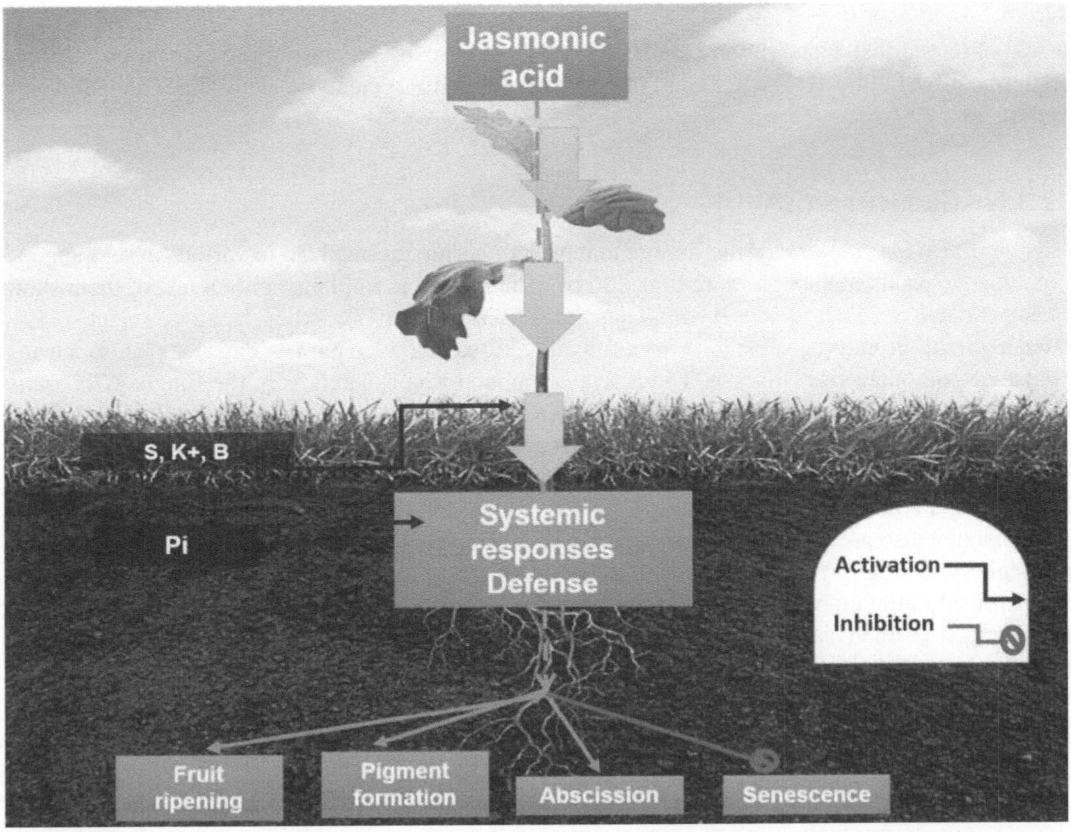

FIGURE 12.5 Factors that can activate/induce (arrows in black) or decrease/inhibit (arrows in red) the activation of the Jasmonic acid (JA) as well as physiological processes in which they participate (arrows in grey).

12.3.5.1 Jasmonic Acid (JA) and Its Influence on Nutrient Homeostasis (Phosphate, Sulphur, Potassium, and Boron)

12.3.5.1.1 Phosphate [Pi]

In Pi-deficient roots, increased expression of LOX2 and OPR3 genes has been found to be responsible for JA accumulation (Khan *et al.*, 2016; Zheng *et al.*, 2016).

12.3.5.1.2 Sulphur (S)

Analysis of the expression profiles of the JA transcriptional response revealed that about 16 genes control sulphur metabolism in response to JA treatment (Jost *et al.*, 2005), showing that JA treatment mainly affects sulphur metabolism positively. In contrast, a large number of JA biosynthetic genes are induced under low-sulphur conditions (Hirai *et al.*, 2003; Maruyama-Nakashita *et al.*, 2004; Nikiforova *et al.*, 2003). However, the fact that JA treatment was not associated with an overall increase in the concentration of sulphur metabolites may indicate that these stimuli, in addition to JA signalling, are necessary to induce full assimilation of the sulphur response.

12.3.5.1.3 Potasio (K^+)

Transcriptome analysis revealed that AOS, AOC1, and LOX2, which catalyse the first step of JA biosynthesis, are positively regulated by low levels of K (Armengaud *et al.*, 2004). Remarkably, these transcriptional responses were mainly observed in shoots rather than in roots, suggesting organic specificity of JA control induced by potassium accumulation. This specificity may be due to increased K deficiency in nutrient cycling in senescing leaves of JA-induced plants (He *et al.*, 2002).

12.3.5.1.4 Boron (B)

It was observed that B starvation enhanced JA biosynthesis by increasing the expression of genes responsible for both early (AOC3) and late (ACX3 and MFP2) catalytic steps (Jia, Giehl and Von Wirén, 2022).

12.3.6 Gibberellins (GA)

Gibberellins (GAs) are tetracyclic diterpenoid growth hormones involved in various developmental processes in plants. Although more than 100 of them are found in plants, only a few of them show biological activity. Their discovery in plants dates back to the 1930s, when Japanese scientists isolated a growth-promoting substance from fungal cultures that were parasitising rice plants, causing "bakanoe" or "plant rise" disease. The active compound was isolated from the fungus *Gibberella fujikoroi* by Eichi Kurosawa in 1926 and was therefore called a "gibberellin." The effect of the fungus on the affected plants consisted of a remarkable increase in height but with a strong decrease in grain yield. The increased growth was due to the high content of this growth factor produced by the fungal attack. The most remarkable physiological effect of this hormone is to induce height growth, but it also promotes sudden inflorescence development and flowering in a number of plants, particularly in long-day plants. They induce germination in seeds under dormancy conditions and are involved in the mobilisation of reserves in cereal grains. They also induce the development of many fruits, parthenocarpy, as well as having a special application in the production of seedless grapes (Gupta and Chakrabarty, 2013; Hedden and Sponsel, 2015) (Figure 12.6).

12.3.6.1 Gibberellins (GA) and Their Influence on Nutrient Homeostasis (Phosphate)

12.3.6.1.1 Phosphate [Pi]

Endogenous GA levels were reduced in *Arabidopsis* plants grown under low Pi conditions, which was correlated with increased accumulation of green fluorescent protein (GFP) fused to RGA, a member of the DELLA protein family. In the same context, *Arabidopsis* mutants showed enhanced DELLA function increasing in response to Pi deprivation. In contrast, several Pi starvation responses

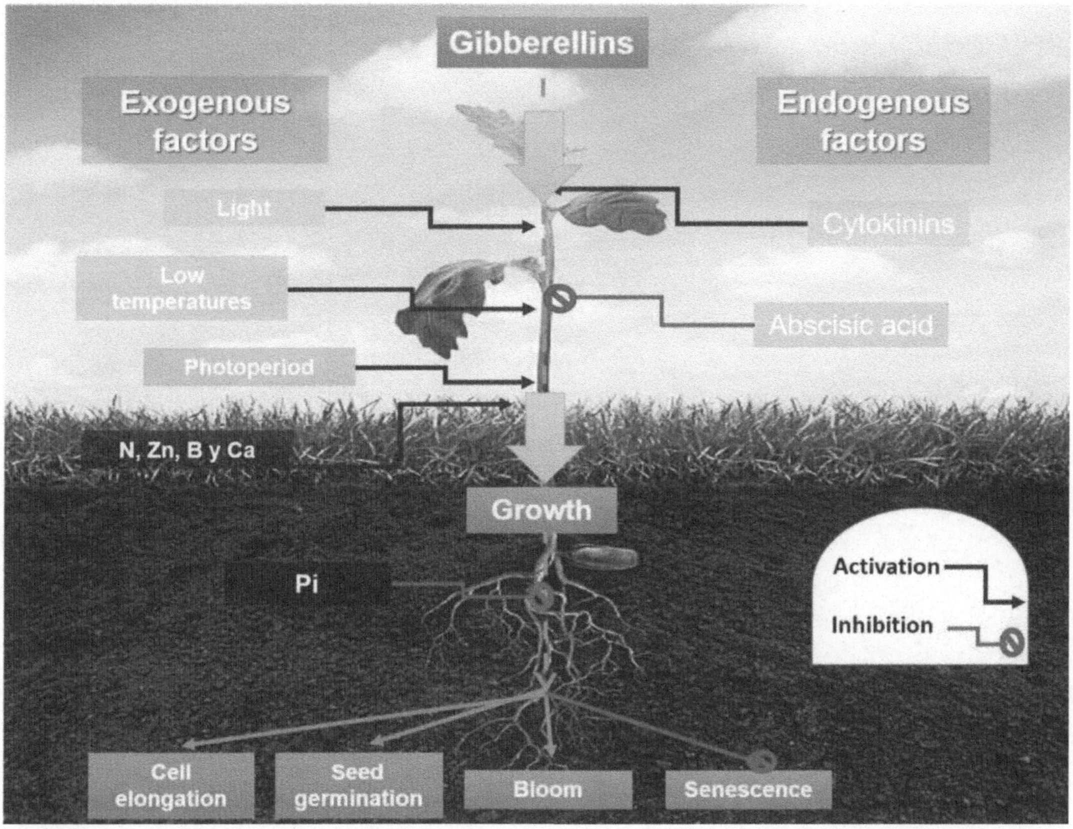

FIGURE 12.6 Factors that can activate/induce (arrows in black) or decrease/inhibit (arrows in red) the activation of the Gibberellins (GA) as well as physiological processes in which they participate (arrows in grey).

were inhibited in GA-treated wild-type plants and in *Arabidopsis* mutants with a lowered DELLA function. Under these conditions, the range of altered responses included anthocyanin accumulation, root hair growth, and changes in root structure. However, GA treatment and altered DELLA function did not affect total Pi accumulation or the expression of representative PSR genes, such as those encoding high-affinity Pi transporters (AtPT1 and AtPT2) or ribose regulatory proteins (At4 and AtIPS1), indicating the contribution of GA (Jiang *et al.*, 2007). Regulation of Pi signalling is only partial and as a result, plants require other signals to respond appropriately to Pi deprivation.

12.3.6.1.2 Other Minerals

Other types of nutrients such as N, Zn, B, and Ca have a broad relationship with their synthesis and action, so they must be at adequate levels (Díaz, 2017).

12.3.7 Brassinosteroids (BR)

In the 1970s, extracts of turnip pollen (*Brassica napus*) were found to promote the elongation of internodes in bean seedlings. From about 40 kg of pollen from *B. napus*, 4 mg of an active crystalline growth-inducing compound could be isolated, leading to the isolation and identification of the first steroidal compound of a hormonal nature: brasinolide (BL) in 1979, castasterone (CS) in 1982, and many other related compounds later. By 2003, about 60 brasinoesteroids (BRs) had been identified in various terrestrial and marine plant species. Different BRs were found in apical shoots

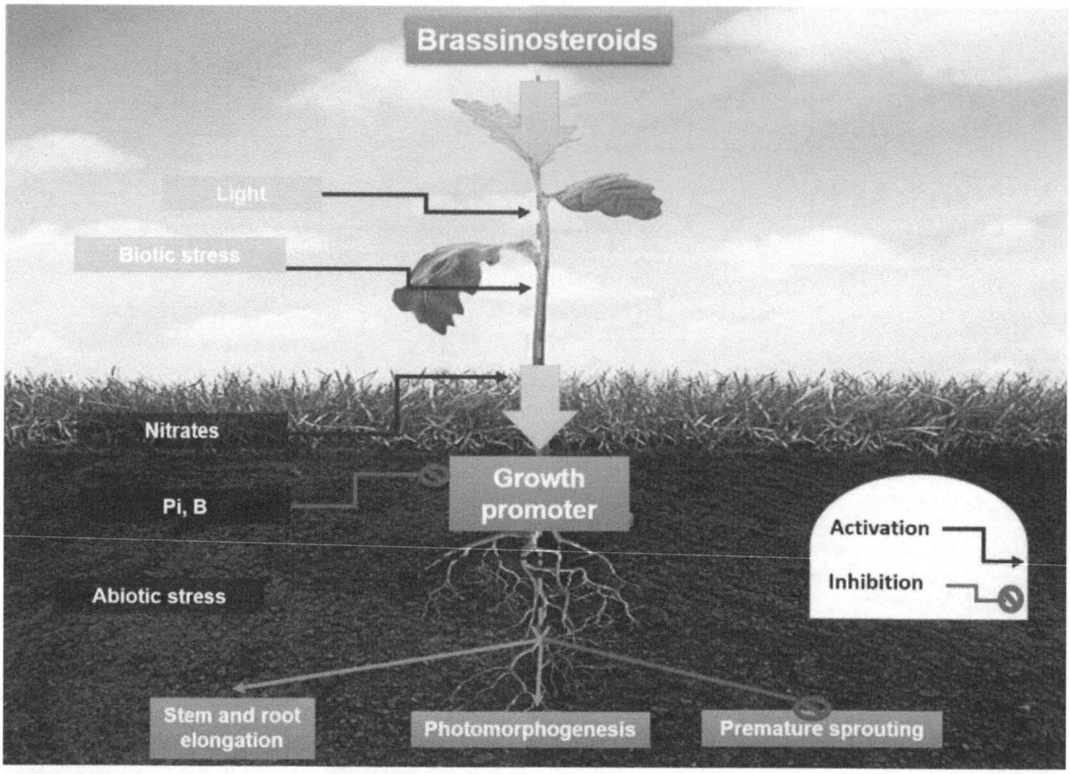

FIGURE 12.7 Factors that can activate/induce (arrows in black) or decrease/inhibit (arrows in red) the activation of the Brassinosteroids (BR) as well as physiological processes in which they participate (arrows in grey).

and active vegetative tissues of young plants as well as in seeds of a large number of herbaceous and shrub species (Bajguz and Tretyn, 2003). A series of parallel studies evidenced a set of effects stimulated by these compounds, among them, activation of proton pumps, reorientation of cellulose microfibrils, xylogenesis, generation of embryogenic tissue, and production of ethylene. Based on these effects, brassinosteroids (BRs) starting with, BRs were considered as new group of endogenous plant hormones, unique for their steroidal structure and essential for normal plant growth, as noted by Nolan *et al.* (2020) (Figure 12.7).

12.3.7.1 Brassinosteroids (BR) and Their Influence on Nutrient Homeostasis (Nitrogen, Phosphate, and Boron)

12.3.7.1.1 Nitrogen

Nitrogen deficiency-induced BR biosynthesis in *Arabidopsis* plants involves a systematic positive regulation of DWF1, CPD, DWF4, and BR6ox2 genes (Jia *et al.*, 2020). Similarly, CPD and DWF4 expression is also increased in maize, and the same effect has been reported for the DWF1 gene in nitrogen-deficient canola roots (Qin *et al.*, 2019; Xing *et al.*, 2022). Nitrogen deficiency-induced BR in biosynthesis is a common physiological response among plant species.

12.3.7.1.2 Phosphate [Pi]

It was observed that the expression of CPD, DWF4, and BR6ox2 was significantly reduced when Pi was deficient in the roots, which coincided with a reduction in BR accumulation (Singh *et al.*, 2014).

12.3.7.1.3 *Boron (B)*

Low B levels reduce BR biosynthesis by negatively regulating DWF4 and BR6ox1/2 (C. Zhang *et al.*, 2021).

12.3.8 Salicylic Acid (SA)

The importance of SA as a growth regulator in plants is reduced to a few processes. In some cases, its presence affects the synthesis of other growth regulators, which directly affect some physiological processes. For example, SA reduces ethylene synthesis, and in some species this causes a delay in flower senescence or induction of flowering (Orozco-Mosqueda *et al.*, 2023) (Figure 12.8).

12.3.8.1 Salicylic Acid (SA) and Their Influence on Nutrient Homeostasis (Nitrogen and Phosphate)

Some studies have shown that SA accumulation increases in roots deprived of N and Pi (Conesa *et al.*, 2020; Silva-Navas *et al.*, 2019). However, the mechanism by which this happens is still unclear.

12.3.9 Strigolactones (SLs)

Strigolactones (SL) are plant secondary metabolites derived from carotenoids. SLs play important roles in regulating plant growth and development through interaction with other plant hormones in the plant, and in chemical communications with beneficial and harmful microorganisms and root-parasitic plants (weeds) in the soil. In the 50 years since the discovery of the first SL, strigol, our knowledge of the chemistry and biochemistry of SLs has advanced explosively, especially during the last two decades. For example, at the end of the 20th century, only six natural SLs were known,

FIGURE 12.8 Factors that can activate/induce (arrows in black) or decrease/inhibit (arrows in red) the activation of the Salicylic acid (SA) as well as physiological processes in which they participate (arrows in grey).

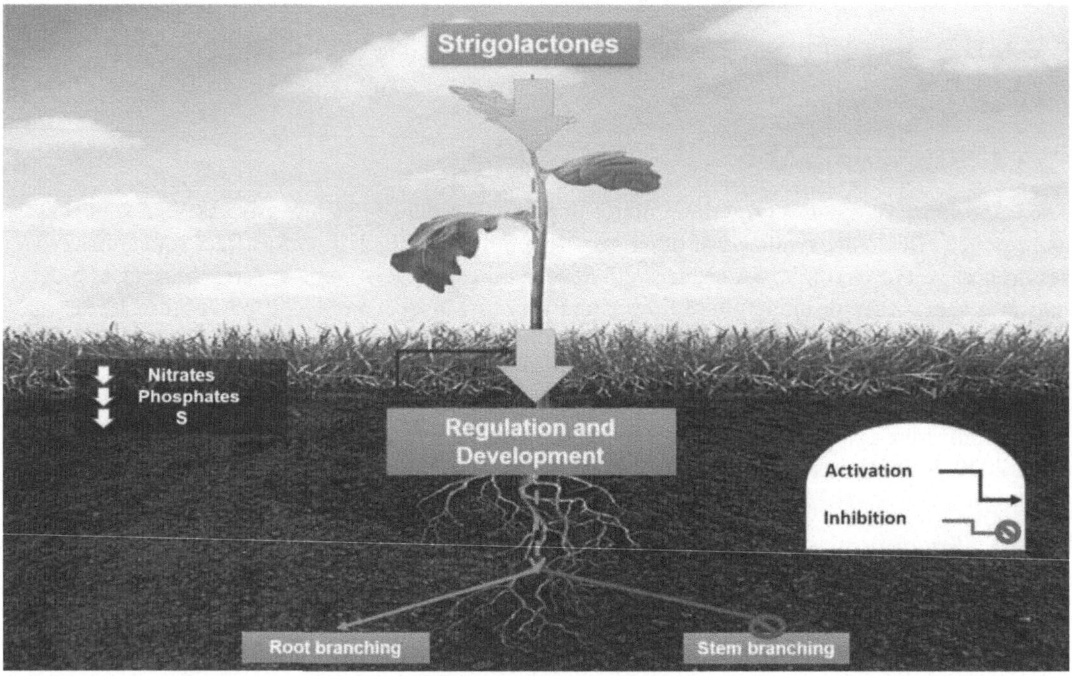

FIGURE 12.9 Factors that can activate/induce (arrows in black) or decrease/inhibit (arrows in red) the activation of the Strigolactones (SLs) as well as physiological processes in which they participate (arrows in grey).

strigol, strigyl acetate, sorgolactone, alectrol, orobankyl acetate, and orobanchol. Now more than 30 natural SLs have been identified. (Yoneyama, 2020) (Figure 12.9).

12.3.9.1 Strigolactones (SLs) and Their Influence on Nutrient Homeostasis (Nitrogen, Phosphate, and Sulphur)

The biosynthesis of strigolactones (SL) has also been linked to N, Pi, and S deficiency. Although deficiencies of these nutrients can increase SL accumulation. Differences have been observed in defined catalytic steps in SL biosynthesis. In *Arabidopsis*, the expression levels of MAX3 and MAX4 genes, and in rice the D17 and D10 genes, respectively, are consistently increased by N and Pi deficiency (according to the studies by Ito *et al.*, 2016; Sun *et al.*, 2014), while the D27 gene is positively regulated by low Pi (Abuauf *et al.*, 2018; Sun *et al.*, 2014). Interestingly, sulphur deficiency-induced SL biosynthesis in rice is specifically depends on on the D27 gene, as indicated by the research of Shindo et al. (2018). This suggests that plants have different response mechanisms in strigolactone synthesis depending on the type of nutrient deficiency, with specific genetic regulation based on the nutrient in question.

12.4 CONCLUSION AND FUTURE REMARKS

Plants have evolved adaptive responses that allow them to grow in soils low in one or more nutrients. These responses involve complex physiological and developmental changes that allow them to increase nutrient uptake and to remobilise and protect the plant from nutrient deficiency stress. Given the role of phytohormones in the control of growth, development, and resistance to stress, it is not surprising that there is an interaction between signalling pathways due to hormone and nutrient deficiency. Given the implication of the role of phytohormones in the control of growth, development, and protection against stress, the existence of interactions between signalling pathways due to the lack of hormones and nutrients is not surprising. As a matter of fact, recent studies have already begun to provide evidence for this interaction. As mentioned above, nearly all hormone signalling pathways are involved in nutrient-dependent processes. This raises the question of how these hormone signalling pathways relate to each other and how signalling layers and networks are established, maintained, and regulated in response to specific nutritional signals. Yet the relationship between nutrients and hormones still needs to be analysed. In this context, single-cell and tissue/cell-type-specific "omics" studies, including transcriptomics, metabolomics, (phospho)proteomics, or ionomics, will be the key to better understand the spatial dimensions of the nutrient-hormone interactions pathway.

REFERENCES

Abuauf, H. *et al.* (2018) 'The *Arabidopsis* DWARF27 gene encodes an all-trans-/9-cis-β-carotene isomerase and is induced by auxin, abscisic acid and phosphate deficiency', *Plant Science*, 277, pp. 33–42. https://doi.org/10.1016/j.plantsci.2018.06.024

Addicott, F.T. and Cams, H.R. (1983) *Abscisic acid.* New York: Praeger Scientific, pp. 1–21.

Armengaud, P., Breitling, R. and Amtmann, A. (2004) 'The potassium-dependent transcriptome of *Arabidopsis* reveals a prominent role of jasmonic acid in nutrient signaling', *Plant Physiology*, 136(1), pp. 2556–2576. https://doi.org/10.1104/pp.104.046482

Arnon, D.I. and Stout, P.R. (1939) 'The essentiality of certain elements in minute quantity for plants with special reference to copper', *Plant Physiology*, 14(2), pp. 371–375. https://doi.org/10.1104/pp.14.2.371

Assmann, S.M. and Shimazaki, K. (1999) 'The multisensory guard cell. Stomatal responses to blue light and abscisic acid', *Plant Physiology*, 119(3), pp. 809–816. https://doi.org/10.1104/pp.119.3.809

Bachheti, A., Deepti, Bachheti, R.K. and Husen, A. (2021) 'Medicinal plants and their pharmaceutical properties under adverse environmental conditions', in Husen A. (ed.) *Harsh environment and plant resilience.* Cham: Springer International Publishing AG, pp. 457–502. https://doi.org/10.1007/978-3-030-65912-7_19

Bajguz, A. and Tretyn, A. (2003) 'The chemical characteristic and distribution of brassinosteroids in plants', *ChemInform*, 34(22). https://doi.org/10.1002/chin.200322270

Bhosale, R. *et al.* (2018) 'Author correction: A mechanistic framework for auxin dependent *Arabidopsis* root hair elongation to low external phosphate', *Nature Communications*, 9(1), p. 1818. https://doi.org/10.1038/s41467-018-04281-x

Brenner, W.G. *et al.* (2005) 'Immediate-early and delayed cytokinin response genes of *Arabidopsis thaliana* identified by genome-wide expression profiling reveal novel cytokinin-sensitive processes and suggest cytokinin action through transcriptional cascades', *The Plant Journal*, 44(2), pp. 314–333. https://doi.org/10.1111/j.1365-313X.2005.02530.x

Brewitz, E., Larsson, C. and Larsson, M. (1995) 'Influence of nitrate supply on concentrations and translocation of abscisic acid in barley (*Hordeum vulgare*)', *Physiologia Plantarum*, 95(4), pp. 499–506. https://doi.org/10.1111/j.1399-3054.1995.tb05515.x

Busk, P.K. and Pagès, M. (1998) 'Regulation of abscisic acid-induced transcription', *Plant Molecular Biology*, 37(3), pp. 425–435. https://doi.org/10.1023/a:1006058700720

Camacho-Cristóbal, J.J. *et al.* (2015) 'Boron deficiency inhibits root cell elongation via an ethylene/auxin/ROS-dependent pathway in Arabidopsis seedlings', *Journal of Experimental Botany*, 66(13), pp. 3831–3840. https://doi.org/10.1093/jxb/erv186

Ciereszko, I. and Kleczkowski, L.A. (2002) 'Effects of phosphate deficiency and sugars on expression of rab18 in Arabidopsis: Hexokinase-dependent and okadaic acid-sensitive transduction of the sugar signal', *Biochimica et Biophysica Acta (BBA) - Gene Structure and Expression*, 1579(1), pp. 43–49. https://doi.org/10.1016/S0167-4781(02)00502-X

Clarkson, D. (1996) 'Marschner H. 1995.Mineral nutrition of higher plants. second edition. 889pp. London: Academic Press, £29.95 (paperback)', *Annals of Botany*, 78(4), pp. 527–528. https://doi.org/10.1006/anbo.1996.0155

Conesa, C.M. *et al.* (2020) 'Alternative polyadenylation and salicylic acid modulate root responses to low nitrogen availability', *Plants*, 9(2), p. 251. https://doi.org/10.3390/plants9020251

Creelman, R.A. and Mullet, J.E. (1997) 'Biosynthesis and action of jasmonates in plants', *Annual Review of Plant Physiology and Plant Molecular Biology*, 48(1), pp. 355–381. https://doi.org/10.1146/annurev.arplant.48.1.355

Dan, H., Yang, G. and Zheng, Z.-L. (2006) 'A negative regulatory role for auxin in sulphate deficiency response in Arabidopsis thaliana', *Plant Molecular Biology*, 63(2), pp. 221–235. https://doi.org/10.1007/s11103-006-9084-0

Díaz, M.D. (2017) 'Las Hormonas Vegetales en las Plantas', *Serie Nutrición Vegetal*, 88, p. 4.

Didi, D.A. *et al.* (2022) 'Effect of plant growth regulators on osmotic regulatory substances and antioxidant enzyme activity of *Nitraria tangutorum*', *Plants (Basel, Switzerland)*, 11(19), p. 2559. https://doi.org/10.3390/plants11192559

Do, D.T., Harbourne, N. and Ellis, A. (2023). 'Anthocyanins: Anthocyanidins, berries, colorants, and copigmentation', in S. M. Jafari, A. Rashidinejad, and J. Simal-Gandara (eds.) *Handbook of food bioactive ingredients.* Cham, Switzerland: Springer Nature Switzerland AG, pp. 341–364. https://doi.org/10.1007/978-3-031-28109-9_9

Ei-D, A.M.S.A., Salama, A. and Wareing, P.F. (1979) 'Effects of mineral nutrition on endogenous cytokinins in plants of sunflower (*Helianthus annuus* L)', *Journal of Experimental Botany*, 30(5), pp. 971–981. https://doi.org/10.1093/jxb/30.5.971

Elawad, S.H., Gascho, G.J. and Street, J.J. (1982) 'Response of sugarcane to silicate source and rate. I. Growth and yield', *Agronomy Journal*, 74(3), pp. 481–484. https://doi.org/10.2134/agronj1982.00021962007400030019x

Franco-Zorrilla, J.M. *et al.* (2002) 'Mutations at *CRE1* impair cytokinin-induced repression of phosphate starvation responses in *Arabidopsis*', *The Plant Journal*, 32(3), pp. 353–360. https://doi.org/10.1046/j.1365-313X.2002.01431.x

Gamalero, E., Lingua, G. and Glick, B.R. (2023) 'Ethylene, ACC, and the plant growth-promoting enzyme ACC deaminase', *Biology*, 12(8), p. 1043. https://doi.org/10.3390/biology12081043

Getnet, Z. *et al.* (2015) 'Growth, water status, physiological, biochemical and yield response of stay green sorghum {*Sorghum bicolor* (L.) Moench} varieties-a field trial under drought-prone area in Amhara regional state, Ethiopia', *Journal of Agronomy*, 14, pp. 188–202.

Ghasemi Pirbalouti, A., Sajjadi, S.E. and Parang, K. (2014) 'A review (Research and patents) on jasmonic acid and its derivatives', *Archiv Der Pharmazie*, 347(4), pp. 229–239. https://doi.org/10.1002/ardp.201300287

Grusak, M.A., Broadley, M.R. and White, P.J. (2016) 'Plant macro- and micronutrient minerals', in *Encyclopedia of life sciences.* 1st ed. Wiley, pp. 1–6. https://doi.org/10.1002/9780470015902.a0001306.pub2

Guilfoyle, T.J. and Hagen, G. (2007) 'Auxin response factors', *Current Opinion in Plant Biology*, 10(5), pp. 453–460. https://doi.org/10.1016/j.pbi.2007.08.014

Gupta, R. and Chakrabarty, S.K. (2013) 'Gibberellic acid in plant: Still a mystery unresolved', *Plant Signaling & Behavior*, 8(9), p. e25504. https://doi.org/10.4161/psb.25504

Hagen, G., Guilfoyle, T.J. and Gray, W.M. (2010) 'Auxin signal transduction', in P. J. Davies (ed.) *Plant hormones.* Dordrecht: Springer Netherlands, pp. 282–307. https://doi.org/10.1007/978-1-4020-2686-7_14

He, Y. *et al.* (2002) 'Evidence supporting a role of jasmonic acid in Arabidopsis leaf senescence', *Plant Physiology*, 128(3), pp. 876–884. https://doi.org/10.1104/pp.010843

Hedden, P. and Sponsel, V. (2015) 'A century of gibberellin research', *Journal of Plant Growth Regulation*, 34(4), pp. 740–760. https://doi.org/10.1007/s00344-015-9546-1

Hermans, C. *et al.* (2006) 'How do plants respond to nutrient shortage by biomass allocation?', *Trends in Plant Science*, 11(12), pp. 610–617. https://doi.org/10.1016/j.tplants.2006.10.007

Hirai, M.Y. *et al.* (2003) 'Global expression profiling of sulfur-starved *Arabidopsis* by DNA macroarray reveals the role of *O* -acetyl- l -serine as a general regulator of gene expression in response to sulfur nutrition', *The Plant Journal*, 33(4), pp. 651–663. https://doi.org/10.1046/j.1365-313X.2003.01658.x

Horgan, J.M. and Wareing, P.F. (1980) 'Cytokinins and the growth responses of seedlings of *Betula pendula* Roth. and *Acer pseudoplatanus* L. to nitrogen and phosphorus deficiency', *Journal of Experimental Botany*, 31(2), pp. 525–532. https://doi.org/10.1093/jxb/31.2.525

Husen, A. (2010) 'Growth characteristics, physiological and metabolic responses of teak (*Tectona grandis* Linn. f.) clones differing in rejuvenation capacity subjected to drought stress', *Silvae Genetica*, 59, pp. 124–136.

Husen, A. (2021a) 'The harsh environment and resilient plants: An overview', in Husen A. (ed.) *Harsh environment and plant resilience.* Cham, Switzerland: Springer International Publishing AG, pp. 1–23. https://doi.org/10.1007/978-3-030-65912-7_1

Husen, A. (2021b) 'Morpho-anatomical, physiological, biochemical and molecular responses of plants to air pollution', in Husen A. (ed.) *Harsh environment and plant resilience.* Cham, Switzerland: Springer International Publishing AG, pp. 203–234. https://doi.org/10.1007/978-3-030-65912-7_9

Husen, A. (2021c) 'Cross talk between autophagy and hormones for abiotic stress tolerance in plants', in Husen A (ed) *Plant performance under environmental stress.* Cham, Switzerland: Springer International Publishing AG, pp. 1–15. https://doi.org/10.1007/978-3-030-78521-5_1

Husen, A. (2021d) *Harsh environment and plant resilience: Molecular and functional aspects.* Cham, Switzerland: Springer Nature Switzerland AG. https://doi.org/10.1007/978-3-030-65912-7

Husen, A. (2021e) *Plant performance under environmental stress: Hormones, biostimulants and sustainable plant growth management*. Cham, Switzerland: Springer Nature Switzerland AG. https://doi.org/10.1007/978-3-030-78521-5

Husen, A., Iqbal, M. and Aref, I.M. (2016) 'IAA-induced alteration in growth and photosynthesis of pea (*Pisum sativum* L.) plants grown under salt stress', *Journal of Environmental Biology*, 37, pp. 421–429.

Husen, A., Iqbal, M. and Aref, M.I. (2014) 'Growth, water status and leaf characteristics of *Brassica carinata* under drought stress and rehydration conditions', *Brazilian Journal of Botany*, 37, pp. 217–227.

Husen, A. *et al.* (2019) 'Modulation of salt-stress tolerance of Niger (*Guizotia abyssinica*), an oilseed plant, by application of salicylic acid', *Journal of Environmental Biology*, 40, pp. 94–104.

Husen, A. (2024a) *Current understanding of biostimulants in plant protection and performance*. Cambridge, MA: Elsevier Inc.

Husen, A. (2024b) *Essential minerals in plant-soil system: Coordination, signalling, and interaction under adverse situations)*. Cambridge, MA: Elsevier Inc.

Hussein, M. *et al.* (2017) 'Salinity-induced modulation of plant growth and photosynthetic parameters in faba bean *(Vicia faba)* cultivars', *Pakistan Journal of Botany*, 49, pp. 867–877.

Ito, S. *et al.* (2016) 'Effects of strigolactone signaling on *Arabidopsis* growth under nitrogen deficient stress condition', *Plant Signaling & Behavior*, 11(1), p. e1126031. https://doi.org/10.1080/15592324.2015.1126031

Jia, Z., Giehl, R.F.H. and Von Wirén, N. (2020) 'The root foraging response under low nitrogen depends on *DWARF1*-Mediated brassinosteroid biosynthesis', *Plant Physiology*, 183(3), pp. 998–1010. https://doi.org/10.1104/pp.20.00440

Jia, Z., Giehl, R.F.H. and Von Wirén, N. (2021) 'Local auxin biosynthesis acts downstream of brassinosteroids to trigger root foraging for nitrogen', *Nature Communications*, 12(1), p. 5437. https://doi.org/10.1038/s41467-021-25250-x

Jia, Z., Giehl, R.F.H. and Von Wirén, N. (2022) 'Nutrient–hormone relations: Driving root plasticity in plants', *Molecular Plant*, 15(1), pp. 86–103. https://doi.org/10.1016/j.molp.2021.12.004

Jiang, C. *et al.* (2007) 'Phosphate starvation root architecture and anthocyanin accumulation responses are modulated by the gibberellin-DELLA signaling pathway in Arabidopsis', *Plant Physiology*, 145(4), pp. 1460–1470. https://doi.org/10.1104/pp.107.103788

Jiang, M. and Zhang, J. (2001) 'Effect of abscisic acid on active oxygen species, antioxidative defence system and oxidative damage in leaves of maize seedlings', *Plant and Cell Physiology*, 42(11), pp. 1265–1273. https://doi.org/10.1093/pcp/pce162

Jost, R. *et al.* (2005) 'Expression profiling of metabolic genes in response to methyl jasmonate reveals regulation of genes of primary and secondary sulfur-related pathways in Arabidopsis thaliana', *Photosynthesis Research*, 86(3), pp. 491–508. https://doi.org/10.1007/s11120-005-7386-8

Khan, G.A. *et al.* (2016) 'Phosphate deficiency induces the jasmonate pathway and enhances resistance to insect herbivory', *Plant Physiology*, 171(1), pp. 632–644. https://doi.org/10.1104/pp.16.00278

Kim, H.J. *et al.* (2006). 'Cytokinin-mediated control of leaf longevity by AHK3 through phosphorylation of ARR2 in *Arabidopsis*', *Proceedings of the National Academy of Sciences*, 103(3), pp. 814–819. https://doi.org/10.1073/pnas.0505150103

Lai, F., *et al.* (2007) 'Cell division activity determines the magnitude of phosphate starvation responses in Arabidopsis', *The Plant Journal*, 50(3), pp. 545–556. https://doi.org/10.1111/j.1365-313X.2007.03070.x

Lawlor, D.W. (2004) 'Mengel, K. and Kirkby, E. A. Principles of plant nutrition', *Annals of Botany*, 93(4), pp. 479–480. https://doi.org/10.1093/aob/mch063

Leung, J. and Giraudat, J. (1998) 'Abscisic acid signal transduction', *Annual Review of Plant Physiology and Plant Molecular Biology*, 49(1), pp. 199–222. https://doi.org/10.1146/annurev.arplant.49.1.199

Li, J., Wu, W.-H. and Wang, Y. (2017) 'Potassium channel AKT1 is involved in the auxin-mediated root growth inhibition in Arabidopsis response to low K+ stress', *Journal of Integrative Plant Biology*, 59(12), pp. 895–909. https://doi.org/10.1111/jipb.12575

Li, Y.-S. *et al.* (2009) 'Phosphorus deficiency-induced reduction in root hydraulic conductivity in *Medicago falcata* is associated with ethylene production', *Environmental and Experimental Botany*, 67(1), pp. 172–177. https://doi.org/10.1016/j.envexpbot.2009.05.013

Lim, P.O., Kim, H.J. and Gil Nam, H. (2007) 'Leaf senescence', *Annual Review of Plant Biology*, 58(1), pp. 115–136. https://doi.org/10.1146/annurev.arplant.57.032905.105316

Liu, M. *et al.* (2023). 'The role of IAA in regulating root architecture of sweetpotato (*Ipomoea batatas* [L.] Lam) in response to potassium deficiency stress', *Plants*, 12(9), 1779. https://doi.org/10.3390/plants12091779

López-Bucio, J., Cruz-Ramírez, A. and Herrera-Estrella, L. (2003) 'The role of nutrient availability in regulating root architecture', *Current Opinion in Plant Biology*, 6(3), pp. 280–287. https://doi.org/10.1016/S1369-5266(03)00035-9

López-Bucio, J. *et al.* (2002) 'Phosphate availability alters architecture and causes changes in hormone sensitivity in the Arabidopsis root system', *Plant Physiology*, 129(1), pp. 244–256. https://doi.org/10.1104/pp. 010934

López-Bucio, J. *et al.* (2005) 'An auxin transport independent pathway is involved in phosphate stress-induced root architectural alterations in Arabidopsis. Identification of *BIG* as a mediator of auxin in pericycle cell activation', *Plant Physiology*, 137(2), pp. 681–691. https://doi.org/10.1104/pp.104.049577

Lucena, C. *et al.* (2006) 'Ethylene could influence ferric reductase, iron transporter, and H+-ATPase gene expression by affecting FER (or FER-like) gene activity', *Journal of Experimental Botany*, 57(15), pp. 4145–4154. https://doi.org/10.1093/jxb/erl189

Ma, W. *et al.* (2014) 'Auxin biosynthetic gene *TAR 2* is involved in low nitrogen-mediated reprogramming of root architecture in *Arabidopsis*', *The Plant Journal*, 78(1), pp. 70–79. https://doi.org/10.1111/tpj. 12448

Ma, Z. *et al.* (2003) 'Regulation of root elongation under phosphorus stress involves changes in ethylene responsiveness', *Plant Physiology*, 131(3), pp. 1381–1390. https://doi.org/10.1104/pp.012161

Malavolta, E. (2006) *Manual de nutrição mineral de plantas*. São Paulo: Editora Agronômica Ceres.

Martín, A.C. *et al.* (2000) 'Influence of cytokinins on the expression of phosphate starvation responsive genes in *Arabidopsis*', *The Plant Journal*, 24(5), pp. 559–567. https://doi.org/10.1046/j.1365-313x.2000. 00893.x

Martin, M. H., & Marschner, H. (1988). 'The mineral nutrition of higher plants,' *The Journal of Ecology*, 76(4), p. 1250. https://doi.org/10.2307/2260650

Maruyama-Nakashita, A., *et al.* (2004) 'A novel regulatory pathway of sulfate uptake in Arabidopsis roots: Implication of CRE1/WOL/AHK4-mediated cytokinin-dependent regulation', *The Plant Journal: For Cell and Molecular Biology*, 38(5), pp. 779–789. https://doi.org/10.1111/j.1365-313X.2004.02079.x

McCabe, M.S. *et al.* (2001) 'Effects of PSAG12- *IPT* Gene expression on development and senescence in transgenic lettuce', *Plant Physiology*, 127(2), pp. 505–516. https://doi.org/10.1104/pp.010244

Mellor, N. *et al.* (2016) 'Dynamic regulation of auxin oxidase and conjugating enzymes *AtDAO1* and *GH3* modulates auxin homeostasis', *Proceedings of the National Academy of Sciences*, 113(39), pp. 11022–11027. https://doi.org/10.1073/pnas.1604458113

Mimmo, T. *et al.* (2018). 'Nutrient availability in the rhizosphere: A review', *Acta Horticulturae*, 1217, pp. 13–28. https://doi.org/10.17660/ActaHortic.2018.1217.2

Miyawaki, K., Matsumoto-Kitano, M. and Kakimoto, T. (2004) 'Expression of cytokinin biosynthetic isopentenyltransferase genes in *Arabidopsis*: Tissue specificity and regulation by auxin, cytokinin, and nitrate', *The Plant Journal*, 37(1), pp. 128–138. https://doi.org/10.1046/j.1365-313X.2003.01945.x

Nacry, P. *et al.* (2005) 'A role for auxin redistribution in the responses of the root system architecture to phosphate starvation in arabidopsis', *Plant Physiology*, 138(4), 2061–2074. https://doi.org/10.1104/pp. 105.060061

Neill, S.O. and Gould, K.S. (2003) 'Anthocyanins in leaves: Light attenuators or antioxidants?', *Functional Plant Biology*, 30(8), p. 865. https://doi.org/10.1071/FP03118

Nikiforova, V. *et al.* (2003) 'Transcriptome analysis of sulfur depletion in *Arabidopsis thaliana*: Interlacing of biosynthetic pathways provides response specificity', *The Plant Journal*, 33(4), pp. 633–650. https://doi.org/10.1046/j.1365-313X.2003.01657.x

Nolan, T.M. *et al.* (2020) 'Brassinosteroids: Multidimensional regulators of plant growth, development, and stress responses', *The Plant Cell*, 32(2), pp. 295–318. https://doi.org/10.1105/tpc.19.00335

Nunes Da Silva, M. *et al.* (2022) 'Non-essential elements and their role in sustainable agriculture', *Agronomy*, 12(4), p. 888. https://doi.org/10.3390/agronomy12040888

Ohkama, N. *et al.* (2002) 'Differential tissue-specific response to sulfate and methionine of a soybean seed storage protein promoter region in transgenic Arabidopsis', *Plant and Cell Physiology*, 43(11), pp. 1266–1275. https://doi.org/10.1093/pcp/pcf149

Ondzighi-Assoume, C.A., Chakraborty, S. and Harris, J.M. (2016) 'Environmental nitrate stimulates abscisic acid accumulation in Arabidopsis root tips by releasing it from inactive stores', *The Plant Cell*, 28(3), pp. 729–745. https://doi.org/10.1105/tpc.15.00946

Orozco-Mosqueda, M.D.C., Santoyo, G. and Glick, B.R. (2023) 'Recent advances in the bacterial phytohormone modulation of plant growth', *Plants*, 12(3), p. 606. https://doi.org/10.3390/plants12030606

Petroni, K. and Tonelli, C. (2011) 'Recent advances on the regulation of anthocyanin synthesis in reproductive organs', *Plant Science*, 181(3), pp. 219–229. https://doi.org/10.1016/j.plantsci.2011.05.009

Phillips, I. D. J., & Wareing, P. F. (1958) 'Studies in dormancy of sycamore. I. Seasonal changes in the growth-substance content of the shoot', *Journal of Experimental Botany*, 9(3), 350–364. https://doi.org/10.1093/jxb/9.3.350

Philippar, K. *et al.* (1999) 'Auxin-induced K^+ channel expression represents an essential step in coleoptile growth and gravitropism', *Proceedings of the National Academy of Sciences*, 96(21), pp. 12186–12191. https://doi.org/10.1073/pnas.96.21.12186

Qin, L. *et al.* (2019) 'Adaption of roots to nitrogen deficiency revealed by 3D quantification and proteomic analysis', *Plant Physiology*, 179(1), pp. 329–347. https://doi.org/10.1104/pp.18.00716

Quamruzzaman, M. *et al.* (2021) 'Improving performance of salt-grown crops by exogenous application of plant growth regulators', *Biomolecules*, 11(6), p. 788. doi:10.3390/biom11060788

Raza, A. *et al.* (2019) 'Impact of climate change on crops adaptation and strategies to tackle its outcome: A review', *Plants*, 8(2), p. 34. https://doi.org/10.3390/plants8020034

Reymond, M. *et al.* (2006) 'Identification of QTL controlling root growth response to phosphate starvation in *Arabidopsis thaliana*', *Plant, Cell & Environment*, 29(1), pp. 115–125. https://doi.org/10.1111/j.1365-3040.2005.01405.x

Rubio, V. *et al.* (2009) 'Plant hormones and nutrient signaling', *Plant Molecular Biology*, 69(4), pp. 361–373. https://doi.org/10.1007/s11103-008-9380-y

Rubio-Wilhelmi, M.D.M. *et al.* (2011) 'Cytokinin-dependent improvement in transgenic P_{SARK}::IPT Tobacco under nitrogen deficiency', *Journal of Agricultural and Food Chemistry*, 59(19), 10491–10495. https://doi.org/10.1021/jf202604k

Rubio-Wilhelmi, M. D. M. *et al.* (2012) 'Response of carbon and nitrogen-rich metabolites to nitrogen deficiency in PSARK::IPT tobacco plants', *Plant Physiology and Biochemistry*, 57, 231–237. https://doi.org/10.1016/j.plaphy.2012.06.004

Sajjad, Y. (2017) 'Application of plant growth regulators in ornamental plants: A review', *Pakistan Journal of Agricultural Sciences*, 54(02), pp. 327–333. https://doi.org/10.21162/PAKJAS/17.3659

Sakakibara, H. (2006) 'Cytokinins: Activity, biosynthesis, and translocation', *Annual Review of Plant Biology*, 57(1), pp. 431–449. https://doi.org/10.1146/annurev.arplant.57.032905.105231

Schachtman, D.P. and Shin, R. (2007) 'Nutrient sensing and signaling: NPKS', *Annual Review of Plant Biology*, 58(1), pp. 47–69. https://doi.org/10.1146/annurev.arplant.58.032806.103750

Schaller, G.E., Bishopp, A. and Kieber, J.J. (2015) 'The Yin-Yang of hormones: Cytokinin and auxin interactions in plant development', *The Plant Cell*, 27(1), pp. 44–63. https://doi.org/10.1105/tpc.114.133595

Scheible, W.-R. *et al.* (2004) 'Genome-wide reprogramming of primary and secondary metabolism, protein synthesis, cellular growth processes, and the regulatory infrastructure of *Arabidopsis* in response to nitrogen', *Plant Physiology*, 136(1), pp. 2483–2499. https://doi.org/10.1104/pp.104.047019

Séguéla, M. *et al.* (2008) 'Cytokinins negatively regulate the root iron uptake machinery in *Arabidopsis* through a growth-dependent pathway', *The Plant Journal*, 55(2), pp. 289–300. https://doi.org/10.1111/j.1365-313X.2008.03502.x

Shao, H.-B. *et al.* (2008) 'Primary antioxidant free radical scavenging and redox signaling pathways in higher plant cells', *International Journal of Biological Sciences*, pp. 8–14. https://doi.org/10.7150/ijbs.4.8

Shin, H. *et al.* (2006) Loss of *At4* function impacts phosphate distribution between the roots and the shoots during phosphate starvation', *The Plant Journal*, 45(5), pp. 712–726. https://doi.org/10.1111/j.1365-313X.2005.02629.x

Shin, R. and Schachtman, D.P. (2004) 'Hydrogen peroxide mediates plant root cell response to nutrient deprivation', Proceedings of the National Academy of Sciences, 101(23), pp. 8827–8832. https://doi.org/10.1073/pnas.0401707101

Shindo, M. *et al.* (2018) 'Upregulation of *DWARF 27* is associated with increased strigolactone levels under sulfur deficiency in rice', *Plant Direct*, 2(4), p. e00050. https://doi.org/10.1002/pld3.50

Signora, L., De Smet, I., Foyer, C. H., Zhang, H. (2001) 'ABA plays a central role in mediating the regulatory effects of nitrate on root branching in Arabidopsis'. *Plant J.*, 28(6): pp. 655–662. doi: 10.1046/j.1365-313x.2001.01185.x. PMID: 11851911.

Silva-Navas, J. *et al.* (2019) 'Role of cis-zeatin in root responses to phosphate starvation', *New Phytologist*, 224(1), pp. 242–257. https://doi.org/10.1111/nph.16020

Singh, A.P. *et al.* (2014) 'Activity of the brassinosteroid transcription factors brassinazole resistant 1 and brassinosteroid insensitive 1-ethyl methanesulfonate-suppressor 1/brassinazole resistant 2 blocks developmental reprogramming in response to low phosphate availability', *Plant Physiology*, 166(2), pp. 678–688. https://doi.org/10.1104/pp.114.245019

Sun, H. *et al.* (2014) 'Strigolactones are involved in phosphate- and nitrate-deficiency-induced root development and auxin transport in rice', *Journal of Experimental Botany*, 65(22), pp. 6735–6746. https://doi.org/10.1093/jxb/eru029

Svistoonoff, S. *et al.* (2007) 'Root tip contact with low-phosphate media reprograms plant root architecture', *Nature Genetics*, 39(6), pp. 792–796. https://doi.org/10.1038/ng2041

Takei, K. *et al.* (2004) 'AtIPT3 is a key determinant of nitrate-dependent cytokinin biosynthesis in *Arabidopsis*', *Plant and Cell Physiology*, 45(8), pp. 1053–1062. https://doi.org/10.1093/pcp/pch119

Ticconi, C.A. *et al.* (2004) '*Arabidopsis pdr2* reveals a phosphate-sensitive checkpoint in root development', *The Plant Journal*, 37(6), pp. 801–814. https://doi.org/10.1111/j.1365-313X.2004.02005.x

Tripathi, R. *et al.* (2022) 'Plant mineral nutrition and disease resistance: A significant linkage for sustainable crop protection', *Frontiers in Plant Science*, 13, p. 883970. https://doi.org/10.3389/fpls.2022.883970

Trull, M.C. *et al.* (1997) 'The responses of wild-type and ABA mutant *Arabidopsis thaliana* plants to phosphorus starvation', *Plant, Cell & Environment*, 20(1), pp. 85–92. https://doi.org/10.1046/j.1365-3040.1997.d01-4.x

Vaishnav, D. and Chowdhury, P. (2023) 'Types and function of phytohormone and their role in stress', in S. Hussain, T. Hussain Awan, E. Ahmad Waraich, and M. Iqbal Awan (eds.) *Plant abiotic stress responses and tolerance mechanisms*. IntechOpen. https://doi.org/10.5772/intechopen.109325

Wagner, B.M. and Beck, E. (1993) 'Cytokinins in the perennial herb *Urtica dioica* L. as influenced by its nitrogen status', *Planta*, 190(4). https://doi.org/10.1007/BF00224790

Walch-Liu, P. *et al.* (2006) 'Nitrogen regulation of root branching', *Annals of Botany*, 97(5), pp. 875–881. https://doi.org/10.1093/aob/mcj601

White, P.J., & Hammond, J.P. (2008) 'Phosphorus nutrition of terrestrial plants', in P.J. White and J.P. Hammond (eds.) *The ecophysiology of plant-phosphorus interactions*, vol. 7, pp. 51–81. Springer Netherlands. https://doi.org/10.1007/978-1-4020-8435-5_4

Xing, J. *et al.* (2022) 'Brassinosteroids modulate nitrogen physiological response and promote nitrogen uptake in maize (*Zea mays* L.)', *The Crop Journal*, 10(1), pp. 166–176. https://doi.org/10.1016/j.cj.2021.04.004

Yoneyama, K. (2020) 'Recent progress in the chemistry and biochemistry of strigolactones', *Journal of Pesticide Science*, 45(2), pp. 45–53. https://doi.org/10.1584/jpestics.D19-084

Zhang, C. *et al.* (2021) 'Boron deficiency-induced root growth inhibition is mediated by brassinosteroid signalling regulation in *Arabidopsis*', *The Plant Journal*, 107(2), pp. 564–578. https://doi.org/10.1111/tpj.15311

Zhang, H., Rong, H. and Pilbeam, D. (2007) 'Signalling mechanisms underlying the morphological responses of the root system to nitrogen in *Arabidopsis thaliana*', *Journal of Experimental Botany*, 58(9), pp. 2329–2338. https://doi.org/10.1093/jxb/erm114

Zhao, Q. *et al.* (2014) 'Sulfur nutrient availability regulates root elongation by affecting root indole-3-acetic acid levels and the stem cell niche', *Journal of Integrative Plant Biology*, 56(12), pp. 1151–1163. https://doi.org/10.1111/jipb.12217

Zhao, Y. (2018) 'Essential roles of local auxin biosynthesis in plant development and in adaptation to environmental changes', *Annual Review of Plant Biology*, 69(1), pp. 417–435. https://doi.org/10.1146/annurev-arplant-042817-040226

Zheng, D. *et al.* (2013) 'The nitrate transporter NRT2.1 functions in the ethylene response to nitrate deficiency in *Arabidopsis*', *Plant, Cell & Environment*, 36(7), pp. 1328–1337. https://doi.org/10.1111/pce.12062

Zheng, H. *et al.* (2016) '*AtOPR3* specifically inhibits primary root growth in *Arabidopsis* under phosphate deficiency', *Scientific Reports*, 6(1), p. 24778. https://doi.org/10.1038/srep24778

13 Perspectives of Nutrient Sensing and Signaling in Crop Plants

Priti Pal, Akhilesh Kumar Singh, and Azamal Husen

13.1 INTRODUCTION

Plants, as stationary organisms, depend heavily on the abundance of nutrients within the soil that surrounds them. The primary route of nutrient absorption involves the roots, where mineral nutrients are absorbed and subsequently transported through vascular systems. Chand *et al.* (2022) delve deeply into the intricate interplay between soil composition and the structural adjustments in plant roots, highlighting their profound influence on the effectiveness of nutrient uptake in the face of limitations. The adaptability of root system architecture (RSA), a trait illuminated by Barbier *et al.* (2023), plays a pivotal role in granting plants access to essential nutrients critical for their overall development, growth, and fertility. Distinct plant species showcase diverse responses to specific nutrient deficiencies. Within the intricate cellular environment, individual organelles assume specific physiological roles. Li *et al.* (2021b) emphasize that comprehending the nutrient concentrations within each organelle is instrumental in unveiling their unique functions and deciphering the metabolic state of the cell. Essential nutrients for crop plants span macronutrients like nitrogen (N), potassium (K), calcium (Ca), magnesium (Mg), phosphorus (P), and sulfur (S). The roster extends to micronutrients, including iron (Fe), zinc (Zn), copper (Cu), boron (B), manganese (Mn), cobalt (Co), molybdenum (Mo), and nickel (Ni), all indispensable for supporting growth and development, as depicted in Figure 13.1. Simultaneously, soils may harbor various other minerals, some with nutritional attributes that, while non-essential (e.g., silicon [Si] and sodium [Na]), can still be beneficial. Conversely, other minerals, such as heavy metals like lead, cadmium, aluminum, or metalloids like arsenic and antimony, can exert detrimental effects (Podar and Maathuis, 2022). The survival and productivity of plants hinge on their ability to sense and respond to fluctuations in nutrient availability. At cellular and molecular levels, intricate processes unfold, enabling crop plants to optimize nutrient uptake and utilization. This chapter embarks on an exploration of the fundamental nutrient sensing mechanisms employed by crop plants.

Recognized as an ancient and fundamental mechanism, nutrient signaling has evolved into a pivotal modulator of cellular activities and organismal development, intricately weaving together intrinsic regulators and environmental inputs across multicellular plants and animals, as highlighted by Kalia *et al.* (2021). Contrary to the previous belief that nutrients merely serve as fuel for cellular metabolism and growth, the complexity of nutrient signaling mechanisms challenges this notion. These mechanisms are integral components of intricate regulatory networks within diverse cells, tissues, and organs, each characterized by specialized physiology, metabolism, and functions. The indispensable roles of nutrient signaling networks span various realms, including plant development, agriculture, human health, disease, and aging, gaining increasing acknowledgment.

In the realm of nutrient dynamics, plants, unencumbered by the organic nutrient limitations observed in animals, play a pivotal role in connecting Earth's inorganic element resources to the synthesis of organic molecules and nutrients. An exemplary feat is their conversion of atmospheric carbon (C), manifested as CO_2, into sugars through photosynthesis in their shoots. Inorganic nitrogen (N) in the form of nitrate (NO_3^-) and phosphorus (P) as phosphate [PO_4^{3-}(Pi)] are absorbed from

DOI: 10.1201/9781032630366-13

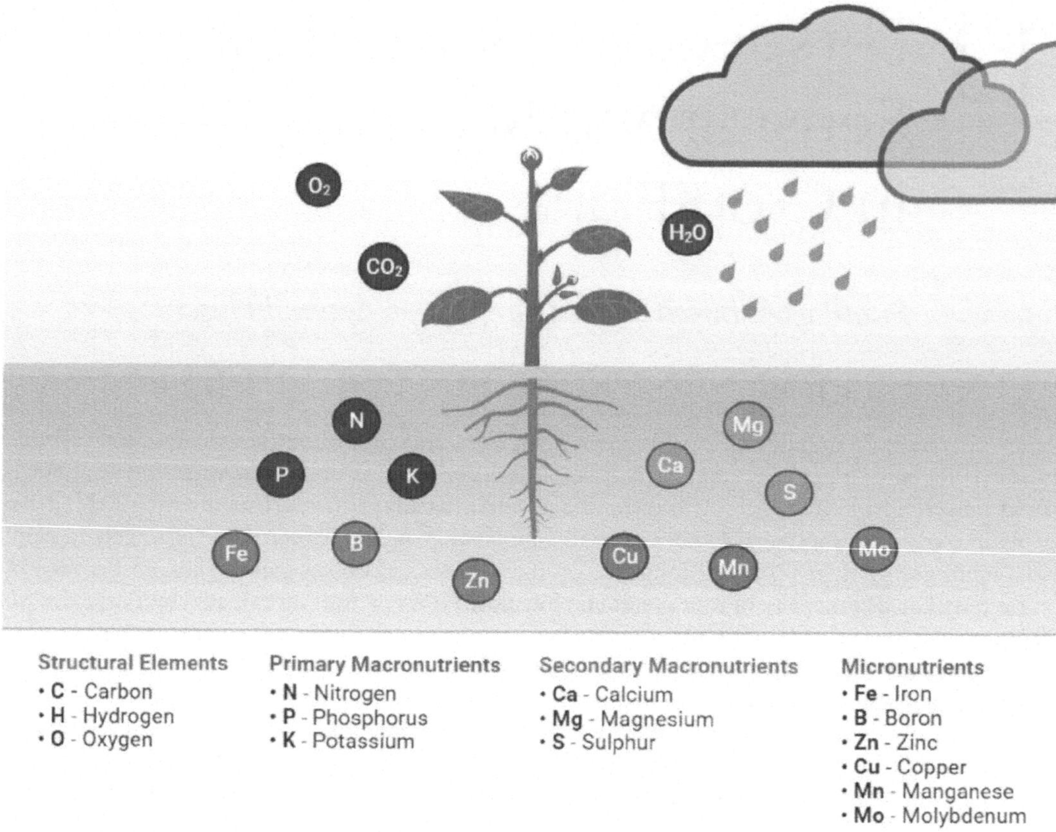

FIGURE 13.1 Vital nutrients for plants: Sustaining growth and health.

the soil by their roots. These nutrients are then harnessed to synthesize crucial components such as nucleic acids, amino acids, lipids, vitamins, and adenosine triphosphate (ATP) (Gudadhe and Thanki, 2021). Photosynthetic plants showcase remarkable adaptability, detecting and responding to a diverse array of inorganic nutrients, whether in the soil or within cells. Ultimately, they serve as the primary source of essential resources, sustaining life forms (see Table 13.1). Considerable progress has been achieved in deciphering the complexities of protein-coding genes that govern the absorption, movement, storage, and integration of macro/micro-elements. The ongoing investigation into the regulatory networks influencing the expression of these genes in response to changing nutritional conditions is underscored by the work of Ichihashi *et al.* (2020). Crop plants frequently encounter imbalances in nutrient availability, which can adversely impact multiple metabolic processes. Nevertheless, plants have evolved strategies to adapt and thrive in the face of nutritional deficiencies.

Globally, the repercussions of nutrient deficiencies cast substantial adverse effects on crop production, manifesting as reduced yields and compromised quality in both food and animal feed. Concurrently, the improper use of fertilizers introduces pollution to terrestrial and aquatic environments, while the energy-intensive nature of fertilizer production contributes significantly to climate change (Dekomah *et al.*, 2022). In natural terrestrial ecosystems, where water availability is paired with nutrient availability, the latter emerges as a pivotal environmental factor, shaping plant responses to climate change and aiding in their resilience against environmental stress, thereby influencing species biodiversity (Naz, 2023).

TABLE 13.1
Transcending Boundaries: How Transgenes Elevate Nitrogen Utilization Efficiency

Genes	Crop	Function	Role in Improving NUE	Reference
GS 1	Rice	Glutamine synthetase (cytosol)	Increase in total N and A.A	Cai *et al.* (2009)
DEP1	Rice	Dense and erect panicles 1	Ammonium uptake and assimilation	Sun *et al.* (2014)
TaNRT2.1	Rice	Uptake of nitrate at post-flowering	High protein and grain yield	Taulemesse *et al.* (2015)
OsNRT2.1	Rice	N-uptake and mobilization	Expressed during low N condition	Chen *et al.* (2016)
TGA1/TGA4 and NTR1.1	*Arabidopsis thaliana*	Root hairs N remobilization	Nitrate increases the density of root hairs which in turn increases the capacity of nitrate uptake	Canales *et al.* (2017)
TaVRN1	Wheat	Vernalization 1	Improve NUE	Lei *et al.* (2018)
AlaAT	Mustard rice	Alanine aminotransferase	Increased in biomass and grain yield	Sisharmini *et al.* (2019)
NRT1.7	*Arabidopsis* Tobacco rice	Loading excess nitrate stored and facilitates nitrate allocation to sink leaves	Enhancing source-to-sink nitrate remobilization	Chen *et al.* (2020)
ASN1	Rice	Asparagine synthetase 1	Improve N content in grain	Lee *et al.* (2020)
ARE1	Rice	N-uptake and mobilization	Increased NUE, delayed senescence, increased biomass	Zhang *et al.* (2021)
NRT2	Cassava (*Manihot esculenta* Crantz)	Developing nitrogen efficient crops	Nitrate uptake and utilization	You *et al.* (2022)

Despite an existing wealth of knowledge on the functional roles of macronutrients in plants, recent years have ushered in a plethora of novel insights into the intricate molecular processes governing a plant's immediate responses to specific nutrient limitations. Researchers have successfully unveiled nutrient starvation networks, shedding light on the intricate interplay and communication among individual nutrients throughout the journey from uptake to integration into plant metabolism (Dreyer *et al.*, 2022).

This chapter aims to illuminate the latest findings regarding the physiological and molecular functions of macronutrients in plants. Emphasis is placed on linking these physiological roles to the distinct symptoms that emerge when nutrient availability dips below the necessary threshold for optimal growth. Acknowledging the correlation between observable deficiency symptoms and the functional aspects of the relevant nutrient(s) is of utmost importance in developing new genotypes with enhanced nutrient efficiency. Such genotypes, capable of enduring adverse climate events and demanding soil conditions, play a vital role in fostering the sustainability and resilience of crop production systems.

Moreover, a comprehensive understanding and interpretation of visible deficiency indications opens avenues for the prospective sustainable enhancement of crop production systems. This involves the creation of cutting-edge technologies that endorse automated and data-driven management approaches. Utilizing instruments like bioimaging, remote sensing, in-field spectrometers, and sensors establishes a sturdy groundwork for the timely administration of nutrients in crop cultivation, promoting effectiveness and environmental responsibility.

13.2 MACRONUTRIENTS

13.2.1 Nitrogen (Nitrate N) Sensing, and Signaling

The term "nitrogen" has its origins in the Greek words "nitron" and "genesis," collectively indicating "saltpeter forming." Typically composing approximately 1.5% of a plant's dry weight, nitrogen primarily enters the plant in the form of nitrate (NO_3^-) and/or ammonium (NH_4^+). Nitrogen plays a fundamental role in supplying amino groups for amino acids, which serve as the fundamental building blocks of proteins. It is also a critical component of nucleotides, forming the ring structure of purines and pyrimidines, the foundational units of nucleic acids. Additionally, nitrogen assumes a crucial role in the biochemistry of photosynthetic pigments and secondary metabolites (Zhang *et al.*, 2020a).

The intricate transport mechanisms associated with nitrogen uptake and distribution have been extensively investigated, particularly in the model species *Arabidopsis thaliana* (Liu et *al.*, 2021a). In *Arabidopsis*, nitrogen uptake predominantly occurs as nitrate in the roots, facilitated by both low-affinity transport systems (LATS, Km 1 mM) and high-affinity transport systems (HATS, Km 25 mM). These systems are encoded by the nitrate transporter 1 (NRT1/NPF) and NRT2 gene families, respectively. Notably, NRT1;1, also recognized as NPF6.3 or CHL1, not only serves as a transporter but also acts as a nitrate receptor, earning it the term "transceptor" (Jia and von Wirén, 2020).

Studies have revealed the independence of NRT1;1's transport and signaling functions, demonstrated by the chl1-9 mutant, which loses transport activity but retains signaling capacity (Yin *et al.*, 2021).

The signaling function of NRT1;1 depends on the phosphorylation and dephosphorylation of a particular site, Thr101 (Wang *et al.*, 2021c), influencing various signaling events in terms of scale and time (Figure 13.2).

Thr101 phosphorylation directly affects NRT1;1 transport. In low external nitrate concentrations, phosphorylation disrupts NRT1;1 dimerization, leading to high-affinity nitrate transport. On the flip side, an abundance of external nitrate triggers dephosphorylation, resulting in the formation of a

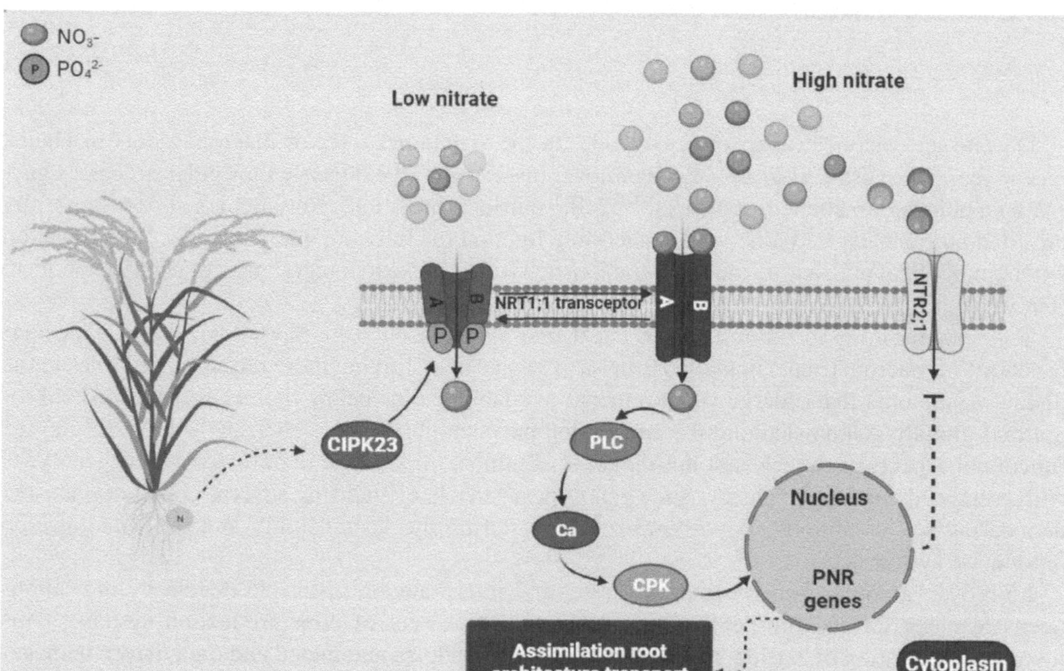

FIGURE 13.2 Nitrate signaling via transceptor-mediated pathways.

dimer that operates as a nitrate transporter with low affinity. Nitrate-induced signaling not only regulates NO_3^- uptake by influencing the affinity of NRT1;1 but also has implications for long-term and subsequent responses, including the primary nitrate response (PNR) (Wang *et al.*, 2021f). The PNR involves the transcriptional control of other NO_3^- transporters and nitrogen-assimilatory genes, depending on calcium (Ca^{2+}) signaling (Rashid *et al.*, 2020) and the stabilization of the dimeric (low-affinity) configuration of NRT1;1. Recent studies propose that crucial Ca^{2+} signals occur through a molecular switch involving NRT1;1 and the Ca-permeable cyclic nucleotide-gated channel 15 (CNGC15) (Wang *et al.*, 2021c). CNGC15 has been demonstrated to interact with NRT1;1 within the membrane, as evidenced by split ubiquitin and BiFC assays. Electrophysiological investigations indicate that the CNGC15:NRT1;1 complex inhibits channel function in a nitrate-dependent manner. This inhibition facilitates complex dissociation under conditions abundant in NO_3^-, allowing CNGC15 to activate and generate a Ca^{2+} signal that initiates downstream processes like the PNR. An illustration of the PNR in action involves the suppression of the high-affinity transporter NRT2;1 under low-affinity NO_3^- conditions. This suppression is triggered by NRT1.1-dependent signaling and involves Ca^{2+}-dependent kinases (CPKs) and the transcription factor NLP7 (NIN-like protein 7) (Wang *et al.*, 2021c).

The third consequence of Thr101 phosphorylation in nitrogen nutrition has implications for root architecture. In addition to its role in nitrate transport, NRT1;1 plays a crucial role in transporting the growth hormone auxin, responsible for lateral root development (Asim *et al.*, 2020). When NO_3^- levels are low, NRT1;1 hinders lateral root development by enhancing basipetal auxin transport, thereby inhibiting lateral root growth (Asim *et al.*, 2020).

Understanding the intricate connection between NO_3^- sensing and signaling through NRT1;1 poses a formidable challenge. Sensing occurs through substrate binding at the external face, while signaling relies on Thr101 phosphorylation on the cytoplasmic side of the protein (Table 13.2).

TABLE 13.2
Revolutionizing Nutrient Sensing and Communication in Crop Plants

Nutrient	Sensor Protein	Role in Nutrient Sensing and Signaling in Crop Plants	Reference
Nitrogen (N)	NRT1/NRT2	Regulates gene expression, root development, and N uptake	Zhang *et al.* (2020)
Phosphorus (P)	PHR1/PHO1	Controls the expression of genes involved in P uptake and usage	Zhou *et al.* (2021)
Potassium (K)	AKT1	Influences the opening and closing of stomata and osmotic balance	Wang *et al.* (2021)
Calcium (Ca)	CAX/CBL/CIPK	Functions as a secondary messenger in various signal pathways	Dong *et al.* (2022)
Magnesium (Mg)	MRS2	Affects photosynthesis and the activation of enzymes in plants	Tao *et al.* (2023)
Sulfur (S)		Regulates the synthesis of sulfur-containing amino acids and coenzymes	Liu *et al.* (2021)
Iron (Fe)	FIT	Impacts gene expression and enzyme activities in iron uptake	Schmidt *et al.* (2020)
Zinc (Zn)	ZIP/IRT	Essential for the activation of various enzymes in nutrient uptake and utilization	Lilay *et al.* (2021)
Copper (Cu)		Involved in various cellular processes and electron transport	Pelegrino *et al.* (2020)
Boron (B)		Essential for cell wall structure and sugar transport in plants	Kohli *et al.* (2023)
Molybdenum (Mo)		Key for nitrogen fixation and nitrate assimilation in legumes	Rana *et al.* (2020)

Structural-functional analyses unveil distinct features in the separate protomers (A and B), forming the NRT1;1 dimer, with varying amino acids responsible for NO_3^- binding. Computational molecular modeling suggests that these differences facilitate allosteric communication between protomer A's NO_3^- binding site and the phosphorylation site, inducing conformational changes (Zhao *et al.*, 2021a). In contrast, protomer B maintains a more rigid structure, impeding this interaction.

High-affinity NO_3^- binding to protomer A induces asymmetric conformational changes, promoting the transition of the transporter from low to high affinity. These alterations impact the transmembrane pathway for NO_3^- transport and expose Thr101 for phosphorylation. The stabilization of the phosphorylated state involves Ca^{2+} signaling, activating the CBL-interacting-protein-kinase 23 (CIPK23) through calcineurin-B-like 9 (CBL9), which is itself activated by Ca^{2+}. In conditions of high external NO_3^- concentrations, the Thr101 residue is sequestered in a hydrophobic pocket, rendering it inaccessible to CIPK23 in either monomer.

The shift between transport modes primarily results from conformational changes that promote Thr101 phosphorylation, and the sustained phosphorylated protein necessitates Ca^{2+} signaling to initiate the CBL9-CIPK23 cascade.

13.2.2 Nitrogen (Ammonium) Sensing and Signaling

The secondary major pathway for nitrogen absorption in plants involves ammonium (NH_4^+). In environments with low oxygen (reductive conditions), such as submerged fields or when nitrate (NO_3^-) availability is restricted, NH_4^+ becomes prevalent. However, when NH_4^+ levels rise significantly, it can swiftly become toxic (Sun *et al.*, 2021). Similar to nitrate, NH_4^+ triggers a broad spectrum of physiological and morphological reactions, including transcriptional adjustments, changes in redox status, shifts in metabolism, and modifications in root structure. In numerous instances, NH_4^+ likely functions as a signaling molecule. The absorption and distribution of NH_4^+ are primarily orchestrated by members of the ammonium transporter (AMT) family (Hao *et al.*, 2020a). While conclusive proof is still pending, it is conceivable that multiple AMT isoforms function as transceptors, akin to the NRT1;1 nitrate transporter discussed earlier. Much like NRT1;1, phosphorylation plays a pivotal role in the allosteric control of AMT transporters (Saini *et al.*, 2021). AMT isoforms, including AMT1;1, form trimeric structures, with each monomer featuring a threonine phosphorylation site (Thr460) in its C-terminal region (Mejias *et al.*, 2021). In conditions of low NH_4^+, the C termini interconnect with the pores of adjacent monomers, facilitating NH_4^+ transport (Hao *et al.*, 2020a).

However, an increase in ambient NH_4^+ concentrations induces the phosphorylation of Thr460 and other conserved residues (Saini *et al.*, 2021), disrupting monomer association and consequently hindering NH_4^+ uptake. Similar effects are observed with mutations at analogous positions, such as Thr460, in other AMT isoforms, suggesting a universal mode of AMT regulation (Wu *et al.*, 2019). Furthermore, when various NH_4^+ assimilation products were examined, none of them influenced Thr460 phosphorylation, indicating that this pathway specifically responds to external NH_4^+ concentrations (Hao *et al.*, 2020b).

An alternative signaling mechanism may involve the putative transceptor AMT1;3, designed to prevent the uptake of toxic NH_4^+ levels (Ding *et al.*, 2021). In conditions of low NH_4^+ supply, AMT1;3 resides in the plasma membrane, facilitating NH_4^+ uptake. However, upon the addition of NH_4^+, the AMT1;3 protein rapidly disappears from the plasma membrane and is internalized via endocytosis (Hao *et al.*, 2020a). While there is intriguing evidence for transceptor functionality in both AMT1;1 and AMT1;3, definitive confirmation of this concept is still required.

13.2.3 Potassium Sensing and Signaling

Potassium (K) emerges as the predominant cation in most plants, functioning as an essential cofactor for numerous enzymes. Cytoplasmic K^+ plays a crucial role in neutralizing negative charges on proteins, while vacuolar K^+ significantly contributes to maintaining turgor pressure and water balance (Wang *et al.*, 2021a). Due to its immobility, bioavailable K^+ levels are typically low in soils, usually in the micromolar range but occasionally reaching millimolar concentrations. Two primary transport systems handle varying external K+ availability. First, inward-rectifying K^+ channels like the *Arabidopsis* K^+ transporter 1 (AKT1), with a millimolar Km, facilitate passive K^+ uptake and primarily function under low-affinity conditions (Sustr *et al.*, 2019). Second, carriers from the high-affinity K transporter (HAK/KUP/KT) family, which are proton-coupled, facilitate active K^+ uptake with a micromolar Km. HAKs become the sole mechanism for K^+ uptake when external K^+ concentrations drop below 10 mM (Sardans and Peñuelas, 2021). Similar to nitrogen, K^+ likely plays a crucial role in maintaining nutritional homeostasis, leading to short-term and long-term adaptations and regulations in response to environmental K^+ fluctuations.

While elevated K^+ levels don't trigger rapid root proliferation as observed with NO_3^- and NH_4^+, low K^+ conditions inhibit primary root growth and first-order lateral roots while promoting higher-order branching (Sustr *et al.*, 2019). The regulation of transport systems in response to low K^+ conditions is well-documented at both the transcriptional and post-transcriptional levels. For instance, low K^+ levels induce the rapid transcription of high-affinity transport systems like HAK5 in *Arabidopsis thaliana* within an hour (Feng *et al.*, 2021). Similar to NRT1;1-mediated NO_3^- uptake, the collaboration of several CBL isoforms activates CIPK23, which phosphorylates HAK5, increasing its affinity for K^+ and enhancing its Vmax (Ma *et al.*, 2020).

In contrast to HAK5, the transcript level of AKT1 remains relatively unaffected by ambient K^+ conditions, but the activity of this channel is upregulated post-translationally in response to low K^+ levels. In *Arabidopsis*, this upregulation involves the interaction of CBL1 and CBL9 with CIPK23. CIPK23-mediated phosphorylation of AKT1 increases K^+ uptake through this channel (Ragel *et al.*, 2019). However, in low K^+ conditions, AKT1 can function as a K^+ leak. A second mechanism down-regulates AKT1 activity and counteracts its CIPK23-based upregulation. This involves K^+ channel 1 (KC1), a "silent" channel subunit transported from the endoplasmic reticulum to the plasma membrane during low K^+ conditions. This leads to the formation of non-functional channels through association with AKT1 subunits, preventing cellular K^+ loss when a steep outward K^+ gradient prevails (Feng *et al.*, 2021; Johnson *et al.*, 2022).

K^+ is implicated in regulatory and signaling roles, with intriguing findings suggesting an intricate mechanism of K^+ sensing. Research utilizing a FRET-based K^+ reporter (GEPII) has revealed that lowering external K^+ from 5 to 0.1 mM results in a rapid reduction of cellular K^+ levels, primarily localized to the cell division zone of the root tip. This localized, cell-specific K^+ signal aligns closely with the dynamics of a low-K^+-induced Ca^{2+} signal, occurring within seconds after reducing external K^+ from millimolar to micromolar levels. The Ca^{2+} signal, essential for the generation of the K^+ signal, suggests that plants sense low K^+ conditions through an initial Ca^{2+} signal activating K^+ channels, leading to K^+ efflux from specific root tip cells. The subsequent drop in internal K^+ levels likely drives signaling events, with CBL isoforms and CIPK23 fine-tuning the activity of HAK and AKT in response to low K^+ conditions (Pantha *et al.*, 2023).

While parallels with nitrogen processes, such as Ca^{2+} and CIPK phosphorylation, exist, there are fundamental differences. Unlike nitrate signaling, K^+ sensing suggests that internal sensing is crucial for subsequent signaling events. This internal sensing, reporting on the intracellular status of nutrients, serves to optimize nutrient homeostasis, including functions like maintaining cytoplasmic K^+ levels and partitioning K^+ into vacuoles (Sardans and Peñuelas, 2021), ultimately influencing nutrient use efficiency. Nevertheless, additional, as-yet-unknown mechanisms for sensing external K^+ levels are likely, potentially involving membrane lipids as primary sensors, as described below for sodium (Na^+).

13.2.4 Phosphorous Sensing and Signaling

Phosphorus (P) is involved in a wide array of physiological functions, serving as a vital component of phospholipids critical for maintaining cell membrane integrity. It is also indispensable in the structure and functioning of nucleic acids by forming the backbone. Additionally, in the form of inorganic phosphate (Pi), P plays an essential role in bioenergetics and serves as a primary mechanism for regulating proteins through phosphorylation and dephosphorylation processes (Zhou *et al.*, 2021). Plants primarily acquire phosphorus (P) in the form of inorganic phosphate (Pi, PO_4^{3-}). The prevalence of phosphate deficiency in various environments often restricts plant development and harvest productivity, emphasizing the need for efficient P utilization, especially considering the finite nature of P fertilizer resources. When confronted with diminished P levels, plants undergo adjustments in their root system architecture to enhance Pi absorption from layers with more abundant Pi concentrations (Hu *et al.*, 2019). This modification commonly entails suppressing primary root elongation in favor of lateral root development, heightened production and extension of root hairs, and the establishment of mycorrhizal symbioses. Certain adaptations are responsive to alterations in external Pi levels (Li *et al.*, 2021), underscoring the importance of primary Pi sensors in proximity to the rhizosphere, while others rely on the cellular phosphorus status.

Exploration of *Arabidopsis thaliana* roots has revealed that the localized detection of external phosphate concentration (Pi) primarily takes place in the root tips, although the precise mechanisms remain incompletely understood (Wang *et al.*, 2021a). Initial elements identified in phosphate sensing include the low phosphate root (LPR) and phosphate deficiency response (PDR) genes (Jinger *et al.*, 2022). The interaction between LPRs and PDRs is governed by the transcription factor STOP1 (sensitive to proton rhizotoxicity 1; Balzergue *et al.*, 2017), prompting malate efflux through the aluminum activated malate transporter 1 (ALMT1). This process leads to the production of reactive oxygen species (ROS) and subsequent callose formation (Prathap *et al.*, 2022). The callose disrupts plasmodesma trafficking of transcription factors, hindering primary root growth (Ham *et al.*, 2018).

While the signaling mechanism has undergone detailed examination, the initial detection of external Pi and how this information links rhizosphere Pi levels to STOP1 activation remain unclear. Phosphate homeostasis involves the high-affinity Pi uptake carrier PHT1, closely resembling the yeast PHO84 transceptor (Wang *et al.*, 2021b). Distinguishing a potential signaling role from its transport function and other pleiotropic effects on phosphate homeostasis has posed challenges (Ham *et al.*, 2018). In contrast, there is more clarity regarding potential primary sensors monitoring internal phosphate status (Wang *et al.*, 2021b), particularly those reporting on inositol polyphosphates (InsPs) like InsP7.

Evidence suggests that SPX (SYG1/PHO81/XPR1) proteins can bind InsPs and serve as primary sensors in various organisms. Involved in phosphate transport and regulation, InsP7 binding to the SPX protein PHO1 catalyzes cellular Pi release into the pericycle during Pi-replete conditions. Binding of InsP7 to other SPX proteins, such as SPX4 in rice, initiates interaction with phosphate starvation response (PHR) transcription factors. This interaction hinders their migration to the nucleus, inhibiting the expression of P-starvation response genes (Wang *et al.*, 2021a). The regulation of vacuolar phosphate may involve more upstream SPX proteins. In rice, OsSPX-MFS1, targeted by miR827 and upregulated by the PHR2 transcription factor, might be initially sensed through PHR2 interaction with a cytoplasmic SPX protein (Wang *et al.*, 2021f). Recent investigations suggest that SPX proteins may control Pi influx through mycorrhizal fungi, with SPX1 and SPX3 in *Medicago truncatula* playing a role in promoting fungal branching and increased root colonization in arbuscule-containing cells of the root (Wang *et al.*, 2021b).

13.2.5 Magnesium Sensing and Signaling

Magnesium (Mg) holds paramount significance in plant physiology, primarily attributed to its essential role in chlorophyll, the green pigment critical for photosynthesis. Beyond its involvement in chlorophyll, Mg serves as a crucial activator for numerous enzymes and plays a pivotal role in neutralizing negative charges, particularly within nucleotides. However, maintaining an optimal Mg level in the soil is crucial, as excessive Mg can prove detrimental, as highlighted by Kleczkowski and Igamberdiev (2021). This study underscores the importance of sequestering Mg into vacuoles as a protective measure against potential toxicity. The mechanism behind this process is potentially mediated by a CBL-CIPK network activated through an Mg-induced calcium signal. Despite these advancements, the primary Mg sensor's identity remains elusive.

Mg plays an essential role in diverse aspects of plant growth and development, including vital functions such as photosynthesis, phloem loading, leaf senescence, stomata opening, and enzyme activation, as underscored by Tian *et al.* (2021). Furthermore, Mg is a significant contributor to sugar storage. Insufficient levels of Mg can lead to detrimental effects in plants, such as weakened stalks, the appearance of yellow and brown spots on leaves, and an overall loss of greenness, as observed by de Bang *et al.* (2021).

13.2.6 Calcium Sensing and Signaling

Calcium (Ca) plays numerous vital physiological roles in plants, acting as a counter-cation for both organic and inorganic anions. It activates enzymes that regulate plant growth, contributes to nitrate assimilation, and is crucial in processes like crosslinking pectin in cell walls and the formation of cell plates during cell division, as outlined by de Bang *et al.* (2021). In addition to these functions, Ca functions as a secondary messenger, playing a central role in signaling during both abiotic and biotic stress, as indicated by the insights shared by Lee and Seo (2021). Its impact on plant physiology extends to root-microbe interactions, especially in the formation of root nodules in legumes and their colonization by nitrogen-fixing bacteria, where Ca^{2+} signaling is essential, as highlighted by Dong *et al.* (2022).

The majority of Ca^{2+} is situated in the apoplasm, typically ranging from 0.3 to 1 mM and, in certain cases, reaching up to 10 mM, as documented by Luan and Wang (2021). Within plant cells, the vacuole and endoplasmic reticulum (ER) function as the primary Ca reservoirs, with concentrations ranging from 200 μM to 50 mM and 50 to 500 μM, respectively, as observed by Choudhury *et al.* (2022). The concentration of free Ca^{2+} ($[Ca^{2+}]$ cyt) in the cytosol is actively maintained at approximately 100 nM, with calcium-binding proteins (Ca-BPs) buffering free Ca^{2+} and actively transporting it to the apoplasm, vacuole, and ER, as highlighted by Plattner and Verkhratsky (2015). Various mechanisms mediate the influx of Ca^{2+} into the cytosol, leading to transient and often oscillatory increases in $[Ca^{2+}]$ cyt, which have been extensively studied for their roles in signaling, as discussed by Costa and Kudla (2015).

Taking a nutrient perspective, calcium deficiencies are less common compared to deficiencies in other nutrients, as indicated by Lee and Seo (2021). However, alterations in calcium availability in the soil can significantly impact plant performance. The intracellular concentration of $[Ca^{2+}]$ can be precisely measured using various Ca^{2+}-sensitive dyes and genetically encoded calcium sensors. Significant progress has been achieved, exemplified by the introduction of a novel series of calcium sensors named NEMO, derived from mNeonGreen. These sensors, as highlighted by Li *et al.* (2022), demonstrate swift kinetics and an exceptionally broad dynamic range. The quantification of intracellular $[Ca^{2+}]$ can be accomplished using ratiometric calcium indicators, although *in vivo* calibration with calcium ionophores or chelators is frequently necessary, as indicated by Sadoine *et al.* (2021). Explorations employing calcium sensors targeted to calcium channels have revealed discrete cytosolic regions termed calcium nanodomains, illustrating the potential of sensors in advancing our comprehension of cellular dynamics and organization. Thus, a well-established mechanism exists for calcium as a secondary signaling molecule, involving numerous proteins that sense cytoplasmic calcium concentrations.

13.2.7 Sulfur Sensing and Signaling

Sulfur (S) stands as an essential element crucial for the vitality of all living organisms, showcasing its remarkable versatility in the plant kingdom. S plays a diverse array of roles in numerous plant processes, contributing to reductive iron assimilation, photosynthesis, and the maintenance of oxidation-reduction homeostasis. It serves as a fundamental building block for crucial amino acids such as methionine and cysteine, as well as essential metabolites like glutathione (GSH), vitamins (e.g., biotin and thiamine), chlorophyll, and coenzyme A. Additionally, sulfur assumes pivotal functions in plant responses to both biotic and abiotic stresses. Notably, GSH plays a crucial role in mitigating the impact of reactive oxygen species during oxidative stress (Li *et al.*, 2020), while glucosinolates, secondary metabolites containing sulfur, significantly contribute to defense mechanisms (Zhang *et al.*, 2020a).

Reactive sulfur species (RSS) encompass molecules that exert significant influence on physiological processes through protein sulfhydration (Liu *et al.*, 2021a). Recently, hydrogen sulfide (H_2S), a gaseous molecule, has emerged as a participant in various physiological activities and distinct signaling pathways (Liu *et al.*, 2021a). Within plants, H_2S actively participates in processes such as seed germination, growth, development, and the enhancement of tolerance to abiotic stressors like salinity, drought, and extreme temperatures. Plants absorb sulfur through two distinct routes: root uptake via sulfur transporters in root epidermal cells and absorption through stomatal openings, facilitating sulfur entry through gas exchange. Subsequently, sulfur is transported into the xylem and distributed throughout the plant, with the option to be stored in vacuoles or transported to chloroplasts (Ristova and Kopriva, 2022). Chloroplasts of young leaves predominantly facilitate sulfur assimilation into amino acids, although assimilation and cysteine production can also occur in seeds and roots (Kopriva *et al.*, 2019).

Measuring intracellular concentrations of sulfur species *in vivo* poses significant challenges, especially when present at low concentrations, highly reactive, and characterized by short lifespans, rendering conventional chemical tools insufficient. For RSS, fluorescence bioimaging provides real-time detection of sulfur species within the cell (Künstler *et al.*, 2020). A recent breakthrough is the innovative sulfane sulfur sensor NIR fluorescent probe, SSNIP, demonstrating remarkable accuracy and reliability in monitoring sulfane sulfur in *Arabidopsis* roots at various developmental stages (Jiang *et al.*, 2019). Additionally, a ratiometric sensor for H_2S (hsFRET) has been developed and tested for monitoring H_2S in living cells (Youssef *et al.*, 2019), although it is important to note that hsFRET requires the exogenous addition of p-azidophenylalanine (pAzF), limiting its application. In summary, S as a versatile and essential nutrient for plants, contributes significantly to various physiological processes, and recent advancements in S species detection offer valuable insights into its dynamic roles within plant cells.

13.2.8 Sodium Sensing and Signaling

Sodium (Na) holds limited essentiality for most plants, with the notable exception of a subgroup of C_4 plants. In specific C_4 plants, sodium ions (Na^+) function as a driving force for pyruvate uptake into chloroplasts, facilitated by a Na^+-pyruvate cotransporter (Keisham *et al.*, 2018). While not universally essential for most plants, sodium has demonstrated a beneficial and "nutritious" role in various plant species, especially when potassium (K^+) is scarce (Hartley *et al.*, 2020). However, elevated external sodium levels can induce salinity stress, negatively impacting agricultural production in multiple regions globally (Wu *et al.*, 2018). The presence of sodium in the soil has dual effects: it diminishes soil water potential, causing osmotic stress, and can accumulate within plant cells, potentially leading to ionic toxicity. The alleviation of osmotic stress often involves the uptake of "cost-effective" osmotic agents in the form of Na^+ and Cl^- ions, crucial to prevent ion toxicity and requiring meticulous regulation. While animal sodium sensing is well-understood and primarily linked to Na^+ channel activity, plant Na^+ sensing follows a distinct mechanism (Hindt *et al.*, 2017). The evidence supporting this mechanism is derived from a mutant screen conducted in *Arabidopsis,* revealing that elevated external NaCl concentrations prompt a swift Ca^{2+} signal. The "monocation-induced-[Ca^{2+}] i-increases-1" (MOCA1) mutant was identified as a key player in salt-induced Ca^{2+} signals. Surprisingly, MOCA1 was recognized as an enzyme responsible for transferring glucuronosyl sugar to glycosyl-inositol-phosphorylceramide (GIPC) sphingolipids in the plasma membrane

(Tay *et al.*, 2012). This glycosylation establishes negative sites to which monovalent cations can bind. The electrostatic binding of cations to these sites diminishes the surface potential, subsequently influencing the transmembrane potential and affecting the activity of voltage-dependent ion channels. This cascade promotes the opening of voltage-dependent Ca^{2+} permeable channels, establishing a connection between the binding of monovalent cations to the plasma membrane's outer leaflet and the gating of voltage-dependent Ca^{2+} channels, resulting in a Ca^{2+} signal and subsequent signaling.

This plant signaling mechanism is distinct, intertwining the electrostatic binding of any monovalent cation to lipids with the gating of voltage-dependent Ca^{2+} permeable channels, differing from the effects observed in prior electrophysiological studies. Although lacking specificity and initiated by the electrostatic binding of any monovalent cation, this mechanism responds to fluctuations in external K^+ concentrations, eliciting a salinity stress response. Moreover, plants may harbor proteins with Na^+ binding sites that contribute to internal sensing and/or signaling, regulating cytoplasmic Na^+ levels, preventing excessive accumulation, and facilitating proper deposition in the vacuole. These proteins may also assist in transporting surplus Na^+ to the xylem for long-distance transport to the shoot. The identification of such proteins in the *Arabidopsis* proteome poses challenges due to significant motif redundancy and a high number of candidate proteins spanning transporters, transcription factors, and kinases (Naz *et al.*, 2023).

13.3 MICRONUTRIENTS

13.3.1 Zinc Sensing and Signaling

Distinguishing itself from iron and copper, zinc (Zn) displays redox-neutrality and acts as a reactive Lewis acid in biological reactions (Zeng *et al.*, 2021). Possessing distinctive characteristics, Zn plays essential roles as a structural, catalytic, and regulatory component in a variety of enzymes (Lilay *et al.*, 2021). Zn's engagement in structural and regulatory functions is evident in its integral role in numerous transcription factors (TFs) featuring zinc finger motifs, crucial for TF-DNA binding (Eide, 2020) and mediating diverse protein-protein interactions (Han *et al.*, 2020). Notably, Zn is the exclusive metal present in all six enzyme classes, namely oxidoreductases, transferases, hydrolases, lyases, isomerases, and ligases (Zlobin, 2021). The regulation of Zn uptake, efflux, and compartmentation is crucial. However, it's not the total zinc content within the cell, typically ranging 0.3–3 mM in plants, but the free cytosolic [Zn^{2+}] and the labile zinc pool (bound to molecules like nicotianamine [NA]), histidine, glutathione, phytochelatins, or phosphate) that significantly influence zinc homeostasis. In most cases, cytoplasmic zinc sensors, typically TFs with zinc-specific binding sites, initiate regulatory processes like transport regulation.

In prokaryotes, TFs, such as ZntR in *E. coli* responding to excess zinc or Zur in *Bacillus subtilis* detecting zinc deficiency, serve as examples of these sensors (Mikhaylina *et al.*, 2018). Recently characterized, plant cellular Zn-sensing mechanisms parallel these known systems. F-group basic leucine-zipper (F-bZIP) TFs oversee the local zinc levels (Zlobin, 2021). When activated, TF dimers form and attach to the promoter regions of pertinent genes. For instance, during zinc deficiency, bZIPs bind to zinc deficiency response elements (ZDREs) situated in gene promoters associated with zinc uptake and distribution. These genes include ZIP/IRT-like proteins like ZIP4 and ZIP9, as well as nicotianamine synthases (NAS2, NAS4) (Zlobin, 2021).

These TFs are integral components of a feedback mechanism, ensuring that sufficient zinc levels lead to the downregulation of F-bZIP activity. In *Arabidopsis*, bZIP19 and bZIP23 play crucial roles in regulating genes responsible for zinc homeostasis (Kalia *et al.*, 2020), and their homologues have been identified in rice, wheat, and barley. Functionally, while bZIP19 and bZIP23 show partial redundancy in single knock-out mutants, double mutants (bzip19/23) display severe growth inhibition and chlorosis in response to zinc deficiency. Characterized by a highly conserved Cys/His-rich motif associated with zinc binding, located in the N-terminal region before the bZIP domain, these zinc-responsive bZIPs function as zinc sensor motifs (ZSMs) (Lilay *et al.*, 2021). *In vitro* binding assays have confirmed direct zinc binding to *Arabidopsis* bZIP19 and bZIP23, with each TF capable of binding two zinc atoms. This was validated through size exclusion chromatography and stable isotope (65Zn) analysis using inductively coupled plasma mass spectrometry (Lilay *et al.*, 2021).

In a living organism, these transcription factors probably interact with free Zn^{2+} in the nucleus, rendering them inactive when zinc levels are satisfactory. This anticipated deactivation is likely facilitated by structural alterations that hinder the binding of transcription factors to DNA (Lilay *et al.*, 2021). Examining crystal structures could offer additional understanding of how these structural modifications influence the functionality of transcription factors. It's important to highlight that alternative zinc-sensing mechanisms, independent of bZIP, might exist, given the observed variations in the promoter regions of ZIP genes. These differences imply potential variations in the number, arrangement, and nucleotide sequence of ZDREs elements. Further investigation is required to determine if these differences directly correlate with ZIP expression patterns under Zn deficiency. Mechanisms limiting excess Zn are less comprehensively understood. Metal tolerance proteins (MTPs) belonging to the cation diffusion facilitator (CDF) family subset predominantly facilitate the extraction of zinc from the cytosol. Tonoplast-localized MTPs, such as AtMTP1 in *A. thaliana*, transport Zn^{2+} into the vacuole, and their activity is likely regulated by cytosolic Zn levels. Another vital facet of zinc homeostasis entails the export of zinc into the apoplast, dependent on the functioning of heavy metal ATPases (HMAs). An exemplar is HMA4 in *A. thaliana*, pivotal for the transportation of zinc from roots to shoots. While these processes remove excess Zn and facilitate Zn translocation, the identity of Zn sensors in these pathways remains unknown. As an illustration, HMA4 is equipped with a C-terminal domain housing cysteine and histidine residues, potentially engaged in the binding and perception of zinc and cadmium ions. Molecular details of these mechanisms and the role of conformational changes in regulating transporter activity necessitate further exploration (Han *et al.*, 2022).

13.3.2 Copper Sensing and Signaling

Copper (Cu) plays a vital role in the redox processes of mitochondria and photosynthesis, serving as a crucial cofactor for enzymes like superoxide dismutases. However, an excess of Cu, similar to iron, can become highly toxic due to the ability of free Cu^{+} ions to catalyze the production of harmful free radicals (Pelegrino *et al.*, 2020). Plants regulate the uptake and distribution of Cu^{2+} through transcription factors belonging to the squamosa promoter binding protein (SBP) family, a unique feature in the plant kingdom. These SBP proteins encompass an SBP domain, consisting of 76 amino acids, along with a nuclear localization signal (NLS) and two atypical zinc finger structures formed by eight conserved Cys and His residues (a C-C-CH motif for the first and C-C-H-C motif for the second zinc finger) (Kumar *et al.*, 2021). SBPs bind to DNA at GTAC promoter sequences present in copper response elements (CuREs) of target genes. The integrity of the GTAC motif is crucial for proper TF binding, and additional sequences beyond the core GTAC are necessary to initiate transcription (Shabbir *et al.*, 2020).

In the genome of the green alga *Chlamydomonas reinhardtii*, 24 SBPs are encoded, including the copper response regulator (CRR1), a pivotal Cu-dependent SBP-TF that activates over 60 genes during Cu deficiency (Rehman *et al.*, 2019). Target genes include Cu^{+} transporters from the CTR family and genes encoding cytochrome c6, a photosynthetic electron carrier that substitutes plastocyanin in Cu deficiency. CRR1 features a Cys-rich C-terminal region similar to the Cu^{+}-sensing domain in the metal-responsive factor-1 of *Drosophila melanogaster* (Giannakoula *et al.*, 2021). Binding studies have revealed that, in addition to the zinc finger regions, CRR1 possesses two highly conserved Cys/His residues crucial for Cu binding. Cu^{+} is favored over Cu^{2+} as the liganded metal, and *in vivo*, Cu^{+} is more likely to be the bound metal due to the protein's higher affinity for Cu^{+}, supported by the greater selectivity for Cu^{+} and reversible high-affinity Cu^{+} binding. This suggests that SBP proteins, such as CRR1, function as primary Cu sensors. When Cu^{+} is bound to

CRR1, it does not interact with DNA; only when Cu^+ is depleted do these TFs bind to the promoter regions of Cu-regulated genes. The mechanism of how changes in Cu status translate into transcriptional adjustments remains to be fully understood, potentially related to nuclear CRR1 localization responding to varying Cu levels or Cu binding causing conformational changes affecting the CRR-DNA interaction.

In higher plants, numerous SBPs, known as squamosa promoter binding protein-like (SPLs), are present (Zhou *et al.*, 2018). Despite minimal amino acid sequence homology, these SPLs share diversity but have the SBP domain in common. *Arabidopsis thaliana*, for instance, possesses approximately 17 SPLs, with SPL7 being the closest ortholog to CRR1. SPL7 promotes high-affinity Cu uptake and optimizes Cu (re)distribution to essential Cu proteins. This regulation is mediated by specific miRNAs that target mRNA encoding non-essential Cu-binding proteins. Although implicated in Cu homeostasis, SPL7 is unlikely to function as the primary Cu sensor, as its transcript levels generally do not respond to Cu availability. Furthermore, in the presence of Cu, SPL7 does not bind to DNA *in vitro*. Notably, SPL7's cellular location in higher plants, associated with the ER via a C-terminal transmembrane domain, differs from the algal CRR1. This suggests that Cu sensing and regulation of SBP-TFs in higher plants differ from the algal system, and an unknown sensor likely operates upstream of SPLs. Alternatively, SPLs may not respond to free ionic Cu, given its exceedingly low concentration, and may instead interact with Cu complexed by organic factors, such as metallothioneins.

13.3.3 Chloride Sensing and Signaling

Chloride (Cl) primarily exists as a highly mobile free anion, demonstrating shoot concentrations within the range of 28 μM to 1.68 mM (Colmenero-Flores *et al.*, 2019). Elevated levels of Cl can lead to adverse effects, such as chlorosis, necrotic lesions, and leaf-tip burning (Geilfus, 2018a). The increasing salinity of soil poses a significant global concern, attributed to irrigation practices and climate change, thereby threatening crop production (van Zelm *et al.*, 2020).

In barley, the vacuolar Cl^- concentration spans 50 to 150 mM, with phloem potentially containing up to 120 mM Cl^- (Colmenero-Flores *et al.*, 2019). Typically, the determination of chloride content involves the use of chloride analyzer devices capable of measuring the total Cl^- content in samples. Alternatively, methods such as inductively coupled plasma mass spectrometry (ICP-MS) (Franco-Navarro *et al.*, 2019) and Cl^- selective microelectrodes are employed. Fluorescent dyes like 6-methoxy-N-(3-sulfopropyl) quinolinium (SPQ), N-(ethoxycarbonylmethyl)-6-methoxyquinolinium bromide (MQAE), and 6-methoxy-N-ethylquinolinium iodide (MEQ) are also available (Geilfus, 2018a). These dyes, characterized as non-ratiometric, exhibit relative insensitivity to changes in bicarbonate concentration and pH while maintaining fast kinetics (Geilfus, 2018b).

Various chloride sensors have been developed, capitalizing on the inherent sensitivity of fluorescent proteins to halides. However, a primary drawback of existing Cl sensors is their susceptibility to alterations in pH (Waadt *et al.*, 2020) and their relatively low affinity for Cl (Lodovichi *et al.*, 2022). The simplest version involves yellow fluorescent protein YFP, where halides directly bind to YFP variants, thereby modulating their fluorescence emission (Franco-Navarro *et al.*, 2019). A sophisticated, ratiometric FRET iteration, identified as Clomeleon, incorporates a CFP that is unaffected by Cl^- and a YFP variant that is responsive to Cl^-, facilitating enhanced accuracy in measurements.

13.3.4 Boron Sensing and Signaling

Boron (B) plays pivotal roles in upholding membrane stability, facilitating sugar transport, contributing to cell wall formation, supporting amino acid production, and regulating the flowering process (Nejad and Etesami, 2020). However, elevated levels of B can become toxic, presenting a global challenge for food production, particularly in arid regions. Conversely, insufficient B in soils is a prevalent issue, resulting in symptoms like wilted growing points, stunted growth, leaf deformation, and inadequate flowering (Singh *et al.*, 2020).

Boron, being an uncharged, small molecule, can passively diffuse across membranes. Nevertheless, when availability is limited, plants employ transporters to ensure a sufficient supply of B (Nejad and Etesami, 2020). The mobility of B in phloem varies among plant species, with B typically binding to sugar alcohols for redistribution from older to younger leaves in most cases (Kohli *et al.*, 2023).

Several techniques, including ICP-MS and laser ablation–ICP-MS, have been employed to gauge B content in *Arabidopsis* roots (Day and Aasim, 2020). Continuous monitoring of cytosolic B concentration in different plant tissues post-B treatment has been carried out to unravel the spatial distribution and temporal dynamics of B levels. This monitoring utilizes sensors based on the uNIP5;1-Venus/Luc system (Kohli *et al.*, 2023).

13.3.5 Manganese Sensing and Signaling

Manganese (Mn) plays a crucial role in a variety of biological processes, impacting respiration, nitrogen assimilation, germination, fruit ripening, and the plant's defense against pathogens (Tao *et al.*, 2023). It serves as an indispensable component of the metalloenzyme cluster within the oxygen-evolving complex (OEC) of photosystem II (PSII), making it vital for photosynthesis. In plants, Mn is predominantly absorbed in the form of Mn^{2+}, and cellular levels typically range from about 20 to 40 ppm (mg kg^{-1}) in dry weight (Alejandro *et al.*, 2020). Consequently, the intricate regulation of Mn homeostasis involves control over its uptake, distribution, allocation, and storage processes (Alejandro *et al.*, 2020).

Recent advancements have led to the identification and characterization of Mn transporters across a broader spectrum of plant species (Alejandro *et al.*, 2020). Roots acquire Mn^{2+} from the soil through various transporters, including NRAMP, ZRT/IRT, and YSL (Alejandro *et al.*, 2020). The primary route for Mn translocation and distribution directs it toward the xylem, subsequently moving to the phloem with relatively low mobility, and ultimately reaching various plant tissues (Li *et al.*, 2017). There is also a suggestion that Mn^{2+} may potentially be taken up by leaf cells in rice (Schmidt and Husted, 2019). Techniques such as atomic emission spectroscopy or ICP-OES have been employed for analyzing Mn levels in plants (Alejandro *et al.*, 2020).

While there are available sensors for Mn detection, such as the fluorescent Mn sensors M1 and iLovU, based on the principles of photo-induced electron transfer (Li *et al.*, 2019), it's noteworthy that these sensors have not yet been utilized in plants.

13.3.6 Molybdenum Sensing and Signaling

Molybdenum (Mo), a transition element with oxidation states ranging from +2 to +6, is an indispensable component for all living organisms (Rana *et al.*, 2020). Functioning as a crucial catalytic center in molybdoenzymes, Mo facilitates the oxidation and reduction of various carbon, nitrogen, and sulfur metabolites (Padhi and Mishra, 2019). It plays a fundamental role in the synthesis and activity of a key enzyme involved in nitrogen fixation, a process essential for symbiotic nitrogen fixation by rhizobacteria (Rana *et al.*, 2020). To maintain its vital functions while preventing potential toxicity, Mo is maintained within cells at low concentrations (Rana *et al.*, 2020).

During plant uptake, Mo is absorbed as molybdate (MoO_4^{2-}), a weak Lewis acid whose availability depends on soil pH (Chanu *et al.*, 2020). It exhibits high mobility in both the xylem and phloem, primarily transported across plant tissues as MoO_4^{2-} (Padhi and Mishra, 2019). In *Arabidopsis*, the vacuole serves as the primary storage compartment for Mo, with the MOT2 transporter responsible for exporting Mo to the cytosol (Banerjee *et al.*, 2021). The import of Mo is governed by MOT1, a high-affinity molybdate transporter localized at the plasma membrane of root epidermis and phloem cells, with a Km of approximately 20 nM (Bossolani *et al.*, 2022).

Traditionally, determining the total Mo concentration in plant tissues relied on the ICP-OES method following acid digestion (Banerjee *et al.*, 2021). However, this method lacks the ability to differentiate between free and protein-bound elements, offering limited cellular and subcellular resolution. One of the initial Mo sensors, MolyProbe, employed a FRET fluorescence protein pair (CFP and YFP) in conjunction with the bacterial-responsive protein ModE (Oliphant *et al.*, 2022). MolyProbe exhibits responsiveness to Mo in the nanomolar range (Kd = 4.7×10^{-8} M), making it suitable for intracellular measurements. It's important to note that MolyProbe also reacts to sulfate (Kd = 2.7×10^{-2} M) (Rana *et al.*, 2020). Recent advancements have enabled the *in vitro* quantification of Mo in cell extracts from *Arabidopsis*, as well as the fungi *Neurospora crassa* and *Saccharomyces cerevisiae* (Oliphant *et al.*, 2022).

13.3.7 Iron Sensing and Signaling

Iron (Fe) plays a pivotal role in the catalytic activity of numerous enzymes, leveraging its properties as a transition metal to serve as a crucial cofactor in various redox reactions. It serves as a vital element within electron transfer chains in both mitochondria and chloroplasts (Pan *et al.*, 2019). Moreover, Fe is crucial for the production of chlorophyll, a pivotal element in the process of photosynthesis. However, it's noteworthy that free Fe^{2+} (similar to free Cu^{+}) can be highly toxic, given its potential to generate harmful oxygen radicals. While being the second most abundant metal in the earth's crust, Fe mainly exists in the state of highly insoluble oxides. In response, plants have evolved effective mechanisms for extracting Fe from the soil (Connorton and Balk, 2019).

To address this challenge, two main strategies have been devised. In dicotyledonous and non-gramineous monocots, Strategy I involves the acidification of the rhizosphere to solubilize iron (Fe), facilitated by H+-releasing ATPases. Additionally, it encompasses the reduction of Fe^{3+} to the more soluble Fe^{2+} through FRO-type Fe (III)-chelate reductases. Subsequently, Fe^{2+} is transported across the plasma membrane using transporters from the iron-regulated transporter 1 (IRT1) family, as depicted in Figure 13.3.

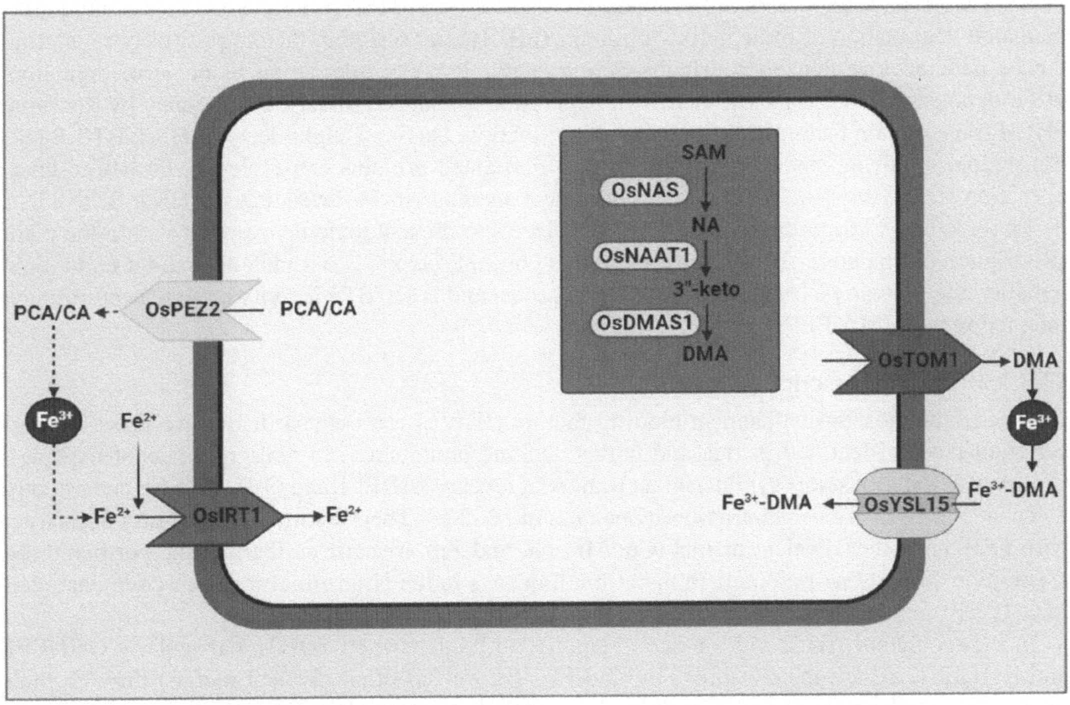

FIGURE 13.3 Iron uptake mechanisms in rice plants.

In contrast, gramineous monocots adopt Strategy II, relying on the release of high-affinity Fe chelators, specifically phytosiderophores like mugineic and deoxymugineic acids (Connorton and Balk, 2019). These phytosiderophores bind to and solubilize Fe^{3+}, and the resulting Fe-chelator complex is transported into the root symplast through yellow stripe (YS) or YS-like (YSL) transporters (Schmidt *et al.*, 2020). The intricate regulation of Fe homeostasis involves complex signaling pathways dependent on the contributions of various classes of proteins.

13.3.7.1 HRZ-BTS-BTSLs Proteins

Proteins containing the haemerythrin motif, known as HRZ/BRUTUS-like (BTSL) within the "really interesting new gene" (RING) and zinc-finger family, play a central role in the regulation of iron (Fe) signaling and are suggested as primary Fe sensors. Although HRZ-BTS proteins are found in all green plants, BTSLs are exclusive to dicotyledonous species such as *A. thaliana* (Rodríguez-Celma *et al.*, 2019). These nuclear proteins (Podar and Maathuis, 2022) consist of an N-terminal haemerythrin (Hr) domain and a C-terminal RING-type E3 ubiquitin ligase (Rodríguez-Celma *et al.*, 2019). Similar to mammals, the Hr domain within HRZ-BTS is proposed to function as a cellular Fe sensor (Rodríguez-Celma *et al.*, 2019). In mammalian cells, Hr domains fold into a bundle of four α-helices containing two Fe atoms, each coordinated by histidine and acidic residues, exhibiting conformational changes based on Fe concentration. In Fe-sufficient conditions, the Hr domain remains stable and resistant to proteolysis, allowing its accumulation. However, in Fe-deficient conditions, the Hr domain's di-iron center integrity is compromised, leading to HRZ ubiquitination and degradation by E3 ligases (Ruiz and Bruick, 2014).

Unlike the behavior of the mammalian Hr Fe sensor, Fe binding to the N-terminal Hr domains of BTS in plant cells induces protein destabilization. The Fe sensing by HRZ-BTS-BTSL and associated regulatory mechanisms may be more intricate in plant cells due to multiple Hr domains and C-terminal E3 ligases catalyzing self-ubiquitination (Selote *et al.*, 2018).

HRZ-BTSs and BTSLs differ in the number of Hr motifs and their distribution, with HRZ-BTSs involved in Fe mobilization, while BTSLs play a role in Fe uptake (Rodríguez-Celma *et al.*, 2019). Serving as negative regulators of Fe homeostasis, HRZ-BTS-BTSL proteins promote the ubiquitin-mediated degradation of basic helix-loop-helix (bHLH) transcription factors positively regulating the Fe deficiency response (Rodríguez-Celma *et al.*, 2019). While acting as negative regulators of Fe-deficiency-induced genes, their own expression is induced under Fe deficiency by the same bHLH transcription factors that facilitate ubiquitination. Due to a slight delay in HRZ-BTS-BTSL translation, requiring small Fe amounts, other Fe-related proteins can replenish Fe before ubiquitination (*Oliveira et al.*, 2022). The subsequent increase in Fe levels causes HRZ-BTS-BTSL destabilization and degradation, fine-tuning Fe levels to prevent toxic Fe overload within the plant (Rodríguez-Celma *et al.*, 2019). Supporting this concept, *bts* knockout mutants exhibit embryonic lethality due to toxic Fe levels accumulating in seeds, and HRZ-BTS activity is influenced through interaction with IMA-FEPs.

13.3.7.2 IDEF Transcription Factors

Iron deficiency-responsive element-binding factors (IDEFs) represent a distinctive class of transcription factors identified in rice and barley, lacking homologues in non-graminaceous species. Contrary to the iron status's influence on transcript levels, OsIDEF1 and OsIDEF2 are consistently detectable across various plant tissues (Zhang *et al.*, 2020a). Their distinctive capability to interact with Fe2+ and other divalent metals (Cu, Mn, Ni, and Zn) arises from the presence of histidine-asparagine and proline-rich tandem metal-binding sites in the N-terminal region, as demonstrated in OsIDEF1 and HvIDEF1 (Connorton and Balk, 2019).

In the regulatory framework of rice's response to Fe deficiency, IDEFs, particularly OsIDEF1 and HvIDEF1, exhibit the capability to bind Fe^{2+} (as well as other divalent metals) through their distinctive metal-binding sites. These domains, abundant in histidine-asparagine and proline, are situated in the N-terminal section of the proteins (Connorton and Balk, 2019). Remarkably, these

IDEFs play a crucial role by identifying CATGC sequences within the Fe deficiency-responsive cis-acting element IDE1 (Kobayashi *et al.*, 2021). Interestingly, the extent of metal binding by IDEFs directly corresponds to the metal concentration in the surrounding solution (Kobayashi *et al.*, 2021). This suggests a potential function for IDEFs as primary sensors, utilizing metal binding through the His-Asn repeats to convey essential information about the Fe status.

13.3.7.3 IMA-FEP Proteins

Iron man-Fe-uptake inducing peptides (IMA-FEPs), found in all angiosperms, form a distinct group of proteins associated with Fe signaling and sensing (Kobayashi *et al.*, 2021). When Fe levels drop, IMA-FEP levels surge in the phloem, particularly within the companion cells of sieve elements, aiding their systemic distribution across the plant (Hirayama *et al.*, 2018). A crucial aspect of IMA-FEP function involves its interaction with HRZ-BTSs. With a conserved BTS interaction domain (BID) in their C-terminal region, IMA-FEPs bind to the C-terminal of BTS. This interaction blocks BTSs from binding to bHLH TFs, which are essential for initiating the plant's response to Fe deficiency (Li *et al.*, 2022). Intriguingly, the relevant bHLH TFs also possess BIDs, allowing IMA-FEPs to stabilize bHLH proteins by competitively interacting with BTS (Li *et al.*, 2021b).

IMA-FEPs demonstrate a direct binding affinity for Fe, Mn, Cu, and Zn, facilitated by a conserved C-terminal aspartate-rich 17 amino acid sequence (Kobayashi *et al.*, 2021). Furthermore, under Fe-rich conditions, IMA-FEPs become destabilized. In laboratory investigations, saturation of the binding sites leads to the precipitation of IMA-FEPs, suggesting their potential role as (phloem) sensors. However, additional experiments are essential to validate this intriguing hypothesis.

13.3.7.4 IRT1 Transporters

IRT1 plays a central role in Fe uptake, serving as the primary transporter for Fe^{2+} located at the plasma membrane of root rhizodermal cells (Dubeaux *et al.*, 2018). While its primary function is the transport of Fe, IRT1 also facilitates the movement of other metals such as Mn, Zn, Co, and Cd across the membrane (Connorton and Balk, 2019). The regulatory mechanisms governing IRT1's activity operate at both the transcriptional and posttranscriptional levels, contingent upon the concentrations of Fe and other metals.

The intricate transcriptional regulation of IRT1 involves a complex network of bHLH transcription factors, specifically from subgroups Ib and IVb. In Fe-deficient conditions, upregulation of IRT1 transcription occurs through the binding of heterodimers, composed of bHLH transcription factors and Fe-deficiency-induced transcription factors (FITs), to the IRT1 promoter (Connorton and Balk, 2019). Both local root signals and systemic signals contribute to the modulation of IRT1 transcriptional regulation (Schmidt *et al.*, 2020).

At the posttranscriptional level, the trafficking of IRT1 between the plasma membrane and endosomes is a crucial regulatory mechanism (Tian *et al.*, 2021). Interestingly, this regulation is more focused on preventing the accumulation of non-iron metals and is largely independent of the Fe status. Elevated levels of non-iron metals, such as Mn, Zn, Co, or Cd, result in their binding to the cytoplasmic histidine-rich loop of IRT1 (Dubeaux *et al.*, 2018). The binding induces phosphorylation of IRT1 at nearby serine and threonine amino acids by the CBL-interacting serine/threonine-protein kinase 23 (CIPK23), followed by multimono-ubiquitination (Schmidt *et al.*, 2020). This ubiquitination initiates the endocytosis of the transporter to the trans-Golgi/early endosomes, marking IRT1 for vacuolar degradation via E3 ligases (Dubeaux *et al.*, 2018).

The multimono-ubiquitination-mediated endocytosis of IRT1 serves as a crucial protective mechanism against the excessive accumulation of non-iron metals like Mn, Zn, Co, or Cd. In this capacity, IRT1 functions as a transceptor, transporting Fe and Zn, Mn, or Co while simultaneously detecting and signaling the surplus presence of Zn, Mn, and Co ions in the cytoplasm (Cointry and Vert, 2019). Despite these insights, the primary sensors for Fe status remain elusive. Existing data propose multiple candidates, including at least one acting as a transceptor (IRT1), while others consist of cytosolic or nuclear proteins with well-defined metal-binding sites (e.g., HRZ-BTS-BTSL proteins, IDEF TFs). Notably, these proteins often exhibit selectivity for multiple metals, and peptides may also play pivotal roles in sensing and signaling the Fe status between root and shoot.

13.4 OTHER MICRONUTRIENTS

The comprehension of plant sensors responsible for monitoring essential nutrients like manganese (Mn), nickel (Ni), boron (B), or molybdenum (Mo) remains limited. Current knowledge, particularly for Mn and Ni, draws largely from bacterial studies. In bacteria, Mn sensors have been extensively studied, revealing a range of protein-based and RNA-based sensors. Protein-centric sensors, illustrated by transcription factors that interact with Mn, and RNA-centric sensors, referred to as riboswitches, play a role in adjusting the expression of manganese transporters (Zhou *et al.*, 2021b). This modulation is critical for the regulation of manganese uptake systems, encompassing ATP-binding cassette (ABC) and NRAMP transporters (Waters, 2020). In the bacterial context, nickel (Ni) import is orchestrated by ABC transporters, with Ni-binding transcription factors like NikR exerting control over the process (Feng *et al.*, 2021). Notably, NikR showcases an extraordinary Ni-binding affinity in the picomolar range. However, the analogous mechanisms in plants are not yet fully elucidated. Studies on bacterial systems provide valuable insights into potential sensing mechanisms, yet it's essential to recognize the complexity of these processes in plant contexts. Further exploration is required to unravel the intricacies of how plants sense and respond to Mn, Ni, and other essential nutrients. Investigating potential homologous systems in plants and understanding the unique adaptations that plant sensors may exhibit will contribute significantly to advancing our knowledge of nutrient sensing in the plant kingdom.

13.5 CHALLENGES AND FUTURE OUTLOOK

The intricate signaling networks orchestrating plant nutrient regulation stand as the linchpin of the global food chain, intricately woven into the fabric of interconnected ecosystems. Photosynthesis, a transformative process capturing CO_2 and releasing O_2, has been a paramount force shaping Earth's evolutionary trajectory. Our evolving comprehension of the intricate cycles of C, N, and P holds the key to revolutionary agricultural advancements, climate change mitigation, and ecosystem restoration (Jiang *et al.*, 2019).

While progress has been achieved in this domain, fundamental inquiries persist. The integration of enigmatic key nutrient, energy, and metabolite sensors, along with signaling regulators, remains a puzzle within the intricate framework of nutrient signaling networks. Diverse sensors are emerging, each with distinct specificity and affinities, unveiling a challenge in delineating their functions from the intricate tapestry of nutrient transport and metabolic processes (Garcia-Molina *et al.*, 2020).

Cutting-edge tools, encompassing chemical modifiers, real-time imaging, and time-series analyses, are indispensable for overcoming challenges posed by mutant lethality, redundancy, and pleiotropic effects. Global profiling experiments and genome-wide screenings offer potential in discovering additional nutrient sensors, unraveling their downstream targets, subcellular actions, and unveiling novel molecular mechanisms. An in-depth exploration of the intricate interactions involving nutrients, hormones, peptides, and metabolites at diverse biological levels holds the promise of unveiling new insights (Asim *et al.*, 2020). The interconnection of nitrate and phosphate signaling pathways with sugars, shrouded in mystery, introduces an additional layer of fascination.

The future brings forth the unveiling of universal sensors, signaling elements, and regulatory mechanisms spanning across multicellular organisms, ranging from plants to humans. Anticipated discoveries include unique and specialized regulators tailored to different plant species. As we plumb the depths of interconnected relationships among C, N, and P within plants and their responses to environmental cues, inspiration for innovative designs to enhance nutrient processes may emerge (Zhang *et al.*, 2020b). This, in turn, holds the potential to decouple stress responses compromising growth and reproduction. In agriculture, strategic approaches focused

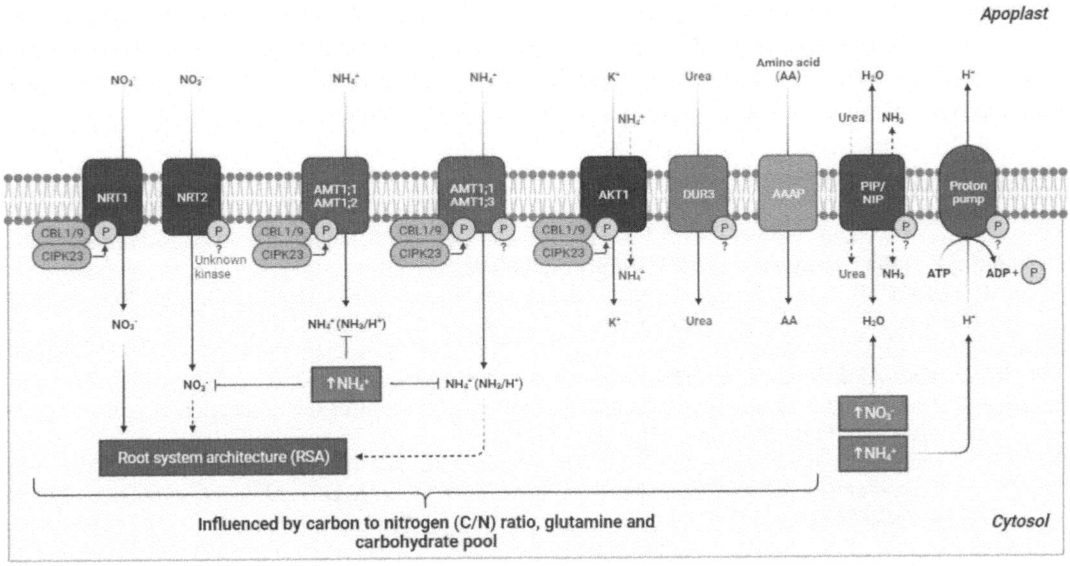

FIGURE 13.4 Nitrogen uptake transporters in *Arabidopsis:* key players in nutrient transport.

on efficient fertilizer acquisition and utilization bear the promise of realizing the ultimate goals of sustainable agriculture and ecosystem preservation (Ródenas and Vert, 2021). In essence, the exploration of plant nutrient regulation unveils a vast frontier with implications spanning agriculture, ecology, and climate. Understanding the interconnected relationships of C, N, and P opens doors to innovative solutions, reshaping our approach to sustainable agriculture and ecosystem management.

Figure 13.4 illustrates the nitrogen uptake transporters in *Arabidopsis thaliana*, highlighting their pivotal role in nutrient transport. These transporters are crucial for facilitating the efficient absorption and distribution of nitrogen, an essential element for plant growth and development.

13.6 CONCLUSION

Exploring the perspectives of nutrient sensing and signaling in crop plants opens a gateway to a future of more sustainable and productive agriculture. Nutrient sensing, a fundamental aspect of plant biology, holds the key to optimizing crop yields, improving nutrient utilization efficiency, and addressing global food security challenges. Through the lens of nutrient sensors, we have gained profound insights into the intricate networks that govern the uptake, transport, and allocation of essential elements in plants. These perspectives have illuminated the potential for precision agriculture, where crops receive the right nutrients, in the right amounts, at the right time. Understanding the molecular mechanisms that underlie nutrient sensing and signaling equips us with tools to develop crop varieties with enhanced nutrient uptake and utilization capabilities. This not only benefits farmers by improving crop productivity but also has far-reaching implications for reducing the environmental footprint of agriculture by minimizing fertilizer usage. Moreover, as we deepen our understanding of how plants perceive and respond to nutrient signals, we gain valuable insights into their ability to adapt to changing environmental conditions, including nutrient-deficient soils and climate challenges. The optimization of nutrient management practices, driven by these perspectives, offers a sustainable path toward meeting the world's increasing demand for food while safeguarding our ecosystems and natural resources. The complex interplay between nutrient sensors, signaling pathways, and environmental cues continues to be a subject of ongoing research.

The future promises the discovery of novel nutrient sensors, signal transduction components, and regulatory mechanisms that will further refine our ability to fine-tune crop nutrient uptake and utilization. Additionally, as we bridge the knowledge gap between nutrient sensing in model plants and crop species, we unlock new opportunities for crop improvement.

REFERENCES

Alejandro, S. *et al.* (2020) 'Manganese in plants: From acquisition to subcellular allocation', *Frontiers in Plant Science*, 11, 300. https://doi.org/10.3389/fpls.2020.00300

Asim, M. *et al.* (2020) 'Nitrate signaling, functions, and regulation of root system architecture: Insights from *Arabidopsis thaliana*', *Genes*, 11(6), p. 633. https://doi.org/10.3390/genes11060633

Balzergue, C. *et al.* (2017) 'Low phosphate activates STOP1-ALMT1 to rapidly inhibit root cell elongation', *Nature Communications*, 8, p. 15300. https://doi.org/10.1038/ncomms15300

Banerjee, P., Das, P. and Sinha, S. (2021) 'Importance of molybdenum for the production of pulse crops in India', *Journal of Plant Nutrition*, 45(2), pp. 300–310.

Barbier, F., Fichtner, F. and Beveridge, C. (2023) 'The strigolactone pathway plays a crucial role in integrating metabolic and nutritional signals in plants', *Nature Plants*, 9(8), pp. 1191–1200. https://doi.org/10.1038/s41477-023-01453-6

Bossolani, J.W. *et al.* (2022) 'Surface liming triggers improvements in subsoil fertility and root distribution to boost maize crop physiology, yield and revenue', *Plant and Soil*, 477(1), pp. 319–341.

Cai, H. *et al.* (2009) 'Overexpressed glutamine synthetase gene modifies nitrogen metabolism and abiotic stress responses in rice', *Plant Cell Reports*, 28(3), pp. 527–537. https://doi.org/10.1007/s00299-008-0665-z

Canales, J. *et al.* (2017) 'Nitrate induction of root hair density is mediated by TGA1/TGA4 and CPC transcription factors in *Arabidopsis thaliana*', *The Plant Journal*, 92(2), pp. 305–316. https://doi.org/10.1111/tpj.13656

Chand, S. *et al.* (2022) 'Agronomical and breeding approaches to improve the nutritional status of forage crops for better livestock productivity', *Grass and Forage Science*, 77(1), pp. 11–32.

Jia, Z. and von Wirén, N. (2020) 'Signaling pathways underlying nitrogen-dependent changes in root system architecture: from model to crop species', *Journal of Experimental Botany*, 71(15), 4393–4404. https://doi.org/10.1093/jxb/eraa033

Chanu, C.K. *et al.* (2020) 'Effect of nitrogen and molybdenum on crop growth, yield and soil properties of pea in acid soil (*Pisum sativum* L.)', *International Journal of Chemical Studies*, 8(5), pp. 2023–2027.

Chen, J. *et al.* (2016) 'Agronomic nitrogen-use efficiency of rice can be increased by driving OsNRT2.1 expression with the OsNAR2.1 promoter', *Plant Biotechnology Journal*, 14(8), pp. 1705–1715. https://doi.org/10.1111/pbi.12531

Chen, K.E. *et al.* (2020) 'Improving nitrogen use efficiency by manipulating nitrate remobilization in plants', *Nature Plants*, 6(9), pp. 1126–1135. https://doi.org/10.1038/s41477-020-00758-0

Choudhury, M.R. *et al.* (2022) 'Detection of calcium, magnesium, and chlorophyll variations of wheat genotypes on sodic soils using hyperspectral red edge parameters', *Environmental Technology & Innovation*, 27, p. 102469.

Cointry, V. and Vert, G. (2019) 'The bifunctional transporter-receptor IRT1 at the heart of metal sensing and signalling', *The New Phytologist*, 223(3), pp. 1173–1178. https://doi.org/10.1111/nph.15826

Colmenero-Flores, J.M. *et al.* (2019) 'Chloride as a beneficial macronutrient in higher plants: New roles and regulation', *International Journal of Molecular Sciences*, 20(19), p. 4686. https://doi.org/10.3390/ijms20194686

Connorton, J.M. and Balk, J. (2019) 'Iron biofortification of staple crops: Lessons and challenges in plant genetics', *Plant & Cell Physiology*, 60(7), pp. 1447–1456. https://doi.org/10.1093/pcp/pcz079

Costa, A. and Kudla, J. (2015) 'Colorful insights: Advances in imaging drive novel breakthroughs in Ca2+ signaling', *Molecular Plant*, 8(3), pp. 352–355. https://doi.org/10.1016/j.molp.2014.11.020

Day, S. and Aasim, M. (2020) 'Role of boron in growth and development of plant: Deficiency and toxicity perspective', in Aftab, T. and Hakeem, K.R. (eds), *Plant Micronutrients: Deficiency and Toxicity Management*. Cham: Springer, pp. 435–453.

de Bang, T. C. *et al.* (2021) 'The molecular-physiological functions of mineral macronutrients and their consequences for deficiency symptoms in plants', *The New Phytologist*, 229(5), pp. 2446–2469. https://doi.org/10.1111/nph.17074

Dekomah, S.D. *et al.* (2022) 'The role of CDPKs in plant development, nutrient and stress signaling', *Frontiers in Genetics 13*, p. 996203. https://doi.org/10.3389/fgene.2022.996203

Ding, B. *et al.* (2021) 'Nitrogen loss through denitrification, anammox and Feammox in a paddy soil', *Science of the Total Environment*, 773, p. 145601. https://doi.org/10.1016/j.scitotenv.2021.145601

Dong, Q. *et al.* (2022) 'Ca^{2+} signaling in plant responses to abiotic stresses', *Journal of Integrative Plant Biology*, 64(2), pp. 287–300. https://doi.org/10.1111/jipb.13228

Dreyer, I. *et al.* (2022) 'Transporter networks can serve plant cells as nutrient sensors and mimic transceptor-like behavior', *iScience*, 25(4), p. 104078. https://doi.org/10.1016/j.isci.2022.104078

Dubeaux, G. *et al.* (2018) 'Metal sensing by the IRT1 transporter-receptor orchestrates its own degradation and plant metal nutrition', *Molecular Cell*, 69(6), pp. 953–964.e5. https://doi.org/10.1016/j.molcel.2018.02.009

Eide, D.J. (2020) 'Transcription factors and transporters in zinc homeostasis: Lessons learned from fungi,' *Critical Reviews in Biochemistry and Molecular Biology*, 55, pp. 88–110. https://doi.org/10.1080/10409238.2020.1742092

Feng, C.Z. *et al.* (2021) 'MYB77 regulates high-affinity potassium uptake by promoting expression of HAK5', *The New Phytologist*, 232(1), pp. 176–189. https://doi.org/10.1111/nph.17589

Franco-Navarro, J.D. *et al.* (2019) 'Chloride as a macronutrient increases water-use efficiency by anatomically driven reduced stomatal conductance and increased mesophyll diffusion to CO_2', *The Plant Journal*, 99(5), pp. 815–831. https://doi.org/10.1111/tpj.14423

Garcia-Molina, A. *et al.* (2020) 'Systems biology of responses to simultaneous copper and iron deficiency in Arabidopsis', *The Plant Journal*, 103(6), pp. 2119–2138. https://doi.org/10.1111/tpj.14887

Geilfus, C.M. (2018a) 'Chloride: From nutrient to toxicant', *Plant & Cell Physiology*, 59(5), pp. 877–886. https://doi.org/10.1093/pcp/pcy071

Geilfus, C.M. (2018b) 'Review on the significance of chlorine for crop yield and quality,' *Plant Science*, 270, pp. 114–122. https://doi.org/10.1016/j.plantsci.2018.02.014

Giannakoula, A., Therios, I. and Chatzissavvidis, C. (2021) 'Effect of lead and copper on photosynthetic apparatus in Citrus (*Citrus aurantium* L.) Plants. The role of antioxidants in oxidative damage as a response to heavy metal stress', *Plants*, 10(1), p. 155. https://doi.org/10.3390/plants10010155

Gudadhe, N.N. and Thanki, J.D. (2021) 'Real time nitrogen management in rice genotypes through leaf colour chart at critical growth stages', *Indian Journal of Fertilizers*, 17, pp. 1040–1046.

Ham, B.K. *et al.* (2018) 'Insights into plant phosphate sensing and signaling', *Current Opinion in Biotechnology*, 49, pp. 1–9. https://doi.org/10.1016/j.copbio.2017.07.005

Han, G. *et al.* (2020) 'C2H2 zinc finger proteins: Master regulators of abiotic stress responses in plants', *Frontiers in Plant Science*, 11, p. 115. https://doi.org/10.3389/fpls.2020.00115

Han, G. *et al.* (2022) 'RING zinc finger proteins in plant abiotic stress tolerance', *Frontiers in Plant Science*, 13, p. 877011. https://doi.org/10.3389/fpls.2022.877011

Hao, D. L. *et al.* (2020a) 'Functional characterization of the *Arabidopsis* ammonium transporter AtAMT1;3 with the emphasis on structural determinants of substrate binding and permeation properties', *Frontiers in Plant Science* 11, p. 571. https://doi.org/10.3389/fpls.2020.00571

Hao, D. L. *et al.* (2020b) 'Function and regulation of ammonium transporters in plants', *International Journal of Molecular Sciences*, 21(10), p. 3557. https://doi.org/10.3390/ijms21103557

Hartley, T.N., Thomas, A.S. and Maathuis, F.J.M. (2020) 'A role for the OsHKT 2;1 sodium transporter in potassium use efficiency in rice', *Journal of Experimental Botany*. 71, pp. 699–706. https://doi.org/10.1093/jxb/erz113

Hindt, M.N. *et al.* (2017) 'BRUTUS and its paralogs, *BTS LIKE1* and *BTS LIKE2*, encode important negative regulators of the iron deficiency response in *Arabidopsis thaliana*', *Metalogics*, 9(7), pp. 876–890. https://doi.org/10.1039/c7mt00152e

Hirayama, T. *et al.* (2018) 'The putative peptide Gene FEP1 regulates iron deficiency response in Arabidopsis', *Plant & Cell Physiology*, 59(9), pp. 1739–1752. https://doi.org/10.1093/pcp/pcy145

Hu, B. *et al.* (2019) 'Nitrate-NRT1.1B-SPX4 cascade integrates nitrogen and phosphorus signalling networks in plants', *Nature Plants*, 5(4), pp. 401–413. https://doi.org/10.1038/s41477-019-0384-1

Ichihashi, Y. *et al.* (2020) 'Multi-omics analysis on an agroecosystem reveals the significant role of organic nitrogen to increase agricultural crop yield', *Proceedings of the National Academy of Sciences of the United States of America*, 117(25), pp. 14552–14560. https://doi.org/10.1073/pnas.1917259117

Jiang, Z. *et al.* (2019) 'Plant cell-surface GIPC sphingolipids sense salt to trigger Ca^{2+} influx', *Nature*, 572(7769), pp. 341–346. https://doi.org/10.1038/s41586-019-1449-z

Jinger, D. *et al.* (2022). 'Co-fertilization of silicon and phosphorus influences the dry matter accumulation, grain yield, nutrient uptake, and nutrient-use efficiencies of aerobic rice', Silicon, *14*, pp. 4683–4697.

Johnson, R. *et al.* (2022) 'Potassium in plants: Growth regulation, signaling, and environmental stress tolerance', *Plant Physiology and Biochemistry* 172, pp. 56–69. https://doi.org/10.1016/j.plaphy.2022.01.001

Kalia, A., Abd-Elsalam, K. A. and Kuca, K. (2020) 'Zinc-based nanomaterials for diagnosis and management of plant diseases: Ecological safety and future prospects', *Journal of Fungi*, 6(4), p. 222. https://doi.org/10.3390/jof6040222

Kalia, V.C. *et al.* (2021) 'Regulation of plant mineral nutrition by signal molecules', *Microorganisms*, 9(4), p. 774. https://doi.org/10.3390/microorganisms9040774

Keisham, M., Mukherjee, S. and Bhatla, S. C. (2018) 'Mechanisms of sodium transport in plants-progresses and challenges', *International Journal of Molecular Sciences*, 19(3), p. 647. https://doi.org/10.3390/ijms19030647

Kleczkowski, L.A. and Igamberdiev, A.U. (2021) 'Magnesium signaling in plants', *International Journal of Molecular Sciences*, 22(3), p. 1159. https://doi.org/10.3390/ijms22031159

Kobayashi, T., Nagano, A. J. and Nishizawa, N. K. (2021) 'Iron deficiency-inducible peptide-coding genes OsIMA1 and OsIMA2 positively regulate a major pathway of iron uptake and translocation in rice', *Journal of Experimental Botany*, 72(6), pp. 2196–2211. https://doi.org/10.1093/jxb/eraa546

Kohli, S. K. *et al.* (2023) 'Boron in plants: Uptake, deficiency and biological potential', *Plant Growth Regulation*, 100(2), pp. 267–282.

Kopriva, S., Malagoli, M. and Takahashi, H. (2019) 'Sulfur nutrition: Impacts on plant development, metabolism, and stress responses', *Journal of Experimental Botany*, 70(16), pp. 4069–4073. https://doi.org/10.1093/jxb/erz319

Kumar, V. *et al.* (2021) 'Copper bioavailability, uptake, toxicity and tolerance in plants: A comprehensive review', *Chemosphere*, 262, p. 127810. https://doi.org/10.1016/j.chemosphere.2020.127810

Künstler, A. *et al.* (2020) 'The versatile roles of sulfur-containing biomolecules in plant defense-A road to disease resistance', *Plants*, 9(12), p. 1705. https://doi.org/10.3390/plants9121705

Lee, H.J. and Seo, P.J. (2021) 'Ca^{2+}talyzing initial responses to environmental stresses', *Trends in Plant Science*, 26(8), pp. 849–870. https://doi.org/10.1016/j.tplants.2021.02.007

Lee, S. *et al.* (2020) 'OsASN1 overexpression in rice increases grain protein content and yield under nitrogen-limiting conditions', *Plant & Cell Physiology*, 61(7), pp. 1309–1320. https://doi.org/10.1093/pcp/pcaa060

Lei, L. *et al.* (2018) 'Nitrogen use efficiency is regulated by interacting proteins relevant to development in wheat', *Plant Biotechnology Journal*, 16(6), pp. 1214–1226. https://doi.org/10.1111/pbi.12864

Li, C. *et al.* (2017) 'Effects of changes in leaf properties mediated by methyl jasmonate (MeJA) on foliar absorption of Zn, Mn and Fe', *Annals of Botany*, 120(3), pp. 405–415. https://doi.org/10.1093/aob/mcx063

Li, J. *et al.* (2019) 'Advances in the mechanisms of plant tolerance to manganese toxicity', *International Journal of Molecular Sciences*, 20(20), p. 5096. https://doi.org/10.3390/ijms20205096

Li, J. *et al.* (2022) 'Engineering of NEMO as calcium indicators with large dynamics and high sensitivity', *Nature Methods*, 20(6), pp. 918–924. https://doi.org/10.1038/s41592-023-01852-9

Li, L., Liu, K. H. and Sheen, J. (2021) 'Dynamic nutrient signaling networks in plants', *Annual Review of Cell Developmental Biology*, 37, pp. 341–367. https://doi.org/10.1146/annurev-cellbio-010521-015047

Li, Q., Gao, Y. and Yang, A. (2020) 'Sulfur homeostasis in plants', *International Journal of Molecular Sciences*, 21(23), p. 8926. https://doi.org/10.3390/ijms21238926

Lilay, G.H. *et al.* (2021) '*Arabidopsis* bZIP19 and bZIP23 act as zinc sensors to control plant zinc status', *Nature Plants*, 7(2), pp. 137–143. https://doi.org/10.1038/s41477-021-00856-7

Liu, H. *et al.* (2021a) 'Hydrogen sulfide (H_2S) signaling in plant development and stress responses', *aBIOTECH*, 2(1), pp. 32–63. https://doi.org/10.1007/s42994-021-00035-4

Liu, Y. *et al.* (2021b) 'Diverse nitrogen signals activate convergent ROP2-TOR signaling in Arabidopsis', *Developmental Cell*, 56(9), pp. 1283–1295.e5. https://doi.org/10.1016/j.devcel.2021.03.022

Lodovichi, C. *et al.* (2022) 'Genetically encoded sensors for chloride concentration', *Journal of Neuroscience Methods*, 368, p. 109455. https://doi.org/10.1016/j.jneumeth.2021.109455

Luan, S. and Wang, C. (2021) 'Calcium signaling mechanisms across kingdoms', *Annual Review of Cell and Developmental Biology* 37, pp. 311–340. https://doi.org/10.1146/annurev-cellbio-120219-035210

Ma, X. *et al.* (2020) 'The CBL-CIPK pathway in plant response to stress signals', *International Journal of Molecular Sciences*, 21(16), p. 5668. https://doi.org/10.3390/ijms21165668

Mejias, J.H. *et al.* (2021) 'Nanofertilizers: A cutting-edge approach to increase nitrogen use efficiency in grasslands', *Frontiers in Environmental Science*, 9, p. 635114.

Mikhaylina, A. *et al.* (2018) 'Bacterial zinc uptake regulator proteins and their regulons', *Biochemical Society Transactions*, 46, pp. 983–1001. https://doi.org/10.1042/bst20170228

Naz, F. (2023) 'Plant nutrition, transport, mechanism and sensing in plants', in Aftab, T. and Hakeem, K.R. (eds.) *Sustainable plant nutrition*. London: Academic Press, pp. 209–228.

Naz, M. *et al.* (2023) 'Sodium and abiotic stress tolerance in plants', in Pandey, S. *et al.* (eds.) *Beneficial chemical elements of plants*: Recent *developments and future prospects.* New York: Wiley, pp. 307–330.

Nejad, S.A.G. and Etesami, H. (2020) 'The importance of boron in plant nutrition', in Deshmukh, R., Tripathi, D.K. and Guerriero, G. (eds.) *Metalloids in plants: Advances and future prospects.* New York: Wiley, pp. 433–449.

Oliphant, K.D. *et al.* (2022) 'Precise quantification of molybdate in vitro by the FRET-based nanosensor 'MolyProbe', *Molecules*, 27(12), p. 3691. https://doi.org/10.3390/molecules27123691

Oliveira, S.L. *et al.* (2022) 'Molybdenum foliar fertilization improves photosynthetic metabolism and grain yields of field-grown soybean and maize', *Frontiers in Plant Science*, 13, p. 887682.

Padhi, P.P. and Mishra, A.P. (2019) 'The role of molybdenum in crop production', *Journal of Pharmacognosy and Phytochemistry*, 8(5), pp. 1400–1403.

Pan, W., Wu, Y. and Xie, Q. (2019) 'Regulation of ubiquitination is central to the phosphate starvation response', *Trends in Plant Science*, 24(8), pp. 755–769. https://doi.org/10.1016/j.tplants.2019.05.002

Pantha, P. *et al.* (2023) 'Living with high potassium: Balance between nutrient acquisition and K-induced salt stress signaling', *Plant Physiology*, 191(2), pp. 1102–1121. https://doi.org/10.1093/plphys/kiac564

Pelegrino, M. T. *et al.* (2020) 'Effects of copper oxide nanoparticles on growth of lettuce (*Lactuca sativa* L.) seedlings and possible implications of nitric oxide in their antioxidative defense', *Environmental Monitoring and Assessment*, 192(4), p. 232. https://doi.org/10.1007/s10661-020-8188-3

Plattner, H. and Verkhratsky, A. (2015) 'The ancient roots of calcium signalling evolutionary tree', *Cell Calcium*, 57(3), pp. 123–132. https://doi.org/10.1016/j.ceca.2014.12.004

Podar, D. and Maathuis, F.J.M. (2022) 'Primary nutrient sensors in plants', *iScience*, 25(4), p. 104029. https://doi.org/10.1016/j.isci.2022.104029

Prathap, V. *et al.* (2022) 'Phosphorus homeostasis: Acquisition, sensing, and long-distance signaling in plants', *Molecular Biology Reports*, 49(8), pp. 8071–8086. https://doi.org/10.1007/s11033-022-07354-9

Ragel, P. *et al.* (2019) 'Regulation of K^+ nutrition in plants', *Frontiers in Plant Science*, 10, p. 281. https://doi.org/10.3389/fpls.2019.00281

Rana, M. *et al.* (2020) 'Molybdenum as an essential element for crops: An overview', *International Journal of Scientific Research and Growth*, 24(5), p. 18535.

Rashid, M. *et al.* (2020) 'Feedforward control of plant nitrate transporter NRT1.1 biphasic adaptive activity', *Biophysical Journal*, 118(4), pp. 898–908. https://doi.org/10.1016/j.bpj.2019.10.018

Rehman, M. *et al.* (2019) 'Copper environmental toxicology, recent advances, and future outlook: A review', *Environmental Science and Pollution Research International*, 26(18), pp. 18003–18016. https://doi.org/10.1007/s11356-019-05073-6

Ristova, D. and Kopriva, S. (2022) 'Sulfur signaling and starvation response in Arabidopsis', *iScience*, 25(5), p. 104242. https://doi.org/10.1016/j.isci.2022.104242.

Ródenas, R. and Vert, G. (2021) 'Regulation of root nutrient transporters by CIPK23: 'One kinase to rule them All', *Plant & Cell Physiology*, 62(4), pp. 553–563. https://doi.org/10.1093/pcp/pcaa156

Rodríguez-Celma, J. *et al.* (2019) 'Hemerythrin E3 ubiquitin ligases as negative regulators of iron homeostasis in plants', *Frontiers in Plant Science*, 10, 98. https://doi.org/10.3389/fpls.2019.00098

Sadoine, M. *et al.* (2021) 'Affinity series of genetically encoded Förster resonance energy-transfer sensors for sucrose', *ACS Sensors*, 6(5), pp. 1779–1784. https://doi.org/10.1021/acssensors.0c02495

Saini, D. K. *et al.* (2021) 'Meta-QTLs, ortho-MQTLs and candidate genes for nitrogen use efficiency and root system architecture in bread wheat (*Triticum aestivum* L.)', *Physiology and Molecular Biology of Plants*, 27(10), pp. 2245–2267. https://doi.org/10.1007/s12298-021-01085-0

Sardans, J. and Peñuelas, J. (2021) 'Potassium control of plant functions: Ecological and agricultural implications', *Plants*, 10(2), p. 419. https://doi.org/10.3390/plants10020419

Schmidt, S. B. and Husted, S. (2019) 'The biochemical properties of manganese in plants', *Plants (Basel, Switzerland*, 8(10), p. 381. https://doi.org/10.3390/plants8100381

Schmidt, W., Thomine, S. and Buckhout, T. J. (2020) 'Editorial: Iron nutrition and interactions in plants', *Frontiers in Plant Science* 10, p. 1670. https://doi.org/10.3389/fpls.2019.01670

Selote, D. *et al.* (2018) 'The E3 ligase BRUTUS facilitates degradation of VOZ1/2 transcription factors', *Plant, Cell & Environment*, 41(10), pp. 2463–2474. https://doi.org/10.1111/pce.13363

Ruiz, J.C. and Bruick R.K. (2014) F-box and leucine-rich repeat protein 5 (FBXL5): Sensing intracellular iron and oxygen. *Journal of Inorganic Biochemistry*, 133, pp. 73–77. https://doi.org/10.1016/j.jinorgbio.2014.01.015

Shabbir, Z. *et al.* (2020) 'Copper uptake, essentiality, toxicity, detoxification and risk assessment in soil-plant environment', *Chemosphere*, 259, p. 127436. https://doi.org/10.1016/j.chemosphere.2020.127436

Singh, A. K., Singh, A. K. and Singh, J. P. (2020) 'Boron in crop production from soil to plant system: A review', *Archives of Agriculture and Environmental Science*, 5(2), pp. 218–222.

Sisharmini, A. *et al.* (2019) 'Expression of a cucumber *alanine aminotransferase2* gene improves nitrogen use efficiency in transgenic rice', *Journal of Genetic Engineering and Biotechnology*, 17(1), p. 9. https://doi.org/10.1186/s43141-019-0010-7

Sun, H. *et al.* (2014) 'Heterotrimeric G proteins regulate nitrogen-use efficiency in rice', *Nature Genetics*, 46(6), pp. 652–656. https://doi.org/10.1038/ng.2958

Sun, X. *et al.* (2021) 'Evaluation of maize root growth and genome-wide association studies of root traits in response to low nitrogen supply at seedling emergence', *The Crop Journal*, 9(4), pp. 794–804.

Sustr, M., Soukup, A. and Tylova, E. (2019) 'Potassium in root growth and development', *Plants*, 8(10), p. 435. https://doi.org/10.3390/plants8100435

Tao, Y. *et al.* (2023) 'Effect of Mn deficiency on carbon and nitrogen metabolism of different genotypes seedlings in maize (*Zea mays* L.)', *Plants*, 12(6), p. 1407. https://doi.org/10.3390/plants12061407

Taulemesse, F. *et al.* (2015) 'Post-flowering nitrate uptake in wheat is controlled by N status at flowering, with a putative major role of root nitrate transporter NRT2.1', *PLoS One*, 10(3), p. e0120291.

Tay, L.H. *et al.* (2012) 'Nanodomain Ca^{2+} of Ca^{2+} channels detected by a tethered genetically encoded Ca^{2+} sensor', *Nature Communications* 3, p. 778. https://doi.org/10.1038/ncomms1777

Tian, X.Y. *et al.* (2021) 'Physiological and molecular advances in magnesium nutrition of plants', *Plant and Soil*, 468, pp. 1–17. https://doi.org/10.1007/s11104-021-05139-w

van Zelm, E., Zhang, Y. and Testerink, C. (2020) 'Salt tolerance mechanisms of plants', *Annual Review of Plant Biology*, 71, pp. 403–433. https://doi.org/10.1146/annurev-arplant-050718-100005

Waadt, R. *et al.* (2020) 'Dual-reporting transcriptionally linked genetically encoded fluorescent indicators resolve the spatiotemporal coordination of cytosolic abscisic acid and second messenger dynamics in Arabidopsis', *The Plant Cell*, 32(8), pp. 2582–2601. https://doi.org/10.1105/tpc.19.00892

Wang, F.L. *et al.* (2021a) 'A potassium-sensing niche in Arabidopsis roots orchestrates signaling and adaptation responses to maintain nutrient homeostasis', *Developmental Cell*, 56(6), pp. 781–794.e6. https://doi.org/10.1016/j.devcel.2021.02.027

Wang, P. *et al.* (2021b) 'Medicago SPX1 and SPX3 regulate phosphate homeostasis, mycorrhizal colonization, and arbuscule degradation', *The Plant Cell*, 33(11), pp. 3470–3486. https://doi.org/10.1093/plcell/koab206

Wang, X. *et al.* (2021c) 'A transceptor-channel complex couples nitrate sensing to calcium signaling in Arabidopsis', *Molecular Plant*, 14(5), pp. 774–786. https://doi.org/10.1016/j.molp.2021.02.005

Wang, Y., Chen, Y. F. and Wu, W. H. (2021d) 'Potassium and phosphorus transport and signaling in plants', *Journal of Integrative Plant Biology*, 63(1), pp. 34–52. https://doi.org/10.1111/jipb.13053

Wang, Y. *et al.* (2021e) 'Phosphate uptake and transport in plants: An elaborate regulatory system', *Plant & Cell Physiology*, 62(4), pp. 564–572. https://doi.org/10.1093/pcp/pcab011

Wang, Z., Kuo, H.-F. and Chiou, T.-J. (2021f) 'Intracellular phosphate sensing and regulation of phosphate transport systems in plants', *Plant Physiology*, 187, 2043–2055. https://doi.org/10.1093/plphys/kiab343

Waters, L.S. (2020) 'Bacterial manganese sensing and homeostasis', *Current Opinion in Chemical Biology* 55, pp. 96–102. https://doi.org/10.1016/j.cbpa.2020.01.003

Wu, H. *et al.* (2018) 'It is not all about sodium: Revealing tissue specificity and signalling roles of potassium in plant responses to salt stress', *Plant and Soil*, 431, pp. 1–17.

Wu, X. *et al.* (2019) 'Ammonium and nitrate regulate NH4+ uptake activity of Arabidopsis ammonium transporter AtAMT1;3 via phosphorylation at multiple C-terminal sites', *Journal of Experimental Botany*, 70(18), pp. 4919–4930. https://doi.org/10.1093/jxb/erz230

Yin, H. *et al.* (2021) 'Soil sensors and plant wearables for smart and precision agriculture', *Advanced Materials*, 33(20), p. e2007764. https://doi.org/10.1002/adma.202007764

You, L. *et al.* (2022) 'Genome-wide identification of nitrate transporter 2 (NRT2) gene family and functional analysis of MeNRT2.2 in cassava (*Manihot esculenta* Crantz)', *Gene*, 809, p. 146038. https://doi.org/10.1016/j.gene.2021.146038

Youssef, S., Zhang, S. and Ai, H.W. (2019) 'A genetically encoded, ratiometric fluorescent biosensor for hydrogen sulfide', *ACS Sensors*, 4(6), pp. 1626–1632. https://doi.org/10.1021/acssensors.9b00400

Zeng, H. *et al.* (2021) 'Molecular regulation of zinc deficiency responses in plants', *Journal of Plant Physiology*, 261, p. 153419. https://doi.org/10.1016/j.jplph.2021.153419

Zhang, J. *et al.* (2021) 'Increasing yield potential through manipulating of an ARE1 ortholog related to nitrogen use efficiency in wheat by CRISPR/Cas9', *Journal of Integrative Plant Biology*, 63(9), pp. 1649–1663. https://doi.org/10.1111/jipb.13151

Zhang, L. *et al.* (2020a) 'Sulfur deficiency-induced glucosinolate catabolism attributed to two β-glucosidases, BGLU28 and BGLU30, is required for plant growth maintenance under sulfur deficiency', *Plant & Cell Physiology*, 61(4), pp. 803–813. https://doi.org/10.1093/pcp/pcaa006

Zhang, Z., Hu, B. and Chu, C. (2020b) 'Towards understanding the hierarchical nitrogen signalling network in plants', *Current Opinion in Plant Biology*, 55, pp. 60–65. https://doi.org/10.1016/j.pbi.2020.03.006

Zhao, L. *et al.* (2021a) 'Genetic effects and expression patterns of the nitrate transporter (NRT) gene family in *Populus tomentosa*', *Frontiers in Plant Science*, 12, p. 661635. https://doi.org/10.3389/fpls.2021.661635

Zhou, J. *et al.* (2021b) 'Mechanism of phosphate sensing and signaling revealed by rice SPX1-PHR2 complex structure', *Nature Communications*, 12(1), p. 7040. https://doi.org/10.1038/s41467-021-27391-5

Zhou, Q. *et al.* (2018) 'Genome-wide identification and characterization of the SBP-box gene family in Petunia', *BMC Genomics*, 19, p. 193. https://doi.org/10.1186/s12864-018-4537-9.

Zlobin, I. E. (2021) 'Current understanding of plant zinc homeostasis regulation mechanisms', *Plant Physiology and Biochemistry*, 162, pp. 327–335. https://doi.org/10.1016/j.plaphy.2021.03.003

14 Functional and Innovative Fertilizers to Improve Crop Plant Growth

Ana Paulina Velasco-Ramírez, Sandra Fabiola Velasco-Ramírez, Armando Rafael Hernández-Pérez, Ana Cristina Ramírez Anguiano, Alejandro Velasco-Ramírez, and Martha Isabel Torres-Morán

14.1 INTRODUCTION

Fertilizers are required in long-term agricultural production systems to sustain or increase crop yields, especially when plants are separated from the production system. The primary reasons for using fertilizers, especially to increase yields, are similar in tropical and temperate climates. In particular, fertilizers are used to complement nutrients naturally present in soils to meet the physiological demands of crops with high yield potential, replenish nutrients lost through leaching or that are removed by plants, maintain acceptable cultivation conditions, or improve unfavorable conditions (Girma *et al.*, 2007). To this end, fertilizers, especially synthetic ones, are essential for increasing agricultural productivity.

Fertilizers are expensive inputs. Given their importance, affordable and ecological fertilizers, especially green ammonia, must be produced with renewable energies, promoted, and supported. Renewable energies, including those powered by solar, wind, and hydrogen sources, may replace traditional energies that consume raw materials, such as gas and coal (FAOSTAT, 2023). Worldwide, nations share a common interest in increasing agricultural production to meet the global demand for food while also lowering production costs and supporting environmental sustainability. Currently, the fertilizer industry is of greater importance to some countries than others (Stewart *et al.*, 2005), although no nation can remain neutral to agricultural fertilization practices, as everyone is inevitably affected, whether they are producers or consumers.

The economic structure of Latin American countries demands that none are unaffected by fertilization practices, although fertilizer use depends on multiple factors that vary in nature and intensity among countries. Indeed, fertilization is one of many practices used to improve agricultural production, and in most cases, it is employed with other methods. However, important factors can be identified that decisively influence the degree of fertilization in Latin American and Asian countries, either directly or indirectly. Identifying these factors is particularly important, as Asian countries will continue to be the largest consumers of fertilizers in the world, depending on imports of the three primary nutrients (i.e., nitrogen, phosphorus, and potassium), even if western Asian nations continue to generate a notable surplus of these nutrients (FAO, 2021; Figure 14.1).

As a whole, Europe will also contain a surplus of these three nutrients due to the high positive balances in Eastern Europe and Central Asia. Fertilizer use through 2023 is expected to remain stable in Western Europe while growing 3.6% percent annually in the Eastern subregion. Latin America and the Caribbean will depend on imports of these three nutrients for the forecasted period, during which the use of fertilizers in the region is expected to grow at a sustained annual rate of 3.3%, according to FAO (2021). An important aspect of the global fertilizer market is the rapid rise in prices in 2021. Indeed, international reference prices for fertilizers increased throughout 2021, with many prices reaching all-time highs. The most notable price increase was for all-important nitrogen

DOI: 10.1201/9781032630366-14

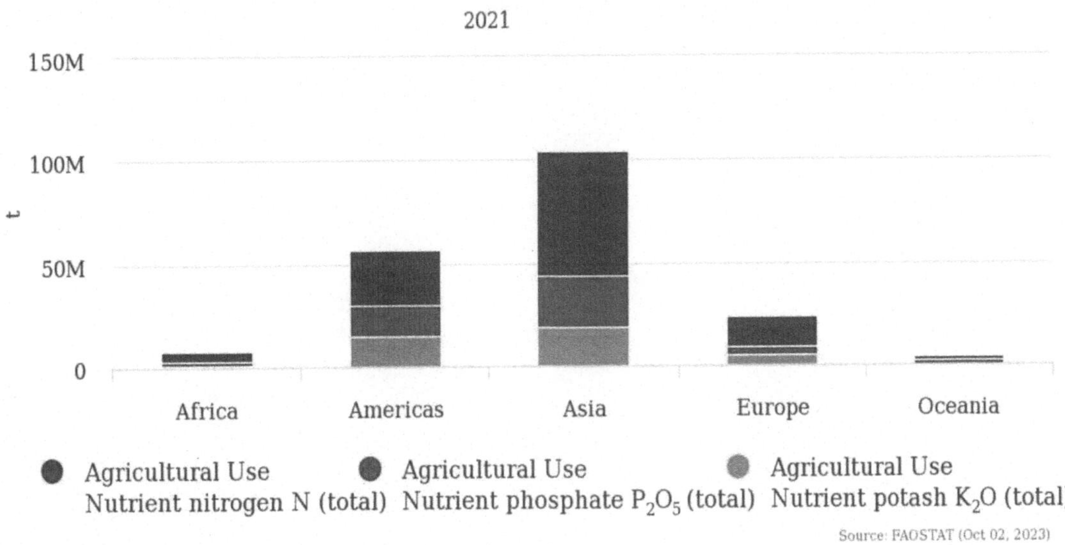

FIGURE 14.1 Fertilizer consumption in nutrients.

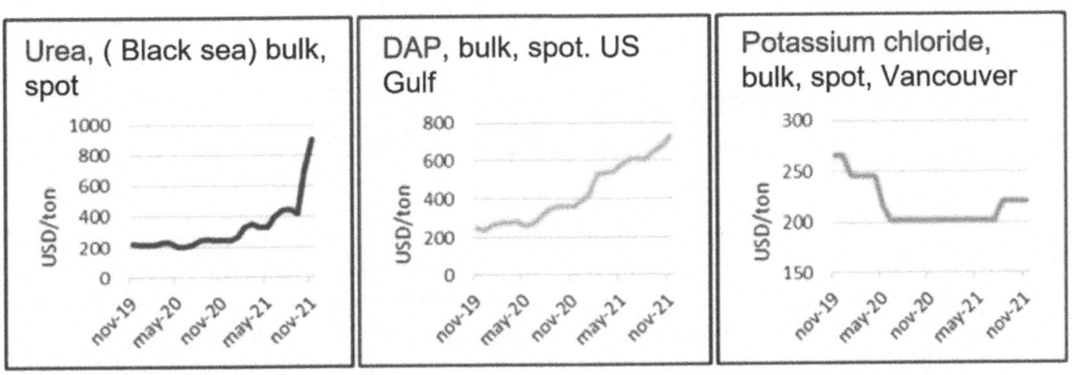

FIGURE 14.2 Spot prices of the main nitrogenous, phosphate, and potassium fertilizers in 2021.

fertilizers, which more than tripled in 2021. For example, nominal spot (bulk) price quotes in the Black Sea rose from USD $245 per ton in November 2020 to USD $901 per ton in November 2021 (Figure 14.2). Furthermore, it should be noted that much of this increase was recorded in the second half of the year.

Phosphate fertilizer prices have risen in parallel to those of nitrogen fertilizer. Prices for diammonium phosphate (DAP), a key compound in phosphate fertilizer, doubled from $360 per ton to $726 per ton in 2021 (Figure 14.1). Undoubtedly, the rise in DAP prices is also due to the price increase of its nitrogen component. Moreover, rising phosphate fertilizer prices are also to blame, accounting for around 50% of the overall increase in DAP prices. On the other hand, potash (potassium fertilizer) prices were less affected in 2021 (Figure 14.2).

14.2 CHEMICAL FERTILIZERS

The primary purpose of chemical fertilizers is to provide plants with the nutritional agents necessary to develop properly. These fertilizers are artificially prepared by combining nutrients, such as nitrogen, phosphorus, and potassium (N, P_2O, K_2O), which improve the fertility and productive capacity of the

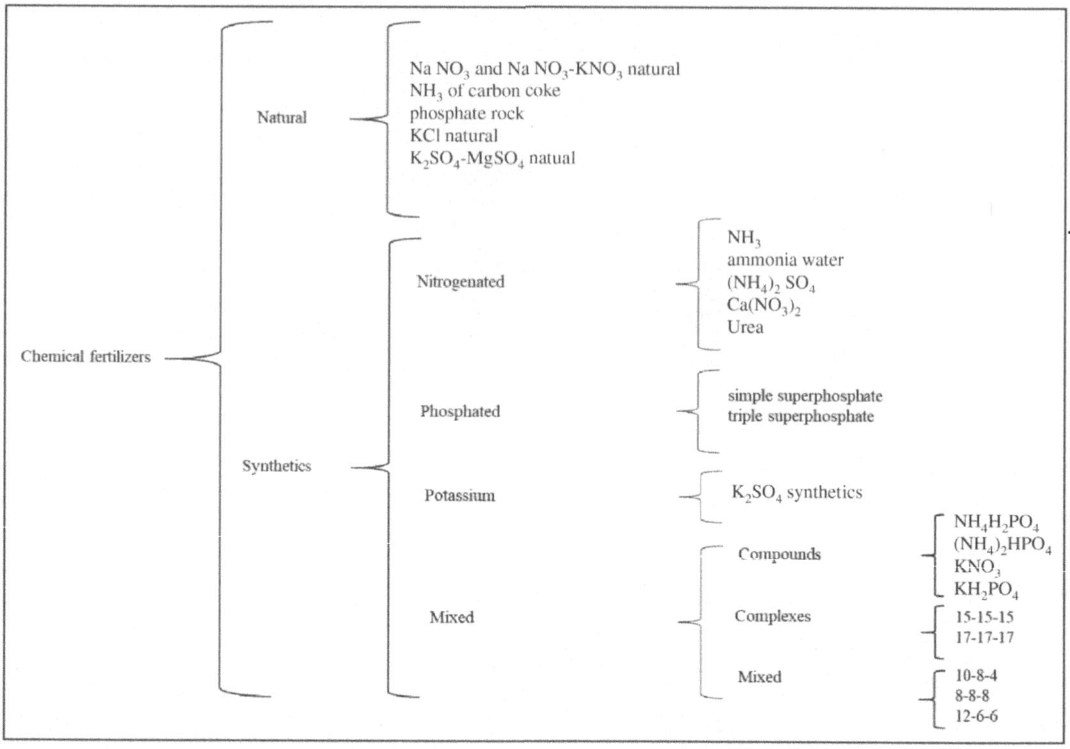

FIGURE 14.3 Classification of chemical fertilizers (Salgado and Núñez, 2012).

land. When evaluating chemical fertilizers, water solubility is the most important characteristic, as this ensures the product can dissolve in soil and passively or actively enter plants through irrigation.

Although farmers widely use fertilizers to maximize harvests and mitigate diseases, it is important to mention that their inappropriate use has negative consequences for both plants and the environment, including the deterioration of microfauna, soil structure, and pH levels. Several fertilizer types are available on the market with different compositions, presentations, and application methods. Therefore, fertilizer selection depends mainly on the characteristics and nutritional demands of the intended crop (Figure 14.3).

14.2.1 Types of Conventional Mineral Fertilizers

Mineral fertilizers dissolve easily in soil. There are many mineral fertilizers available, each with its own particular uses. Potassium fertilizers are ideal for sensitive plants. Nitrogen fertilizers come in multiple forms, although high nitrogen levels characterize all nitrogen fertilizers. Sodium nitrate fertilizers and ammonium sulfate fertilizers are useful in acidic soils. Ammonium nitrate fertilizers are used with multiple crops because of their low risk. In contrast, ammonia fertilizers are not suitable for home gardeners because they contain approximately 80% nitrogen. Urea fertilizers also contain high nitrogen concentrations and are absorbed by plants rapidly. These fertilizers require extreme care when applied, as they will raise the pH of the soil, causing burns and damaging crops if applied inappropriately (IFA, 2023).

14.2.1.1 Nitrogen Fertilizers

Nitrogen fertilizers contain nitrogen in chemical forms that plants can assimilate after being absorbed by different organs. When applied to the soil or leaves, nitrogen fertilizers support plant growth,

TABLE 14.1
Largest Ammonia-Producing Companies Worldwide

Company	Tons	Company	Tons
Yara (Norway)	8.5	Sinopec (China)	3
CF Industries (USA)	8.1	IFFCO (India)	2.8
Agrium (Canada)	4.8	EuroChem (Russia)	2.8
Group DF (Ukraine)	4.2	Uralchem (Russia)	2.8
PotashCorp (Canada)	4.1	Koch (USA)	2.7
TogliattiAzot (Russia)	3.5	OCI (Egypt)	2.2

development, and production (FAO-IFA, 2000). The primary solid nitrogenous fertilizers derived from ammonia are urea, ammonium sulfate, and ammonium nitrate. Urea is the most widely used due to its high nitrogen concentration (46%) and low cost. This fertilizer comes in the form of white granules or pearls that are converted in the soil into ammonium (NH_4^+) after interacting with water and into nitrate (NO_3^{--}) by microbiological decomposition, with both forms being usable by plants (Erisman et al., 2008). However, urea shows notable nitrogen loss due to volatilization, especially when applied as a surface fertilizer. For this reason, urea should be used in already-established crops and preferably applied in small doses. Table 14.1 lists the companies with the highest ammonia production worldwide.

The high consumption of nitrogen fertilizers worldwide is because nitrogen is a vital element for proper plant growth and development, along with phosphorus and potassium. This necessity is reflected in the large quantities consumed throughout cultivation cycles. On the other hand, soils worldwide are nitrogen-deficient, which implies a direct dependence between the use of these products and yields, especially in highly specialized crops such as hybrids and transgenics (Salgado and Núñez, 2012).

Plants absorb nitrogen in its most soluble forms, namely nitrate (NO_3), ammonium (NH_4), and low-molecular-weight nitrogenous compounds (e.g., amines and amino acids). Nitrogen is a highly mobile element in soil and can be lost through different processes such as denitrification, leaching, and volatilization (Mengel and Kirkby, 2000). Table 14.2 shows the different types of nitrogenous fertilizers and how they are obtained.

TABLE 14.2
Types of Nitrogen Fertilizers

Nitrogen Fertilizer	Method	Chemical Reaction
Ammonium sulfate $(NH_4)_2SO_4$	Obtained from the reaction of ammonia with sulfuric acid.	$2NH_3 + H_2SO_4 \rightarrow (NH_4)_2SO_4$
Potassium nitrate (KNO_3)	Obtained from nitric acid (produced industrially or from atmospheric nitrogen) and potassium compounds or by chemically treating sodium nitrate.	$2NHO_3 + K_2O \rightarrow 2KNO_3 + H_2O$
Ammonium nitrate (NH_4NO_3)	Obtained from ammonia and nitric acid.	$HNO_3 + NH_3 \rightarrow NH_4NO_3$
Urea $(NH_2)_2$ CO	Nitrogen compound of animal origin. Obtained by chemical synthesis; ammonia reacts with carbon dioxide (CO_2) or carbon anhydrous.	$2NH_3 + CO_2 \rightarrow NH_2COONH_4\ (NH_2)_2CO + H_2O$
Sodium nitrate ($NaNO_3$)	Obtained synthetically from sodium carbonate and nitric acid and from natural sources.	$Na_2CO_3 + 2HNO_3 \rightarrow 2NaNO_3 + H_2CO_3$
Anhydrous ammonia (NH_3)	Obtained from the combination of nitrogen (N_2) and hydrogen (H_2) through the Claude-Haber process.	

TABLE 14.3
Types of Phosphorus Fertilizers

Phosphorus Fertilizer	Method	Chemical Reaction
Phosphoric rock	Obtained from geological deposits located around the world.	$(Ca3(PO_4)2)_3CaF_2$
Orthophosphoric acid synthesis	Obtained by treating phosphate rocks with diluted sulfuric acid; the phosphoric acid obtained contains 55% P_2O_5.	$[Ca_3(PO_4)_2]_3$. $CaF_2 + 10\ H_2SO_4 + 20H_2O$ 10 $\rightarrow$ $CaSO_4$. $2H_2O + 6H_3PO_4 + 2HF$
Sulfuric acid synthesis	Obtained from sulfur (S) and the desulfurization of natural gas.	$S + \rightarrow SO_2$ + energy $2\ SO_2 + O_2 \rightarrow SO_3$ + 22,600 cal/mol $SO_3 + H_2O \rightarrow H_2SO_4$
Nitric acid synthesis	Obtained from the oxidation of ammonia.	$4NH_3 + 5\ O_2 \rightarrow 4NO + 6\ H_2O$ $2NO + O \rightarrow 2NO_2$ $3\ NO_2 + H_2O \rightarrow 2\ HNO_3 + NO$
Ammonium phosphate	Obtained through the saponification of phosphoric acid or a mixture of the phosphoric and sulfuric acids monoammonium phosphate (MAP) 11–46–0 and diammonium phosphate (DAP) 18–46–0.	

14.2.1.2 Phosphorus Fertilizers

In soil, phosphorus is found in organic and inorganic forms. Soil inorganic phosphorus can be found adsorbed to the mineral or organic fraction of the soil, in solution, or as a constituent of phosphate minerals (Salgado and Núñez, 2012). Once plants absorb phosphorus, it circulates as monobasic phosphate. Internally, monobasic phosphate is mobile and essential to nucleic acids and nucleotides. It also stimulates root development and promotes flowering and seed formation. Plants also require phosphorus to fix nitrogen biologically. The practical importance of phosphorus in agriculture in terms of frequency of use and the amounts applied to correct deficiencies is only surpassed by that of nitrogen, especially when considering the volumes of both fertilizer types. Table 14.3 shows the types of phosphorus fertilizers.

14.2.1.3 Potassium Fertilizers

Potassium is the seventh most abundant element on earth. The arable layer of agricultural soils contains 4–60 t ha^{-1} of potassium, although mostly in soluble minerals. In addition, 2.6% of the earth's crust is potassium, which is present in primary minerals, such as feldspar ($KAlSi_3O_8$); orthoclase and microline in sand and silt; or micas, including muscovite [$H2KA13(SiO_4)_3$] or biotite [$(H, K)_2 (Mg, Fe)_2\ Al_2(SiO_4)_3$]. Weathering susceptibility follows the order of biotite > muscovite > feldspar (Malavolta, 2006).

In plants, potassium maintains the ionic balance and water status and participates in sugar production and transport, enzyme activation, and protein synthesis. Potassium is also necessary for synthesizing pigments, particularly lycopene. When potassium enters the metabolic system of a cell, it forms unstable salts, such as organic and inorganic acids, which regulate the cellular osmotic potential (Núñez, 1990). Table 14.4 shows the types of potassium fertilizers.

14.2.2 Chemical Fertilizer Application

14.2.2.1 Foliar Fertilization

Foliar fertilization is commonly used to complement traditional soil fertilization. Indeed, adding foliar fertilization to existing fertilization schemes is often employed to overcome sub-optimal

TABLE 14.4
Type of Potassium Fertilizers

Potassium Fertilizer	Method	Chemical Reaction
Potassium chloride (KCl)	Obtained through the reaction of phosphoric acid with potassium hydroxide	
Potassium sulfate (K_2SO_4)	Obtained from potassium salts containing potassium and magnesium chlorides.	(KCl +Cl_2 Mg, etc.) 2 KCl + H_2SO_4 K_2SO_4 + → 2HCl
Monopotassium phosphate (KH_2PO_4)	Obtained through the reaction of phosphoric acid with potassium hydroxide.	KH_2PO_4

yields from exclusive edaphic fertilization. Moreover, foliar fertilization causes less environmental pollution than traditional fertilization methods, while still increasing crop productivity and quality. According to El-Fouly (2002), foliar fertilizers are classified as mono- and multi-macronutrient formulations containing one or more macronutrients, respectively; mono- and multi-micronutrient formulations containing one or more micronutrients in concentrations greater than 2% for chelates and 5% for other salts; and mixed formulations of macro- and micronutrients.

14.2.2.2 Soil Fertilization

Soil fertilization occurs directly at the base of the plant, the substrate, or within the soil to ensure nutrients are delivered as closely as possible to the root absorption zone to be taken up by the plant. Typically, granulated solid fertilizers are applied in low-frequency application regimens to ensure that the nutrients are available over long periods while being absorbed (IFA, 1992). Edaphic fertilization management greatly impacts the efficiency of fertilization schemes, with efficiency being measured by the amount of nutrients taken up by the plant with respect to the total amount applied. A thorough analysis should be conducted of the variables that directly affect this efficiency, including soil type, climate, plant type, crop age, application technique, agronomic behavior, economic components, and product availability, to achieve maximum fertilization efficiency (IFA, 1992; Laegreid *et al.*, 1999).

14.2.2.3 Fertigation

Fertigation involves applying fertilizers to crops through irrigation water. This agricultural technique provides an excellent opportunity to maximize yields while reducing environmental pollution by increasing fertilizer use efficiency, minimizing fertilizer application, and increasing the economic benefits of fertilizer investments. With fertigation, the timing, quantities, and concentrations of fertilizers can be controlled easily (Kafjafi and Tarchitzky, 2012).

14.2.3 Advantages of Chemical Fertilizers

Chemical fertilizers are crucial to meeting the high demands of intensive agriculture. When used properly, chemical fertilizers notably improve the spaces and surfaces set aside for agricultural activities, resulting in greater production per hectare. Due to their great nutritional contributions, chemical fertilizers can be employed to optimize production, facilitate plant development, and optimize harvest results. In addition to improving nutrient absorption, chemical fertilizers can increase root depth by up to one meter, improving plant nutrition, water supply, and structure.

Soil components and characteristics, such as pH, can be adjusted by applying chemical fertilizers. Thus, the moderate and conscious use of chemical fertilizers makes soils more fertile and improves their quality to meet the requirements of upcoming production cycles. When critical situations arise and plants do not develop correctly, chemical fertilizers can be applied as an excellent means to improve plant health. Indeed, the foliar application of N-P-K fertilizers, which involves diluting them with water and applying the mixture directly to plant leaves, improves plant health and the outcomes of advanced cultivation stages. The chemical properties and amounts of specific nutrients in N-P-K fertilizers can be adjusted to meet the needs of each product. For this reason, chemical fertilizers available on the market can meet the specific agricultural needs that arise during particular phases of the production cycle (Chemical Fertilizers, 2023).

14.2.4 Disadvantages of Chemical Fertilizers

Despite their usefulness, when used in a disorganized manner, chemical fertilizers produce negative consequences (Chemical Fertilizers, 2023). In the long term, soil can become infertile due to the inappropriate application of chemical fertilizers. This infertility is primarily because chemical fertilizers substantially increase soil acidity while altering macronutrient levels, even to the point of saturation, so that the soil loses its sensitivity and capacity to absorb different nutrients. In addition, the improper application of chemical fertilizers can result in the accumulation of excess nutrients in the soil and subsoil that are not absorbed by plants, leading to groundwater contamination. This form of contamination results in inevitable losses in the immediate production cycle and compromises future agricultural activities in the affected area.

In addition, salt burns indicate a deficient or inappropriate use of chemical fertilizers that has resulted in the dehydration of plant tissue. Fertilizers with a high salt index, mainly those made with compounds such as sodium nitrate, must be used cautiously to avoid salt burns that hinder development. Far from increasing productivity, excessive plant growth due to poorly applied chemical fertilizers compromises plant survival and the yield at harvest. Unexpectedly large plant size indicates that foliar and root systems are under intense pressure, and the soil may be degraded. For this reason, the soils and crops must be analyzed continually to detect any negative consequences to correct them in time. To avoid the drawbacks associated with chemical fertilizers, their use should be complemented with the application of organic fertilizers and compounds.

14.3 ORGANIC FERTILIZERS

It must be clarified that in some countries, organic fertilizers are only those approved for organic farming by an international standard (e.g., European, North American, and Japanese) and certified by an accredited external company. More generally, as recently defined in the European Commission working groups, organic fertilizers are those whose nutrients are contained in organic materials of animal, plant, or other natural organic origin, consisting of compounds or materials in which the primary nutrients are chemically linked or part of these organic matrices (Ferreira de Oliveira *et al.*, 2023). In Spain, an organic fertilizer is a product whose main function is to provide nutrients from carbonaceous materials of animal or plant origin to plants. Thus, not all organic fertilizers can be used in organic farming. Organic fertilizers may be solids or liquids, with each form having different types classified according to their origins and specific properties.

14.3.1 Types of Organic Fertilizers

Composting is based on the controlled decomposition of different solid elements of natural origin, resulting in an earthy product that condenses many nutrients, favoring plant growth (Makhkamova and Kamilov, 2023). Aided by favorable atmospheric conditions, different microorganisms metabolize compounds through decomposition, converting them into compost loaded with varying quantities of sulfate, nitrogen, carbon, phosphorus, and other elements. Vermicompost differs from other organic fertilizers because of its higher quality. Humus, a form of vermicompost, is rich in minerals and produced by detritivorous worms that actively degrade solid, organic compost that is mostly of plant origin (Makhkamova and Kamilov, 2023). To create favorable conditions for humus-producing worms, geomembranes that protect worms and facilitate leachate drainage should cover vermiculture systems. Liquid fertilizers produced by natural sources, such as animal manure and plant waste from irrigation systems, are known collectively as slurry (Makhkamova and Kamilov, 2023). Thanks to the physical qualities of slurry, large amounts of methane can be extracted and returned as soil nutrients to improve plant growth.

14.3.2 Advantages of Organic Fertilizers

Organic fertilizers are easy, ecological, and efficient alternatives that confer multiple benefits to plants. For example, organic fertilizers release nutrients slowly, gradually, and naturally in soils, which is one of the main advantages of their use. This natural release of nutrients allows for greater control over the amount of fertilizer delivered to plants, preventing excessive fertilizer application that can seriously and negatively affect soil quality. This degree of control also reduces the application frequency and, in turn, considerably reduces crop costs (Alzamel *et al.*, 2022). Inevitably, chemical fertilizers must be used more and more frequently, hastening the loss of soil fertility. Organic fertilizers that naturally regenerate soils by improving the quantities of beneficial, stabilizing compounds combat this problem by extending soil fertility (Mayer *et al.*, 2022). Given that organic fertilizers are 100% of natural origin, they do not produce any chemical residues that could greatly harm the environment. Thus, the use of organic fertilizers with few dangerous compounds or consequences provides peace of mind to producers and results in products that are closer to nature. Indeed, foods grown with organic fertilizers are free of harmful chemicals, ensuring those who consume them are unlikely to be exposed to diseases or health problems.

14.4 SLOW-RELEASE FERTILIZERS

The intention behind fertilization is to ensure nutrients are available to crops as they grow. Insufficient nutrients produce nutritional deficiencies, resulting in losses in crop yield and quality. On the other hand, when excess nutrients are present, nutritional imbalances that arise from wasted fertilizers can affect the appearance of crops and, in certain cases, result in phytotoxicity and dangerous environmental consequences.

To avoid these situations, nutrients must always be present in assimilable forms and necessary quantities throughout plant development. However, this is difficult to achieve in practice. Fertilizers must be used in a manner that is balanced by the demands of crops. In this way, high fertilizer efficiency and performance can be obtained along with high yields while minimizing costs and risks. As such, slow- and controlled-release and stabilized fertilizers (SCRSFs), also known as slow-release fertilizers, controlled-release fertilizers, or stabilized fertilizers, have been developed. These fertilizers exhibit enhanced efficiency, which stems from releasing nutrients at a certain rate beyond their application time, reducing losses, improving yields, and supporting environmental health (Kiplangat *et al.*, 2019).

Slow-release fertilizers release nutrients at a rate that reflects the demands of a crop. As such, these fertilizers are highly versatile. However, the behavior of a fertilizer in the laboratory or under pre-established test conditions will only sometimes align with that of the fertilizer under variable growing conditions in an open field. Generally, to achieve controlled nutrient release, the action rates of the most soluble and active fertilizers must be reduced to limit losses. After a single application, slow-release fertilizers can cover all of the nutritional needs of a crop. For these reasons, slow-release fertilizers are often synonymous with controlled-release fertilizers (Larios-González *et al.*, 2021). Indeed, there is no official difference between slow-acting and controlled-release fertilizers (Trenkel, 1997), as the European Committee for Standardization has proposed that they be considered as such if the nutrients contained in the fertilizers appear under defined conditions (e.g., 25°C).

14.4.1 Nitrogen-stabilized SCRSF Fertilizers

Some SCRSF fertilizers have nitrogen stabilizers, which extend the time that the nitrogen component remains in the soil as either urea or ammonia. In general, nitrogen-stabilized SCRSF fertilizers can be categorized as nitrification or urease inhibitors. Nitrification inhibitors, as their name suggests, inhibit the biological oxidation of ammoniacal N into nitrite or nitrate. Urease inhibitors inhibit the hydrolysis of urea into its ammonium and CO_2 components by urease.

Following the application of ammoniacal and urea fertilizers, principal nitrogen losses occur after nitrogen-containing compounds are converted to nitrates. Certain materials are toxic to nitrifying bacteria. When added to soil, these compounds can temporarily inhibit nitrification. Thus, these compounds reduce nitrate losses through leaching and denitrification and increase the yields of ammoniacal fertilizers, as well as the quantity of ammoniacal nitrogen that originates from the decomposition of organic matter in soil (Bolio-López *et al.*, 2008). However, nitrification should not be inhibited completely. As such, nitrogen-stabilized SCRSF fertilizers must be selective to ensure they only act on nitrifying microorganisms. In general, these products are very effective in preventing nitrate washing in sandy soils and denitrification in waterlogged soils.

Slow-release fertilizers ensure the slow hydrolysis of low-molecular-weight, water-soluble compounds. Semipermeable coatings, occlusion, protein materials, and other chemical forms ontrol solubility. Although slow-release fertilizers imply that they release nutrients at slower rates than usual, the release rates, patterns, and durations are not well-controlled (Trenkel, 1997). Slow-release fertilizers include both organic and inorganic products, including isobutylidene diurea (IBDU), methylene urea (MU), urea-formaldehyde (UF), and sulfur-coated urea (SCU).

14.4.2 Controlled-Release SCRSF Fertilizers

Controlled-release fertilizers are characterized by containing nutrients in their centers, which are then coated by polymer resins of varying thickness that control their release. Once controlled-release fertilizers are manufactured, the nutrient release rate only depends on the presence of water surrounding the granule and ambient temperature. The dependence on microorganism degradation, which is relevant in fertilizers coated with sulfur, is not important in this case. Nutrient release is achieved in two stages. Once the product is applied to the soil or substrate and exposed to moisture, water vapor infiltrates the fertilizer granule containing the completely soluble salt, creating a concentrated solution with high osmotic pressure. This elevated pressure within the granule drives the fertilizer solution through the micropores of the coating and into the soil or substrate (Cahill *et al.*, 2008).

14.5 BIOFERTILIZERS

Biofertilizers provide plants with organic nutrients necessary for their development while improving soil quality and helping to create an optimal and natural microbiological environment. Biofertilizers are essential to organic farming because they improve agricultural production and support large harvests without harming the environment. The use of biofertilizers aligns with organic farming guidelines, ensuring these compounds are respectful of the soil and nature and compatible with sustainable development.

Bacteria and fungi produce biofertilizers, allowing plants to naturally acquire nutrients to improve their performance by providing them with beneficial elements, mainly nitrogen and phosphorus (Khalil and Agah, 2017). These microorganisms transform atmospheric nitrogen into organic nitrogen, which can then be incorporated into plants. This process is natural, as the elements are extracted from the soil itself, and following mass production, they are returned to the plants and the earth. In this way, a cycle is closed in which no artificial or harmful products participate.

In general, biofertilizer quality depends on the raw materials used for production or, as indicated by Restrepo (2007), on the physical characteristics of color and smell. However, other variables determine biofertilizer quality with greater certainty. Soto and Meléndez (2004) proposed seven parameters to evaluate the quality of a biofertilizer, including safety, which refers to eliminating the possibility that a biofertilizer is damaging to human health. The main risks come from pathogenic microorganisms such as *Salmonella* spp., *Shigella* spp., and *Escherichia coli.*

14.5.1 Types of Biofertilizers

When rainwater circulates through accumulated organic matter, leachate forms due to fermentation and decomposition, extracting chemical compounds and biological materials from the organic matter. Some examples of fermented leachate come from compost and worm castings.

To investigate the effects of compost made from tobacco leaves on soil microorganisms during potting, Yu-Xin *et al.* (2024) applied the compost alone and in combination with a chemical fertilizer. The authors concluded that the simultaneous application of compost and fertilizer altered the ecological functions of the soil microbial community, enhancing the stochasticity of community formation. Hernández *et al.* (2017) evaluated the impacts of organic fertilization on the chemical and physical fertility of Luvisol soil found on a hillside used for agricultural purposes, concluding that adding compost improved soil fertility. Alves *et al.* (2023) studied the emergence speed index and morphological parameters of Wanda hybrid cherry tomato seedlings (*Solanum lycopersicum* var. pimpinellifolium) after applying worm humus, highlighting that morphological characteristics and dry biomass improved following humus application, especially when mixed with commercial substrates.

Mycorrhiza is a natural fertilization system based on mutually beneficial associations between mycorrhizal fungi and plants. By wrapping around plant roots, mycorrhizal fungi can receive sugars from plants while metabolizing atmospheric nitrogen, which they provide to plants. Worldwide, different environmental conditions likely affect mycorrhiza (Deng *et al.*, 2023); thus, mycorrhizal ecologists should develop specific hypotheses for each region to identify possible differences in mycorrhizae functioning between subtropical and temperate regions to better understand this form of symbiosis.

The *Trichoderma* genus of fungi is composed of beneficial, versatile, and multifaceted species that are found abundantly and naturally in many agricultural soils. *Trichoderma* is used in foliar applications and in seed and soil treatments to control various diseases caused by fungi and is also used to manufacture enzymes. Indeed, *Trichoderma* is commonly used as a fungicide because it is environmentally friendly, with benefits and minimal to no negative impact on humans, wildlife, pollinators, or insects. After colonizing plant roots, *Trichoderma* attacks, parasitizes, and feeds on other fungi while improving plant growth. In soil, *Trichoderma* behaves as a natural inhabitant, respecting the natural flavors of edible crops. In tomato plants, *Trichoderma* has been proposed as a sustainable tool and bioagent to promote plant growth, improve soil health, and support the antioxidant defense system of tomato plants (Kariuki *et al.* 2020). Moreover, *Trichoderma* has been found to be an efficient microbial antagonist to manage bacterial wilt in field-grown tomato plants (Wamani *et al.*, 2023).

14.5.2 Advantages of Biofertilizers

The use of biofertilizers is advantageous for multiple reasons. In addition to utilizing organic waste and improving crop productivity, biofertilizers require less energy to produce than traditional fertilizers, thus supporting sustainable resource management. Given that biofertilizers recover organic matter from the soil and promote carbon fixation, they improve the capacity of the soil to absorb water. By improving soil texture, biofertilizers also increase its useful life, ensuring optimal growing conditions and safeguarding the overall quality of the environment. In fact, biofertilizers can often replace chemical fertilizers entirely, with their use resulting in no environmental harm. Economically, biofertilizers cost less than chemical fertilizers. Indeed, biofertilizers cost approximately 10% of the average cost of chemical fertilizers.

14.6 BIOSTIMULANTS

In general terms, a biostimulant is defined as a mixture of two or more plant regulators and other substances, including amino acids, nutrients, and vitamins. Biostimulants can be obtained from plant extracts or compost containing humic, fulvic, and phytormonium acids (e.g., auxins, cytokinins, and gibberellins). Biostimulants are ecologically benign, as most are biodegradable, non-toxic, non-polluting, and non-hazardous to organisms (González-Morales, 2021; Husen, 2024; Yakhin *et al.*, 2017). Understandably, biostimulants for agricultural purposes have garnered ecological and commercial importance. Interestingly, the definition of "biostimulant" is not derived from the definition of "biostimulation" but rather precedes it due to the need for their regulation in the agricultural sector, which contains an important niche for biostimulants.

Europe has led the way in developing a working definition for agricultural biostimulants and a regulatory and legislative framework for their use. According to the European Union Fertilising Products Regulation, which came into force in July 2019, a plant biostimulant is a product that stimulates plant nutrition processes regardless of the nutrient content of the product with the sole objective of improving one or more characteristics of the plant or plant rhizosphere (EBIC, 2019). According to this regulation, biostimulants include humic and fulvic acids, amino acids and peptide mixtures, algae and plant extracts, chitosan and other biopolymers, inorganic compounds (e.g., Si, Al, Se, Co, and Na), beneficial mushrooms, hormones, and organic acids. The terms "plant biostimulant" and "agricultural biostimulant" encompass a diverse group of product technologies, including bacterial or microbial inoculants, biochemical materials, amino acids, humic acids, fulvic acids, and seaweed and plant extracts.

Biostimulation has been described as a general biological phenomenon that depends on the interactions between cellular molecular structures and external physical, chemical, and biological stimuli or impulses. Biostimulation alters metabolic processes and allows plants to use environmental resources more efficiently, substantially increasing growth and yields and improving tolerance to adverse environmental factors (Juárez-Maldonado *et al.*, 2019). Biostimulants can mitigate the negative effects of water stress by helping plants maintain physiological processes and recover quickly. They can help mitigate the harmful effects of climatic stressors when applied before or during stressful conditions (Soares da Silva *et al.*, 2022; Figure 14.4).

It is important to mention that biostimulants are perceived by cells because they contain signaling, regulatory, and second-messenger substances. Biostimulants trigger signaling cascades that reach transcription factors, activating genes responsible for forming defense compounds and antioxidants. Indeed, antioxidant compounds have been found to increase within 24 to 48 hours after a biostimulant is applied and can persist for seven to ten days following application. Thus, biostimulant application activates signaling without inducing plant stress as the plant prepares for future stressful conditions. The mixture of several classes of biostimulants in a product will generate greater signaling and, consequently, better plant tolerance to different stressors, in addition to substantially improving crop yield and quality (du Jardin, 2015).

Various studies have evaluated the consequences of applying biostimulants. Tosacano *et al.* (2024) applied natural biostimulants to sustainably mitigate damage due to water stress in melon (*Cucumis melo* L.). Chaoucha *et al.* (2023) examined the biostimulant effects of microalgae and thyme essential oil during *in vitro* and *ex vitro* strawberry rooting. Szparaga *et al.* (2023) explored the agricultural potential of *Levisticum officinale* WDJ Koch extract for soybean cultivation. Villa e Vila *et al.* (2023) reported that *Ascophyllum nodosum* algae extract improved the growth, yield, and soil fertility of tomato crops when applied as a biostimulant under subtropical conditions.

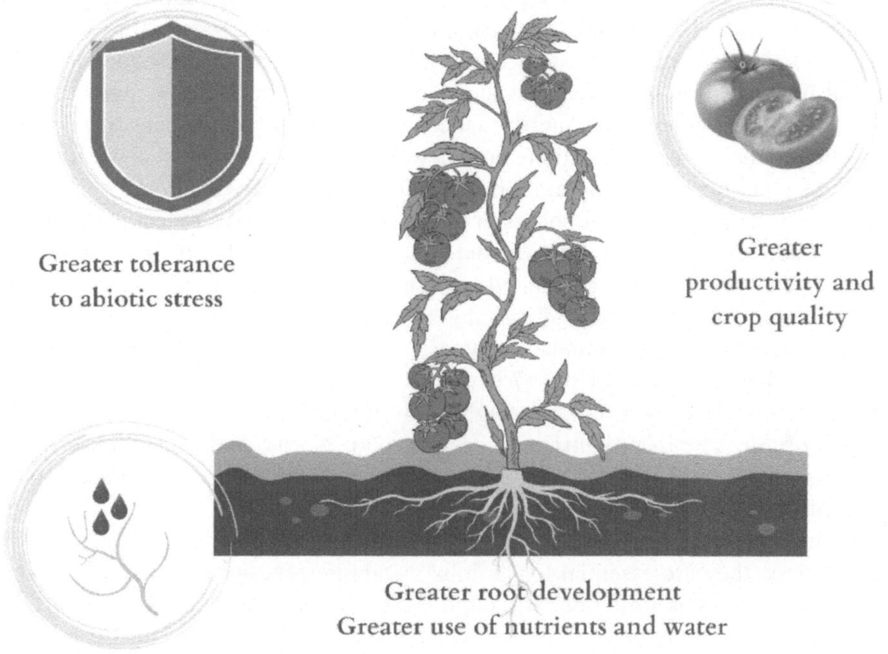

FIGURE 14.4 Effects of agricultural biostimulants on plants.

The main difference between biostimulants and fertilizers is that the latter provides nutrients to plants, whereas the former improves the efficiency of nutrient absorption mechanisms and tolerance to environmental stressors that alter physiological processes in plants and negatively affect development. However, without a doubt, the most important consequence of employing agricultural biostimulants is the reduction in fertilizer use, which supports one of the main objectives of the EU strategy "From Farm to Fork," namely, reducing fertilizer use by at least 20% by 2030.

14.7 NANOFERTILIZERS

Nanoscience is the study of phenomena and material manipulation at the scale of nanometers (nm; 10^{-9} m). The working range at the nano level spans a few tenths to a few hundred nanometers. Thus, nanotechnology involves producing and applying substances or structures smaller than 100 nanometers, which can be classified as well-defined or undefined nanoparticles. Class I nanoparticles are well-defined and include more than 90% monodisperse forms in terms of composition, size, mass, shape, chemical surface, and flexibility. Class II nanoparticles are undefined and only available in polydisperse forms, depending on composition, mass, size, and shape.

Nanoparticles are increasingly used in commercial and consumer products (Carrillo and González-Chávez, 2009). Some nanoparticles show rigid, latex-like, inflexible, and inorganic metallic structures and are referred to as hard nanoparticles. Other nanoparticles resemble biological structures and are relatively flexible, non-conductive, and insulating, forming covalent bonds similar to organic structures (Tomalia, 2009). These nanoparticles have non-crystalline, rather amorphous chemical structures composed of carbon, hydrogen, oxygen, and nitrogen (Tomalia, 2009). Nanoparticles exhibit varying morphologies, including spherical, cubic, and octahedral shapes. The morphology and chemical composition (e.g., Fe, Cu, Zn, or Ag) of a nanoparticle determines its physical, chemical, and biological properties.

Research on the applications of nanoparticles in agriculture (i.e., agro-nanotechnology) has gained momentum in the last decade. Agro-nanotechnology has the potential to improve conventional crop production systems while supporting sustainable agricultural production with reduced environmental impact. Nanoparticles can be used as controlled-release agrochemicals (e.g., nanofertilizers, nanopesticides, and nanoherbicides), nanosensors (i.e., devices that detect biotic or abiotic stress before they affect production), or in new genetic manipulation techniques that improve the efficiency of plant breeding programs. In the next section, we will provide examples of how nanoferilizers have helped minimize losses and reduce adverse environmental effects due to the excessive use of agricultural inputs (Carrillo and González-Chávez, 2009).

For agroecosystems to perform adequately and sustainably, the proper balance of sunlight and nutrients must be provided to plants and included in cultivation practices and technologies. Plants require 17 nutrients containing chemical elements in different concentrations and compounds to develop correctly. The first three of the 17 essential elements are carbon, oxygen, and hydrogen. These three elements are of non-mineral origin and obtained from air and water, while the rest are of mineral origin. The macronutrients nitrogen, phosphorus, and potassium are required in relatively high quantities and included in traditional (e.g., manure, guano, and compost) or industrial (chemical origin) fertilizer formulations.

Plants use the remaining 11 elements in small quantities; thus, they are referred to as micronutrients. Nonetheless, they are essential for various metabolic functions, including chlorophyll formation and macronutrient assimilation. Although these micronutrients are commonly available in soil, agricultural soils often require more than what is available naturally. For this reason, mineral supplements are added, including traces contained in irrigation water. In the case of hydroponics, nutrient solutions containing these micronutrients are applied to crops.

The nutritional needs of plants vary depending on the type of plant, developmental stage, and product to be obtained (e.g., tubers, flowers, fruits, or foliage). Nanostructured formulations applied to agricultural crops have been shown to increase fertilizer use efficiency and nutrient absorption from the soil, thereby reducing production costs and negative environmental impacts (Naderi and Danesh-Shahraki, 2013; Srilatha, 2011).

14.7.1 What Are Nanofertilizers and How Do They Work?

Nanofertilizers are nutrients encapsulated within nanoparticles. Nanofertilizers are created by encapsulating nutrients within porous materials, which are then covered by a thin layer of polymer (Rai *et al.*, 2012). Nanofertilization occurs when the encapsulated nutrients are released as particles or emulsions at the nanometric scale (Rai *et al.*, 2012). In different plant species, essential micronutrients (e.g., Cu, Zn, and Fe) in the form of nanoparticles of their oxides function as nanofertilizers. When these are applied to the root or foliage, plant growth is positively affected.

14.7.2 Nanofertilizers with Macronutrients

The disadvantages of nitrogen fertilizers include low utilization and the loss of nitrogen to the environment, increasing eutrophication and greenhouse gas emissions (Kahrl *et al.*, 2010). Due to rapid volatilization and leaching, commercially available nanofertilizers containing nitrogen (e.g., urea) disappear after 75% of the fertilizer has been applied. However, an advanced nitrogen nanofertilizer has been developed recently that employs urea-coated hydroxyapatite nanoparticles (i.e., rod-shaped structures with an average aspect ratio of 10) for targeted delivery and slow release (Kahrl *et al.*, 2010). These composite nanoparticles were selected for their rich sources of nitrogen and phosphorous and chemical compatibility.

Phosphorus forms part of enzymes, amino acids, proteins, nucleic acids (DNA and RNA), chlorophyll, adenosine triphosphate (ATP), phospholipids (membranes), and phytin. Adequate nutrition that includes phosphorus is essential for root growth and nodulation in leguminous plants while also improving plant photosynthesis, nitrogen fixation, flowering, and fruiting (Soetan *et al.*, 2010). Nanofertilizers containing phosphorous have been prepared and evaluated in a greenhouse experiment. For this, hydroxyapatite [$Ca_5(PO_4)_3OH$] nanoparticles (~16 nm) were applied to soybeans (*Glycine max*) in an inert culture medium (50% perlite and 50% peat moss). The application of hydroxyapatite nanoparticles increased the growth rate and seed yield by 33% and 20%, respectively, when compared to the effects obtained by applying a standard phosphorous fertilizer [$Ca(H_2PO_4)_2$]. Thus, soybean roots can efficiently absorb hydroxyapatite nanoparticles as phosphorous sources, maintaining healthy growth rates and high yields. The use of hydroxyapatite nanoparticles as a novel phosphate fertilizer could improve agronomic performance and increase the efficiency of phosphorous application. Hydroxyapatite nanoparticles have much weaker interactions with soil components than charged ions of PO_4^{3-}· HPO_4^{2-}, [H_2PO_4]$^-$, or Ca_2^+. Therefore, a notable portion of nanoparticles remains in the soil for root uptake, while most charged phosphate ions from conventional phosphorous fertilizers are adsorbed by soil particles and, therefore, unavailable for plants (Liu and Lal, 2014).

14.7.3 Nanofertilizers with Micronutrients

The effects of different concentrations of a CuO nanoparticle on aquatic algae (*Elodea canadensis*) have been studied. Following nanofertilizer application, photosynthesis was either stimulated with less than 0.25 mg/L or inhibited with more than 1 mg/L (Johnson *et al.*, 2011). The same biphasic effect was observed when ZnO nanoparticles were applied to peanut plants (*Arachis hypogaea*). When the nanofertilizer was applied in a dose of 1000 ppm (parts per million, relatively equivalent to micro-units µg/g or µL/L), seed germination and seedling vigor were promoted. However, when applied in a dose of 2000 ppm, the nanofertilizer resulted in negative and toxic effects on crop growth and yield (Abir and Muhittin, 2022). Recently, it was found that ZnO nanoparticles can act as cofactors (or auxiliary molecules) of enzymes that mobilize or solubilize phosphorus, which would help plant roots absorb compounds with this element (Al Jabri *et al.*, 2022).

Despite considerable amounts of iron in soil, many crops are iron-deficient due to the rapid conversion of this element into unavailable forms. Application of inorganic iron fertilizers to iron-deficient soils is usually ineffective. On the other hand, iron that is dynamically released from iron oxide nanoparticles (e.g., magnetite, maghemite, hematite, and ferrihydtrite) is suitable for plants. In this regard, the foliar application of Mg and Fe nanoparticle solutions has been tested in cowpea (*Vigna unguiculata*) plants. A combination of 0.5 g L^{-1} Mg nanoparticles and 0.5 g L^{-1} Fe nanoparticles increased the weight of 1000 seeds by 7% from 216 g (without Mg or Fe) to 232 g. The highest yield (245 g, reflecting an increase of 13.5% compared to the control) was obtained for the application of a combination of 0.5 g L^{-1} regular Fe salt and 0.5 g L^{-1} Mg nanoparticles, suggesting that foliar applications of these two elements increased photosynthetic efficiency. However, the exclusive application of nanoparticle Mg decreased the yield to 202 g (~6% compared to the control). The authors observed that the application of nanoparticle Mg improved Mg uptake in plant stems and leaves compared to the application of regular Mg salt, suggesting greater availability and mobility of the nanoparticle Mg (Delfani *et al.*, 2014). In addition, maghemite nanoparticles, a form of iron (III) oxide ($Fe_2O_3^-$), promoted the growth of peanut plants, suggesting that interactions were present between the regulation of phytohormone content and the activities of enzymes and antioxidants. Something similar was found to occur in wheat. Magnetite nanoparticles were found to promote the growth of this crop and its ability to prevent oxidative stress (Abir and Muhittin, 2022; Pariona *et al.*, 2017).

Finally, nanoparticles used as nanofertilizers can be applied by spraying foliage or supplying them to the soil for the roots to absorb, increasing the efficiency of micronutrient distribution, given that nutrient release is controlled. Micronutrient absorption can occur in several ways. For example, studies have suggested that nanoparticles can enter plant cells by binding to transport proteins, passing through aquaporins and ion channels, creating new pores (this pathway has been described for carbon nanotubes, with the nanotubes creating new pores to enter cells), or inducing endocytosis. In addition, nanoparticles can form complexes with membrane transporters or root exudates and subsequently be transported. Once inside cells, nanoparticles are transported apoplastically or symplastically. However, the exact mechanisms and reasons explaining why only some plant species readily absorb nanoparticles remain unknown (Kurepa *et al.*, 2010; Rico *et al.*, 2011).

14.8 CONCLUSIONS AND FUTURE REMARKS

Continual population growth, the increasing demand for food, and the ongoing reduction of cultivated land per capita have intensified the use of fertilizers worldwide. To provide sufficient food for the global population and reduce the harmful environmental effects of chemical fertilization and inappropriate disposal or reuse of agricultural waste, novel and modern technologies must be implemented in agroecosystems. Together, biotechnology and nanotechnology have the potential to transform agricultural practices and provide answers to immediate and long-term problems. We must look toward intelligent agricultural practices for the good of the environment and human health.

REFERENCES

Abir, D.O. and Muhittin, D. (2022) 'Physiological effect of ZnO nanoparticles on peanut (*Arachis hypogaea* L.)', *International Journal of Academic Multidisciplinary Research*, 6(3), pp. 192–197.

Al Jabri, H. *et al.* (2022) 'Zinc oxide nanoparticles and their biosynthesis: Overview', *Life (Basel)*, 12(4), p. 594.

Alves, S.E.E. *et al.* (2023) 'Emergence and structural characteristic of the *Solanum pimpinellifolium* in trays under different levels and types of substrates', *Horticulturae*, 9, pp. 2–15.

Alzamel, N.M. *et al.* (2022) 'Effect of organic and inorganic fertilizers on soil properties, growth yield, and physiochemical properties of sunflower seeds and oils', *Sustainability*, 14, pp. 2–18.

Bolio-López, G.I. *et al.* (2008) 'Dinámica de potasio en suelos cañeros', *Terra Latinoamericana*, 26(3), pp. 252–263.

Cahill, S. et al. (2008) 'Response of corn and cotton to starter phosphorus on soils testing very high in phosphorus', *Agronomy Journal*, 100, pp. 537–542.

Carrillo, G.R. and González-Chávez, M.C.A. (2009) 'La nanotecnología en la agricultura y rehabilitación de suelos contaminados', *Mundo Nano. Revista Interdisciplinaria En Nanociencias y Nanotecnología*, 2(2), pp. 50–63. Available at: http://www.scielo.org.mx/scielo.php?script=sci_arttext&pid=S2448-56912009000200050&lng=es&tlng=es (Accessed on 18 October 2023).

Chaoucha, R. *et al.* (2023) 'Assessing the biostimulant effect of micro-algae and thyme essential oil during in-vitro and ex-vitro rooting of strawberry', *South African Journal of Botany*, 162, pp. 120–128.

Chemical Fertilizers (2023) Available at: https://isam.education/en/advantages-and-disadvantages-of-chemical-fertilizers/ (Accessed: 29 July 2023)

Delfani, M. *et al.* (2014) 'Some physiological responses of black-eyed pea to iron and magnesium nanofertilizers', *Communications in Soil Science and Plant Analysis*, 45(4), pp. 530–540.

Deng, M. *et al.* (2023) 'Tree mycorrhizal association types control biodiversity productivity relationship in a subtropical forest', *Science Advances*, 9, pp. 1–11.

du Jardin, P. (2015) 'Plant biostimulants: Definition, concept, main categories and regulation', *Scientia Horticulturae*, 196, pp. 3–14.

EBIC (2019) Consorcio Europeo de la Industria de Bioestimulantes. Disponible en línea. Available at: https://www.ferbainternacional.com/consorcio-europeo-de-la-industria-de-bioestimulantes-ebic/ (Accessed: 10 July 2023).

El-Fouly, M. M. (2002) 'Quality of foliar fertilizers', *Acta Horticulturae*, 594, pp. 277–281.

Erisman, J.W. *et al.* (2008) 'How a century of ammonia synthesis changed the world', *Nature Geoscience*, 1, pp. 636–639.

FAO (2021) Uso óptimo de los fertilizantes para los cereales. FAO. Roma (Boletín de fertilizantes No. 3), p. 31.

FAO-IFA (2000) *Fertilizers and their use. A pocket guide for extensión officers.* 4th ed. FAO-International Fertilizer Industry Association (www.fertilizer.org), p. 70.

FAOSTAT (2023) Available at: https://www.fao.org/faostat/en/#home (Accessed: 2 October 2023).

Ferreira de Oliveira, O. et al. (2023) 'Application of organic and mineral fertilizers increases carbon fractions in two classes of aggregates in an integrated crop-livestock system', *Revista Brasileira De Ciencia Do Solo*, 47, p. e0220044.

Fertilizantes químicos (2023) Available at: https://bloglatam.jacto.com/fertilizantes-quimicos/ (Accessed: 17 November 2023).

Girma, K.L.E. *et al.* (2007) 'The magruder plots: Untangling te puzzle', *Agronomy Journal*, 99, pp. 1191–1198.

González-Morales, G. Y. (2021) *Uso de microorganismos como bioestimulantes sobre el crecimiento y desarrollo en Solanum betaceum* Cav, *Zea mays* L, *Solanum lycopersicum* Mill. Bajo invernadero. Documento de Maestría. Mexico: Universidad Veracruzana, p. 89.

Hernández, C.B.O. *et al.* (2017) 'Composts use to improve fertility of a slope Luvisol soil', *Revista Mexicana De Ciencias y Agrícolas*, 8(6), pp. 1273–1285.

Husen, A. (2024) *Current understanding of biostimulants in plant protection and performance*. Cambridge, MA: Elsevier Inc.

IFA, International Fertilizer Industry (1992) Urea Statistics January-June 1993 (Accessed: 20 September 2023). https://www.fertilizer.org/wp-content/uploads/2023/01/NOUVEAU-Fertiliser-Report.pdf

IFA, International Fertilizer Industry (2023) Urea Statistics January-June 2023 (Accessed: 5 September 2023). https://www.fertilizer.org/wp-content/uploads/2024/05/2024_IFA_Annual_Report_2023.pdf

Johnson, M.E. et al. (2011) 'Study of the interactions between Elodea canadensis and CuO nanoparticles', *Russian Journal of General Chemistry*, p. 1350.

Juárez-Maldonado, A. *et al.* (2019) 'Nanoparticles and nanomaterials as plant biostimulants', *International Journal of Molecular Sciences*, 20(162), pp. 2–19.

Kafjafi, U. and Tarchitzky, J. (2012) Fertirrigación Una herramienta para una eficiente fertilización y manejo del agua. Asociación Internacional de la Industria de Fertilizantes (IFA) Instituto Internacional de la Potasa (IIP) París, Francia y Horgen, Suiza.

Kahrl, F. et al. (2010) 'Greenhouse gas emissions from nitrogen fertilizer use in China', *Environmental Science and Policy*, 13, pp. 688–694.

Kariuki, C.K., Mutitu, E.W. and Muiru, W.M. (2020) 'Effect of *Bacillus* and *Trichoderma* species in the management of the bacterial wilt of tomato (*Lycopersicum esculentum*) in the field', *Egypt Journal of Biological Pest Control*, 30, pp. 1–8.

Khalil, N.H. and Agah, R.J. (2017) 'Effect of chemical, organic and bio fertilization on growth and yield of strawberry plant', *International Journal of Advances in Chemical Engineering and Biological Sciences*, 4(1), pp. 2349–2507.

Kiplangat, R. *et al.* (2019) 'Evaluating the effects of formulated nano-NPK slow-release fertilizer composite on the performance and yield of maize, kale and capsicum', *Annals of Agricultural Sciences*, 64, pp. 9–19.

Kurepa, J. *et al.* (2010) 'Uptake and distribution of ultrasmall anatase TiO_2 alizarin red S nanoconjugates in *Arabidopsis thaliana*', *Nano Letters*, 10(7), pp. 2296–302.

Laegreid, M., Bockman, O.C. and Kaarstad, E.O. (1999) *Agriculture fertilizers and environment*. Wallingford: CAB, p. 294.

Larios-González, R.C. *et al.* (2021) 'Pérdidas de nitrógeno por volatilización a partir de dos fuentes nitrogenadas y dos métodos de aplicación', *Siembra*, 8(2), p. e2475.

Liu, R. and Lal, R. (2014) 'Synthetic apatite nanoparticles as a phosphorus fertilizer for soybean (*Glycine max*)', *Scientific Reports*, 4, p. 5686.

Makhkamova, A. and Kamilov, B. (2023) Importance of humine preparation and organic fertilizers in improving the fertility of eroded typical gray soils. Web of Conferences, 37, pp. 1–6.

Malavolta, E. (2006) *Manual de nutrición mineral de plantas*. Sao Paola, Brasil: Editora agronómica CERES ltda, p. 631

Mayer, M. *et al.* (2022) 'Fertilizer quality and labile soil organic matter fractions are vital for organic carbon sequestration in temperate arable soils within a long-term trial in Switzerland', *Geoderma*, 426, p. 116080.

Mengel, K. and Kirkby, E.A. (2000) *Principios e nutrición vegeta*. Switzerland: International Potash Institute, p. 665.

Naderi, M.R. and Danesh-Shahraki, A. (2013) Nanofertilizers and their roles in sustainable agriculture.

Núñez, E.R. (1990) Aplicación de fertilizantes. Sección técnica de la Revista Fertimundo. No. 70 fertilizantes mexicanos, S.A.

Pariona, N. *et al.* (2017) 'Effect of magnetite nanoparticles on the germination and early growth of *Quercus macdougallii*', *Science Total Environment*, 575, pp. 869–875.

Rai, V., Acharya, S. and Dey, N. (2012) 'Implications of nanobiosensors in agriculture', *Journal of Biomaterials and Nanobiotechnology*, 3(2A), pp. 315–324.

Restrepo, R. J. (2007) 'Biofertilizantes preparados y fermentados a base de heces de vaca', in *Manual práctico ABC de la agricultura orgánica y panes de piedra*. 1st ed. Colombia: Cali, vol. 2, pp. 1–103.

Rico, C.M. *et al.* (2011) 'Interaction of nanoparticles with edible plants and their possible implications in the food chain', *Journal of Agricultural and Food Chemistry*, 59(8), pp. 3485–3498.

Salgado, G.S. and Núñez, E.R. (2012) *Manejo de fertilizantes químicos y orgánicos*. 2nd ed. Biblioteca básica de agricultura bba, Texcoco Edo. of Mexico: Mexico p. 145.

Soares da Silva, C.W. *et al.* (2022) 'Use of biostimulants in cotton seed treatment', *Research, Society and Development*, 11(5), p. e31211528396.

Soetan, K.O., Olaiya, C.O. and Oyewole, O.E. (2010) 'The importance of mineral elements for humans, domestic animals and plants: A review', *African Journal of Food Science*, 4(5), pp. 200–222.

Soto, G. and Meléndez, G. (2004) *Cómo medir la calidad de los abonos orgánicos*. Costa Rica: Manejo Integrado de Plagas y Agroecología, vol. 72, pp. 91–97.

Srilatha, B. (2011) 'Nanotechnology in agriculture', *Journal of Nanomedicine and Nanotechnology*, 2, p. 123.

Stewart, M.W. et al. (2005) 'The contribution of commercial fertilizer nutrients to food production', *Gronomy Journal*, 97(1), pp. 1–6.

Szparaga, A. et al. (2023) 'Exploring the agro-potential of extract from *Levisticum officinale* WDJ Koch in soybean cultivation', *Industrial Crops and Products*, 203, p. 117235 1–15.

Tomalia, D.A. (2009) 'In quest of a systematic framework for unifying and defining nanoscience', *Journal of Nanoparticles Research*, 11(6), pp. 1251–1310.

Tosacano, A.M.L. *et al.* (2024) 'Natural biostimulants foliar application as sustainable mitigation strategy of drought stress damage on the melon crop (*Cucumis melo* L.)', *Scientia Horticulturae*, 323, p. 112471.

Trenkel, M.E. (1997) *Controlled release and stabilized fertilisers in agriculture*. Paris: IFA.

Villa e Vila, V. *et al.* (2023) 'Seaweed extract of *Ascophyllum nodosum* applied in tomato crop as a biostimulant for improving growth, yield and soil fertility in subtropical condition', *Journal of Applied Psychology*, 35, pp. 1–11.

Wamani, A.O. et al. (2023) 'Efficacy of microbial antagonists in the management of bacterial wilt of field-grown tomato', *Journal of Natural Pesticide Research*, 6(2023), p. 100051.

Xu, G., Fan, X. and Miller, A.J. (2012) 'Plant nitrogen assimilation and use efficiency', *Annual Review of Plant Biology*, 63, pp. 153–82.

Yakhin, O.I. et al. (2017) 'Biostimulants in plant science: A global perspective', *Frontiers in Plant Science*, 7, pp. 1–32.

Yu-Xin, W. et al. (2024) 'The mechanism of microbial community succession and microbial co-occurrence network in soil with compost application', *Science of The Total Environment*, 906, p. 167409.

15 Perspectives of Soil Enzymes in Sustainable Crop Production

Mohamed A. Amin, Hossam M. Fouda, and Mahmoud S. Osman

15.1 INTRODUCTION

Atmospheric temperature will rise by an additional 2.7°C by 2100, after rising by more than 1°C during the 1900s. Being the largest sink of carbon (C) in terrestrial ecosystems, soils are among the most significant problems raised by several international studies about the impact of global warming on soil C storage (Masson-Delmotte *et al.*, 2021; Shukla *et al.*, 2019). As microbes play a major role in the cycling of organic matter and long-term C stabilization in soil (García-Palacios *et al.*, 2021), it is critical to evaluate the extent and direction of the effects of global warming on microbial communities and soil C cycling rates (Allison *et al.*, 2010). Rising temperatures may boost soil microbial activity in a range of soil types and habitats, according to evidence (Chen, 2015; Fanin *et al.*, 2022). A positive feedback loop on global warming can be created by increased microbial activity, leading to faster breakdown rates, which can then release CO_2 produced from the soil into the atmosphere and reduce soil C storage (Bardgett *et al.*, 2008; García-Palacios and Chen, 2022). However, there is little confidence in the estimates of carbon-climate feedbacks in global models due to the extremely erratic reactions of microbial communities to warming (Sulman *et al.*, 2018).

One of the primary limiting mechanisms preventing the decomposition of soil organic matter is soil enzymes, which are mostly generated by microorganisms (Burns *et al.*, 2013). High-molecular-weight organic chemicals are broken down into smaller oligomers or monomers by soil enzymes, which are then taken up by cell-wall receptors and entered by microorganisms. Since it is sometimes required to comprehend the breakdown of organic matter at a very tiny scale in order to predict ecosystem functioning at greater spatial scales (Allison *et al.*, 2010; Bradford *et al.*, 2021), monitoring variations in total enzyme activity may be used to forecast biogeochemical reactions related to the cycling of sulfur, nitrogen, phosphorus, and carbon in terrestrial environments. This chapter evaluates soil enzyme functional diversity within major nutrient cycles, providing an overview of their past uses and suggesting methodologies for this purpose and the role of soil enzymes in sustainable crop production.

15.2 ENZYMES AND SOIL QUALITY

Soil quality is an important aspect in agricultural plant development and determines plant nutrient availability. To meet the nutritional needs of the rising population by 2030, food grain output must be increased by 280 million tons (Singh *et al.*, 2016). Since the green revolution, the use of chemical fertilizers has increased food grain output by a factor of ten (Dotaniya *et al.*, 2019; Dotaniya and Kushwah, 2013). Conversely, owing to the overmining of soil minerals beyond those provided by fertilizers, loss of soil organic matter, and changing climate conditions, this pace of grain production cannot be sustained indefinitely (Meena *et al.*, 2016). The functional state of soil, including its ability to support plant, animal, and human health as well as crop productivity, is referred to as soil quality and health. Biological soil quality refers to the biological role of soil in development and production (Doran and Zeiss, 2000). Soil health refers to the state of the soil as a dynamic living system, influenced by interactions with other living things like symbiosis or disease. It emphasizes

DOI: 10.1201/9781032630366-15

the importance of conservation and upkeep of the soil ecosystem to preserve its biological activity and biodiversity (Allen *et al.*, 2011). The Soil Science Society of America defines soil quality as soil's capacity to support biological production, preserve environmental integrity, and advance the health of plants and animals while remaining within ecological bounds (Herrick, 2000). It may be enhanced with thoughtful management strategy formulation and is responsive to management practices (Rao *et al,* 2014).

Factor efficiency partially declines with time, accompanied by a rise in fertilizer requirement for a comparable level of production. Farmers are frequently misled by the idea that using excessive amounts of fertilizer and pesticides increases crop yields proportionately, consequently diminishing soil biological quality. The rapidity of soil organic matter decomposition influences the progress of nutrient transformation along with the accessibility of plant nutrients in the soil solution (Omar *et al.*, 2017). Higher nitrogen fertilizer quantities or a split application are required due to the increased volatilization and leaching loss of nitrogenous fertilizers (Saka *et al.*, 2017). In sandy soils, pesticide and fertilizer residues contaminating groundwater are a global problem. In fields, overuse of fertilizers such as phosphatic can build up (Dotaniya *et al.*, 2013). In today's agriculture, adjustments to the pace, timing, and climate are essential for crop production that is sustainable (Dotaniya and Meena, 2015; Verma *et al.*, 2017).

Many variables, including changes in land use, govern extracellular enzymes, which are susceptible to environmental changes. Additionally, many factors have an impact on the catalytic activity of these enzymes (Acosta-Martínez *et al.*, 2019; Adetunji *et al.*, 2020) to connect to agricultural yields (Castro Lopes *et al.*, 2018; de Carvalho Mendes *et al.*, 2019). Enzyme activity is therefore one of the most important biological markers of soil health and quality. Variations in land utilization and managing soils have a substantial impact on the soil's microbial population, enzyme activity, and nutritional quality. For example, the substrate that the microbial community accesses is changed when forests are cleared for annual crops (Malik *et al.*, 2018; Sobucki *et al.*, 2021). Compared to annual crop residues with lower lignin ratios, which directly impact the rates at which plant cell wall polymers degrade and soil carbon sequestration plant residues in natural forest ecosystems have a greater lignin content, which protects cellulose from enzyme assault (Puttaso *et al.*, 2011).

A greater exposure of substrate to enzymatic assault and accessible organic carbon results from soil disturbance in agricultural situations, which causes organic carbon loss (Ghimire *et al.*, 2019). Nevertheless, owing to a decrease in substrate supply, soil enzymatic activity falls off following fast degradation. Additionally, disturbance of the soil limits substrate and enzyme diffusion and catalysis by increasing evaporation and decreasing soil moisture (Manzoni *et al.*, 2014). Higher soil temperatures can quickly boost enzyme activity and, as a result, accelerate the breakdown and mineralization of organic carbon due to increased exposure to sun radiation (Hou *et al.*, 2016; Liu *et al.*, 2019). In light of the current global warming scenario and the amplified greenhouse effect, this is a very undesirable outcome (Hassan *et al.*, 2015; Looby and Treseder, 2018). Temperature increases over acceptable limits can decrease the half-life, substrate affinity, and catalytic effectiveness of enzymes because of conformational changes and denaturation of proteins (Alvarez *et al.*, 2018).

15.3 SOIL ENZYMES' FUNCTION IN NUTRIENTS' AVAILABILITY AND YIELD PRODUCTION

The primary plant nutrients, C, H, O, N, P, and K, comprise the structural and functional molecules of the plant body. In smaller levels, micronutrients like B, Cl, Cu, Fe, Mn, Mo, and Zn as well as secondary nutrients like Ca, Mg, and S are crucial for plant growth. Of these, photosynthesis and soil absorption provide C, H, and O, which are obtained from CO_2 and H_2O, respectively. The earth provides additional nutrients in the form of minerals. The biogeochemical cycles of these nutrients are initiated and maintained in large part by soil enzymes, which also supply the fundamental fertility support for plants to grow healthily (Fang *et al.*, 2010). By facilitating the contact area of the reacting agents, such as the substrate and water during hydrolysis, soil enzymes speed up the breakdown of organic crop residue and liberate plant-derived nutrients toward the soil solution (Dotaniya *et al.*, 2017). The unique habitat known as the rhizosphere (the soil zone in contact with the root and impacted by root activities or exudates) is where microorganisms are selectively enhanced. Because of this, the rhizosphere zone has more abundant enzyme activity than the bulk soil when it comes to facilitating the biogeochemistry of minerals, which benefits soil-plant ecosystems more (Gobran *et al.*, 1998). The soil enzymes that are essential for the cycling of C, N, P, and S are among them. Although enzymes that regulate C cycling are unable to supply minerals for growing plants, they're necessary for the growth of soil microorganisms, which in turn promote plant growth indirectly. They break down complicated organic carbon compounds into simpler, more usable forms, including sugars and organic acids. Microorganisms and plants can use the inorganic compounds of N, P, and S enzymes, which mineralize the corresponding nutrients from the organic components in the soil. The physical characteristics of soil, such as pH, EC, soil texture, and structure, have an impact on the composition, activity, and persistence of enzymes because these variables regulate the availability of nutrients for microbes. Thus, microbial respiration reflects their activity (Fang *et al.*, 2010) as well as microbiological diversity. Although sandy soils have a lower organic matter concentration than clay soils, they are more aerated and hence restrict the growth of microorganisms. These types of soils exhibit limited variety and population density of microorganisms. In high pH saline and alkaline soils (Tabatabai, 1994), the process by which microbes engaged in the metabolism of plant wastes produce soil enzymes (Saiya-Cork *et al.*, 2002) and other substrates (Fang *et al.*, 2010; Rao *et al.*, 2014) is greatly reduced in a similar vein. Soils that have been severely degraded by chemical crop management techniques have lower soil fertility and less potential for crop development (Dotaniya *et al.*, 2016). The microbial count and soil fertility are increased by adding crop residue, organic manure, farmyard manure (FYM), and biofertilizers to the soil as well as by reclaiming damaged land with specific amendments, such as gypsum in alkaline soils, lime in acidic soils, and FYM in heavy metal-contaminated soils (Gianfreda and Rao, 2014). Plant yield and soil enzymatic activity have been found to significantly correlate in several research studies, although there is no clear association seen in others. For instance, the dehydrogenase activity (DHA), arylsulfatase activity (ASA), and urease activity (UA) in soil were favorably linked with the mean grain yields of several crops grown in rainfed production systems (Mandal *et al.*, 2007). While groundnut monocropping and pearl millet with three enzymes were not significantly and positively connected with DHA, ASA, and UA activities, the grain yields of groundnut-finger millet, finger millet monocropping, winter sorghum, soybean, and upland rice were (Saha *et al.*, 2008). Kanchikerimath and Singh (2001) reported improvement correlation between yields and all the soil enzymes studied. Similarly, using linear regression models, Alves de Castro Lopes *et al.* (2013) analyzed the enzymatic activity (ASA, acid phosphatase, β-glucosidase, and cellulase) in relation to the relative cumulative yields of soybean and corn. Numerous factors, including soil quality, plant genotype, and biotic and abiotic stressors have an impact on plant productivity. Furthermore, the relationship between plant productivity and enzyme activity may be influenced by a wide range of other variables (external inputs) in controlled ecosystems.

15.4 SOIL ENZYMES

Soil plays a crucial role in maintaining terrestrial ecosystems through chemical, physical, and biological processes (Dick and Burns, 2011). It filters water, absorbs environmental pollutants, and provides over 98% of human food and the world's supply of grain, fiber, oil, and wood (Kopittke *et al.*, 2019). Plant nutrients are stored in the organic matter of the soil until they are released or changed into forms that the plants can use. Nutrient transformations are driven by microflora, which mineralizes nutrients through growth and activity. Extracellular or endogenous microbial enzymes are essential to this process. Plants can use nutrients that are inorganic or organic because extracellular enzymes break them down or change their forms (Meena et al., 2016). Together, these enzymes are referred to as soil enzymes, and they serve a substantial function in keeping the health and fertility of the soil as well as safeguarding the environment by breaking down pollutants (Stirling *et al.*, 2016).

The microbial population and its composition of active soil enzymes determine the availability of nutrients and, consequently, the health of the soil at any given moment. Organic matter is essential for soil fertility, the global carbon cycle, and heavy metal complexation (Gmach *et al.*, 2019). Soil organisms rely on enzyme-mediated catalytic activity, and soil nutrients are crucial for soil fertility and the global carbon cycle (Dotaniya *et al.*, 2019). For example, Badzinski *et al.* (2021) and Baldoni *et al.* (2020) have shown a direct correlation between soil enzymatic activity and the transformation of organic matter and cycling of nutrients, as well as the biological degradation of harmful organic pollutants. The rhizosphere is a distinct soil zone impacted by root activities that has selective enrichment of microorganisms, resulting in more dominant enzyme activities and improved soil-plant ecosystem feeding (Fang *et al.*, 2010).

The cytoplasm, membranes of living cells, excretion into soil solutions, complexing in the soil matrix, and microbial detritus are all places where soil enzymes can be found (Dick and Kandeler, 2005). Intracellular enzymes (IE) are enzymes that function on particular metabolic pathways within the cellular milieu. Their primary mode of action is on tiny substrates that are absorbed by microbial cells in a regulated environment. To hydrolyze high-molecular-weight organic substrates into oligomers or monomers, extracellular enzymes (EE) are released into the extracellular environment or bound to cell walls (Burns *et al.*, 2013). Indicators of alterations in the soil environment, soil IE, and soil EE can give information on nutrient release in the soil via microbial activity and the breakdown of organic residue (Kumar *et al.*, 2013). Since enzyme-mediated catalysis is essential for many important biological soil processes, soil quality is defined as soil's capacity to operate within the bounds of the ecosystem, support biological productivity, preserve environmental quality, and enhance plant and animal health (Lehmann *et al.*, 2020).

15.5 SOIL EXTRACELLULAR ENZYME ECOLOGY

Temperature, pH, moisture content, and substrate all affect the activity of extracellular soil enzymes. Since enzymes cannot express their catalytic activity without the substrate, it is the primary limiting factor. Several enzyme groups, including cellulose, lignin, hemicellulose, starch, and pectin, must operate simultaneously or sequentially to break down major organic compounds (Wallenstein and Burns, 2011). Plant cell wall polymers are the main substrate that microbes use to break down materials in soil. Enzymes that are hydrolytic (like cellulases) and oxidative (like laccases and peroxidases) contribute to their breakdown. Plant cell walls are made up of cellulose polymers linked to hemicellulose, lignin, pectin, and other proteins. The lignin forms a barrier around the cellulose, creating a compact and stiff structure (Chen and Chen, 2014). Oxidative enzyme activity, which is an energetically unfavorable process, is necessary to depolymerize lignin before soil cellulases can effectively access cellulose (Chauhan, 2020; Moorhead *et al.*, 2013). Because of this structural variability, a variety of different enzymes generated by microbes from various groups must work together to break down these polymers, even if they do so in a cooperative manner (Janusz *et al.*, 2017). Soil moisture has a significant role in determining the community of microbes and, in turn, enzyme catalysis. Only watery media can support hydrolytic enzyme reactions (Lupatini *et al.*, 2019). Guber *et al.* (2018) state that transport channels and diffusion activities within an enzyme are necessary for contact between it and a substrate. According to Baveye *et al.* (2018) and Schimel (2018), extracellular enzymes diffuse across the water content of the micro- and macropores in the direction of the substrate and away from the parent cells. Temperature variations above or below the optimal temperature at which an enzyme reaches its highest efficiency might slow down a process until the enzyme is completely inactivated (Yang *et al.*, 2019). In order to sustain extracellular enzyme activity in the face of temperature fluctuations in the environment, microorganisms have devised biochemical methods. They secrete highly thermally stable enzymes that can catalyze stable reactions across a broad temperature range, or they can synthesize numerous isoenzymes that have comparable substrate affinities but differ in optimum temperature. For instance, protease enzyme always may retain their thermal stability and affinity within the whole temperatures range 0 -40°C (Razavi *et al.*, 2016).

When it comes to interfacial attraction and repulsion, an enzyme's pH has a direct impact on its electrostatic behavior (Datta *et al.*, 2017). Under the microorganism's control, the pH of live cells is around 7.2 (Wade *et al.*, 2021). Extracellular enzymes are dependent on their pH environment when they are discharged into soil; increased activity of certain enzymes is indicated by soil pH values close to neutrality (pH 6.0–7.0) (Burns *et al.*, 2013). For the purpose of quantifying soil enzyme activity, tested procedures that have been established for a range of soil conditions are needed, such as colorimetric measurement techniques (Wang *et al.*, 2023). Data comparability and appropriate laboratory conditions are guaranteed by these guidelines. Because it is more prone to mistakes, minimal dirt amounts like in the microplate method are essential. By adhering to these procedures, the chance of mistakes is reduced, and accurate results are guaranteed (Nannipieri *et al.*, 2018).

15.6 SOIL ENZYMES AND THEIR FUNCTIONS

The oxidoreductase, hydrolase, and lyase enzyme classes include the significant soil enzymes. The enzymes engaged in the oxidation of various substrates, as well as laccases and dehydrogenases specialized for substrates, are all considered oxidoreductases. There are no external enzymes associated with oxidation, such as dehydrogenases, because they are confined to cells. It is demonstrated that laccases are extracellular. Hydrolases, which disassemble macromolecules into smaller forms and eliminate inorganic groups or ions, make up most soil enzymes. Among these are amidases, urease, sulfatases, phosphoesterases, cellulases, glucosidases, and amylases. On the other hand, lyases like ammonia lyases, dehydratases, and carboxylases eliminate specific chemical species without requiring hydrolysis. While other kinds of amino acids are less significant ecologically and in the soil environment, ammonia lyases deaminate amino acids without hydrolysis (Figure 15.1).

15.6.1 Dehydrogenase Activity (DHA), Amylase, Cellulase, and b-Glucosidase

Dehydrogenases are extracellular enzymes that are present in all living microorganisms, indicating the microbial activity present in soil. As they mediate the oxidation-reduction processes of the cells, they are crucial carriers of electrons and hydrogen ions (protons). Dehydrogenases are

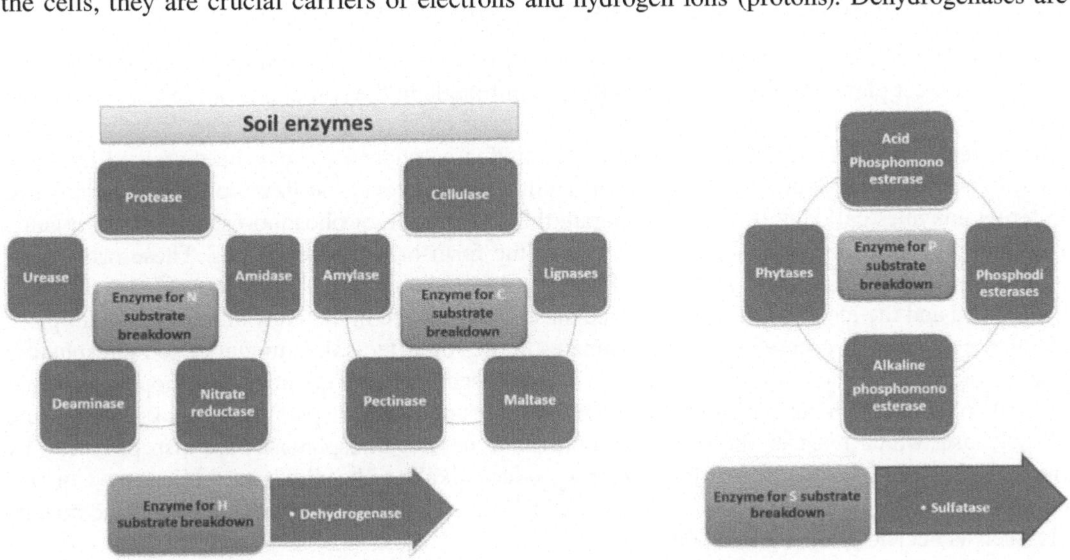

FIGURE 15.1 Different types of soil enzymes.

essential components of respiration and electron transport chains, transferring hydrogen to produce reductants through the oxidation of various substrates. Depending on whether nicotinamide adenine dinucleotide is phosphorylated, these dehydrogenases transmit H^+ ions on it (Rao *et al,* 2014). As an excellent indication of biological activities in soil, soil DHA reflects the whole oxidative metabolic activity of soil bacteria (Gu *et al.*, 2009). Soil type, moisture content, and the redox state (aeration) of the soil all have an impact on the overall DHA of soils (Lisowska *et al.*, 2023). A rapid transition from aerobic to anaerobic activity is encouraged by raising the water volume beyond the ground's potential to submerged conditions. It is employed as a more accurate measure of oxidative activity in microorganisms (Kumar and Garkoti, 2022). A class of enzymes known as amylases breaks down starch into short-chain polysaccharides known as dextrins during the process of starch hydrolysis (Neemisha and Sharma, 2022). Microorganisms and plants both manufacture these enzymes, although seldom do both α and β amylases come from the same organism. Amylases have an impact on vegetation types, agronomic techniques, and the carbon cycle. Amylase is a soil enzyme; however, its substrate starch is not a significant component of soil organic inputs. The origins of α and β amylases might differ, and they are seldom produced in the same organism (Paul and Sahoo, 2022). Cellulase, a set of enzymes with β (1 → 4) glycosidic linkages, is also involved in the breakdown of cellulose, a glucose polymer. Cellodextrins are produced when endocellulase hydrolyzes internal glycosidic linkages, while cellobioses and cellotrioses are eliminated from the nonreducing terminal of cellulose by exocellulase hydrolyzing this same structure. The enzyme called β-glucosidase hydrolyzes the β (1 → 4) glycosidic bond found in cellobioses and cellotrioses, releasing glucose molecules. The soil microbial population needs to break down the high cellulose content of the organic matter inputs, which include plant wastes and litter, before they can be used. Soil enzymes that are significant are cellulases and β-glucosidase, the latter being especially important for breaking down low-molecular-weight carbohydrates found in soil organic matter (Ajwa and Tabatabai, 1994). Similar to soil DHA, β-glucosidase, an enzyme produced by soil bacteria, is becoming an important indication of microbial activity. Since glucose is the most common carbon source for soil microorganisms, this extracellular enzyme influences their proliferation by releasing glucose during the breakdown of organic materials (Acosta-Martínez and Tabatabai, 2011; Rao *et al,* 2014). In contrast to dehydrogenase, β-glucosidase exhibits stability and less seasonal variation (Turner *et al.*, 2002).

15.6.2 Phosphatases, Protease, Urease, and Arylsulfatase

With 15–20% utilization efficiency, phosphorus (P) is a vital growth-limiting nutritional element that is scarce for plants. More than 80% of P stays immobile in the soil and is not readily accessible during crop development, even with artificial fertilizers. Ores contain bound phosphorus, which is only released in acidic environments. Small quantities of phosphates can be dissolved by some microorganisms and plant roots, but higher acidity also prevents the development of organisms. External enzymes called phosphates accelerate the dehydration of phospho-ester bonds in organic P-containing substrates, releasing inorganic P in the form of orthophosphates. These orthophosphates are used by plants and soil biota. Phosphates are classified based on the amount of bonds hydrolyzed and the ideal pH for action. Since these enzymes are inductively adaptable (Nannipieri, 1994), the quantity of accessible orthophosphates in the soil affects the amount of soil phosphatase that is available. The activity of phosphatases is decreased when the quantity of orthophosphates in the soil solution is increased (Šarapatka, 2003). In times of P deficit, plant roots also generate acid phosphatase, which releases the soil's immobilized P in situ. Phosphatases are also produced by fungi, mainly acid phosphatases. Most bacteria produce alkaline phosphatases, while some can produce both. Phosphodiesterase could liberate orthophosphates that are connected to organic materials (such as nucleic acids) by two bonds (Figure 15.2).

Protease is a soil enzyme that is crucial for the mineralization of organic nitrogen in soil. It is associated with colloidal substances in soil, both biological and inorganic (Chen and Arai, 2023).

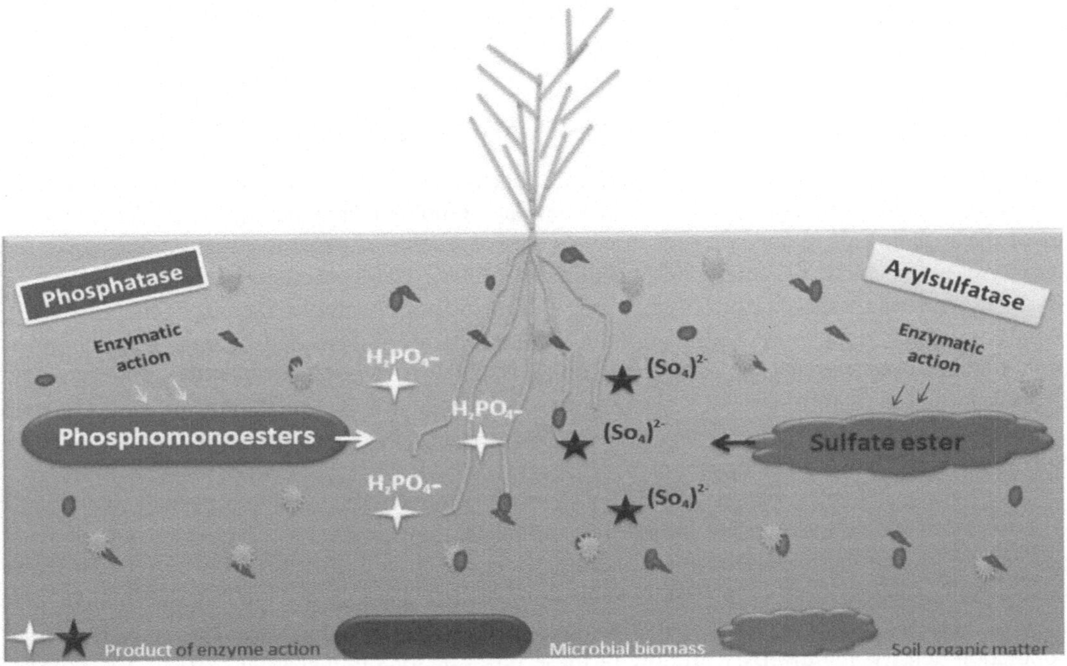

FIGURE 15.2 Enzyme phosphatase and arylsulfatase in soil.

Being extracellular, it is essential to preserve ecosystem function and provide nitrogen nutrition for plant development (Burns, 1982). Proteins are broken down by protease, which hydrolyzes their peptide bonds and releases small peptides that are then broken down even more to reveal their component amino acids. For the soil microorganisms, the amino acids serve as sources of nitrogen. They might either be ingested by the bacteria or undergo more mineralization, which would produce ammonia. During crop production, protease plays a critical role in the nitrogen mineralization of soils receiving organic crop residual inputs, helping to satisfy plant nitrogen needs.

Urea is hydrolyzed into ammonium and carbon dioxide by urease enzymes that are secreted by the soil microbiota (Fazekašová, 2012). Nearly every type of soil microorganism, including bacteria, fungus, yeast, and algae, as well as the roots of certain plants, generates urease (Follmer, 2008). In microorganisms, this soil enzyme can be found both extracellularly and intracellularly (Mobley and Hausinger, 1989). The urease enzyme raises the ammonical nitrogen concentration, which affects the nitrogen dynamics in soil. When urea is used as a fertilizer, ammonium is released into the soil, raising its pH. The atmosphere is negatively impacted by the volatilization of ammonia, which can lead to the loss of released nitrogen (Rahul *et al.*, 2022). It mostly lowers the effectiveness of nitrogen fertilizers that are applied. The emphasis of research is on how urease activities are regulated during crop growth seasons. Soil urease's action is lessened by slow-release urea, such as polymer-coated urea (Das and Varma, 2011).

Organic matter is the main source of sulfur in soil; bacteria mineralize it into inorganic sulfate. The hydrolysis of sulfur esters in soil is mostly dependent on soil enzymes called sulfatases, which liberate inorganic sulfur from the organic substrate into the soil solution (Kertesz and Mirleau, 2004). According to McGill and Cole (1981), soils with low or restricted sulfur content cause the bacterial population to produce sulfatases. Arylsulfatase concentration in soil and the external application of S sources are mostly influenced by soil bacterial populations (Vong *et al.*, 2003). Arithmetic sulfatase is the sole enzyme that can hydrolyze aromatic sulfate esters into phenols (R-OH) and sulfate sulfur (SO_4^{-2}) by breaking the oxygen-sulfur (O-S) bond and converting inorganic S to plant absorption (Tabatabai, 1994). The amount of inorganic S present in the soil influences the activity of this enzyme, particularly while crops are actively growing.

15.7 THE KINETICS OF SOIL ELEMENTS AND ENZYMES

Molecules of carbon, nitrogen, sulfur, and phosphorus make up organic compounds; these molecules are shaped by biological activity. The interplay of multiple enzyme types affects the dynamics of these chemicals in soil. Key reactions in the breakdown of organic compounds involve the activity of extracellular enzymes (Sinsabaugh *et al.*, 2008). These reactions include the hydrolysis of organic nitrogen reservoirs such as proteins, chitin, and peptidoglycan; the biogeochemical cycle of carbon (which involves the degradation of cellulose, hemicellulose, and lignin); the mineralization of phosphorus from nucleic acids, phospholipids, and ester phosphates; and the catalysis of the hydrolysis of organic sulfur esters. Soil enzymes such as cellulases (C cycle), protease and urease (N cycle), phosphatase (P cycle), and arylsulfatase (S cycle) have been studied in relation to soil nutrition and health (Henry, 2013; Turner *et al.*, 2016).

The physical characteristics of soil – pH, EC, texture, and structure – as well as their interactions with microbes to regulate nutrient availability, all have an impact on the composition, activity, and persistence of enzymes. Consequently, microbial respiration and microbial diversity both exhibit their activity (Abdel Latef *et al.*, 2021; Fang *et al.*, 2010). Though the low organic matter level inhibits the development of microorganisms, sandy soils are more aerated than clay soils. Microbial population density and diversity are low in these types of soils (Omer *et al.*, 2022). Soil enzyme production from microorganisms involved in the utilization of plant residues (Saiya-Cork *et al.*, 2002) and other substrates (Das and Varma, 2011; Fang *et al.*, 2010) is greatly reduced in saline and alkaline, which are high-pH soils (Tabatabai, 1994). Similarly, soils degraded as a result of heavy chemical crop management methods have low soil fertility and crop production potential (Dotaniya and Meena, 2015). Organic manure, crop residue, biofertilizers, and soil reclamation with specific amendments increase microbial count and soil fertility (Rao *et al,* 2014).

15.8 PRINCIPAL CYCLES-RELATED SOIL ENZYMES

15.8.1 Carbon Cycle

Glycosidic linkages hold together a lengthy chain of glucose molecules to form cellulose, the most prevalent natural polymer. Heterotrophic soil bacteria use it as their main energy source; however, they are unable to directly access the glucose molecules found in cellulose structures. This access is dependent on the cellulose being broken down by three extracellular enzymes collectively referred to as cellulases (Sobucki *et al.*, 2021).

Entoglucanase, an enzyme of the glucanase family, is responsible for the initial phase of cellulose depolymerization. According to Wallenstein and Burns (2011) and Chen (2015), endoglucanase (endo-1, 4-β-glucanase; EC 3.2.1.4) breaks lengthy cellulose chains by randomly cutting β-1, 4 glycosidic linkages. The disrupted molecular chains have one reducing end and one non-reducing end. Later, the enzyme cellobiosidase acts on the short cellulose chains. It is made up of two parts that cut glucose and cellobiose from the reducing end of the long chain, respectively (exo-1, 4-β-glucanase; EC 3.2.1.91) (Thapa *et al.*, 2020).

Following the action of cellobiosidase, two glucose molecules connected by β $(1 \rightarrow 4)$ links produce cellobiose fragments ($C_{12}H_{22}O_{11}$), which are disaccharides. According to Wallenstein and Burns (2011), β-d-glucosidase (EC 3.2.1.21) catalyzes the reduction of cellobiose and other water-soluble cellodextrins to glucose molecules, which microbial cells can subsequently take up. Owing to organisms' unique cellulose-degradation methods, glucose is widely available despite the complexity of depolymerization (Figure 15.3). The primary sources of cellulase are fungus, specifically ascomycetes and basidiomycetes. *T. reesei* is the fungus that produces a variety of enzymes that

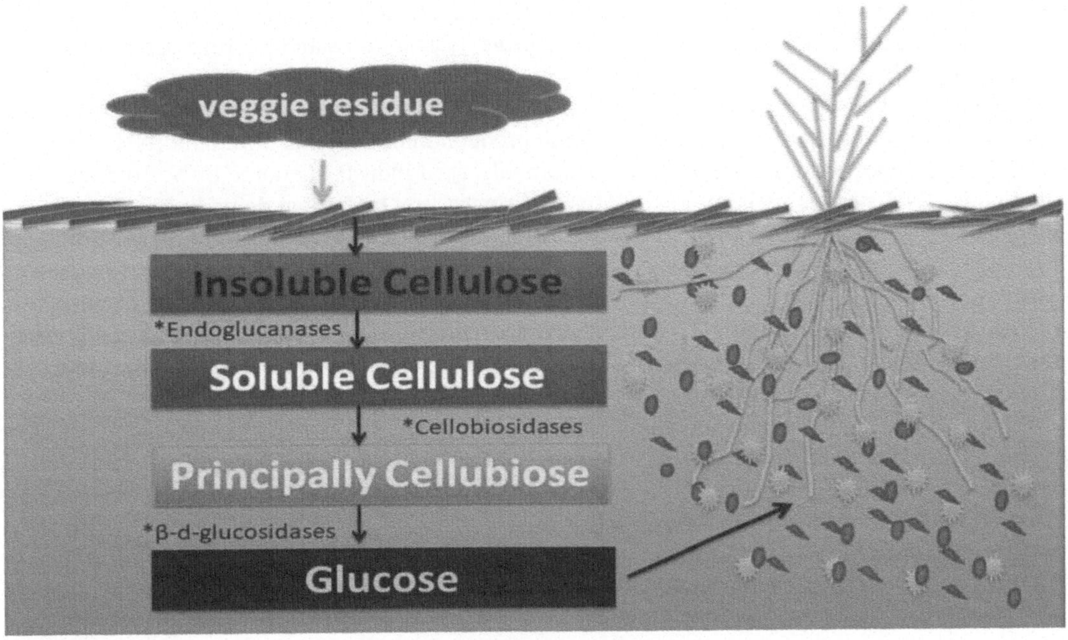

FIGURE 15.3 Glucose formation from vegetable residue through different reactions in the soil.

break down cellulose, according to (Martinez *et al.*, 2008). Although glucose produced by cellulase enzymes is still required for the growth of soil microorganisms, which then in turn regulate the availability of N, P, S, and other nutrients and further stimulate plant growth through other channels, cellulase enzymes do not provide nutrients to plants (Dotaniya *et al.*, 2019). An appropriate metric to evaluate soil health and microbial activity is the activity of β-glycosidase, since it is the final enzyme in the cellulose depolymerization process.

15.8.2 Nitrogen Cycle

Humic materials, proteins, chitin, cell walls, and nucleic acids are examples of organic sources of nitrogen in soil (da Silva *et al.*, 2023). These materials come from burlap, rhizodeposition, or the necromass of soil micro-, meso-, and macroorganisms. By adding plant and microbial residues, the primary source of nitrogen to soils is protein (Greenfield *et al.*, 2020). The two processes of protein breakdown in soil are proteolysis and protease enzymes (Cox and Nelson, 2008). Proteins are polymers of amino acids bound together by peptide bonds. Smaller peptides are created during proteinolysis and then liberated from the protein chain by peptidase activity. The class of enzymes known as proteases is responsible for catalyzing reactions in proteins. They vary in terms of their molecular makeup, type of reaction, and affinity for substrates (Alef and Nannipieri, 1995). Protease enzyme activity is thought by some to occur in free proteins in solution, but it is also thought by other people to occur in proteins absorbed on soil mineral surfaces. Still, not everything about it is known (Greenfield *et al.*, 2020).

Amino acid consumption by the soil microbial community as a source of nitrogen, carbon, and energy is the second stage. These molecules go through deamination processes within the microbial cells, producing ammonium (NH^{4+}), which can be released into the soil solution. A portion of the carbon in the amino acid structure mineralizes as CO_2 during these processes. Plant roots can absorb either form of nitrite (NO_2^-), which is created when part of the NH^{4+} is nitrified in solution (Nakatani *et al.*, 2014). Proteases are proteins that play a crucial role in the degradation of soil enzymes, such as those produced by plant roots, which can help resist pathogens. Microorganisms produce protease inhibitor molecules to regulate proteolysis in roots and the rhizospheric environment (Wang *et al.*, 2023).

Urease (EC 3.5.1.5), in addition to protease, is a crucial enzyme in the nitrogen cycle, as it facilitates the hydrolysis of urea, resulting in the production of CO_2 and ammonia (NH_3). In agriculture around the world, urea is the most commonly used industrial source of N. To release the ammonium (NH_3) or ammonium ions (NH_4^+), urea must first be broken down because plants are unable to absorb it (Qin and Cabral, 2002). However, nitrogen loss via NH_3 ammonia volatilization may occur from high urease enzyme activity. As a result, many tactics can be devised to lessen its effects, such as employing soil urease inhibitors and creating urea granules coated with substances that restrict dissolution, and urea derivatives that are hydrolyzed by soil urease more slowly when urea is used. By reducing nitrogen losses through volatilization, these methods enhance fertilizer efficiency determined by urea (Lourenço *et al.*, 2016).

Enzymes mediate the nitrification and denitrification processes, which are essential to nitrogen dynamics. Because these processes release nitrous oxide (N_2O), a greenhouse gas, and nitrate leach, they are significant for the quality of the environment. Microbial oxidation during nitrification converts reduced forms of nitrogen (NH_3) into less reduced forms, such as NO_3^- (Hu *et al.*, 2015). This process consists of two steps. Ammonia monooxygenase, also known as hydroxylamine reductase (EC 1.7.1.10), is the enzyme that first facilitates the oxidation of NH_3 to hydroxylamine. Following this, the enzyme hydroxylamine oxidoreductase (EC 1.7.3.4) converts hydroxylamine to NO_3. The first step involves bacteria from various phylogenetic groups, whereas the β subclass of proteobacteria primarily performs the second step. The reduction of NO^{3-}, which is either added by external sources (such as fertilizers) or results from nitrification, is the first step in denitrification. The nitrate reductase enzyme converts the NO_3^- to nitrogen dioxide (NO_2^-) through reduction. Nitrite reductase enzymes then convert NO_2^- to nitric oxide (NO). The microbes that perform these conversions have only one enzyme, not both. Nitric oxide reductase subsequently reduces NO to N_2O, and nitrous oxide reductase subsequently converts N_2O to N_2 gas (Kandeler *et al.*, 2011), depending on the state of the environment.

15.8.3 Phosphorus Cycle

Phosphorus is crucial for energy transport and early plant growth, making its high concentration in soil hazardous to water and plants. Soil fertilizers supply phosphorus, which is catalyzed by phosphatase class enzymes, which convert insoluble organic phosphorus compounds into inorganic molecules that are easily absorbed by plants. This process is essential for the health of both plants and water.

Enzymes known as phosphomonoesterases (EC 3.1.3.) catalyze the hydrolysis of orthophosphoric acid esters and anhydrides. They are in charge of changing organic phosphorus forms in soil into inorganic phosphates, which are forms that plants may directly access (Eivazi and Tabatabai, 1988). From an agricultural perspective, this is a crucial activity since, according to estimates, 40% of Polish soils are deficient in phosphorus. Hoffmann (1968) suggested a separation of phosphatases into two categories based on the distinct optimal pH: alkaline (EC 3.1.3.1) and acidic (EC 3.1.3.2). Phosphomonoesterases are synthesized in soil by both plant roots and microbes. Nevertheless, research indicates that microbiological acid phosphatases break down organic forms of phosphorus more quickly than plant-derived phosphatases. The amount of phosphatases present in the soil can serve as a proxy for the biological activity of the soil as well as the ability of organic phosphorus to mineralize. Research has indicated that the biological activity of soil is influenced by several factors, such as soil moisture, pH, land use, variety and richness of soil microorganisms, and fertilization level (Furtak and Gałązka, 2019). Both the United States and Sweden track the activity of soil phosphatases. It has been demonstrated that these enzymes are susceptible to petroleum pollutants and heavy metals. Soil from organic farming had high acid phosphatase activity (Tautges *et al.*,

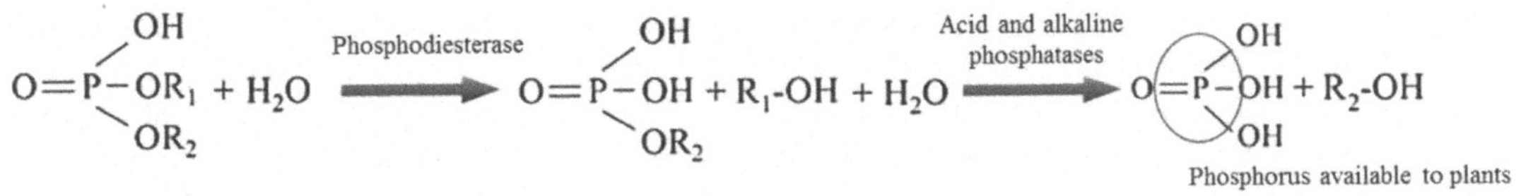

FIGURE 15.4 Showing changing phosphorus compounds in soil using phosphodiesterase and phosphomonoesters enzymes. R_1 and R_2 represent alcohol groups, phenolic groups, and nucleosides. (Acosta-Martínez and Tabatabai [2011].)

2016). The absence of accessible phosphorus forms in the soil causes an increase in phosphatase activity (Lemanowicz, 2018).

Phosphodiesterase (EC 3.1.4.1) contributes to the soil's organic phosphorus supply by breaking down phospholipids and nucleic acids. Microorganisms, plants, and animals all create it. Phosphodiesterase-catalyzed processes provide substrates for phosphomonoesterases in the soil, which in turn catalyze reactions that liberate phosphates that are vulnerable to plant attack. The pH of the soil is correlated with the activity of phosphodiesterase (Figure 15.4). Liming the soil and raising its pH even caused this enzyme's activity to double (Acosta-Martínez and Tabatabai, 2011). Manure-fertilized soils are also shown to have high phosphodiesterase activity (Parham *et al.*, 2002).

15.8.4 Sulfur Cycle

Plant development requires sulfur, a macronutrient that ranges in content in plant tissues 0.1–0.5 g kg-1 according to the species. Amino acids like cysteine and methionine include this element (Lucheta and Lambais, 2012). According to Brady and Weil (2009), sulfur in SOM can be highly reduced when bound to carbon in compounds (such as sulfides and thiols, which include amino acids like cysteine, cysteine, and methionine) and highly oxidized when bound to carbon and hydrogen in sulfoxides and sulfonates.

Numerous sulfatase group (EC 3.1.6.) enzymes, including glucosulfatases, arylsulfatases, alkyl sulfatases, and steroid sulfatases, catalyze the hydrolysis of ester sulfates in the natural environment. But because arylsulfatase (EC 3.1.6.1.) was the first sulfatase found in nature, it has been the subject of most research. It is an enzyme that breaks the O-S link to catalyze the hydrolysis of sulfate esters in soil, and it is generated by both bacteria and fungi. Research conducted by Siwik-Ziomek and Koper (2008) revealed that the type of soil fertilization technique affects the activity of arylsulfatase in agricultural soils. According to Siwik-Ziomek and Koper (2008), the use of mineral fertilizers inhibits its function. According to Cregut *et al.* (2009), various plant species may have an impact on soil sulfur fluctuations and microbes. Temperature, the percentage of clay in the soil, the amount of organic matter present, soil moisture, and the average annual rainfall all affect arylsulfatase activity (Akhtar *et al.*, 2019). The activity of arylsulfatase is enhanced by the addition of organic carbon to the soil through the retention of plant leftovers following harvesting.

Cregut *et al.* (2009) revealed that the kind of plant growing in a certain setting affects the makeup of the soil's arylsulfatase-producing bacteria. Compared to the rhizosphere of barley, the researchers discovered that the rape rhizosphere had a greater activity of this enzyme. Because of its sensitivity to environmental circumstances, arylsulfatase has been shown to be useful for analyzing soil biological activity and evaluating its quality (Zwikel *et al.*, 2007). Potassium p-nitrophenyl sulfate (PNS) is used as the substrate in the conventional colorimetric approach to measure the arylsulfatase activity (Pandey *et al.*, 2022).

The enzyme rhodanase, also known as thiosulfate sulfurtransferase (EC 2.8.1.1), catalyzes the reaction between thiosulfate and cyanide to produce thiocyanate and sulfites. One intermediate product that is created when basic sulfur is oxidized is thiocyanate. The tissues of plants, animals, and humans all contain rhodanase. Rhodanase is also produced by a few fungi (*Fusarium* sp., *Trichoderma* sp.) and bacteria (*Escherichia coli*, *Thiobacillus* sp., *Chromatium* sp.). In soils, rhodanase activity was found. It was also shown that there was a link between the amount of sulfuric amino acids in the soil and the activity of rhodanase. The phosphorus ion concentration, soil texture, soil water content, and soil carbon biomass are all correlated with rhodanase activity. Because SO_4^{2-} ions are the end product of sulfur oxidation, the administration of sulfur fertilizer in the form of K_2SO_4 or Na_2SO_4 caused a reduction in rhodanase activity. When compared to crop rotations, rhizospheric soil and soil under continuous cultivation showed higher rhodanase activity (Klose *et al.*, 2011).

15.9 CONCLUSIONS

The functional variety of soil enzymes may greatly improve our comprehension of the relationships among microbial community structure, resource availability, and ecosystem processes. The particular questions being answered will determine how soil enzyme functional diversity is measured and interpreted. Utilizing certain inhibitors to identify various reaction pathways and particular substrates to investigate diversity both within and across nutrition cycles are examples of potential components. By using selective inhibitors, pH optimization, or glycosylation to identify the wide taxonomic origins of certain soil enzyme activity, more research should be conducted. Ratios of different activity of the enzymes that process nutrients can reveal information on how the soil community reacts physiologically to alterations in the nutritional environment.

REFERENCES

Abdel Latef, A.A.H., *et al.* (2021) 'Strategy of salt tolerance and interactive impact of *Azotobacter chroococcum* and/or *Alcaligenes faecalis* inoculation on canola (*Brassica napus* L.) plants grown in saline soil', *Plants*, 10(1), p. 110.

Acosta-Martínez, V., Pérez-Guzmán, L. and Johnson, J.M.F. (2019) 'Simultaneous determination of β-glucosidase, β-glucosaminidase, acid phosphomonoesterase, and arylsulfatase activities in a soil sample for a biogeochemical cycling index', *Applied Soil Ecology*, 142, pp. 72–80.

Acosta-Martínez, V. and Tabatabai, M.A. (2011) 'Phosphorus cycle enzymes', *Methods of Soil Enzymology*, 9, pp. 161–183.

Adetunji, A.T., *et al.* (2020) *Potential use of soil enzymes as soil quality indicators in agriculture: Frontiers in soil and environmental microbiology*. Boca Raton: CRC Press.

Ajwa, H.A. and Tabatabai, M.A. (1994) 'Decomposition of different organic materials in soils', *Biology and Fertility of Soils*, 18, pp. 175–182.

Akhtar, K., *et al.* (2019) 'Integrated use of straw mulch with nitrogen fertilizer improves soil functionality and soybean production', *Environment International*, 132, p. 105092.

Alef, K. and Nannipieri, P. (1995) *Methods in applied soil microbiology and biochemistry*. London San Diego: Academic Press, c1995.

Allen, D.E., Singh, B.P. and Dalal, R.C. (2011) 'Soil health indicators under climate change: A review of current knowledge', in Singh, B., Cowie, A., Chan, K. (eds), Soil *health and climate change*, Soil Biology, vol [29]. Berlin, Heidelberg: Springer, pp. 25–45. https://doi.org/10.1007/978-3-642-20256-8_2

Allison, S.D., Wallenstein, M.D. and Bradford, M.A. (2010) 'Soil-carbon response to warming dependent on microbial physiology', *Nature Geoscience*, 3(5), pp. 336–340.

Alvarez, G., *et al.* (2018) 'Catalytic power of enzymes decreases with temperature: New insights for understanding soil C cycling and microbial ecology under warming', *Global Change Biology*, 24(9), pp. 4238–4250.

Alves de Castro Lopes, A., *et al.* (2013) 'Interpretation of microbial soil indicators as a function of crop yield and organic carbon', *Soil Science Society of America Journal*, 77(2), pp. 461–472.

Badzinski, C., *et al.* (2021) 'Soil acidification and impacts over microbial indicators during attenuation of soybean biodiesel (B100) as compared to a diesel-biodiesel blend (B8)', *Fuel*, 289, p. 119989.

Baldoni, D.B., *et al.* (2020) 'Chitinase production by *Trichoderma koningiopsis* UFSMQ40 using solid state fermentation', *Brazilian Journal of Microbiology*, 51, pp. 1897–1908.

Bardgett, R.D., Freeman, C. and Ostle, N.J. (2008) 'Microbial contributions to climate change through carbon cycle feedbacks', *The ISME Journal*, 2(8), pp. 805–814.

Baveye, P.C., *et al.* (2018) 'Emergent properties of microbial activity in heterogeneous soil microenvironments: Different research approaches are slowly converging, yet major challenges remain', *Frontiers in Microbiology*, 9, p. 1929.

Bradford, M.A., *et al.* (2021) 'Quantifying microbial control of soil organic matter dynamics at macrosystem scales', *Biogeochemistry*, 156(1), pp. 19–40.

Brady, N.C. and Weil, R.R. (2009) *Elementos da natureza e propriedades dos solos*. Porto Alegre, Rio Grande do Sul: Bookman Editora.

Burns, R.G. (1982) 'Enzyme activity in soil: Location and a possible role in microbial ecology', *Soil Biology and Biochemistry*, 14(5), pp. 423–427.

Burns, R.G., *et al.* (2013) 'Soil enzymes in a changing environment: Current knowledge and future directions', *Soil Biology and Biochemistry*, 58, pp. 216–234.

Castro Lopes, A.A., *et al.* (2018) 'Temporal variation and critical limits of microbial indicators in oxisols in the Cerrado, Brazil', *Geoderma Regional*, 12, pp. 72–82.

Chauhan, P.S. (2020) 'Role of various bacterial enzymes in complete depolymerization of lignin: A review', *Biocatalysis and Agricultural Biotechnology*, 23, p. 101498.

Chen, A. and Arai, Y. (2023) 'A review of the reactivity of phosphatase controlled by clays and clay minerals: Implications for understanding phosphorus mineralization in soils', *Clays and Clay Minerals*, 71(2), pp. 119–142.

Chen, H. (2015) 'Lignocellulose biorefinery feedstock engineering', in Lignocellulose *biorefinery engineering*. Amsterdam, Netherlands: Elsevier, pp. 37–86.

Chen, H. and Chen, H. (2014) 'Chemical composition and structure of natural lignocellulose', in *Biotechnology of lignocellulose: Theory and practice*. Dordrecht: Springer, pp. 25–71. https://doi.org/10.1007/978-94-007-6898-7_2

Cox, M.M. and Nelson, D.L. (2008) *Lehninger principles of biochemistry*. 5th ed, New York: Freeman.

Cregut, M., *et al.* (2009) 'Density, structure, and diversity of the cultivable arylsulfatase-producing bacterial community in the rhizosphere of field-grown rape and barley', *Soil Biology and Biochemistry*, 41(4), pp. 704–710.

da Silva, M.B., *et al.* (2023) 'Residual effects of composted sewage sludge on nitrogen cycling and plant metabolism in a no-till common bean-palisade grass-soybean rotation', *Frontiers in Plant Science*, 14, p. 1281670. https://doi.org/10.3389/fpls.2023.1281670

Das, S.K. and Varma, A. (2011) 'Role of enzymes in maintaining soil health', in Shukla, G., Varma, A. (eds), *Soil enzymology*. Berlin, Heidelberg: Springer, pp. 25–42. https://doi.org/10.1007/978-3-642-14225-3_2

Datta, R., *et al.* (2017) 'How enzymes are adsorbed on soil solid phase and factors limiting its activity: A review', *International Agrophysics*, 31(2), pp. 287–302.

de Carvalho Mendes, I., *et al.* (2019) 'Critical limits for microbial indicators in tropical *Oxisols* at post-harvest: The FERTBIO soil sample concept', *Applied Soil Ecology*, 139, pp. 85–93.

Dick, R.P. and Burns, R.G. (2011) 'A brief history of soil enzymology research', *Methods of Soil Enzymology*, 9, pp. 1–34.

Dick, R.P. and Kandeler, E. (2005), 'Enzymes in soils', In Hillel, D. (ed.) *Encyclopedia of Soils in the Environment*. Oxford, U.K.: Elsevier Ltd, pp. 448–455, 2005.

Doran, J.W. and Zeiss, M.R. (2000) 'Soil health and sustainability: Managing the biotic component of soil quality', *Applied Soil Ecology*, 15(1), pp. 3–11.

Dotaniya, M.L., *et al.* (2013) 'Effect of solution phosphorus concentration on the exudation of oxalate ions by wheat (*Triticum aestivum* L.)', *Proceedings of the National Academy of Sciences, India Section B: Biological Sciences*, 83, pp. 305–309.

Dotaniya, M.L., *et al.* (2016) 'Use of sugarcane industrial by-products for improving sugarcane productivity and soil health', *International Journal of Recycling of Organic Waste in Agriculture*, 5, pp. 185–194.

Dotaniya, M.L., *et al.* (2017) 'Influence of chromium contamination on carbon mineralization and enzymatic activities in vertisol', *Agricultural Research*, 6, pp. 91–96.

Dotaniya, M.L., *et al.* (2019) 'Role of soil enzymes in sustainable crop production', in *Enzymes in food biotechnology*. Amsterdam, Netherlands: Elsevier, pp. 569–589.

Dotaniya, M.L. and Kushwah, S.K. (2013) 'Nutrients uptake ability of various rainy season crops grown in a vertisol of central India', *African Journal of Agricultural Research*, 8(44), pp. 5592–5598.

Dotaniya, M.L. and Meena, V.D. (2015) 'Rhizosphere effect on nutrient availability in soil and its uptake by plants: A review', in *Proceedings of the National Academy of Sciences, India Section B: Biological Sciences*, 85, pp. 1–12.

Eivazi, F. and Tabatabai, M.A. (1988) 'Glucosidases and galactosidases in soils', *Soil Biology and Biochemistry*, 20(5), pp. 601–606.

Fang, S., *et al.* (2010) 'Enzymatic activity and nutrient availability in the rhizosphere of poplar plantations treated with fresh grass mulch', *Soil Science & Plant Nutrition*, 56(3), pp. 483–491.

Fanin, N., *et al.* (2022) 'Soil enzymes in response to climate warming: Mechanisms and feedbacks', *Functional Ecology*, 36(6), pp. 1378–1395.

Fazekašová, D. (2012) 'Evaluation of soil quality parameters development in terms of sustainable land use', in *Sustainable development-authoritative and leading edge content for environmental management.* London, UK: Intech open, pp. 435–458.

Follmer, C. (2008) 'Insights into the role and structure of plant ureases', *Phytochemistry*, 69(1), pp. 18–28.

Furtak, K. and Gałązka, A. (2019) 'Enzymatic activity as a popular parameter used to determine the quality of the soil environment', *Polish Journal of Agronomy*, 37, pp. 22–30.

García-Palacios, P., *et al.* (2021) 'Evidence for large microbial-mediated losses of soil carbon under anthropogenic warming', *Nature Reviews Earth & Environment*, 2(7), pp. 507–517.

García-Palacios, P. and Chen, J. (2022) *Emerging relationships among soil microbes, carbon dynamics and climate change.* Wiley Online Library.

Ghimire, R., Bista, P. and Machado, S. (2019) 'Long-term management effects and temperature sensitivity of soil organic carbon in grassland and agricultural soils', *Scientific Reports*, 9(1), p. 12151.

Gianfreda, L., Rao, M.A. (2014) 'Enzymes in Agricultural Sciences'. OMICS Group eBooks, USA.

Gmach, M.R., *et al.* (2019) 'Processes that influence dissolved organic matter in the soil: A review', *Scientia Agricola*, 77, p. e20180164.

Gobran, G.R., Clegg, S. and Courchesne, F. (1998) 'Rhizospheric processes influencing the biogeochemistry of forest ecosystems', in Van Breemen, N. (eds.), *Plant-induced soil changes: Processes and feedbacks*, Developments in Biogeochemistry, vol 4. Dordrecht: Springer, pp. 107–120.

Greenfield, L.M., *et al.* (2020) 'Is soluble protein mineralisation and protease activity in soil regulated by supply or demand?, *Soil Biology and Biochemistry*, 150, p. 108007.

Gu, Y., Wang, P. and Kong, C.H. (2009) 'Urease, invertase, dehydrogenase and polyphenoloxidase activities in paddy soil influenced by allelopathic rice variety', *European Journal of Soil Biology*, 45(5–6), pp. 436–441.

Guber, A., *et al.* (2018) 'Quantitative soil zymography: Mechanisms, processes of substrate and enzyme diffusion in porous media', *Soil Biology and Biochemistry*, 127, pp. 156–167.

Hassan, W., *et al.* (2015) 'Temperature sensitivity and soil organic carbon pools decomposition under different moisture regimes: Effect on total microbial and enzymatic activity', *CLEAN–Soil, Air, Water*, 43(3), pp. 391–398.

Henry, H.A.L. (2013) 'Reprint of 'Soil extracellular enzyme dynamics in a changing climate', *Soil Biology and Biochemistry*, 56, pp. 53–59.

Herrick, J.E. (2000) 'Soil quality: An indicator of sustainable land management?, *Applied Soil Ecology*, 15(1), pp. 75–83.

Hoffmann, G. (1968) 'Phosphatasen im Enzymsystem unserer Kulturböden und Möglichkeiten zur Bestimmung ihrer Aktivität', *Zeitschrift für Pflanzenernährung und Bodenkunde*, 118(3), pp. 153–160.

Hou, R., *et al.* (2016) 'Lasting effect of soil warming on organic matter decomposition depends on tillage practices', *Soil Biology and Biochemistry*, 95, pp. 243–249.

Hu, H.-W., Chen, D. and He, J.-Z. (2015) 'Microbial regulation of terrestrial nitrous oxide formation: Understanding the biological pathways for prediction of emission rates', *FEMS Microbiology Reviews*, 39(5), pp. 729–749.

Janusz, G., *et al.* (2017) 'Lignin degradation: Microorganisms, enzymes involved, genomes analysis and evolution', *FEMS Microbiology Reviews*, 41(6), pp. 941–962.

Kanchikerimath, M. and Singh, D. (2001) 'Soil organic matter and biological properties after 26 years of maize–wheat–cowpea cropping as affected by manure and fertilization in a Cambisol in semiarid region of India', *Agriculture, Ecosystems & Environment*, 86(2), pp. 155–162.

Kandeler, E., *et al.* (2011) 'Nitrogen cycle enzymes', *Methods of Soil Enzymology*, 9, pp. 211–245.

Kertesz, M.A. and Mirleau, P. (2004) 'The role of soil microbes in plant sulphur nutrition', *Journal of Experimental Botany*, 55(404), pp. 1939–1945.

Klose, S., *et al.* (2011) 'Sulfur cycle enzymes', *Methods of Soil Enzymology*, 9, pp. 125–159.

Kopittke, P.M., *et al.* (2019) 'Soil and the intensification of agriculture for global food security', *Environment International*, 132, p. 105078.

Kumar, S., Chaudhuri, S. and Maiti, S.K. (2013) 'Soil dehydrogenase enzyme activity in natural and mine soil-a review', *Middle-East Journal of Scientific Research*, 13(7), pp. 898–906.

Kumar, S. and Garkoti, S.C. (2022) 'Rhizosphere influence on soil microbial biomass and enzyme activity in banj oak, chir pine and banj oak regeneration forests in the central Himalaya', *Geoderma*, 409, p. 115626.

Lehmann, J., *et al.* (2020) 'The concept and future prospects of soil health', *Nature Reviews Earth & Environment*, 1(10), pp. 544–553.

Lemanowicz, J. (2018) 'Dynamics of phosphorus content and the activity of phosphatase in forest soil in the sustained nitrogen compounds emissions zone', *Environmental Science and Pollution Research*, 25(33), pp. 33773–33782.

Lisowska, A., *et al.* (2023) 'Changes in soil sulfate sulfur content as an effect of fertilizer granules containing elemental sulfur, halloysite and phosphate rock', *Agronomy*, 13(5), p. 1410.

Liu, Q., *et al.* (2019) 'Dominant extracellular enzymes in priming of SOM decomposition depend on temperature', *Geoderma*, 343, pp. 187–195.

Looby, C.I. and Treseder, K.K. (2018) 'Shifts in soil fungi and extracellular enzyme activity with simulated climate change in a tropical montane cloud forest', *Soil Biology and Biochemistry*, 117, pp. 87–96.

Lourenço, K.S., *et al.* (2016) 'Addition of urease inhibitor has no effect on ammonia volatilization following soil application of poultry litter or organomineral fertilizer, unlike urea', *Revista Brasileira De Ciência Do Solo*, 40, p. e0150031.

Lucheta, A.R. and Lambais, M.R. (2012) 'Sulfur in agriculture', *Revista Brasileira De Ciência Do Solo*, 36, pp. 1369–1379.

Lupatini, M., *et al.* (2019) 'Moisture is more important than temperature for assembly of both potentially active and whole prokaryotic communities in subtropical grassland', *Microbial Ecology*, 77, pp. 460–470.

Malik, A.A., *et al.* (2018) 'Land use driven change in soil pH affects microbial carbon cycling processes', *Nature Communications*, 9(1), p. 3591.

Mandal, A., *et al.* (2007) 'Effect of long-term application of manure and fertilizer on biological and biochemical activities in soil during crop development stages', *Bioresource Technology*, 98(18), pp. 3585–3592.

Manzoni, S., *et al.* (2014) 'A theoretical analysis of microbial eco-physiological and diffusion limitations to carbon cycling in drying soils', *Soil Biology and Biochemistry*, 73, pp. 69–83.

Martinez, D., *et al.* (2008) 'Genome sequencing and analysis of the biomass-degrading fungus *Trichoderma reesei* (syn. *Hypocrea jecorina*)', *Nature Biotechnology*, 26(5), pp. 553–560.

Masson-Delmotte, V., *et al.* (2021) 'Climate change 2021: The physical science basis', *Contribution of working group I to the sixth assessment report of the intergovernmental panel on climate change* 2(1), p. 2391.

McGill, W.B. and Cole, C.V. (1981) 'Comparative aspects of cycling of organic C, N, S and P through soil organic matter', *Geoderma*, 26(4), pp. 267–286.

Meena, B.L., *et al.* (2016) 'Response of compaction with sulphur fertilization to nutrient content, uptake and economics of barley on highly permeable soil', *International Journal of Agriculture Sciences*, 8(34), pp. 1719–1722.

Mobley, H.L. and Hausinger, R.P. (1989) 'Microbial ureases: Significance, regulation, and molecular characterization', *Microbiological Reviews*, 53(1), pp. 85–108.

Moorhead, D.L., *et al.* (2013) 'Calculating co-metabolic costs of lignin decay and their impacts on carbon use efficiency', *Soil Biology and Biochemistry*, 66, pp. 17–19.

Nakatani, A.S., *et al.* (2014) 'Effects of the glyphosate-resistance gene and of herbicides applied to the soybean crop on soil microbial biomass and enzymes', *Field Crops Research*, 162, pp. 20–29.

Nannipieri, P. (1994) 'The potential use of soil enzymes as indicators of productivity, sustainability and pollution', in Pankhurst, C.E., Double, B.M., Gupta, V.V.S.R., Grace, P.R., (eds.), *Soil biota: Management in sustainable farming systems*. East Melbourne: CSIRO, pp. 238–244.

Nannipieri, P., Trasar-Cepeda, C. and Dick, R.P. (2018) 'Soil enzyme activity: A brief history and biochemistry as A basis for appropriate interpretations and meta-analysis', *Biology and Fertility of Soils*, 54, pp. 11–19.

Neemisha, P. and Sharma, S. (2022) 'Soil enzymes and their role in nutrient cycling', in *Structure and functions of pedosphere*. Dordrecht: Springer, pp. 173–188.

Omar, Z., *et al.* (2017) 'Soil property and soil organic carbon pools and stocks of soil under oases in arid regions of Tunisia', *Environmental Earth Sciences*, 76(12), p. 415.

Omer, A.M., Osman, M.S. and Badawy, A.A. (2022) 'Inoculation with *Azospirillum brasilense* and/or *Pseudomonas geniculata* reinforces flax (*Linum usitatissimum*) growth by improving physiological activities under saline soil conditions', *Botanical Studies*, 63(1), p. 15.

Pandey, B., Suthar, S. and Chand, N. (2022) 'Effect of biochar amendment on metal mobility, phytotoxicity, soil enzymes, and metal-uptakes by wheat (*Triticum aestivum*) in contaminated soils', *Chemosphere*, 307, p. 135889.

Parham, J.A. S.P., *et al.* (2002) 'Long-term cattle manure application in soil: I. Effect on soil phosphorus levels, microbial biomass C, and dehydrogenase and phosphatase activities', *Biology and Fertility of Soils*, 35, pp. 328–337.

Paul, R. and Sahoo, S. (2022) 'Clay–enzyme interactions and their implications', in *Soil management for sustainable agriculture: New research and strategies*, edited by Nintu Mandal, Abir Dey, Rajiv Rakshit. New York, NY: Apple Academic Press, p. 49.

Puttaso, A., *et al.* (2011) 'Relationship between residue quality, decomposition patterns, and soil organic matter accumulation in a tropical sandy soil after 13 years', *Nutrient Cycling in Agroecosystems*, 89, pp. 159–174.

Qin, Y. and Cabral, J.M.S. (2002) 'Review properties and applications of urease', *Biocatalysis and Biotransformation*, 20(1), pp. 1–14.

Rahul, R., *et al.* (2022) 'Soil enzymes and their role in soil health improvement', in *Advances in agricultural and industrial microbiology: Volume 1: Microbial diversity and application in agroindustry*, edited by Suraja Kumar Nayak, Bighneswar Baliyarsingh, Ilaria Mannazzu, Ashutosh Singh, Bibhuti Bhusan Mishra. Dordrecht: Springer, pp. 39–61.

Rao, M.A., Scelza, R., and Gianfreda, L. (2014). 'Soil enzymes', in *Enzymes in agricultural sciences.* Foster City: OMICS Group eBooks, pp. 10–43.

Razavi, B.S., Blagodatskaya, E. and Kuzyakov, Y. (2016) 'Temperature selects for static soil enzyme systems to maintain high catalytic efficiency', *Soil Biology and Biochemistry*, 97, pp. 15–22.

Saha, S., *et al.* (2008) 'Soil enzymatic activity as affected by long term application of farm yard manure and mineral fertilizer under a rainfed soybean–wheat system in NW Himalaya', *European Journal of Soil Biology*, 44(3), pp. 309–315.

Saiya-Cork, K.R., Sinsabaugh, R.L. and Zak, D.R. (2002) 'The effects of long term nitrogen deposition on extracellular enzyme activity in an Acer saccharum forest soil', *Soil Biology and Biochemistry*, 34(9), pp. 1309–1315.

Saka, H.A., *et al.* (2017) 'Dynamics of soil nitrogen availability indices in a sandy clay loam soil amended with animal manures', *International Journal of Recycling of Organic Waste in Agriculture*, 6, pp. 167–178.

Šarapatka, B. (2003) *Phosphatase activities (ACP, ALP) in agroecosystem soils*, Department of Ecology and Crop Production Science, Uppsala, Sweden.

Schimel, J.P. (2018) 'Life in dry soils: Effects of drought on soil microbial communities and processes', *Annual Review of Ecology, Evolution, and Systematics*, 49, pp. 409–432.

Shukla, P.R., *et al.* (2019) 'IPCC, 2019: Climate Change and Land: An IPCC special report on climate change, desertification, land degradation, sustainable land management, food security, and greenhouse gas fluxes in terrestrial ecosystems,' CGIAR.

Singh, V. K., *et al.* (2016). 'Soil physical properties, yield trends and economics after five years of conservation agriculture based rice-maize system in north-western India,' *Soil and Tillage Research*, 155, 133–148.

Sinsabaugh, R.L., *et al.* (2008) 'Stoichiometry of soil enzyme activity at global scale', *Ecology Letters*, 11(11), pp. 1252–1264.

Siwik-Ziomek, A. and Koper, J. (2008) 'Arylsulphatase activity and sulphur content in a soil after modification of fertilization', *Roczniki Gleboznawcze (Poland)*, 1, pp. 198–202 (in Polish).

Sobucki, L., *et al.* (2021) 'Contribution of enzymes to soil quality and the evolution of research in Brazil', *Revista Brasileira De Ciência Do Solo*, 45, p. e0210109.

Stirling, G., *et al.* (2016) *Soil health, soil biology, soilborne diseases and sustainable agriculture: A guide.* Clayton South, VIC, Australia: Csiro Publishing.

Sulman, B.N., *et al.* (2018) 'Multiple models and experiments underscore large uncertainty in soil carbon dynamics', *Biogeochemistry*, 141, pp. 109–123.

Tabatabai, M.A. (1994) 'Soil enzymes', *Methods of Soil Analysis: Part 2 Microbiological and Biochemical Properties*, 5, pp. 775–833.

Tautges, N.E., *et al.* (2016) 'Soil microbial diversity and activity linked to crop yield and quality in a dryland organic wheat production system', *Applied Soil Ecology*, 108, pp. 258–268.

Thapa, S., *et al.* (2020) 'Microbial cellulolytic enzymes: Diversity and biotechnology with reference to lignocellulosic biomass degradation', *Reviews in Environmental Science and Bio/Technology*, 19, pp. 621–648.

Turner, B.L., *et al.* (2002) 'β-glucosidase activity in pasture soils', *Applied Soil Ecology*, 20(2), pp. 157–162.

Turner, B.L., *et al.* (2016) 'Sulfur dynamics during long-term ecosystem development', *Biogeochemistry*, 128, pp. 281–305.

Verma, R., *et al.* (2017) 'Enhancing production potential of cabbage and improves soil fertility status of Indo-Gangetic plain through application of bio-organics and mineral fertilizer', *International Journal of Current Microbiology and Applied Sciences*, 6(3), pp. 301–309.

Vong, P.-C., *et al.* (2003) 'Immobilized-S, microbial biomass-S and soil arylsulfatase activity in the rhizosphere soil of rape and barley as affected by labile substrate C and N additions', *Soil Biology and Biochemistry*, 35(12), pp. 1651–1661.

Wade, J., *et al.* (2021) 'Prescribed pH for soil β-glucosidase and phosphomonoesterase do not reflect pH optima', *Geoderma*, 401, p. 115161.

Wallenstein, M.D. and Burns, R.G. (2011) 'Ecology of extracellular enzyme activities and organic matter degradation in soil: A complex community-driven process', *Methods of Soil Enzymology*, 9, pp. 35–55.

Wang, G., *et al.* (2023) 'Benefits of biological nitrification inhibition of *Leymus chinensis* under alkaline stress: The regulatory function of ammonium-N exceeds its nutritional function', *Frontiers in Plant Science*, 14, p. 1145830.

Yang, Z., *et al.* (2019) 'Temperature sensitivity of mineral-enzyme interactions on the hydrolysis of cellobiose and indican by β-glucosidase', *Science of the Total Environment*, 686, pp. 1194–1201.

Zwikel, S., Lavee, H., and Sarah, P. (2007). 'Temporal evolution of salts in Mediterranean soils transect under different climatic conditions', *Catena*, 70(3), 282–295.

16 Rhizosphere Biology for Higher Growth and Yield

Pankaj Mandal, Pronobesh Ghosh, Rittick Mondal, Harshita Shand, and Amit Kumar Mandal

16.1 INTRODUCTION

Food deficiency is becoming a worldwide problem as the population grows at an alarming rate. Crop production is experiencing several problems as a result of biotic and abiotic stressors such as soil fertility, climate change, environmental degradation, urbanization, and rising food demand. In the contemporary period, agricultural production for a growing population on limited land presents a big problem for academics and farmers (Viana *et al.*, 2022). Population growth-mediated stress in farming has resulted in the excessive use of synthetic formulations and wastewater, all of which pose major ecological hazards and induce soil deterioration in order to maximize productivity (Pathak *et al.*, 2022; Tudi *et al.*, 2021).

Apart from various environmental risk factors, environmental pollution has a great impact on overall crop production. Production of grains has become challenging under various stresses. We cannot expect a strain-free environment in today's world. In spite of the various risk factors, sustainable food production for an increasing population has become the top priority for researchers, minimizing adverse climatic effects and improving soil health (Serebrennikov *et al.*, 2020). Therefore, the future of agro-production requires comprehensive policies that can boost crop productivity in an eco-friendly manner. Strategies should be modified and developed in such a way as to avoid the further decay of natural resources. Among various approaches, the use of plant growth-promoting microorganisms (PGPMs) has been gaining momentum in plant growth and yield. Rhizospheric bacteria and fungi have a beneficial effect on crop development (Khoshru *et al.*, 2020; Mitra *et al.*, 2021b). Rhizobacteria have a broad array of direct and indirect mechanisms to enhance crop health in various agricultural settings (Chandran *et al.*, 2021).

Plant growth-promoting rhizobacteria (PGPR) converts essential plant nutrients into soluble forms. In the rhizosphere, PGPR produces certain phytohormones, enzymes, antibiotics, stress-reducing agents, chelating agents, and quorum-sensing signaling agents (e.g., AHLs) to react with the other organisms (Fadiji *et al.*, 2022)

16.2 PGPMS

The three primary elements of the rhizosphere are soil, plants, and microorganisms. It comprises an intricate mesh of plant roots, soil, and various bacteria, fungi, eukaryotes, and archaea. These diverse microorganisms support plant growth and production, soil regeneration, increased nutrient levels, and disease protection. The capacity to monitor and alter the whole rhizosphere microbiome is limited because of its complexity. The inoculation of artificially reproduced microorganisms is the most environmentally benign method of changing the rhizosphere. The rhizosphere contains bacteria known as PGPR that have a progressive association with the plant and endorse their development (Basu *et al.*, 2021; Kong and Liu, 2022). These microorganisms encourage plant development and production by increasing nutrient availability, modifying the soil structure, and blocking disease development. Furthermore, they also engaged in soil bioremediation (Lopes *et al.*, 2021; Park *et al.*, 2021).

DOI: 10.1201/9781032630366-16

16.3 PGPR

PGPR interacts well with plants, fostering their development and survival. PGPRs (e.g. *Sinorhizobium*, *Mesorhizobium*, *Azorhizobium*, *Rhizobium*, and *Bradyrhizobium*) have the aptitude to settle plant roots and improve plant growth when injected into seeds; this phenomenon was first identified by Schroth and Kloepper (Kloepper and Schroth, 1978). Actinomycetes and free-living rhizobacteria (Azospirillum and Azotobacter) are another significant category of microorganisms that subsist in the rhizosphere and exhibit a variety of advantageous characteristics to increase plant production (Mitra *et al.*, 2022a; Muleta and Assefa, 2018). By creating a variety of defensive chemicals, PGPR has a significant impact on plant development and helps to reduce biotic and abiotic stressors. Additionally, compared to *Rhizobium* inoculation alone, the combinatorial inoculation of *B. megaterium* and *P. polymyxa* with *Rhizobium* increased *Phaseolus vulgaris* crop biomass (Korir *et al.*, 2017). Similarly, using *Rhizobium* and *Pseudomonas* boosted plant nitrogen (N), potassium (K), and sodium (Na) concentrations and mung bean biomass and yield (Ahmad *et al.*, 2012).

16.4 DIRECT AND INDIRECT MECHANISMS OF PGPR IN PLANT GROWTH PROMOTION

PGPR influences plant development through various direct processes, like nutrient mobilization, phytohormone synthesis, siderophore synthesis, secondary metabolite production, and lytic enzyme production, as well as indirect mechanisms like abiotic stress tolerance (Figure 16.1).

16.5 MOBILIZATION OF NUTRIENTS

Nitrogen, a fundamental limiting component in agriculture that maintains crop productivity, has been shown to be insufficient, owing to a variety of environmental circumstances. Almost 40–70 Tg N/year are contributed by biological nitrogen fixation (BNF) in agricultural systems (Goyal *et al.*, 2021). BNF aids almost 65% of the nitrogen used in farming. PGPR strains exhibit a crucial role in BNF. The nitrogen-fixing organisms during BNF convert plant-inaccessible nitrogen into an available form (i.e., ammonia). BNF can occur in three ways: symbiotic nitrogen fixation, associative nitrogen fixation, and free-living nitrogen fixation. Symbiotic nitrogen fixers such as *Rhizobium*, *Bradyrhizobium*, and *Sinorhizobium* undergo BNF through the production of root nodules in conjunction with legume crops. Non-symbiotic nitrogen fixers such as *Azospirillum* fix nitrogen through close or loose connections with *Poaceae* families such as wheat, rice, oats, and maize, whereas *Azotobacter*, *Klebsiella*, and *Bacillus* exist independently in soil and fix nitrogen without the affiliation of others (Wagner, 2011). Nitrogen-fixing bacteria boost soil NH_4^+ concentrations, plant growth, and N absorption.

Phosphorus (P) is a notable element of phospholipids and nucleic acids and is engaged in vital physiological and biochemical processes. The prime role of P in cells is energy storage and transfer (Kaviyarasi *et al.*, 2011). P deficiency shows chlorosis and lowered plant growth (Malhotra *et al.*, 2018). Although phosphorus is plentiful in the soil, only a small percentage of it is soluble (Sharma *et al.*, 2013). Phosphorus is solubilized by different phosphate-solubilizing microbes, including *Bacillus*, *Klebsiella*, *Pseudomonas*, and *Xanthomonas*, via H^+ exudation and organic acid generation (Hanif *et al.*, 2020; Naqqash *et al.*, 2020).

Potassium (K) is responsible for activating the enzymes responsible for photosynthesis and nitrate reduction (Adams and Shin, 2014; Wang and Wu, 2017; Zahoor *et al.*, 2017). Potassium-solubilizing microorganisms are thought to be an important donor of K dissolution in soil (Sindhu *et al.*, 2014). *B. mucilaginosus*, *B. edaphicus*, and *B. circulans* are the major effective K solubilizers in soil bacterial groups.

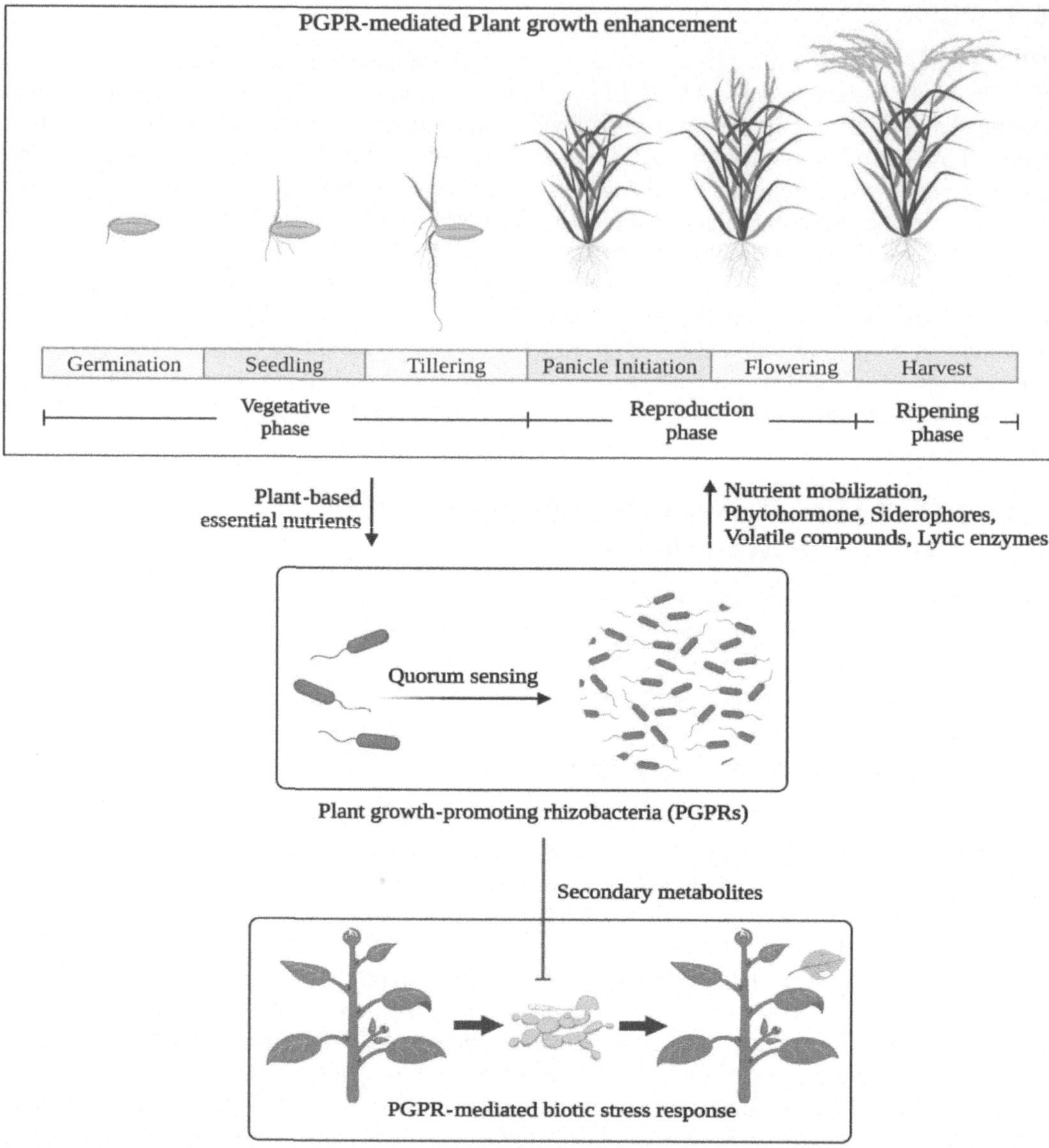

FIGURE 16.1 PGPR-mediated direct and indirect mechanisms for plant growth promotion.

16.6 PLANT HORMONES AND PGPR

Plant hormones that act as chemical signals are essential for metabolic activity, plant health, and morphogenesis in plants. The major hormones essential for plants are auxins, cytokinins, and gibberellins, that are deeply involved in plant development. Numerous rhizosphere-dwelling bacteria have the ability to produce phytohormones, which aid in the growth of plants (Maheshwari *et al.*, 2015).

The first phytohormone to be researched was auxin. Due to their multifaceted developmental roles (apical dominance, cell division, germination, and root growth in plants), it is now often applied in agriculture as growth regulators and herbicides. Additionally, it aids in the manufacture of metabolites, stress resistance, and photosynthesis (Maheshwari *et al.*, 2015). The majority of the

PGPMs (e.g., *Agrobacterium, Erwinia, Azospirillum, Pseudomonas, Bradyrhizobium, Enterobacter, Klebsiella*, and *Rhizobium*) have the ability to synthesize auxin, as they comprise iaaM and iaaH genes (Olenska *et al.*, 2020). Earlier studies revealed that auxin synthesized by the *Ignatzschineria* sp. has a beneficial impact on apple mint and chrysanthemum root initiation, number, length, and growth (Park *et al.*, 2021). Raheem *et al.* (2018) suggested that the application of auxin-producing microbes such as *Bacillus, Enterobacter, Moraxella*, and *Pseudomonas* can improve the yield of wheat under drought conditions.

Cytokinins have been proven to include changes in apical dominance, apical meristem growth and activity, gametophytic development, vascular development, and bud dormancy breaking (Osugi and Sakakibara, 2015). Additionally, cytokinins are engaged in the interaction of plants with salt and drought stress, macronutrients, N_2-fixing bacteria, and mycorrhizal fungi. Cytokinins can be synthesized by bacteria belonging to the genera *Bacillus, Pseudomonas, Escherichia, Agrobacterium, Proteus, Methylobacterium*, and *Klebsiella* (Maheshwari *et al.*, 2015). Additionally, cytokinin-generated *Bacillus subtilis* minimizes drought stress and boosts the growth of the *Platycladus orientalis* plant (Liu *et al.*, 2013). *P. syringae* infection was successfully controlled by the cytokinins produced by *P. fluorescens* (Großkinsky *et al.*, 2016).

Gibberellins (GAs) play a role in the intrinsic biological activity of plants. Gibberellins' promotion of inter-node elongation in dwarf seedlings is one of their most noticeable effects in encouraging cell elongation. GAs can also encourage fruit development, pollen development, pollen tube growth, germination, blooming, and flowering (Binenbaum *et al.*, 2018). The first bacterial species to characterize gibberellins was *Rhizobium meliloti* (Atzorn *et al.*, 1988). GAs-producing bacteria include *Bacillus, Acetobacter, Herbaspirillum, Azospirillum*, and *Pseudomonas* (Maheshwari *et al.*, 2015). The growth of radishes and cucumbers was accelerated by the application of GAs producing strain *Leifsonia soli* (Kang *et al.*, 2014).

Under heat stress, the GAs-producing strain *Bacillus tequilensis* can improve the shoot length, photosynthetic pigment, biomass, and leaf area of the soybean (Kang *et al.*, 2019).

16.7 PGPR AS BIOCONTROL AGENTS

To stop the expansion of bad microbes in the soil, harmful chemical formulations have been vigorously applied. These chemicals can negatively affect soil texture, fertility, indigenous flora, and fauna. To ensure soil fertility which helps plant development, the application of PGPR has become popular In order to stop pathogenicity, the promotion and manipulation of several PGPRs are in use. PGPRs can act as a biocontrol agent as well as an alternative agent of harmful agrochemicals by producing lytic enzymes, siderophores, ACC-deaminase, induced systemic resistance (ISR), and quorum sensing. These biocontrol agents are essential for controlling diseases in plants with the enhancement in agricultural productivity and soil health (Liu *et al.*, 2017).

Antibiotics produced by rhizobacterial species are the most effective defenses and inhibiting mechanisms against harmful microorganisms. Antimicrobial substances produced by *Streptomyces, Bacillus* sp., and *Stenotrophomonas*, such as kanosamine, zwittermicin A, xanthobaccin, and oligomycin, were effective in inhibiting the proliferation of phytopathogens (Procópio *et al.*, 2012). Phenazine synthesized by pseudomonas can prevent the growth of *Fusarium oxysporum* and *Gaeumannomyces graminis* (Biessy and Filion, 2018). *Bacillus* species also generate a variety of antibiotics like polymyxin and colistin that are active against phytopathogens (Boro *et al.*, 2022).

Iron existing in the soil is not easily accessible by plants. PGPR creates siderophores, an iron-chelating molecule with the ability to supply iron to plant cells or tissues (Hider and Kong, 2010). Pyoverdines, phenol catecholate, hydroxamates, and carboxylates are the four major classes of bacterial siderophores. Moreover, siderophores can reduce harmful microorganisms. *Pseudomonas* sp. generated siderophores in green gram, which improved plant development and slowed the progress of the disease (Sahu and Sindhu, 2011). Siderophores pyochelin and pyoverdine are generated by *Pseudomonas aeruginosa* and *Pseudomonas fluorescens.* Microbes that produce siderophores enhance Fe uptake and inhibit pathogenic growth by competing for iron. Siderophore-producing bacteria like *Klebsiella*, *Bacillus*, *Bradyrhizobium*, *Streptomyces*, *Serratia*, and *Rhizobium* have been identified.

Plants produce ethylene in response to pathogenic infection, which causes nutrient deprivation, oxidative stress, and root hair curling, mortality, and stunted growth (Premachandra *et al.*, 2016). The detrimental impacts of phytopathogens' enhancement are reduced by PGPR. Soil bacteria produce ACC deaminase enzymes, which lessen the quantity of ethylene caused by pathogens as well as its consequent detrimental consequences. Numerous microorganisms such as *Pseudomonas, Rhizobium, Azospirillum lipoferum, Ralstonia solanacearum*, and *Bacillus* are known to make ACC-deaminase and cope with stress (Ali *et al.*, 2020). In a state of homeostasis, the plant produces a sizable amount of ACC and controls ethylene production and improves plant growth (Gamalero and Glick, 2015; Nadeem *et al.*, 2010). Additionally, the microbes that consume it transform it back into ammonia and α-ketobutyrate, which controls the ethylene synthesis. As a result, PGPR, which has also produced ACC deaminase, acts as a biocontrol agent against a variety of pathogenic attacks, including those caused by *Phytophthora sp., Xanthomonas oryzae, Fusarium oxysporum, Sclerotium rolfsii, Rhizoctonia solani, Botrytis cinerea*, and *Pythium ultimum,* among others. Determining the precise mechanism underlying ACC deaminase activity is crucial, though, Scientists are fascinated by this ACC deaminase-producing-PGPR, because they can molecularly alter them and develop a vision for a more targeted usage.

Induced immunity is a defense-oriented mechanism that strengthens and stimulates a plant's innate defense system and gets it ready for potential pathogenic attack. The plants develop protective immunity against plant diseases when PGPR is used to bio-prime them. ISR-producing beneficial bacteria can directly inhibit root immunity. Pathogen infection changes the exudation pattern in roots, which in turn uses some specific microbiota to induce ISR. This strengthening of the cellular membrane shows how stress and immunity tolerance in plants interact, enabling plants to withstand abiotic and biotic challenges (Kesten *et al.*, 2019). The herbivory of insects above ground is impacted by the recruitment of plant jasmonate and defense signals that are controlled by various microbial cultures in the rhizosphere (Yuan *et al.*, 2018). ISR is brought on by a variety of bacterial elements, including siderophores, cyclic lipopeptides (CLPs), 2, 4-diacetylphloroglucinol flagella, LPS, homoserine lactones, and acetoin (Torres-Cortés *et al.*, 2018). Thus, it is possible to build a plant's immune system to draw in microorganisms and give it the resilience it needs to survive for multiple generations.

Biological control agents produce lytic enzymes to prevent the spread of pathogenic microorganisms. Lytic enzymes work by compromising the specific pathogen's cell wall's integrity and maintainability. Different lytic enzymes, including chitinases, dehydrogenases, lipases, proteases, b-glucanases, and phosphatases, are produced by PGPR. The main way that PGPR contributes to plant development through the activity of these lytic enzymes is by defending the plant against pathogenic fungi such as *Athelia rolfsii, Botrytis cinerea, Phytophthora* sp., *Fusarium oxysporum*, and *Rhizoctonia solani.* Studies have shown that *Serratia marcescens* has antifungal and chitinolytic properties that can suppress pathogenic fungi (i.e., *Rhizoctonia solani* and *Fusarium oxysporum*) (Moon *et al.*, 2017). When *Serratia marcescens* was applied to the mycelia of a pathogenic fungus, the hyphae curl, the hyphal tip ruptures, and some of the hyphae expand. β-1,3-Glucanase and cellulase, enzymes produced by *Streptomyces* sp. and *Paenibacillus*, degrade the cell walls of pathogenic fungi such as *Sclerotinia sclerotiorum* and *Fusarium oxysporum.*

16.8 RHIZOBACTERIA STRESS TOLERANCE

Studies have documented plants' tolerance to different abiotic conditions due to the presence of rhizobacteria. The capacity of soil bacteria to enhance plant systems for the expression of stress-response EPS, transcription factors, scavenging of reactive oxygen species (ROS), synthesis of proline, and stabilization of biomass determines the tolerance levels. The better exploitation of these microbes to reduce various abiotic stresses is made possible by understanding and exploring these routes.

Plant growth and development are enhanced by drought-tolerant rhizobacterial strains during water deficit conditions. To overcome reduced water potential, PGPR exhibits several tolerance mechanisms, including thickening of walls, maintaining dormancy, accumulating osmolytes, forming spores, and producing EPS. PGPR provides optimal environmental and nutritional requirements for continued plant growth. Plant hormones like ACC deaminase, ISR, and EPS are involved in PGPR-induced drought tolerance (Orozco-Mosqueda *et al.*, 2020). Indole-3-acetic acid (IAA) is produced when PGPR is applied during drought stress, and it is responsible for regulating the growth of shoots, cell division, lateral root differentiation, and vascular tissue differentiation. Abscisic acid (ABA) production in plants can be triggered by co-inoculation with rhizobacterial strains, enhancing the physiological processes of plants by controlling stress-associated transcription factors and root hydraulic activity to promote against water deficit conditions. High production of ABA by *Azospirillum brasilense* lessens water scarcity stress in *Arabidopsis thaliana*. The inoculation of favorable microorganisms diminishes antioxidant activity and increases sugar and free amino acid production to stimulate abiotic stress tolerance. In maize plant rhizobacteria strains such as *Pseudomonas aeruginosa*, *Alcaligenes faecalis* produces sugar, proline, and proteinaceous compounds which help reduce loss of water to induce water potential, growth, and tolerance against drought conditions (Naseem and Bano, 2014). *P. Putida* application stimulates photosynthetic machinery in drought-affected soybean plants with minimized chlorophyll content and decreased photosynthesis.

High-saline soils can support halo-tolerant rhizobacterial strains, which also have a few traits that make it easier for plants to flourish there (Abbas *et al.*, 2019) with the formation of biofilm, absorption of osmolyte in the cytoplasm, and maintaining turgor pressure in salt-affected cells. By generating antioxidants, phytohormones, and ISR signaling, PGPR reduces oxidative stress. PGPR characteristics like solubilization of phosphate, HCN formation, production of IAA, and salt stress resistance were present in *Bacillus subtilis* and *Bacillus pumilus* strains isolated from saline soil. When *Pseudomonas pseudoalcaligenes* and *B. pumilus* are co-injected into rice, glycine betaine levels rise, and the plant's tolerance to salinity stress is improved (Jha *et al.*, 2011). During saline stress, *Pseudomonas* and *Acinetobacter* store 1-aminocyclopropane-1-carboxylate deaminase (ACC) deaminase and IAA to enhance the development of cereal crops like oats and barley. Salinity makes PGPR cause physical and chemical changes that boost growth and induced systemic resistance (ISR) signaling. Wheat plants under salt stress grew better and produced heavier grains thanks to salt-tolerant *Azospirillum* strains. *Azospirillum* bio-priming of lettuce seeds enhanced the quality of products, shelf life, and growth under stressful circumstances (Fasciglione *et al.*, 2015). The rhizosphere of the tomato crop contained *Pseudomonas* strains, which improved the tricarboxylic acid cycle, the storage of growth regulators, and ACC deaminase, which leads to increased tolerance to salinity. By generating auxin and regulated expression of numerous salt-stressed genes, *Bacillus amyloliquefaciens* inoculation in rice increases salinity tolerance and boosts production.

Living species are seriously threatened by climate warming, which is turning into a significant problem on a global scale. The Intergovernmental Panel on Climate Change (IPCC) study from 2007 predicted that due to changes in climatic circumstances, global temperatures would increase by 1.8–3.6°C later this century. The largest obstacle to crop development and microbial colonization is thought to be the rise in temperature, which causes protein breakdown and serious cell damage. For instance, an organism may create heat-shock proteins (HSPs) against high temperatures. Because PGPR has multiple ways of establishing tolerance against heat stress, various genes in microorganisms are regulated for responding to stress. According to various researchers, heat stress can be managed by PGPR manipulation. By producing various phytohormones, improving biofilm formation, and raising HSP levels, these PGPR promote tolerance. The use of *Pseudomonas putida* in the chickpea plant demonstrated thermotolerance through the development of a thick biofilm and the production of transcription factors linked to stress (Kumar *et al.*, 2016). *Bacillus cereus* has the ability to create a number of metabolites, including organic acids, IAA, GAs, vital amino acids, and HSPs, which improve heat tolerance in a wide variety of plants (Khan *et al.*, 2020). However, more research is required to fully understand how PGPR-mediated production of HSPs contributes to thermotolerance.

16.9 MANIPULATION OF RHIZOSPHERE BY QUORUM SENSING SIGNALS

Quorum sensing (QS) systems are a successful agricultural strategy to intervene against plant pathogens. QS is a method of intercellular communication that links bacterial cell density and gene expression in microbes. It is controlled by the diffusion of certain signaling molecules, including acyl-homoserine lactones (AHLs). These phenotypic features are the major cause of microbial pathogenesis in many significant agricultural diseases, and this mechanism regulates their expression. For instance, *Ralstonia solanacearum*, *Agrobacterium tumefaciens*, *Pseudomonas syringae*, and *Erwinia amylovora* are some examples of the organisms in which QS governs diverse phenotypes. Another intriguing method for combating the pathogenicity of pathogenic organisms in agriculture is to interfere with this sensing system. This can be achieved by quorum quenching (QQ), halting the QS system that engages in enzyme-based hydrolysis or destruction of signal molecules (AHLs). AHLs are modified and degraded by enzymes called oxidoreductases, lactonases, and analyses. Studies have shown that this strategy is effective in reducing the pathogenicity of numerous phytopathogens. QS-based defense reduces virulence and infection without impairing bacterial pathogens' growth and without the use of antibiotics. This technique is more effective over the long term than antibiotics since it neither affects the vital genes of microorganisms nor leads to the development of resistance, though further research is required to support this assertion. The development of resistance to QS interference has been documented in several studies, and it has also been suggested that this QS obstructing method can inhibit the resistance in other microbes (Garcia-Contreras *et al.*, 2016).

16.10 MYCORRHIZAL FUNGI AND NUTRIENT ACQUISITION

Mycorrhizae define the symbiotic collaboration between a soil fungus and a plant root (Mitra *et al.*, 2022). Mycorrhizae are now found almost everywhere in the plant world because the potential for plants to create symbiotic relationships with them emerged earlier in the history of plants (about 450 mya) than it did in the evolution of legumes (around 60 mya). Although some of these relationships are neutral or parasitic, the majority are advantageous to both the host plant and the fungi that are colonizing it.

Mycorrhiza aids plants in drawing water and essential minerals from the soil (such as Zn, Cu, etc.) in return for food from the plant (Mitra *et al.*, 2021). Mycorrhizal relationships with plants fall into two general categories: ectomycorrhiza (EM) and endomycorrhiza, which are distinguished based on the interaction with plants. The EM are primarily found in the roots of large trees and develop a fungal sheath or mantel on the lateral root tip. Hyphae develop in the intercellular spaces of the outer cortex to form a net of hyphae known as Hartignet without penetrating cell walls. In contrast, the plant plasma membrane serves as a barrier between the cytoplasm and the endomycorrhizal fungal hyphae as they penetrate the root cortex and, in the cells, create an arbuscular, which resembles a fan and is highly branched (Harrison, 2015). The majority of mycorrhizal connections are found in the arbuscular mycorrhizae (AM) fungi. The Hartignet and the arbuscules expand the area between the plant and fungi, facilitating carbon transport for fungi and supporting plant nutrients. Endomycorrhiza is completely reliant on the plant for its carbon, in contrast to ectomycorrhiza. To capture nutrients for plants, ecto-endo mycorrhizae expand hyphae deep into the soil, which increases the root surface area and enhances nutrient uptake compared to non-mycorrhizal plants.

The interweaved fungal hyphae form a common mycorrhizal network (CMN), in which interconnection of hyphae from several mycorrhizae infected plants occurs, which helps to share nutrients between plants that are spaced distantly (Selosse *et al.*, 2006).

16.11 CURRENT CHALLENGES AND FUTURE PERSPECTIVE

As a substitute for inorganic fertilizers, PGPR inoculants have a high potential for viable agriculture. The majority of PGPR studies are conducted in greenhouses and laboratories. Therefore, PGPR's role in the natural ecosystem should be explored. Microbe screening, production, marketing, and commercialization are also critical challenges in implementing microbial inoculants. Metaproteomics, nanotechnology, and bioengineering are some of the recent and advanced technologies that need to be employed in the future for producing eco-friendly and effective inoculants. A bioformulation must emphasize PGPR inoculant variety and its ecological and physiological acceptance. PGPR can be injected along with ice-nucleating agents which act as chilling stress. Through the development of smart biosensors, different pollutants and nutrients can be assessed, allowing precision farming to be achieved. Nanoparticles can be inoculated as nano-biofertilizers. A nanobiofertilizer can be encapsulated and delivered to a specific cell, regulating fertilizer transfer and preventing unintentional losses. PGPR strains can help control plant stresses; these manipulated PGPR can be eco-friendly, low-input, and sustainable.

REFERENCES

Abbas, R., *et al.* (2019) 'Halotolerant PGPR: A Hope for cultivation of saline soils', *Journal of King Saud University - Science*, 31, pp. 1195–1201.

Adams, E. and Shin, R. (2014) 'Transport, signaling, and homeostasis of potassium and sodium in plants', Journal of Integrative Plant Biology, 56, pp. 231–249.

Ahmad, M., *et al.* (2012) 'The combined application of rhizobial strains and plant growth promoting rhizobacteria improves growth and productivity of mung bean (*Vigna radiata* L.) under salt-stressed conditions', *Annals of Microbiology*, 62, pp. 1321–1330.

Ali, S., *et al.* (2020) 'Functional characterization of potential PGPR exhibiting broad-spectrum antifungal activity', *Microbiological Research*, 232, p. 126389.

Atzorn, R., *et al.* (1988) 'Production of gibberellins and indole-3-acetic acid by rhizobium phaseoli in relation to nodulation of *Phaseolus vulgaris* roots', *Planta*, 175, pp. 532–538.

Basu, A., *et al.* (2021) 'Plant growth promoting rhizobacteria (PGPR) as green bioinoculants: Recent developments, constraints, and prospects', *Sustainability*, 13(3), p. 1140.

Biessy, A. and Filion, M. (2018) 'Phenazines in plant-beneficial *Pseudomonas spp.*: Biosynthesis, regulation, function and genomics', *Environmental Microbiology*, 20(11), pp. 3905–3917.

Binenbaum, J., Weinstain, R. and Shani, E. (2018) 'Gibberellin localization and transport in plants', *Trends in Plant Science*, 23, pp. 410–421.

Boro, M., *et al.* (2022) 'Microorganisms in biological control strategies to manage microbial plant pathogens: A review', *Archives of Microbiology*, 204(11), p. 666.

Chandran, H., Meena, M. and Swapnil, P. (2021) 'Plant growth-promoting rhizobacteria as a green alternative for sustainable agriculture', *Sustainability*, 13(19), p. 10986.

Fadiji, A.E., *et al.* (2022) 'Efforts towards overcoming drought stress in crops: Revisiting the mechanisms employed by plant growth-promoting bacteria', *Frontiers in Microbiology*, 13, p. 962427.

Fasciglione, G., *et al.* (2015) 'Azospirillum inoculation effects on growth, product quality and storage life of lettuce plants grown under salt stress', *Scientia Horticulturae*, 195, pp. 154–162.

Gamalero, E. and Glick, B.R. (2015) 'Bacterial modulation of plant ethylene levels', *Plant Physiology*, 169, pp. 13–22.

García-Contreras, R., Maeda, T. and Wood, T.K. (2016) 'Can resistance against quorum-sensing interference be selected?, *The ISME Journal*, 10, pp. 4–10.

Goyal, R.K., Schmidt, M.A. and Hynes, M.F. (2021) 'Molecular biology in the improvement of biological nitrogen fixation by rhizobia and extending the scope to cereals', *Microorganisms*, 9(1), p. 125.

Großkinsky, D.K., *et al.* (2016) 'Cytokinin production by *Pseudomonas fluorescens* G20-18 determines biocontrol activity against *Pseudomonas syringae* in Arabidopsis', *Scientific Reports*, 6, p. 23310.

Hanif, M.K., *et al.* (2020) 'Growth stimulatory effect of AHL producing *Serratia spp.* from potato on homologous and non-homologous host plants', *Microbiological Research*, 238, p. 126506.

Harrison, M.J. (2015) 'Signaling in the arbuscular mycorrhizal symbiosis', *Annual Review of Microbiology*, 59, pp. 19–24.

Hider, R.C. and Kong, X. (2010) 'Chemistry and biology of siderophores', *Natural Product Reports*, 27, pp. 637–657.

Jha, Y., Subramanian, R.B. and Patel, S. (2011) 'Combination of endophytic and rhizospheric plant growth promoting rhizobacteria in *Oryza sativa* shows higher accumulation of osmoprotectant against saline stress', *Acta Physiologiae Plantarum*, 33, pp. 797–802.

Kang, S.M., *et al.* (2014) 'Gibberellin production by newly isolated strain *Leifsonia* soli SE134 and its potential to promote plant growth', *Journal of Microbiology and Biotechnology*, 24(1), pp. 106–112.

Kang, S.-M., *et al.* (2019) 'Integrated phytohormone production by the plant growth-promoting rhizobacterium *Bacillus tequilensis* SSB07 induced thermotolerance in soybean', *Journal of Plant International*, 14, pp. 416–423.

Kaviyarasi, K., *et al.* (2011) 'Isolation, identification and molecular characterization of phosphate solubilizing actinomycetes isolated from the coastal region of Manora, Thanjavur (Dt.)', *Asian Journal of Pharmaceutical Sciences*, 1, pp. 119–122.

Kesten, C., *et al.* (2019) 'Pathogen-induced pH changes regulate the growth defense balance in plants', *The EMBO Journal*, 38, p. e101822.

Khan, M.A., *et al.* (2020) 'Extending thermotolerance to tomato seedlings by inoculation with SA1 isolate of *Bacillus cereus* and comparison with exogenous humic acid application', *PLoS ONE*, 15, p. e0232228.

Khoshru, B., *et al.* (2020) 'Role of soil rhizobacteria in utilization of an indispensable micronutrient zinc for plant growth promotion', *Journal of Critical Reviews*, 21, pp. 4644–4654.

Kloepper, J.W. and Schroth, M.N. (1978) 'Plant growth-promoting rhizobacteria on radishes', in Végétale et Phyto-Bactériologie, (eds.) Station de Pathologie, Proceedings of the 4th International Conference on Plant Pathogenic Bacteria; Tours, France. pp. 879–882

Kong, Z. and Liu, H. (2022) 'Modification of rhizosphere microbial communities: A possible mechanism of plant growth promoting rhizobacteria enhancing plant growth and fitness', *Frontiers in Plant Science*, 13, p. 920813.

Korir, H., *et al.* (2017) 'Co inoculation effect of rhizobia and plant growth promoting rhizobacteria on common bean growth in a low phosphorus soil', *Frontiers in Plant Science*, 8, p. 141.

Kumar, M., *et al.* (2016) 'Synergistic effect of *Pseudomonas putida* and *Bacillus amyloliquefaciens* ameliorates drought stress in chickpea (*Cicer arietinum* L.)', *Plant Signaling & Behavior*, 11(1), p. e1071004.

Liu, F., *et al.* (2013) 'Cytokinin-producing, plant growth-promoting rhizobacteria that confer resistance to drought stress in *Platycladus orientalis* container seedlings', *Applied Microbiology and Biotechnology*, 97, pp. 9155–9164.

Liu, Y., *et al.* (2017) 'Identification of root-secreted compounds involved in the communication between cucumber, the beneficial *Bacillus amyloliquefaciens*, and the soilborne pathogen *Fusarium oxysporum*', *Molecular Plant-Microbe Interactions*, 30, pp. 53–62.

Lopes, M.J., Bernardino, M. and Gurgel, E.S. (2021) 'Successful plant growth-promoting microbes: Inoculation methods and abiotic factors', *Frontiers in Sustainable Food Systems*. 5, p. 606454.

Maheshwari, D.K., Dheeman, S. and Agarwal, M. (2015) 'Phytohormone producing PGPR for sustainable agriculture', in Maheshwari, D. (ed.) *Bacterial metabolites in sustainable agroecosystem*. Cham: Springer, pp. 159–182.

Malhotra, H., Sharma, S. and Pandey, R. (2018) 'Phosphorus nutrition: Plant growth in response to deficiency and excess', in Hasanuzzaman, M., Fujita, M., Oku, H., Nahar, K. and Hawrylak-Nowak, B. (eds.) *Plant nutrients and abiotic stress tolerance*. Singapore: Springer, pp. 171–190.

Mitra, D., *et al.* (2021a) 'Impacts of arbuscular mycorrhizal fungi on rice growth, development, and stress management with a particular emphasis on strigolactone effects on root development', *Communications in Soil Science and Plant Analysis*, 52, pp. 1591–1621.

Mitra, D., *et al.* (2021b) 'Rhizobacteria mediated seed bio-priming triggers the resistance and plant growth for sustainable crop production', *Current Research in Microbial Sciences*, 2, p. 100071.

Mitra, D., *et al.* (2022a) 'Actinobacteria-enhanced plant growth, nutrient acquisition, and crop protection: Advances in soil, plant, and microbial multifactorial interactions', *Pedosphere*, 32, pp. 149–170.

Mitra, D., *et al.* (2022b) 'Arbuscular mycorrhizal fungal association boosted the arsenic resistance in crops with special responsiveness to rice plant', *Environmental and Experimental Botany*, 193, p. 104681.

Moon, C., *et al.* (2017) 'Antifungal activity and patterns of N-acetyl-chitooligosaccharide degradation via chitinase produced from *Serratia marcescens* PRNK-1', *Microbial Pathogenesis*, 113, pp. 218–224.

Muleta, A. and Assefa, F. (2018) 'Isolation and screening of antibiotic producing actinomycetes from rhizosphere and agricultural soils', *African Journal of Biotechnology*, 17, pp. 700–715.

Nadeem, S.M., *et al.* (2010) 'Rhizobacteria capable of producing acc-deaminase may mitigate salt stress in wheat', *Soil Science Society of America Journal*, 74, pp. 533–542.

Naqqash, T., *et al.* (2020) 'First report of diazotrophic *Brevundimonas* spp. As growth enhancer and root colonizer of potato', *Scientific Reports*, 10, pp. 1–14.

Naseem, H. and Bano, A. (2014) 'Role of plant growth-promoting rhizobacteria and their exopolysaccharide in drought tolerance of maize', *Journal of Plant Interactions*, 9(1), pp. 689–701.

Olenska, E., *et al.* (2020) 'Beneficial features of plant growth-promoting rhizobacteria for improving plant growth and health in challenging conditions: A methodical review', *Science of the Total Environment*, 743, p. 140682.

Orozco-Mosqueda, M.D.C., Glick, B.R. and Santoyo, G. (2020) 'ACC deaminase in plant growth-promoting bacteria (PGPB): An efficient mechanism to counter salt stress in crops', *Microbiological Research*, 235, p. 126439.

Osugi, A. and Sakakibara, H. (2015) 'Q&A: How do plants respond to cytokinins and what is their importance?, *BMC Biology*, 13, p. 120.

Park, S., *et al.* (2021) 'A highly efficient auxin-producing bacterial strain and its effect on plant growth', *Journal of Genetic Engineering & Biotechnology*, 19, p. 1–9.

Pathak, V.M., *et al.* (2022) 'Current status of pesticide effects on environment, human health and it's eco-friendly management as bioremediation: A comprehensive review', *Frontiers in Microbiology*, 13, p. 962619.

Premachandra, D., Hudek, L. and Brau, L. (2016) 'Bacterial modes of action for enhancing of plant growth', *Journal of Biotechnology and Biomaterials*, 6, pp. 1–8.

Procópio, R.E., *et al.* (2012) 'Antibiotics produced by streptomyces', *The Brazilian Journal of Infectious Diseases: An Official Publication of the Brazilian Society of Infectious Diseases*, 16(5), pp. 466–471.

Raheem, A., *et al.* (2018) 'Auxin production by rhizobacteria was associated with improved yield of wheat (*Triticum aestivum* L.) under drought stress', *Archives of Agronomy and Soil Science*, 64(4), pp. 574–587.

Sahu, G. and Sindhu, S. (2011) 'Disease control and plant growth promotion of green gram by siderophore producing *Pseudomonas* sp.', *Research Journal of Microbiology*, 6, p. 735.

Selosse, M.A., *et al.* (2006) 'Mycorrhizal networks" des liaisons dangereuses', *Trends in Ecology & Evolution*, 21(11), pp. 621–628.

Serebrennikov, D., *et al.* (2020) 'Factors influencing adoption of sustainable farming practices in Europe: A systemic review of empirical literature', *Sustainability*, 12(22), p. 9719.

Sharma, S.B., *et al.* (2013) 'Phosphate solubilizing microbes: Sustainable approach for managing phosphorus deficiency in agricultural soils', *SpringerPlus*, 2, p. 587.

Sindhu, S.S., Parmar, P. and Phour, M. (2014) 'Nutrient cycling: Potassium solubilization by microorganisms and improvement of crop growth', in *Geomicrobiology and biogeochemistry*. Berlin/Heidelberg, Germany: Springer, pp. 175–198.

Torres-Cortés, G., *et al.* (2018) 'Functional microbial features driving community assembly during seed germination and emergence', *Frontiers in Plant Science*, 9, p. 902.

Tudi, M., *et al.* (2021) 'Agriculture development, pesticide application and its impact on the environment', *International Journal of Environmental Research and Public Health*, 18(3), p. 1112.

Viana, C.M., *et al.* (2022) 'Agricultural land systems importance for supporting food security and sustainable development goals: A systematic review', *Science of The Total Environment*, 806, p. 150718.

Wagner, S.C. (2011) 'Biological nitrogen fixation', *Nature Education Knowledge*, 3(10), p. 15.

Wang, Y. and Wu, W.H. (2017) 'Regulation of potassium transport and signaling in plants', *Current Opinion in Plant Biology*, 39, pp. 123–128.

Yuan, J., *et al.* (2018) 'Root exudates drive the soil-borne legacy of aboveground pathogen infection', *Microbiome*, 6, pp. 1–12.

Zahoor, R., *et al.* (2017) 'Potassium improves photosynthetic tolerance to and recovery from episodic drought stress in functional leaves of cotton (*Gossypium hirsutum* L.)', *Plant Physiology and Biochemistry*, 119, pp. 21–32.

17 Nutrient Management on Soil Organic Carbon Storage and Crop Production under Changing Environments

Naeem Ahmad, Ahmad Latif Virk, Muhammad Rizwan Shoukat, Noreen Zahra, Iqra Arshad, Xiaoli Wang, Jun Li, and Muhammad Bilal Hafeez

17.1 INTRODUCTION

Nutrient management is vital for soil organic carbon (SOC) storage and crop production, particularly under changing environmental conditions (Lin *et al.*, 2023). The relationship between SOC storage and crop yield mainly depends on different nutrient management practices and diverse climatic conditions (Singh *et al.*, 2023a). Nutrient management practices (including appropriate fertilization, organic amendments, tillage systems, and crop rotation with the inclusion of legume/cover crops) have a substantial impact on soil fertility and SOC contents, thereby improving crop productivity (Chowdhury *et al.*, 2021; Hussan *et al.*, 2021). Use of appropriate nitrogen (N), phosphorus (P), and potassium (K) fertilizers have a crucial role in improving soil fertility and also contribute to crop yield (Pahalvi *et al.*, 2021). Appropriate fertilization can enhance microbial activities, which play an important role in nutrient cycling and decomposition of the organic material into the soil. The breakdown of the organic material releases essential minerals and promotes the stable organic compounds that increase the SOC accumulation (Keiblinger *et al.*, 2023). Organic materials (likewise manure, compost, biochar, biofertilizer, and crop residue) increase the levels of soil organic matter (SOM) into the soil. The SOM acts as a binding agent to form stable soil aggregates, which improve the soil structure, SOC stability, water infiltration, and porosity, and thus increase crop yield (Brichi *et al.*, 2023). Similarly, the inclusion of legumes in crop rotation has potential to improve soil fertility and nutrient cycling by enhancing SOM and carbon-to-nitrogen (C/N) ratio (Virk *et al.*, 2021b). Legumes have the ability to fix the atmospheric N in the soil, converting atmospheric N_2 into NH_4^+ by a symbiotic N-fixing-relationship, which increases N availability to the subsequent crop (Virk *et al.*, 2022). The extensive root system of the legume crops facilitates breakage of the soil compaction, improving soil porosity and water retention, which contributes more towards root biomass and their conversion to soil C inputs; it also enriches the soil with SOC (Kumar *et al.*, 2018). Moreover, a combination of organic and inorganic practices, called integrated nutrient management (INM), is effective in improving soil properties, nutrient holding capacity, SOC storage, and crop yield (Singh *et al.*, 2023b). The INM decreases the risks of nutrient leaching and thus provides a comprehensive nutrient supply to the crops by optimizing nutrient and use efficiency (Darjee *et al.*, 2023).

Therefore, it is imperative to highlight the potential knowledge gaps regarding different nutrient management practices for improving SOC storage and crop yield. There is a great debate regarding the individual nutrient source to increase SOC sequestration and crop productivity. However, to date, there is a paucity of information on how organic and inorganic sources and their combinations can influence SOC storage and crop yield under diverse environments. Therefore, this chapter

DOI: 10.1201/9781032630366-17

focuses on developing a critical understanding of nutrient dynamics, soil health, and crop productivity under changing climatic conditions.

17.2 NITROGEN MANAGEMENT FOR SOC STORAGE AND CROP YIELD IMPROVEMENT

Nitrogen (N) is an essential nutrient, highly required by plants in metabolic pathways, and has a significant impact on crop productivity and soil fertility (Plett *et al.*, 2020). The N fertilizer can influence SOC and crop yield by improving nitrogen-use efficiency (NUE) and biomass production and reducing the loss of N through leaching and volatilization (Liu *et al.*, 2021; Shiade *et al.*, 2023). An adequate supply of N can contribute to SOC storage by chemically stabilizing C in the soil and increasing dry matter production (Xu *et al.*, 2023). It can increase the microbial activities in the soil, resulting into greater organic matter breakdown, nutrient release, and availability. Microbial activities in the soil mostly depend on N cycling, and it has a critical role in the regulation of microbial growth and different biochemical processes in the soil (Hu *et al.*, 2023a). The N cycle is the process by which N is changed into both organic and inorganic molecules. The breakdown of organic matter under the process of ammonification releases N-containing organic substances (protein, nucleic acid) and inorganic substance ammonium (NH_4^+). Microorganisms, including bacteria and fungi, mediate these transformations in the soil (Zhang *et al.*, 2015). After ammonification, NH_4^+ can be further converted into nitrate (NO_3^-) via a two-step process of nitrification. NH_4^+ is converted to nitrite (NO_2^-) by *Nitrosomonas* bacteria, which is further converted to nitrate (NO_3^-) by *Nitrobacter* bacteria. The cycling of N in soil is closely linked to microbial activity. The availability and cycling of N in the soil ecosystem are greatly impacted by the several processes, including N fixation, ammonification, nitrification, and denitrification. These activities are mediated by microbial communities (Van Groenigen *et al.*, 2015; Zhang *et al.*, 2015). Variations in soil fertility due to N can have adverse effects on plant growth, microbial populations, and soil fertility status (Li *et al.*, 2023).

Root exudate and rhizodeposition can influence N availability and microbial activity and thus contribute to soil N cycling. Root exudates are the organic substances that plants release through their roots. Sugars, amino acids, organic acids, and other substances are examples of these substances (Jing *et al.*, 2023). It can attract or stimulate the N-regulating microbes in the soil, which enhances the N cycling and provides the readily available form of N to plants (Gong *et al.*, 2023). Moreover, the root exudates and other organic substances released by plant roots are referred as rhizodeposition (Mondal *et al.*, 2023). The rhizosphere or the soil around the roots in which more carbon-rich chemicals are released can stimulate the microbial activity and thus nitrogen transformations into the soil. The microbes in the rhizosphere play an important role in nitrification (conversion of NH_4^+ to NO_3^-) and denitrification (conversion of NO_3^- to N_2). The root exudates are strongly influenced by microbial activity and composition, which can impact the soil's nitrogen cycle (Mondal *et al.*, 2023).

Appropriate N management can significantly contribute to the SOC storage by enhancing the microbial growth, which accelerates the decomposition of organic material (Xun *et al.*, 2016). The rate of decomposition of organic material is greatly depending on nutrient cycling and C/N ratio. Low C/N ratio suggests that the substance is high in N and has a tendency to decompose more quickly. The amount of N released into the soil during decomposition can alter the C/N ratio and influence the breakdown of other organic material (Qiao *et al.*, 2016). Moreover, the availability and uptake of N can promote root development, and their nutritional decomposition after subsequent harvest can strongly influence the formation of soil aggregates. These aggregates can improve the soil structure and SOC storage by providing physical protection to the aggregates and protecting it from microbial decomposition (Wang *et al.*, 2018).

N is a limiting factor in crop productivity. Optimum N is essential for vegetative growth, photosynthetic efficiency, and crop productivity (Shoukat *et al.*, 2023; Zhang *et al.*, 2015). Several others factors, including N application rate, time, and method can also influence the crop yield and NUE considerably (Hirel *et al.*, 2011). The primary aim of NUE is to maximize plant N uptake while reducing losses from leaching, volatilization, and denitrification. Optimizing NUE requires applying N at the right time, rate, and place (Haile *et al.*, 2012). Appropriate N application, according to crop requirements, facilitates nutrient availability/uptake and thus improves crop yield (Ahmad *et al.*, 2022; Zhang *et al.*, 2015).

17.3 PHOSPHORUS MANAGEMENT FOR SOC STORAGE AND CROP YIELD IMPROVEMENT

Phosphorus (P) is an essential nutrient for plant growth and contributes to energy transfer, photosynthesis, and nucleic acid synthesis (Malhotra *et al.*, 2018). It is an essential component of macromolecular structures, including nucleic acids, and plays a pivotal role in the synthesis of energy transfer organic compounds such as adenosine triphosphate (ATP) in plants (Hawkesford *et al.*, 2012; Malhotra *et al.*, 2018). P is often present in soil in the forms that plants can't readily uptake. The two major forms of P are organic and inorganic. Organic material and animal waste contain organic P, whereas inorganic P is present in mineral form, likewise $H_2PO_4^-$ and HPO_4^{2-}. The P cycle includes different transformations. The weathering of rock, mineralization of organic matter, plant absorption, and breakdown and decomposition of organic matter can add P to the soil system (Shen *et al.*, 2011). The transformation of P in the soil undergoes mineralization and immobilization. Mineralization involves the conversion of organic P into inorganic forms through microbial activities (Spohn and Kuzyakov, 2013). However, immobilization is the binding or incorporating of inorganic P by microbes into their biomass, which temporarily reduces P availability in the soil and makes it unavailable for plant uptake. Moreover, certain soil microbes have the ability to solubilize P from insoluble compounds, making it readily available to plants. This process frequently produces the organic acids or enzymes that help to break down P-containing minerals in the soil (Bilyera and Loginova, 2023). Similarly, microbial diversity strongly affects the soil P cycle. Diverse microbial communities play a key role in P transformation and conversion. For example, mycorrhizal fungi have a symbiotic relationship with plant roots, increasing the root surface area and nutrient absorption, especially P uptake. These fungi can transfer P from soil particles to plant roots in a more soluble form (Sarmah *et al.*, 2022; Wu *et al.*, 2022).

Appropriate P fertilization can enhance nutrient absorption and translocation within plant roots (Elhawat *et al.*, 2023). P contains phospholipids that are integral parts of cell membrane, provide membrane integrity, and facilitate nutrient absorption and overall root functions (Zhang *et al.*, 2013). For example, root architecture in response to phosphate availability shows a phosphate-sensing mechanism, which triggers various root development signaling pathways, thereby affecting the nutrient uptake (Zhang *et al.*, 2023). The roots can uptake P via active absorption and passive diffusion. Active absorption is energy-dependent, in which P is actively absorbed through their root cells. Passive diffusion is that when P concentration in the soil solution is high, uptake of P can occur through passive diffusion (Dalir, 2023). Optimal P application can enhance root biomass, as extensive root systems may contribute to organic matter input into the soil (Kätterer *et al.*, 2011). Furthermore, the root exudates contain organic acids and enzymes that increase the solubility and mobilization of P in the soil. P can chelate with organic acids (e.g., citric and oxalic acid), which facilitates the P uptake by plant roots, even in insoluble forms (Pantigoso *et al.*, 2023). The decomposition of organic material by microbes releases carbon into the soil's organic carbon pools (Brichi *et al.*, 2023). Thus, root architecture and growth can influence soil structure and SOC pools (Souza *et al.*, 2023). Variations in soil structure can influence the transport of water and nutrients, especially P in the soil. Increased root development may result in the formation of channels that help nutrients and water pass through the soil, affecting the distribution of phosphorus and thereby SOC storage (Wang *et al.*, 2023).

P is a key component of ATP, which is the major energy source in cell. During the process of photosynthesis, plants transform light energy into a chemical form of energy, which is stored in ATP; then glucose and organic compounds are synthesized using this energy, and this process is known as phosphorylation (Khan *et al.*, 2023). P is an essential component of DNA and RNA. Genetic information is carried by DNA, whereas RNA serves in many biological functions, such as protein synthesis, cell division, and elongation. These mechanisms are important for the development and functioning of chloroplast, where photosynthesis takes place (Shah *et al.*, 2023). Appropriate P application supports reproductive structure and seed formation, and it ultimately improves crop yield (Khan *et al.*, 2023).

17.4 POTASSIUM MANAGEMENT FOR SOC STORAGE AND CROP YIELD IMPROVEMENT

Potassium (K) is an essential macronutrient for plant growth and quality yield formation (Hawkesford *et al.*, 2012). It is essential for several physiological and biochemical functions in plants, including the absorption of water, photosynthesis, nutrient uptake and translocation, photosynthesis, and enzyme activation (Dalir, 2023). The availability of K in the soil comes through different sources. Mineral sources: The mineral forms of K can be found in the soil as mica, feldspar, and other minerals that contain K. Organic sources: This includes the decomposition of organic material that releases K into the soil. Clay particles: This negatively charges sites in the soil, which can retain K ions and protect them from leaching. Weathering of rocks: This can also release K into the soil (Soumare *et al.*, 2023). The K uptake mechanism in the soil involves root interception and mass flow. Root uptake is an energy-dependent process, also known as active uptake, that involves transport proteins (potassium channels and transporters) in their root cell membranes, which actively absorb K+ ions through the plant roots and transport to the other plant parts against the concentration gradient. Mass flow or passive uptake involves the movement of soil solution containing dissolved K in the direction of roots along the concentration gradient. Water intake by plant roots and transpiration from leaves propel this movement (Jose, 2022; White *et al.*, 2010).

Appropriate K application enhances root growth that supports greater root volume to absorb more nutrient and water from the soil (Soumare *et al.*, 2023). The extensive root system can occupy a larger soil area that can improve the roots' access to minerals and water uptake, thereby enhancing the rhizodeposition. This improved root activity can facilitate the formation of root exudates and rhizodeposition in the rhizosphere soil (Yang *et al.*, 2023). Adequate K enhances the nutrient uptake efficiency in the rhizosphere soil, which can affect the nutritional composition and quality of root exudates. These increments in root growth and exudation resulting from K application can promote microbial association and SOC concentration (Yuan *et al.*, 2021). The exudates which are released by the plant roots provide a carbon source of soil microbes. These microbes break down organic material and promote SOC storage (Yang *et al.*, 2023).

The higher amount of K uptake by plant roots can enhance photosynthetic efficiency. It facilitates light energy to convert it into chemical forms of energy and enhances the production of starch, sugar, and biomass (Imtiaz *et al.*, 2022). It plays a crucial role in osmotic potential, enzymatic activities, and stomatal functioning; it also facilitates the translocation of assimilates. K application controls the opening and closing of stomata, which facilitates plants to use water more effectively, even under drought stress conditions (Mulet *et al.*, 2023). Consequently, this improves resistance to drought and provides resilience in a changing environment (Mulet *et al.*, 2023). It activates enzymes, maintains cell integrity, and has a strong influence on different metabolic pathways related to plant growth and seed formation, which ultimately leads to higher crop productivity, even under stressful environments (Bons and Sharma, 2023).

17.5 CONSERVATION AGRICULTURAL PRACTICES FOR THE OPTIMIZATION OF CROP NUTRIENT REQUIREMENTS, SOC STORAGE, AND CROP YIELD

Conservation agriculture (CA) is a climate-smart and sustainable farming approach that aims to improve soil health, increase crop yields, and minimize the environmental impact of agriculture (Kovács *et al.*, 2023). It involves four main principles:

1. Minimum or no tillage and residue retention for soil structure improvement.
2. Appropriate crop rotation with focus on economic benefits as well as soil fertility improvement.
3. Permanent soil coverage through residue/mulch and cover cropping for nutrient replenishes and carbon (C) enrichment.
4. Inclusion of legumes/cover crops for soil fertility improvement.

These CA practices reduce soil erosion, which helps to retain nutrients in the soil and also enhances the organic matter (OM) (Zhang *et al.*, 2022). The CA prevents the oxidation and decomposition of SOC compared to conventional and C-depleting farming practices, in which soil is turned over and OM exposed to the outer environment, and rapid decomposition and C loss occurs (Kong *et al.*, 2005). Moreover, CA facilitates the optimization and uptake of essential macronutrients (N, P, and K) (Arshad *et al.*, 2023). The minimum soil disturbance and residue retention under CA facilitate the breakdown of organic material and promote the buildup of OM, which may serve as a nutrient reservoir and provide nutrients to the crops over a long period of time (Fiorini *et al.*, 2020). The CA practices also promote microbial abundance, which plays a crucial role in nutrient cycling and making nutrients easily available to the crops (Hussain *et al.*, 2021). Besides this, CA strategies also enhance the soil moisture and nutrient status, and improve its uptake and availability during the critical crop growth periods. The protective layer of residue mulch under CA helps reduce evaporation, conserves soil moisture, facilitates the root penetration, and thus produces higher crop yield (Fiorini *et al.*, 2020; Wang *et al.*, 2021) (Figure 17.1).

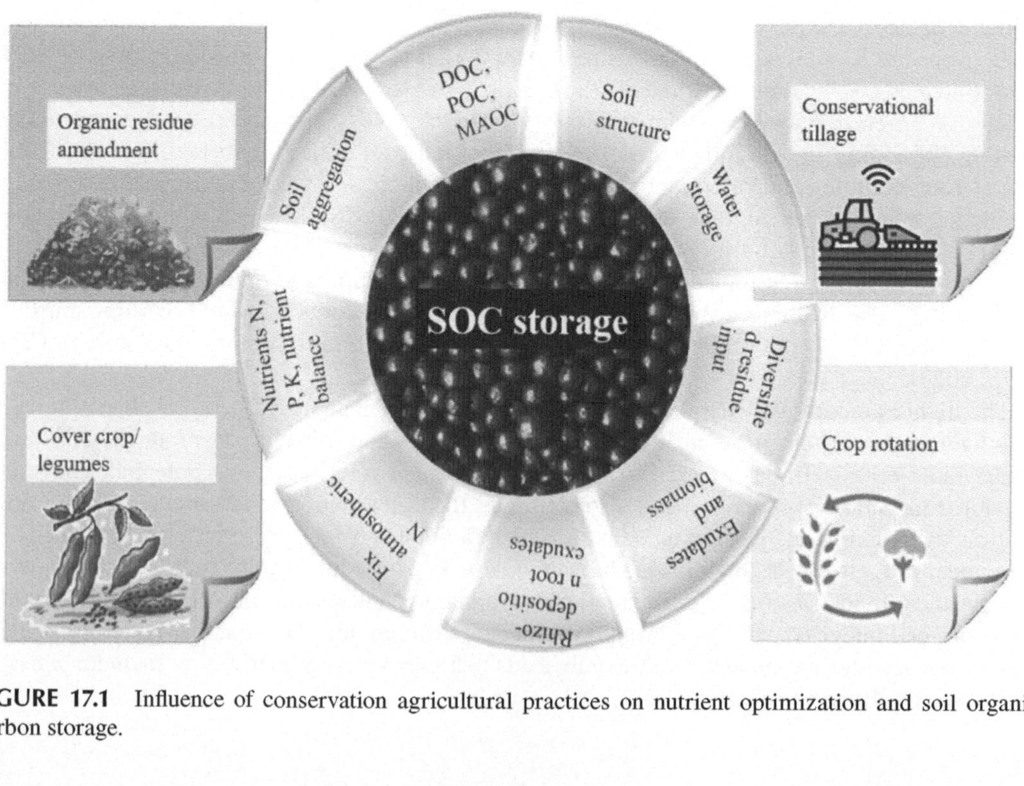

FIGURE 17.1 Influence of conservation agricultural practices on nutrient optimization and soil organic carbon storage.

17.5.1 Conservation Tillage on SOC Storage and Crop Yield

The primary aim of the CA is to increase the soil quality and crop productivity through the increments in SOM (Bhattacharyya *et al.*, 2015). Conventional tillage practices may deplete the SOM, because it destroys the soil structure and accelerates soil erosion. The SOM is oxidized under intensive tillage operations and thus decreases SOM contents unless the additional organic C inputs in the form of crop residues were added into the soil (Kan *et al.*, 2022). The conventional tillage breaks the soil aggregates and causes rapid decomposition and losses in the SOM (Kan *et al.*, 2020).

Conservation tillage (CT) refers to the tillage practice that involves minimum soil disturbance and organic C inputs to improve the soil structural stability through soil aggregation and reduce the impact of soil and water erosion (Hussain *et al.*, 2021; Iqbal *et al.*, 2024). The organic C input (including crop residues, root biomass, and rhizodeposition) under CT enhance the soil aggregation, which interacts as a binding agent between the substrate and soil microbes (Ahmad *et al.*, 2024a). They protect the macro-aggregates from microbial decomposition and thus increase SOC accumulation (Ahmad *et al.*, 2023; Kong *et al.*, 2005). CT enhances the SOC concentration, soil macro-aggregation, availability of N, and SOC sequestration in the soil matrix (Ahmad *et al.*, 2024a,b; Kan *et al.*, 2020). A previous study reported that rotary and no-tillage with residue retention increased SOC concentration by 17.6% and 18% at upper soil layers (0–5 cm) compared to the conventional tillage (Zhu *et al.*, 2023). It can facilitate SOM buildup by retaining crop residue in the soil. SOM serves as a food source for the microbes and improves soil structure (Nugroho *et al.*, 2023). Several studies have indicated that CT had a strong impact on soil physico-chemical and biological properties (Lv *et al.*, 2023; Sun *et al.*, 2020), soil respiration rates, and microbial diversity (Nugroho *et al.*, 2023), and thus it increased SOC storage (Das *et al.*, 2022). Residue retention under CT systems stabilizes soil pH. The continuous soil coverage with crop residues has a buffering effect which can stabilize the soil pH (Liu *et al.*, 2023). CT with residue retention improves soil structure, water infiltration, and SOC storage. It also reduces soil pH and bulk density (Zhang *et al.*, 2022). The improved soil structure contributes to nutrient cycling, conserves soil moisture, retains nutrients, and provides a favorable environment for root growth (Guo *et al.*, 2022). It prevents water loss through evaporation and runoff, and increases the water availability in the plant roots, leading to improved crop yield (Zhang *et al.*, 2022).

17.5.2 Crop Rotation on SOC Storage and Crop Yield

Crop rotation is an agricultural practice in which different crops are grown on the same piece of land in a predetermined sequence over the years to maintain soil fertility and improve crop yield in a sustainable manner (Demirdogen *et al.*, 2023). Crop rotation has a significant effect on SOC dynamics, and the variations in SOC flux mainly depend on the input of biomass in the soil (Almagro *et al.*, 2023). Due to the increase in cropping intensity via multiple cropping over the monoculture, the higher biomass and residue produced under diversified cropping systems results in higher SOC accumulation (Wang *et al.*, 2015). As different crops have varying nutrient requirements, the rotation of different crops (e.g., legume inclusion in rotation with cereals, shallow-rooted crops with deep-rooted crops) helps to maintain the soil nutrient balance and improve crop yield (Iheshiulo *et al.*, 2023; Virk *et al.*, 2021b). Similarly, a field study reported by Chahal *et al.* (2021) found that SOC concentration was considerably higher under diversified rotation compared to monoculture corn. This suggests that crop rotation using various crops may be helpful in sequestering C into the soil.

The rotation of legume with non-legume crops can enhance biomass productivity and C sequestration, which ultimately increases crop yield (Virk *et al.*, 2021b). Legumes have higher conversion efficiency of residue C input to soil C storage compared to monoculture (Liu *et al.*, 2023). The nodules on the roots of legumes fix atmospheric N, which can be utilized by the succeeding crop, produce higher biomass, and ultimately release more C sequestration via root exudation and rhizodeposition (Virk *et al.*, 2022). Legumes can add more in the labile organic C pools compared to the monoculture system. This is primarily due to the better microbiological activity, NUE, and biomass productivity of the next crop (Li *et al.*, 2023). Moreover, the intercropping system supports the biomass production and crop yield because it can maximize crop architecture's spatial arrangement and increase light utilization efficiency. Intercropping is divided into four categories: mixed cropping, relay intercropping, row intercropping, and strip intercropping, depending on the aim of cultivation and the spatial distribution of distinct crops (Xiao *et al.*, 2023). Perhaps in soybean-maize intercropping, soybeans provide supplementary N to maize crops and improves their yield, while maize serves as a windbreaker for the protection of soybeans from strong winds (Nasar *et al.*, 2023). Likewise, soybean/pigeon pea intercropping improves the nutrient contents, especially N, which increases root growth and crop yield (Ghosh *et al.*, 2006). In addition to improving resilience against climatic variations, mixed cropping optimizes the soil nutrient status, improves crops yield, and decreases weed and insect infestation (Virk *et al.*, 2021a).

Rotation of the existing cropping system with cover crops can increase the potential of SOC sequestration and crop yield. Cover crops improve soil health through the addition of the different nature of C inputs through root exudation and residue decomposition (Hu *et al.*, 2023b). Cover crops release organic substances through root exudation. These substances (e.g., amino acid, sugars, and organic acid) act as a source of food for microbes (Engedal *et al.*, 2023). The proliferation of soil microorganisms is encouraged by the consistent supply of organic matter that is provided by cover crops. These soil microbes decompose residues, increase soil microbial biomass, and form a stable SOC in the soil (Anuo *et al.*, 2023). In conclusion, rotations of cereals with cover/leguminous crops can increase SOC storage and crop productivity through the interaction of diversified C inputs, microbial activities, C and N cycling, soil aggregation, and other ecological processes (Hernanz *et al.*, 2009).

17.5.3 Organic Amendments on SOC Storage and Crop Yield

Organic amendments such as farmyard manure, residues, compost, biochar, and bio-solid waste have a significant impact on soil nutrients, SOC storage, and crop yield improvement (Bai *et al.*, 2023). Organic amendments slowly release nutrients into the soil upon decomposition. The slow release provides a balanced nutrient supply which facilitates the availability and uptake of essential nutrients to multiple field crops, compared to inorganic fertilizers (Diacono and Montemurro, 2011). Organic amendments are richer in organic matter, which facilitates the formation of humus, a more stable form of organic carbon that persists in the soil for a longer period of time (Diacono and Montemurro, 2011). Its addition can promote the physico-chemical and biological functions of the soil that can lead to more SOC storage and crop production (Agegnehu *et al.*, 2016). The organic matter provides the soil structural stability via two mechanisms elaborated by Luo and Xu (2023):

1. By promoting the inter-particle cohesion within the soil aggregates, and
2. By improving their hydrophobicity, thus decreasing their breakdown.

Soil microorganisms play a crucial role in the breakdown of organic matter, nutrient cycling, and other bio-geochemical transformations in the soil (Cui *et al.*, 2023). A long-term field study reported that compost application significantly improves the soil biological functionality, including soil microbial biomass C, basal respiration, and soil enzymatic activities (Wang *et al.*, 2024). Composts are typically breakdown slowly into the soil by microorganisms, which slowly release nutrients for longer duration compared to inorganic fertilizers (Siedt *et al.*, 2021). Similarly, another study indicated that the application of manure and compost has a long-duration residual effect in soil, which results in the increments of microbial biomass carbon of 20% and 40% after four years of the treatment application (Ginting *et al.*, 2003). Moreover, a long-term experiment of 17 years of organic addition of solid waste in Mediterranean conditions reported 70% increments in SOM content (Bastida *et al.*, 2008). In fact, the quantity and quality of the organic material have strongly influenced the decomposition rate and microbial and enzymatic activities in the soil (Cui *et al.*, 2023). A long-term study of 11 years comprises the addition of different organic amendments that indicated that soil respiration followed the order as 755.1 > 682.1 > 640.9 C-CO_2 kg^{-1} for straw, cattle slurry, and farmyard manure (Monaco *et al.*, 2008).

A study over five years indicated that organic amendments increase SOC by 143% and 54% under dairy waste compost and liquid dairy waste compared to inorganic fertilizer (Habteselassie *et al.*, 2006). Likewise, addition of solid waste and olive pomace compost increased total SOC by 24% and 43.2% in cocksfoot and alfalfa fields, respectively, compared to the control (Montemurro *et al.*, 2006). Another long-term (32 years) experiment showed that farmyard manure application increases SOC content about 17% compared with inorganic fertilizers in a rice-wheat cropping system (Kukal *et al.*, 2009). Organic amendments provide the major soil nutrients, including N, P, and K, to the crops. Previous studies regarding long-term use of compost showed that N and other nutrients became available for plants during the first year, and their residual effects remained in the soil for several years (Diacono and Montemurro, 2011). As Tittarelli *et al.* (2007) reported, compost releases 30–35% of total N content during the first year of the application. Regular inputs of organic amendments (via compost or manures) for more than 10 years promote the C and N stocks and N buildup within the macro-aggregates through physical protection (Mallory and Griffin, 2007). Also, compost application can enhance the available P contents in the soil. Application of solid waste at the rate of 0–200 t ha^{-1} has the potential to increase extractable P concentration 7.2–86 mg kg^{-1} (Zhang *et al.*, 2006). Additionally, a five-year study reported that available K content increased by 26% under co-compost of organic household waste compared to the control (Hartl *et al.*, 2003). More specifically, the availability of the nutrients and their release in the soil depend on multiple factors such as type of organic material, composition, and specific soil conditions (Pahalvi *et al.*, 2021).

17.5.4 Integrated Nutrient Management

INM is a diverse system that involves different organic and inorganic sources collectively to enhance nutrient availability and crop productivity while improving soil fertility (Wu and Ma, 2015). INM is a key approach to enhance SOC sequestration by maximizing the organic matter in the soil, nutrient holding capacity, water retention, and microbial activities (Wu and Ma, 2015). Long-term use of balanced organic and inorganic fertilizers can promote soil fertility and SOC pools. The combination of synthetic fertilizers (N, P, and K) and organic manures (such as farm yard manure, compost, vermicompost, biosolids, biochar, and crop residues) are examples of INM techniques in agricultural farmlands that have been effective in increasing SOC accumulation and lowering CO_2 emissions from the soil (Darjee *et al.*, 2023; Wu and Ma, 2015; Zhang *et al.*, 2012) (Figure 17.2).

INM promotes nutrient availability in the soil, which has potential to increase biomass carbon generation, which results in higher root exudation and rhizodeposition input in the soil (Irfan *et al.*, 2023). It increases organic matter, which is a key component of SOC. It acts as a substrate for soil microbes, which helps in the breakdown of organic substances and facilitates the formation of stable organic carbon in the soil (Patra *et al.*, 2023). The mechanism of binding the stable carbon in macro-aggregates involves physical protection, which protects the carbon from microbial decomposition and outer environmental exposure (Kan *et al.*, 2020). Chemical interactions such as sorption with clay particles also play a role in stabilizing SOC (Das *et al.*, 2023). Overall, these physical, chemical, and biological transformations in the soil can improve crop production through effective nutrient management and long-term SOC sequestration in the soil (Mazumdar *et al.*, 2023).

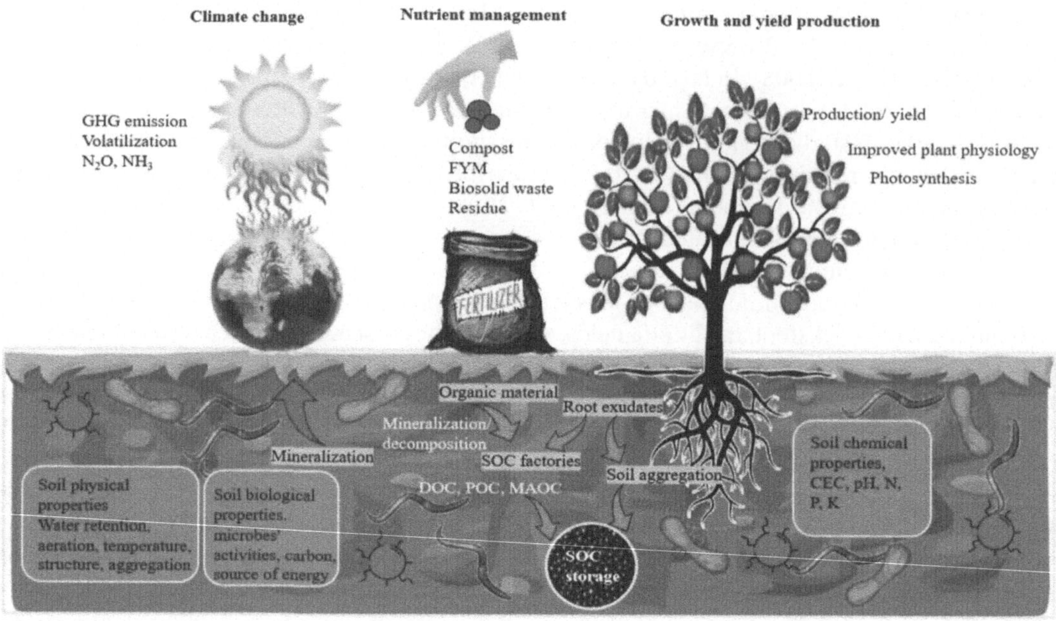

FIGURE 17.2 Role of different organic and inorganic nutrient management practices on soil organic carbon storage and crop yield.

17.6 NUTRIENT MANAGEMENT UNDER CLIMATE CHANGE AND THE IMPACT ON SOC AND CROP YIELD

Amid the present environmental crises that exert enormous pressure on crop production, there is a need to adopt appropriate nutrient management that plays a crucial role in sustainable crop production (Kiran *et al.*, 2023). Climate change is vastly influencing the temperature and precipitation patterns and soil fertility levels (Bibi and Rahman, 2023). Appropriate nutrient management through organic and inorganic sources can enhance soil fertility, SOC sequestration, and sustainable crop production under a changing environment (Kiran *et al.*, 2023).

17.6.1 Climate Change and Crop Nutrient Requirements

Climate change refers to long-term fluctuations in the patterns of various typical weather elements, such as temperature and precipitation (Bibi and Rahman, 2023). The rising temperature and changing precipitation patterns are direct consequence of increased greenhouse gas (GHG) concentration. The concentration of GHGs such as carbon dioxide (CO_2), methane (CH_4), and nitrous oxide (N_2O) has increased greatly in the atmosphere, and it is the main cause of current climate change. These emissions are mostly caused by human activities, mainly burning of fossil fuels, deforestation, and improper agricultural and nutrient management practices (Vijayavenkataraman *et al.*, 2012).

Rising atmospheric temperature due to climate change is a direct result of increased GHGs concentrations (Bibi and Rahman, 2023). Global warming affects weather elements on a colossal level, leading to an increase in the frequency and severity of extreme weather events such as droughts, heat waves, floods, and storms. The occurrence of such disasters potentially disrupts sustainable agriculture, impacting nutrient cycles, crop growth, water availability, and soil fertility (Diffenbaugh *et al.*, 2017; Shoukat *et al.*, 2021). Elevated temperature affects nutrient cycling by impacting crop growth rates, nutrient uptake, and translocation within the plant's body (Krishnan *et al.*, 2011). On the other hand, severe heat stress might impede the absorption of nutrients and unbalance the nutrients' status in the soil (Vimal *et al.*, 2017). Moreover, the changes in precipitation patterns can also attributed to change in climate, as it can cause variations in rainfall intensity, frequency, and distribution at a global and regional scale (Vijayavenkataraman *et al.*, 2012). Some regions are facing extreme rainfall and flooding conditions while others endure severe droughts. Such fluctuations strongly impact the availability of water and nutrients in the soil, reducing soil fertility status and crop yield. An inadequate supply of water restricts nutrient availability and transport to plant roots, influencing crop growth and nutrient utilization efficiency (Bardgett and Wardle, 2010). Several studies report that climate change is a major challenge that vastly impacts soil nutrients (Clair and Lynch, 2010; Vimal *et al.*, 2017). There are two diverse strategies that can enhance the soil fertility status, water retention, and nutrient availability and also mitigate the adverse effects of climate change (Smith and Gregory, 2013).

1. Appropriate and real-time use of inorganic fertilizers in agricultural farmlands.
2. Adopting conservation agricultural practices such as, no-tillage, cover crops, and organic farming.

17.6.2 Climate Change Impact on SOC Storage

Climate change and fluctuations in the atmospheric temperature are key factors influencing SOC storage (Lavallee *et al.*, 2020). Increased global temperatures have a direct impact on soil microbial activity, accelerating the breakdown of organic materials. High temperature stimulates the activity of microorganisms, which are responsible for decomposing organic molecules into simpler substances, which consequently leads to an increase in SOC turnover (Classen *et al.*, 2015; Sierra *et al.*, 2015). This temperature-driven microbial interaction is particularly noticeable in high temperature, and SOC is more vulnerable to these conditions (Sierra *et al.*, 2015). However, variations in the climate due to lower temperature cause permafrost thaw in polar and high-latitude areas, releasing higher amounts of SOC, which was earlier frozen into the soil. Permafrost soils can store significant amounts of SOC that have been accumulated over the millennium (Burke *et al.*, 2012). Climate warming causes permafrost soils to gradually thaw, which restricts the microbial activity and slowly breakdown of ancient carbon reserves. Notably, permafrost thaw soils can contributed to SOC storage; however, the released carbon may potentially exacerbate global warming (Feng *et al.*, 2020; Koven *et al.*, 2011).

Similarly, changes in the precipitation patterns also have profound effects on SOC storage. The intensity of extreme weather events such as as heavy rainfall and dry weather conditions can disrupt the water and nutrient balance in the soil (Truman *et al.*, 2007). Dry conditions reduce microbial activities and organic matter breakdown and potentially contribute to SOC buildup in the soil (Singh *et al.*, 2021). Heavy rainfall can cause soil erosion and transport carbon-rich soil particles and essential nutrients from their original place, which eventually decreases SOC reserves (Truman *et al.*, 2007).

17.7 CONCLUSIONS

This chapter discusses the impact of various nutrient management practices, including organic and inorganic sources and their combinations, on soil organic carbon storage and crop production in changing environments. Climate change potentially disrupts sustainable agriculture, impacting nutrient cycles, crop growth, water availability, and soil fertility status. It can mitigate through the adoption of appropriate and real-time use of organic and inorganic sources and their combinations. Inorganic fertilizers such as nitrogen (N), phosphorus (P), and potassium (K) promote the formation of organic acids, which stimulate microbial activities, nutrient cycling and transformation, physio-biochemical mechanisms, vegetative growth, and ultimately crop productivity. Adoption of conservation farming practices (no-tillage, residue retention, crop rotation, cover/legumes cropping) enhance water retention and soil organic matter, which facilitate the binding and encapsulation of organic matter within soil aggregates and promote soil structural stability, SOC storage, and crop yield. Organic materials, such as farmyard manure, residue, compost, biochar, and bio-solid waste, slowly release nutrients into the soil upon decomposition and facilitate the formation of humus, a more stable form of organic carbon that persists in the soil for a longer period of time. It promotes the availability of NPK, which facilitates the formation of stable organic carbon and thus improves SOC storage. A combination of organic and inorganic sources known as integrated nutrient management (INM) facilitates the optimization and uptake of essential macro-nutrients (N, P, and K) and improves soil physico-chemical and biological properties. INM improves the nutrient holding capacity, soil aggregation, and SOC storage, which ultimately increases crops yield. Thus, appropriate nutrient management through organic and inorganic sources is essential to improve SOC storage and crop yield while reducing the negative consequences of climate change.

REFERENCES

Agegnehu, G., Nelson, P.N. and Bird, M.I. (2016) 'Crop yield, plant nutrient uptake and soil physicochemical properties under organic soil amendments and nitrogen fertilization on nitisols', *Soil Tillage Research*, 160, pp. 1–13.

Ahmad, N. *et al.* (2022) 'Integrated application of plant bioregulator and micronutrients improves crop physiology, productivity and grain biofortification of delayed sown wheat', *Environmental Science and Pollution Research*, 29(35), pp. 52534–52543.

Ahmad, N. *et al.* (2023) 'Soil carbon mineralization and aggregate distribution in various tillage practices of rice-wheat cropping system. A field and laboratory study', *Journal of Soil Science and Plant Nutrition*. https://doi.org/10.1007/s42729-023-01555-2

Ahmad, N. *et al.* (2024a) 'Effects of different tillage and residue management systems on soil organic carbon stock and grain yield of rice-wheat double cropping system', *Ecological Indicators*, 158, 111452.

Ahmad, N. *et al.* (2024b) 'Carbon tradeoff and energy budgeting under conventional and conservation tillage in a rice-wheat double cropping system', *Journal of Environmental Management*, 351, 119888.

Almagro, M. et al. (2023) 'Crop diversification effects on soil organic carbon and nitrogen storage and stabilization is mediated by soil management practices in semiarid woody crops', *Soil Tillage Research*, 233, p. 105815.

Anuo, C.O. et al. (2023) 'Effect of cover cropping on soil organic matter characteristics: Insights from a five-year field experiment in Nebraska', *Agriculture Ecosystem and Environment*, 347, p. 108393.

Arshad, T. et al. (2023) 'Conservation tillage for sustainable agriculture', *Climate Resilient Agriculture*, 1, pp. 313–327.

Bai, X. et al. (2023) 'Organic amendment effects on cropland soil organic carbon and its implications: A global synthesis', *CATENA*, 231, p. 107343.

Bardgett, R.D. and Wardle, D.A. (2010) *Aboveground-belowground linkages: Biotic interactions, ecosystem processes, and global change.* Oxford, UK: Oxford University Press.

Bastida, F. *et al.* (2008) 'Long-term effect of municipal solid waste amendment on microbial abundance and humus-associated enzyme activities under semiarid conditions', *Microbial Ecology*, 55, pp. 651–661.

Bhattacharyya, R. et al. (2015) 'Conservation agriculture effects on soil organic carbon accumulation and crop productivity under a rice–wheat cropping system in the western Indo-Gangetic plains', *European Journal of Agronomy*, 70, pp. 11–21.

Bibi, F. and Rahman, A. (2023) 'An overview of climate change impacts on agriculture and their mitigation strategies', *Agriculture*, 13, p. 1508.

Bilyera, N. and Loginova, I. (2023) 'Phosphorus in soils—Biological interactions', *Encyclopedia of Soils and Environment*, 1(2), pp. 293–302.

Bons, H.K. and Sharma, A. (2023) 'Impact of foliar sprays of potassium, calcium, and boron on fruit setting behavior, yield, and quality attributes in fruit crops: A review', *Journal of Plant Nutrition*, 46, pp. 3232–3246.

Brichi, L. *et al.* (2023) 'Organic residues and their impact on soil health, crop production and sustainable agriculture: A review including bibliographic analysis', *Soil Use Management*, 39, pp. 686–706.

Burke, E.J., Hartley, I.P. and Jones, C.D. (2012) 'Uncertainties in the global temperature change caused by carbon release from permafrost thawing', *The Cryosphere*, 6, pp. 1063–1076.

Chahal, I. *et al.* (2021) 'Long-term effects of crop rotation, tillage, and fertilizer nitrogen on soil health indicators and crop productivity in a temperate climate', *Soil Tillage Research*, 213, p. 105121.

Chowdhury, S. *et al.* (2021) 'Role of cultural and nutrient management practices in carbon sequestration in agricultural soil', *Advance in Agronomy*, 166, pp. 131–196.

Clair, S.B. and Lynch, J.P. (2010) 'The opening of Pandora's box: Climate change impacts on soil fertility and crop nutrition in developing countries', *Plant and Soil*, 335, pp. 101–115.

Classen, A.E. *et al.* (2015) 'Direct and indirect effects of climate change on soil microbial and soil microbial-plant interactions: What lies ahead?, *Ecosphere*, 6, pp. 1–21.

Cui, J. *et al.* (2023) 'Investigating the effects of organic amendments on soil microbial composition and its linkage to soil organic carbon: A global meta-analysis', *Science of the Total Environment*, 894, p. 164899.

Dalir, N. (2023) 'Nutrients uptake and transport in plants', In: *Plant ionomics: Sensing, signaling and regulation*. Hoboken, NJ: John Wiley & Sons, pp. 180–190.

Darjee, S. *et al.* (2023) 'Integrated nutrient management reduced the nutrient losses and increased crop yield in irrigated wheat', *Archives of Agronomy and Soil Science*, 69, pp. 1298–1309.

Das, A. *et al.* (2023) 'Influence of clay mineralogy on soil organic carbon stabilization under tropical climate, India', *Journal of Soil Science and Plant Nutrition*, 23, pp. 1003–1018.

Das, S., Chatterjee, S. and Rajbanshi, J. (2022) 'Responses of soil organic carbon to conservation practices including climate-smart agriculture in tropical and subtropical regions: A meta-analysis', *Science of the Total Environment*, 805, p. 150428.

Demirdogen, A., Guldal, H.T. and Sanli, H. (2023) 'Monoculture, crop rotation policy, and fire', *Ecological Economics*, 203, p. 107611.

Diacono, M. and Montemurro, F. (2011) 'Long-term effects of organic amendments on soil fertility', *Sustainable Agriculture*, 2, pp. 761–786.

Diffenbaugh, N.S. *et al.* (2017) 'Quantifying the influence of global warming on unprecedented extreme climate events', *Proceedings of the National Academy of Sciences*, 114, pp. 4881–4886.

Elhawat, N. *et al.* (2023) 'Phosphorus fertilization rate and dynamics in early development of three giant reed (*Arundo donax*) ecotypes', *Biomass and Bioenergy*, 173, p. 106805.

Engedal, T. *et al.* (2023) 'Legume-based cover crop mixtures can overcome trade-offs between C inputs, soil mineral N depletion and residual yield effects', *Agriculture Ecosystem and Environment*, 349, p. 108408.

Feng, J. *et al.* (2020) 'Warming-induced permafrost thaw exacerbates tundra soil carbon decomposition mediated by microbial community', *Microbiome*, 8, pp. 1–12.

Fiorini, A. *et al.* (2020) 'May conservation tillage enhance soil C and N accumulation without decreasing yield in intensive irrigated croplands? Results from an eight-year maize monoculture', *Agriculture Ecosystem and Environment*, 296, p. 106926.

Ghosh, P.K. *et al.* (2006) 'Growth, competition, yields advantage and economics in soybean/pigeonpea intercropping system in semi-arid tropics of India: II. Effect of nutrient management', *Field Crops Research*, 96, pp. 90–97.

Ginting, D. *et al.* (2003) 'Greenhouse gas emissions and soil indicators four years after manure and compost applications', *Journal Environmental Quality*, 32, pp. 23–32.

Gong, X. *et al.* (2023) 'Linkages of microbial community structure and root exudates: Evidence from microbial nitrogen limitation in soils of crop families', *Science of the Total Environment*, 881, p. 163536.

Guo, X. *et al.* (2022) 'Subsoiling and plowing rotation increase soil C and N storage and crop yield on a semi-arid loess plateau', *Soil Tillage Research*, 221, p. 105413.

Habteselassie, M.Y. *et al.* (2006) 'Soil nitrogen and nutrient dynamics after repeated application of treated dairy-waste', *Soil Science Society of America Journal*, 70, pp. 1328–1337.

Haile, D., Nigussie, D. and Ayana, A. (2012) 'Nitrogen use efficiency of bread wheat: Effects of nitrogen rate and time of application', *Journal of Soil Science and Plant Nutrition*, 12, pp. 389–410.

Hartl, W., Putz, B. and Erhart, E. (2003) 'Influence of rates and timing of biowaste compost application on rye yield and soil nitrate levels', *European Journal of Soil Biology*, 39, pp. 129–139.

Hawkesford, M. *et al.* (2012) 'Functions of macronutrients', In: *Marschner's minerals nutrition of higher plants*. Amsterdam, Netherlands: Academic Press, pp. 135–189.

Hernanz, J.L., Sánchez-Girón, V. and Navarrete, L. (2009) 'Soil carbon sequestration and stratification in a cereal/leguminous crop rotation with three tillage systems in semiarid conditions', *Agriculture Ecosystem Environment*, 133, pp. 114–122.

Hirel, B. *et al.* (2011) 'Improving nitrogen use efficiency in crops for sustainable agriculture', *Sustainability*, 3, pp. 1452–1485.

Hu, Q. *et al.* (2023a) 'Effects of nitrogen fertilizer on soil microbial residues and their contribution to soil organic carbon and total nitrogen in a rice-wheat system', *Applied Soil Ecology*, 181, p. 104648.

Hu, Q. *et al.* (2023b) 'Soil organic carbon fractions in response to soil, environmental and agronomic factors under cover cropping systems: A global meta-analysis', *Agriculture Ecosystem and Environment*, 355, p. 108591.

Hussan, M.U. *et al.* (2021) 'Impact of soil applied humic acid, zinc and boron supplementation on the growth, yield and zinc translocation in winter wheat', *Asian Journal of Agriculture and Biology*, 10, pp. 1–14.

Hussain, S. *et al.* (2021) 'Carbon sequestration to avoid soil degradation: A review on the role of conservation tillage', *Plants*, 10, p. 2001.

Iheshiulo, E.M.A. *et al.* (2023) 'Do diversified crop rotations influence soil physical health? A meta-analysis', *Soil Tillage Research*, 233, p. 105781.

Imtiaz, H. *et al.* (2022) 'Impact of potassium starvation on the uptake, transportation, photosynthesis, and abiotic stress tolerance', *Plant Growth Regulation*, 99, pp. 429–448.

Iqbal, J. *et al.* (2024) 'Productivity, profitability and energy use efficiency of wheat maize cropping under different tillage systems', *Farming System*, 2(3), p. 100085.

Irfan, M. *et al.* (2023) 'Optimum wheat productivity under integrated plant nutrient management is associated with improved root system and high nutrient efficiency', *Eurasian Journal of Soil Science*, 12, pp. 111–121.

Jing, H. *et al.* (2023) 'The mechanism effects of root exudate on microbial community of rhizosphere soil of tree, shrub, and grass in forest ecosystem under N deposition', *ISME Communications*, 31(3), pp. 1–13.

Jose, J.V. (2022) 'Physiological and molecular aspects of macronutrient uptake by higher plants', In: Sustainable *plant nutrition*, Cambridge, MA: Academic Press, pp. 1–21.

Kan, Z.R. *et al.* (2020) 'Carbon sequestration and mineralization in soil aggregates under long-term conservation tillage in the North China plain', *CATENA*, 188, p. 104428.

Kan, Z.R. *et al.* (2022) 'Mechanisms of soil organic carbon stability and its response to no-till: A global synthesis and perspective', *Global Change Biology*, 28, pp. 693–710.

Kätterer, T. *et al.* (2011) 'Roots contribute more to refractory soil organic matter than above-ground crop residues, as revealed by a long-term field experiment', *Agriculture Ecosystem and Environment*, 141, pp. 184–192.

Keiblinger, K., Wichern, F. and Cong, W.F. (2023) 'Interplay between living or dead plant carbon input and soil organic matter – Key drivers and agricultural management for soil carbon sequestration', *Plant and Soil*, 488, pp. 1–8.

Khan, F. *et al.* (2023) 'Phosphorus plays key roles in regulating plants' physiological responses to abiotic stresses', *Plants*, 12, p. 2861.

Kiran, K.K. *et al.* (2023) 'Carbon sequestration potential of sustainable agricultural practices to mitigate climate change in Indian agriculture: A meta-analysis', *Sustainable Production and Consumption*, 35, pp. 697–708.

Kong, A.Y.Y. *et al.* (2005) 'The relationship between carbon input, aggregation, and soil organic carbon stabilization in sustainable cropping systems', *Soil Science Society of America Journal*, 69, pp. 1078–1085.

Kovács, G.P. *et al.* (2023) 'Conservation tillage improves soil quality and crop yield in Hungary', *Agronomy*, 13(3), p. 894.

Koven, C.D. *et al.* (2011) 'Permafrost carbon-climate feedbacks accelerate global warming', *Proceedings of the National Academy of Sciences*, 108(36), pp. 14769–14774.

Krishnan, P. *et al.* (2011) 'High-temperature effects on rice growth, yield, and grain quality', *Advance in Agronomy*, 111, pp. 87–206.

Kukal, S.S., Rehana, R. and Benbi, D.K. (2009) 'Soil organic carbon sequestration in relation to organic and inorganic fertilization in rice–wheat and maize–wheat systems', *Soil Tillage Research*, 102, pp. 87–92.

Kumar, S. *et al.* (2018) 'Role of legumes in soil carbon sequestration', In: *Legumes for Soil Health and Sustainable Managemant*, Berlin, Germany: Springer, pp. 109–138.

Lavallee, J.M., Soong, J.L. and Cotrufo, M.F. (2020) 'Conceptualizing soil organic matter into particulate and mineral-associated forms to address global change in the 21st century', *Global Change Biology*, 26, pp. 261–273.

Li, G. *et al.* (2023) 'Response of soil organic carbon fractions to legume incorporation into cropping system and the factors affecting it: A global meta-analysis', *Agriculture Ecosystem and Environment*, 342, p. 108231.

Liu, B. *et al.* (2021) 'Combined application of organic and synthetic nitrogen fertilizers for improving crop yield and reducing reactive nitrogen losses from China's vegetable systems: A meta-analysis', *Environmental Pollution*, 269, p. 116143.

Lin, B.J. *et al.* (2023) 'Changes in cropland soil carbon through improved management practices in China: A meta-analysis', *Journal of Environmental Management*, 329, p. 117065.

Liu, C. *et al.* (2023) 'Legume-based rotation enhances subsequent wheat yield and maintains soil carbon storage', *Agronomy for Sustainable Development*, 43, pp. 1–13.

Liu, W.S. *et al.* (2023) 'Conservation tillage increases surface soil organic carbon stock by altering fungal communities and enzyme activity', *Environmental Science and Pollution Research*, 30, pp. 80901–80915.

Luo, Y. and Xu, J. (2023) 'Soil organic matter dynamics', In: Encyclopedia of soils and environment, Oxford, UK: Elsevier, pp. 387–395.

Lv, L. *et al.* (2023) 'Impact of conservation tillage on the distribution of soil nutrients with depth', *Soil Tillage Research*, 225, p. 105527.

Malhotra, H. *et al.* (2018) 'Phosphorus nutrition: Plant growth in response to deficiency and excess', In: *Plant nutrition and abiotic stress tolerence*, Singapore: Springer, pp. 171–190.

Mallory, E.B. and Griffin, T.S. (2007) 'Impacts of soil amendment history on nitrogen availability from manure and fertilizer', *Soil Science Society of America Journal*, 71, pp. 964–973.

Mazumdar, S.P. *et al.* (2023) 'Trade-off between soil aggregate stability and carbon decomposition under 44 years long-term integrated nutrient management in rice-wheat-jute system', *Archives of Agronomy and Soil Science*, 69, pp. 417–430.

Monaco, S. *et al.* (2008) 'Changes in chemical and biochemical soil properties induced by 11-yr repeated additions of different organic materials in maize-based forage systems', *Soil Biology and Biochemistry*, 40, pp. 608–615.

Mondal, S. *et al.* (2023) 'Multifaceted roles of root exudates in light of plant-microbe interaction', In: *Unravelling plant-microbe synergy*. Cambridge, MA: Academic Press, pp. 49–76.

Montemurro, F. *et al.* (2006) 'Compost organic amendments in fodder crops: Effects on yield, nitrogen utilization and soil characteristics', *Compost Science Utilization*, 14, pp. 114–123.

Mulet, J.M., Porcel, R. and Yenush, L. (2023) 'Modulation of potassium transport to increase abiotic stress tolerance in plants', *Journal of Experimental Botany*, 74, pp. 5989–6005.

Nasar, J. *et al.* (2023) 'Maize-soybean intercropping at optimal N fertilization increases the N uptake, N yield and N use efficiency of maize crop by regulating the N assimilatory enzymes', *Frontiers in Plant Science*, 13, p. 1077948.

Nugroho, P.A. *et al.* (2023) 'Long-term conservation tillage results in a more balanced soil microbiological activity and higher nutrient supply capacity', *International Soil Water Conservation Research*, 11, pp. 528–537.

Pahalvi, H.N. *et al.* (2021) 'Chemical fertilizers and their impact on soil health', In: Microbiota *biofertilizers,* Vol 2: Ecofriendly *tools reclamation degraded soil environs*, Cham: Springer, pp. 1–20.

Pantigoso, H.A. *et al.* (2023) 'Root exudate-derived compounds stimulate the phosphorus solubilizing ability of bacteria', *Scientific Reports*, 13(3), pp. 1–12.

Patra, B., Jena, S. and Phonglosa, A. (2023) 'Date of planting, spacing and integrated nutrient management practices improves soil microbial populations and soil fertility status under rice -arhar cropping system in acidic soils of Eastern India', *Communications in Soil Science and Plant Analysis*, 54, pp. 1397–1414.

Plett, D.C. *et al.* (2020) 'The intersection of nitrogen nutrition and water use in plants: New paths toward improved crop productivity', *Journal of Experimental Botany*, 71, pp. 4452–4468.

Qiao, N. *et al.* (2016) 'Carbon and nitrogen additions induce distinct priming effects along an organic-matter decay continuum', *Scientific Reports.*, 61(6), pp. 1–8.

Sarmah, R. *et al.* (2022) 'Phosphate solubilizing microorganisms: A Review', *Communications in Soil Science and plant Analysis*, 54(10), pp. 1306–1315.

Shah, A.N. *et al.* (2023) 'Genetic and molecular factors modulating phosphorus use efficiency in plants', In: *Sustainable agriculture in the era omics and revolution*, Cham: Springer, pp. 343–356.

Shen, J. *et al.* (2011) 'Phosphorus dynamics: From soil to plant', *Plant Physiology*, 156, pp. 997–1005.

Shiade, S.R.G. *et al.* (2023) 'Nitrogen contribution in plants: Recent agronomic approaches to improve nitrogen use efficiency', *Journal of Plant Nutrition*, 47(2), pp. 1–18.

Shoukat, M.R. *et al.* (2021) 'Effects of foliar applied thiourea on maize physiology, growth and yield (*Zea Mays* L.) under shaded conditions', *Journal of Plant Nutrition*, 9, pp. 1312–1321.

Shoukat, M.R. *et al.* (2023) 'Growth, yield, and agronomic use efficiency of delayed sown wheat under slow-release nitrogen fertilizer and seeding rate', *Agronomy*, 13(7), p. 1830.

Siedt, M. *et al.* (2021) 'Comparing straw, compost, and biochar regarding their suitability as agricultural soil amendments to affect soil structure, nutrient leaching, microbial communities, and the fate of pesticides', *Science of the Total Environment*, 751, p. 141607.

Sierra, C. A. *et al.* (2015) 'Sensitivity of decomposition rates of soil organic matter with respect to simultaneous changes in soil temperature and moisture', *Journal of Advances in Modeling Earth Systems*, 7(1), pp. 335–356.

Singh, P., Nazir, G. and Dheri, G.S. (2023a) 'Influence of different management practices on carbon sequestration of agricultural soils – A review', *Archives of Agronomy and Soil Science*, 69, pp. 2471–2492.

Singh, P., Singh, B. and Farmaha, B.S. (2023b) 'Nutrient management impacts on organic carbon pool in soils under different cropping systems in the Indo-Gangetic Plains in South Asia', *Proceedings of Indian National Science Academy*, 893(89), pp. 520–559.

Singh, S. *et al.* (2021) 'Soil organic carbon cycling in response to simulated soil moisture variation under field conditions', *Scientific Reports*, 111(11), pp. 1–13.

Smith, P. and Gregory, P.J. (2013) 'Climate change and sustainable food production', *Proceedings Nutritional Society*, 72, pp. 21–28.

Soumare, A., Sarr, D. and Diedhiou, A.G. (2023) 'Potassium sources, microorganisms and plant nutrition: Challenges and future research directions', *Pedosphere*, 33, pp. 105–115.

Souza, L.F.T. *et al.* (2023) 'Root distributions, precipitation, and soil structure converge to govern soil organic carbon depth distributions', *Geoderma*, 437, p. 116569.

Spohn, M. and Kuzyakov, Y. (2013) 'Phosphorus mineralization can be driven by microbial need for carbon', *Soil Biology and Biochemistry*, 61, pp. 69–75.

Tittarelli, F. et al. (2007) 'Quality and agronomic use of compost', *Waste Management Series*, 8, pp. 119–157.

Truman, C.C. et al. (2007) 'Variable rainfall intensity and tillage effects on runoff, sediment, and carbon losses from a loamy sand under simulated rainfall', *Journal of Environmental Quality*, 36, pp. 1495–1502.

Van Groenigen, J.W. *et al.* (2015) 'The soil N cycle: New insights and key challenges', *SOIL*, 1, pp. 235–256.

Vijayavenkataraman, S., Iniyan, S. and Goic, R. (2012) 'A review of climate change, mitigation and adaptation', *Renewable and Sustainable Energy Reviews*, 16, pp. 878–897.

Vimal, S.R. *et al.* (2017) 'Soil-plant-microbe interactions in stressed agriculture management: A review', *Pedosphere*, 27, pp. 177–192.

Virk, A.L. *et al.* (2021a) 'Effects of diversified cropping sequences and tillage practices on soil organic carbon, nitrogen, and associated fractions in the North China plain', *Journal of Soil Science and Plant Nutrition*, 21, pp. 1201–1212.

Virk, A.L. *et al.* (2021b) 'Role of legumes in managing soil organic matter and improving crop yield', In: *Soil organic matter and feeding the future*. Boca Raton, FL: CRC Press, pp. 259–277.

Virk, A.L. *et al.* (2022) 'Simultaneous effects of legume cultivation on carbon and nitrogen accumulation in soil', *Advance in Agronomy*, 171, pp. 75–110.

Wang, H. et al. (2021) 'Ploughing/zero-tillage rotation regulates soil physicochemical properties and improves productivity of erodible soil in a residue return farming system', *Land Degradation and Development*, 32, pp. 1833–1843.

Wang, J. *et al.* (2015) 'Contributions of wheat and maize residues to soil organic carbon under long-term rotation in north China', *Science Reports*, 51(5), pp. 1–12.

Wang, Y. *et al.* (2018) 'Long-term effects of nitrogen fertilization on aggregation and localization of carbon, nitrogen and microbial activities in soil', *Science of the Total Environment*, 624, pp. 1131–1139.

Wang, Y. *et al.* (2023) 'Changes in rhizosphere phosphorus fractions and phosphate-mineralizing microbial populations in acid soil as influenced by organic acid exudation', *Soil Tillage Research*, 225, p. 105543.

Wang, Y.X. *et al.* (2024) 'The mechanism of microbial community succession and microbial co-occurrence network in soil with compost application', *Science of the Total Environment*, 906, p. 167409.

White, P.J. *et al.* (2010) 'Potassium', In *Cell Biology of Metals and Nutrients*. Plant Cell Monographs, vol 17. Berlin, Heidelberg: Springer. https://doi.org/10.1007/978-3-642-10613-2_9

Wu, W. and Ma, B. (2015) 'Integrated nutrient management (INM) for sustaining crop productivity and reducing environmental impact: A review', *Science of the Total Environment*, 512–513, pp. 415–427. https://doi.org/10.1016/j.scitotenv.2014.12.101

Wu, W. et al. (2022) 'Meta-analysis of the impacts of phosphorus addition on soil microbes', *Agriculture Ecosystem and Environment*, 340, p. 108180.

Xiao, X. *et al.* (2023) 'Intercropping enhances microbial community diversity and ecosystem functioning in maize fields', *Frontiers in Microbiology*, 13, p. 1084452.

Xu, W. *et al.* (2023) 'Long-term partial substitution of chemical nitrogen fertilizer with organic fertilizers increased SOC stability by mediating soil c mineralization and enzyme activities in a rubber plantation of Hainan Island, China', *Applied Soil Ecology*, 182, p. 104691.

Xun, W. *et al.* (2016) 'Significant alteration of soil bacterial communities and organic carbon decomposition by different long-term fertilization management conditions of extremely low-productivity arable soil in South China', *Environmental Microbiology*, 18, pp. 1907–1917.

Yang, X. *et al.* (2023) 'Contribution of roots to soil organic carbon: From growth to decomposition experiment', *CATENA*, 231, p. 107317.

Yuan, G. *et al.* (2021) 'Effects of straw incorporation and potassium fertilizer on crop yields, soil organic carbon, and active carbon in the rice–wheat system', *Soil Tillage Research*, 209, p. 104958.

Zhang, F. *et al.* (2012) 'Integrated nutrient management for food security and environmental quality in China', *Advance in Agronomy*, 116, pp. 1–40.

Zhang, M. *et al.* (2006) 'A four-year study on influence of biosolids/MSW cocompost application in less productive soils in Alberta: Nutrient dynamics', *Compost Science Utilization*, 14, pp. 68–80.

Zhang, Q. *et al.* (2022) 'Conservation tillage improves soil water storage, spring maize (*Zea mays* L.) yield and WUE in two types of seasonal rainfall distributions', *Soil Tillage Research*, 215, p. 105237.

Zhang, X. *et al.* (2015) 'Managing nitrogen for sustainable development', *Nature*, 528, pp. 51–59.

Zhang, Y. et al. (2023) 'Phosphorus/nitrogen sensing and signaling in diverse root–fungus symbioses', *Trends in Microbiology*, 32(2), pp. 200–215.

Zhang, Y.K. *et al.* (2013) 'Genetic improvement of root growth contributes to efficient phosphorus acquisition in maize (*Zea mays* L.)', *Journal of Integrative Agriculture*, 12, pp. 1098–1111.

Zhu, C. *et al.* (2023) 'Driving factors of soil organic carbon sequestration under straw returning across China's uplands', *Journal of Environmental Management*, 335, p. 117590.

18 Traditional Agricultural Crop Plants Practices and Food Security

Eksha Guliani, Jaya Tuteja, Fouad Damiri, Mithilesh Yadav, and Arpit Sand

18.1 INTRODUCTION

In today's scenario, due to a burgeoning population, there has been a tremendous increase in the needs and demands of people. As a result, natural resources are depleting at an alarming rate, which is posing a challenge towards sustainability (Russelle *et al.*, 2007). According to a survey conducted by the Food and Agricultural Organization, the population of the globe will be about 10 billion by 2050, and it will simultaneously increase demand by 50%. Also, about 815 million citizens on Earth are malnourished, and this will increase to 2 billion in the upcoming years (Yadav *et al.*, 2023). Traditional agriculture evolved about 10,000 years ago. It is one of the most sustainable methods, and it has led to a steady system of farming. An incredible variety of crops and livestock could be produced by practicing traditional farming. In the developing world, many conventional farmers also practice traditional farming methods to maintain an equilibrium with the nature. Traditional farming is one of the oldest methods of farming which is still practiced by farmers in today's developing world, as depicted in Figure 18.1 (Strock *et al.*, n.d.). It basically includes the utilization of traditional tools, organic fertilizers, indigenous knowledge and natural resources, which is an advantage in maintaining a balance between the ecosystem and biodiversity. It eventually enhances

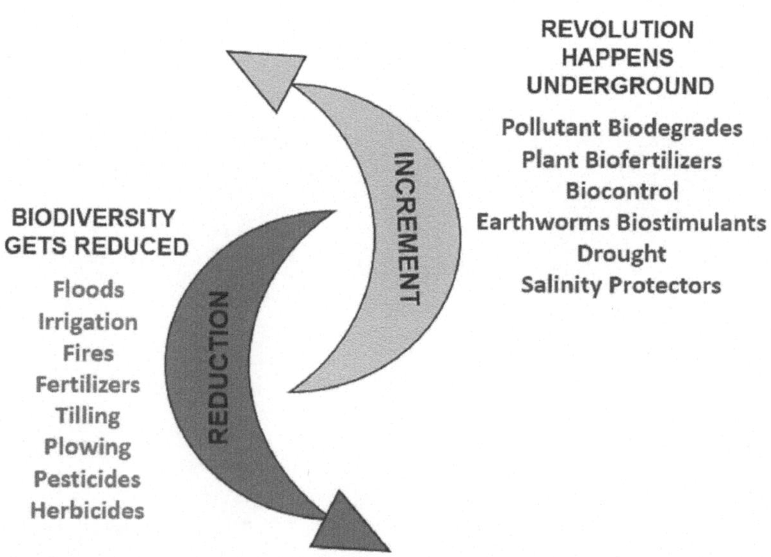

FIGURE 18.1 A mutual relation among traditional farming and increased crop production.

DOI: 10.1201/9781032630366-18

food security and protects natural resources, but there are certain negative impacts as well, which should be considered (Alam *et al.*, 2014).

About 80% of fertile land is cultivated with crops like wheat, corn, soyabean and rice, which reduce the genetic diversity of agriculture globally (Barros-Rodríguez *et al.*, 2021). Due to overpopulation, there has been an increase in the demand of water for irrigation and land for cultivation, which is leading to the overexploitation and deterioration of available natural resources and agricultural systems. Productivity has been hampered by climatic conditions, and the inappropriate usage of fertilizers, agrochemicals and fossil fuels has triggered degradation to the environment. Traditional agriculture possesses an adaptive and mitigation ability for anthropogenic threats, which can lead to sustainability and environment security (Gupta, Rai and Risam, n.d.). It is one of the most dynamic tools utilized for farming, which eventually leads to the conservation of natural resources. Its productivity is sustained by an appropriate utilization of all the local resources that are available and induces those kinds of practices which are site-specific in nature, making it compatible with the climatic conditions (Lincoln, 2019).

Food security has now become a major concern to work on (Guliani and Jeyaseelan, n.d.). The land which is available for cultivation and grazing for animals is gradually depleting because of urbanization. This predicament has evolved over time because of low income, climate change, lack of knowledge and techniques regarding modern tools (Wayman *et al.*, 2017). Also, greenhouse gases will increase by 50% by 2050, with population expanding up to 66%, which will cause a drop of 20% in food production, posing a danger to food security and nutrition. Food security is one of the most vital aspects for fulfilling the needs of human beings, supporting the regional economic prosperity and promoting nutrition (Tyczewska *et al.*, 2023). Global food security has been recognized as an important aspect of sustainable development goals (SDGs) by the United Nations (Alam *et al.*, 2023).

18.2 HISTORY BEHIND TRADITIONAL FARMING

Talking about history, agricultural development could be categorized into three stages: the traditional stage, green revolution era and post-green revolution phase. Agriculture has transformed itself gradually through these stages along with the passage of time due to an increase in the demands of people. The technology implemented in supporting agriculture was very rough for farming operations to work out (Khmelevtsova *et al.*, 2022). As a result, advancements took place, and it stimulated the need for technology required for farming operations that improved with modernization. Traditional agriculture could be further divided into two stages: early stage and pre-modern stage. The time period in which traditional agriculture was widespread up to the beginning of the 19th century symbolizes the early stage of traditional agriculture, and after that, the pre-modern stage of traditional agriculture began (Temesgen, 2021). Traditional farming practices governed a sustainable interaction among animals, plants, food, soil and water, which eventually led to the retention of soil fertility and maintained consistency in crop production. In earlier times, livestock rearing was combined with agriculture by the farmers in order to improve productivity. All the input in traditional farming were organic and naturally available or produced by the farms themselves. Farmers made a common practice of recycling waste produced in farms, which was utilized as compost to maintain soil fertility (Kolanthasamy *et al.*, 2020). The fundamentals of farming process involve the selection of appropriate varieties in order to maintain diversity. Traditional farming has been practiced over generations and has been the most popular method to date on the Asian subcontinent (Hamid Dar *et al.*).

18.3 TRADITIONAL AGRICULTURAL PRACTICES

Traditional agricultural practices the best methods for healthy crop production. Some of these practices are explained in this section (Cheema *et al.*, 2022).

18.3.1 Agroforestry

Agroforestry is basically an agricultural practice that involves planting trees along with some crops. It is one of the most practiced techniques for climate change, food security and crop productivity. Trees basically act as a sink for carbon dioxide. They have the ability to fix carbon dioxide via the process of photosynthesis and even store excess carbon in the form of biomass. This enhances carbon sequestration, water retention and organic matter present in the soil. Agroforestry is influenced by certain factors like species of trees, geographical location and environmental factors. The presence of trees in the farming system has the ability to reduce the severity of weather conditions like hurricanes, floods and tropical storms. Agroforestry is widespread in the tropical regions and is very common in Southeast Asia. A huge portion of the population depends upon agroforestry for agriculture productivity along with their livelihood. Integrating animals into farming methodology helps in recycling the feed into useful manure, which leads directly to the enhancement of carbon sequestration. These methods are highly preferred in drought conditions, as the roots of trees possess a larger volume of soil for nutrients and water. In today's scenario, there has been an increase in the adoption of agroforestry as the prices are rising alarmingly for agrochemicals and fossil fuels (Singh and Singh, 2017).

18.3.2 Shifting Cultivation

Shifting cultivation is also known as slash-and-burn cultivation, in which a particular portion of the land is cleared for variable cultivation. It is one of the most practiced systems of agriculture through which the fertility of soil is restored instead of using artificial fertilizers. It is an example of an extensive agricultural system which depends upon the nature and type of soil, climate, terrain and vegetation. Shifting agriculture is quite complex yet diverse in nature, which includes cultivation of ten or more crops along with the combination of livestock and trees. The replenishment of soil is done through cycling of the nutrients via biological nitrogen fixation. It therefore maintains the structure of soil at a certain level, which eventually minimizes the physical degradation of soil. It controls the excessive runoff and manages the pests by enhancing biodiversity by inducing a balance among parasites and predators. Therefore, it increases the income of farmers by generating value-added products, which creates off-farm employment (Lal, 2015).

18.3.3 Crop Rotation

Crop rotation is another traditional agricultural practice, which involves the growth of plants in a sequence on that particular land. It can be described by monitoring over a certain period of time. Consider an example of a sugarcane plantation, which it utilizes a lot of nitrogen from the soil (Dury *et al.*, 2012). Therefore, to replenish the soil with nitrogen, farmers grow beans on that same portion of land in order to enhance and balance soil fertility. It is a very useful technique when it comes to practicing sustainable agriculture. Crop rotation helps in the reduction of trade-offs, maintaining the fertility of soil and disrupting disease formation via intrinsic recycling of nutrients to sustain productivity (Shah *et al.*, 2021). Pathogens have the ability to add up on the farm, and if crop rotation isn't practiced, these infections will incorporate in the soil, which would lead to a severe outbreak of disease in plants. Crop rotation aids in combating erosion forces, and it even requires less fertilizers, which will result in economic benefits (Shah *et al.*, 2021).

18.3.4 Cover Cropping

A cover crop is defined as that crop which isn't harvested but could be grown in order to provide nourishment to the soil and other crops in various methods. It includes improved quality of soil, less soil erosion and less presence of insects. These crops are versatile in nature and fix atmospheric nitrogen in soil (Strock *et al.*). One of the most amazing benefits of practicing cover cropping is long-term protection of soil and its stabilization, which leads to sustainable agriculture. The residue of cover crops leads to control of the germination of weeds. Practicing cover cropping leads to the reduction in usage of herbicides, which eventually reduces the impact on the environment (Teasdale, 2013). Cover cropping is done more efficiently when there is use of manure in the farming. The cover crops have the ability to store and hold nutrients until another cash crop uses it (Mitchell *et al.*, 2015). The planting is done in such a way that the dates of establishment of cover crops and application of manure are within the days of one another. The cover crops therefore improve the efficiency and productivity of the cropping system along with the minimization of environmental impacts (Groff, 2015).

18.3.5 Integrated Crop-Livestock Farming

Integrated crop-livestock systems, referred to as ICLs, are basically planned systems which have the ability to interact spatially with crops and animals in the same area in succession (De Moraes *et al.*, 2014). These are man-made agricultural systems which combine the rate of production with animal husbandry in order to get a large variety of products. This type of farming has the ability to reduce the loss of yield from any sort of diseases. For example, *Arachis hypogaea* (peanut disease) and *Rhizoctonia solani* Kuhn (limb rot) were reduced by *Paspalum notatum* (bahiagrass), which even increased production to 30% (Franzluebbers, 2007). Both environmental and economic benefits are observed when this type of farming is processed. If cattle graze in the farmland of legume-grain crops, the accumulation of soil is doubled, which even recycles the nutrients upon addition of manure. ICLs result in some biogeochemical cycles, which mimic the natural ecosystem. Therefore, it is regarded as a sustainable ecosystem, which is implied throughout the world (Sekaran *et al.*, 2021).

18.4 TRADITIONAL AGRICULTURE AND CLIMATE CHANGE

Climate change has now become an important issue to be discuss with respect to the economic and socio-ecological outlook. Anthropogenic activities are mainly responsible for inducing environmental problems like pollution, climatic change and degradation of natural resources like biodiversity loss and soil degradation (Duffy *et al.*, 2021). Agriculture is one of the most vulnerable systems and is hampered by even minute changes in climatic conditions. Recently, climatic changes have become a great threat to food security globally. They cause a decline in the production of food, which eventually leads to inflation, and it is intensified by overpopulation (Singh and Singh, 2017). Traditional agriculture provides livelihood to about 1 billion people. Climatic change hampers agriculture in various ways. In many situations, farmers face erratic rainfall and uneven changes in seasonal patterns, causing floods, droughts, etc. These sort of stresses in the environment lead to a direct impact on the productivity of land, which eventually affects agriculture. It has an indirect impact on soil and water, which leads to soil erosion (Mijatović *et al.*, 2013).

Global warming has become a threat to crop growth, which compromises the production of agriculture and food security (Ahmed *et al.*, 2020). The changes in climate have an impact on the yields of crops, and they vary from region to region, which leads to certain aftereffects in the developing world, as depicted in Figure 18.2. These effects include severity of droughts, floods in lowland areas and excessive heat conditions, which pose a hazard to agriculture by limiting the growth of crops (Perotti *et al.*, 2020). As a result, poor farmers will be most affected because of their low income, limited ability to seek livelihood and limited geographic exposure. About 370 million people in the world are poor and live in low-resource areas, which are very much prone to severe risks, and that makes them vulnerable (Gururani *et al.*, 2021). Even a minute change in the climatic condition can

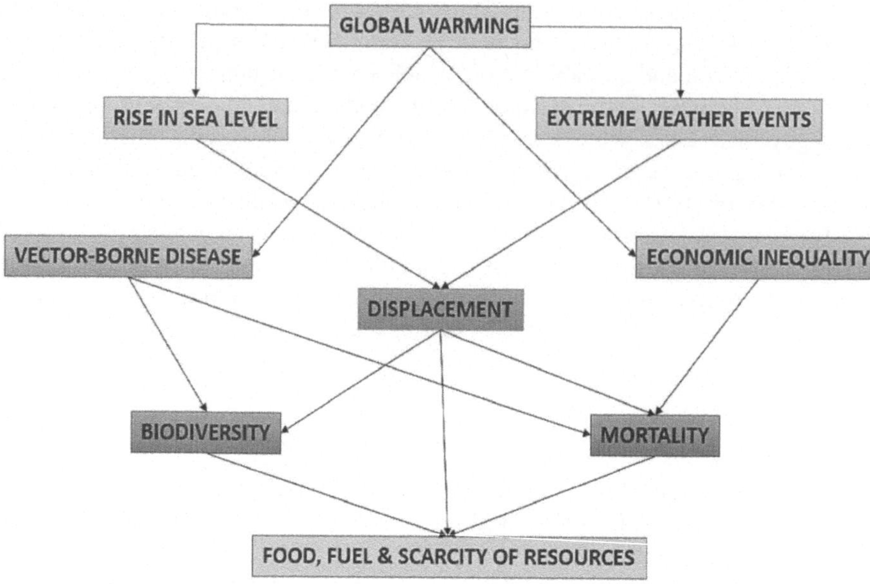

FIGURE 18.2 A cascading view of global warming impacting the ecosystem.

impact these vulnerable groups disastrously, ultimately posing a challenge for food security. Based on the facts stated above, experts recommend that combining traditional agriculture practices with agroecological strategies would increase productivity level, resilience and sustainability of agricultural production. Managing risk exposure is a vital preoccupation of households, and it's the only insurance available to the farmers.

An archaeological record gives a vital data source on agricultural systems that is sustainable and resilient in nature. Many systems consist of raised fields in highland basins. Therefore, a raised bed is a cultivation system which is an ancient technique practiced by farmers (Jehn *et al.*, 2022). Its use and origin are linked with the water management techniques by inducing various opportunities in order to cause a reduction in overflow of water while irrigating crops. It also provides methods to irrigate fields during scarcity of water. Among the most widely used methods are chinampa and waru. Chinampa is an artificial cultivation system that comprises a raised platform which is about 10 m wide and 100 m long. It is constructed with some scraped mud obtained from shallow lakes or swamps. Farmers grow some basic food crops on them. They are also known as a floating gardens and are widely used for agriculture on shallow beds of lakes. These have the ability to produce bumper crops even in floods or droughts. The combination of canals and raised beds helps eliminate the distressing effects of frost. During droughts, the moisture from the canals ascends slowly by capillary action to the roots, whereas during floods, troughs have the ability to drain the runoff in excess. Heat emitted by the sun is absorbed by the water in canals, and it is radiated back during night, which protect the crops from frost. It has a higher productivity than waru when compared to the use of chemical fertilizers on crops. Another vital technique is rainwater harvesting, which comprises water collection in small catchments. Also, earthen basins are a simple way of capturing rainwater and retaining the capacity of soil. There are many farmers who follow traditional farming, as it helps to understand the integrity of agroecosystems, which depends upon the interactions of plant diversity, with soil microbial community having a great percentage of organic matter. Several researches have concluded that incorporating organic matter into farming enriches the underground biological diversity that empowers plant roots, leading to the production of healthy and strong crops (Altieri and Nicholls, 2017).

18.5 SUSTAINABILITY AND FOOD SECURITY

Sustainable agriculture can also be regarded as sustainable intensification, which is basically a system that enhances crop yields with negligible impact on the environment. It aims to increase the production of food by regenerating the same land used for cultivation and the goal is to end hunger by keeping in mind the environment protection (Rehman *et al.*, 2022). It is applicable mainly to those developing countries in which a high yield gap is observed because of the low yields obtained from the existing production system. The major cause is the usage of inadequate fertilizers, which leads to soil degradation. Adopting sustainable agriculture practices like precision agriculture, integrated pest control, conservation tillage and reforestation will definitely have the potential to enhance food security by reducing the usage of synthetic fertilizers and agrochemicals, as well as avoiding non-renewable energy, by increasing the yields and maintaining a healthy environment (Thomson *et al.*, 2019). Income is one of the most important outcomes and is directly proportional to the rate of production. It can be increased by using a high-quality and innovative mechanization of agricultural techniques. To improve efficiency, multiple practices become an important aspect in traditional farming, as it involves the judicial use of resources which eventually improves sustainability. Considering the usage of water, there are certain approaches that farms can adopt in order to curb wasting of water. Some activities like mixed-crop rotation and crop husbandry can greatly impact water usage. Optimal irrigation planning and maintenance influence the efficiency of system (Jürkenbeck *et al.*, 2019).

Crops are required to be stored under cool conditions in order to maintain proper nutrition and freshness. To obtain a high value for the food, it is transported to processing facilities which are responsible for cutting, washing, centrifuging and removing the excess water that was accumulated in the products. To maintain an appropriate inertia status for food, it is stored in low temperatures. But farmers are unable to maintain cold storage conditions, which makes it more possible for pathogens to attack and cause infection. Food security is a major concern to maintain food hygiene and nutrition. Over the past few years, there has been a constant and increasing demand for the food products by a growing population. Therefore, to increase the rate of production, farmers use chemicals to fight pests which are hampering the nutrition and freshness of the crops. After a heavily ploughed farming, tremendous irrigation and monocrop selection, soil is degraded and its nutrients are depleted. Even if there are technological automations and innovations to avoid all the food-borne diseases, there are certain rules and regulations which are governed by Global Food Safety Initiatives (GFSI). This organization was established in 2000 to control and reduce the risks linked with food production. There are certain certifications that are provided to the farmers in order to maintain high levels of food quality: Hazard Analysis Critical Control Points (HACCP) and Sage Quality Food (SQF) (Avgoustaki and Xydis, 2020).

18.6 CHALLENGES

Groups that manage traditional agricultural along with biodiversity on various scales create a dynamic landscape of orchards, pastures, fields, gardens and ecosystem patches. Minimizing the biodiversity loss and simultaneously increasing the sustainability of agricultural systems have been recognized as major challenges. One of the greatest challenges in adopting the traditional agricultural practices is a huge decline in production matrix (Farrokhi and Pellegrina, 2023). This mainly includes the work done manually, such as weeding, grain threshing, milling, post-harvest processing and seed sowing, which slows the production rate. Another challenge is that traditional agriculture leads to low yield in the absence of scientific techniques. These crops can develop in a very stressful environment where the main crops fail to grow. Some small millets are grown by farmers who practice traditional agriculture in order to develop stress resistance from biotic and abiotic factors (Obirikorang *et al.*, 2021).

Fulfilling the requirements of populations by providing proper dietary requirements and sustainable food poses to be another major challenge in current scenario. After the green revolution, production has multiplied greatly, but there are certain implications to it (Phan *et al.*, 2021). With issues in the environment, which include population growth, degradation of natural resources, loss of biodiversity and climate change, food security has become necessary. The traditional agriculture crop system still remains largely unexplored while it is discussed at a molecular level. Therefore, if coordination, funding and communication improve, these can be overcome (Patel *et al.*, 2020).

18.7 CONCLUSION

Traditional farming has always been the backbone in terms of the economy in many countries and largely in India. Farmers have utilized various traditional techniques in order to enhance the yield of crops for a long time, but in the present day, they do not completely fulfil needs due to increasing demand for food. This type of farming possesses several drawbacks, and one of the most severe ones is pest control. Therefore, this type of farming becomes less beneficial in terms of the economy. To overcome it, new techniques are emerging to increase crop yield for the health of the environment. Traditional farming has therefore gained a great deal of attention for contributing to food production in the era of environmental crises. Climate changes have always influenced crop production, which could impact food security, too. Therefore, adoption of traditional agricultural practices has an impact on the quality and productivity of crop production.

REFERENCES

Ahmed, N., Thompson, S. and Turchini, G.M. (2020) 'Organic aquaculture productivity, environmental sustainability, and food security: Insights from organic agriculture', *Food Security*, 12, pp. 1253–1267. https://doi.org/10.1007/s12571-020-01090-3

Alam, A., *et al.* (2014) 'The changing scenario in Indian agriculture: A review', *International Journal of Scientific Research in Agricultural Sciences*, 2(7), pp. 118–127. https://doi.org/10.12983/ijsras-2014-p0118-0127

Alam, M.F.B., *et al.* (2023) 'Analysis of the drivers of agriculture 4.0 implementation in the emerging economies: Implications towards sustainability and food security', *Green Technologies and Sustainability*, 1(2), p. 100021. https://doi.org/10.1016/j.grets.2023.100021

Altieri, M.A. and Nicholls, C.I. (2017) 'The adaptation and mitigation potential of traditional agriculture in a changing climate', *Climatic Change*, 140(1), pp. 33–45. https://doi.org/10.1007/s10584-013-0909-y

Avgoustaki, D.D. and Xydis, G. (2020) 'How energy innovation in indoor vertical farming can improve food security, sustainability, and food safety?', in *Advances in food security and sustainability.* Elsevier Ltd., Vol. 5, pp. 1–51. https://doi.org/10.1016/bs.af2s.2020.08.002

Barros-Rodríguez, A., *et al.* (2021) 'Impacts of agriculture on the environment and soil microbial biodiversity', *Plants*, 10(11). https://doi.org/10.3390/plants10112325

Cheema, S.M., *et al.* (2022) 'IoAT enabled smart farming: Urdu language-based solution for low-literate farmers', *Agriculture (Switzerland)*, 12(8). https://doi.org/10.3390/agriculture12081277

De Moraes, A., *et al.* (2014) 'Integrated crop-livestock systems in the Brazilian subtropics', *European Journal of Agronomy*, 57, pp. 4–9. https://doi.org/10.1016/j.eja.2013.10.004

Duffy, C., *et al.* (2021) 'Agroforestry contributions to smallholder farmer food security in Indonesia', *Agroforestry Systems*, 95(6), pp. 1109–1124. https://doi.org/10.1007/s10457-021-00632-8

Dury, J., *et al.* (2012) 'Models to support cropping plan and crop rotation decisions. A review', *Agronomy for Sustainable Development*, 32(2), pp. 567–580. https://doi.org/10.1007/s13593-011-0037-x

Farrokhi, F. and Pellegrina, H.S. (2023) 'Trade, technology, and agricultural productivity', *Journal of Political Economy*, 131(9), pp. 2509–2555. https://doi.org/10.1086/724319

Franzluebbers, A.J. (2007) 'Integrated crop-livestock systems in the southeastern USA', *Agronomy Journal*, 99(2), pp. 361–372. https://doi.org/10.2134/agronj2006.0076

Groff, S. (2015) 'The past, present, and future of the cover crop industry', *Journal of Soil and Water Conservation*, 70(6), pp. 130A–133A. https://doi.org/10.2489/jswc.70.6.130A

Guliani, E. and Jeyaseelan, C. (n.d.) 'The biochemistry of food colouring agents', *Journal of Agricultural Engineering and Food Technology*, 7.

Gururani, K., *et al.* (2021) 'Mainstreaming *Barahnaja* cultivation for food and nutritional security in the Himalayan region', *Biodiversity and Conservation*, 30(3), pp. 551–574. https://doi.org/10.1007/s10531-021-02123-9

Jehn, F.U., *et al.* (2022) 'Focus of the IPCC assessment reports has shifted to lower temperatures', *Earth's Future*, 10(5). https://doi.org/10.1029/2022EF002876

Jürkenbeck, K., Heumann, A. and Spiller, A. (2019) 'Sustainability matters: Consumer acceptance of different vertical farming systems', *Sustainability (Switzerland)*, 11(15). https://doi.org/10.3390/su11154052

Khmelevtsova, L.E., *et al.* (2022) 'Influence of agricultural practices on bacterial community of cultivated soils', *Agriculture (Switzerland)*, 12(3). https://doi.org/10.3390/agriculture12030371

Kolanthasamy, E., *et al.* (2020) 'Traditional agricultural practices as a tool for management of insects and nematode pests of crops: An overview', *Journal of Entomology and Zoology Studies*, 8(3), pp. 237–245. http://www.entomoljournal.com

Lal, R. (2015). 'Shifting cultivation versus sustainable intensification', in *Reference module in earth systems and environmental sciences*. Elsevier. https://doi.org/10.1016/b978-0-12-409548-9.09295-2

Lincoln, N.K. (2019) 'Learning from indigenous agriculture', *Nature Sustainability*, 2(3), pp. 167–168. https://doi.org/10.1038/s41893-019-0234-1

Mijatović, D., *et al.* (2013) 'The role of agricultural biodiversity in strengthening resilience to climate change: Towards an analytical framework', *International Journal of Agricultural Sustainability*, 11(2), pp. 95–107. https://doi.org/10.1080/14735903.2012.691221

Mitchell, J.P., Shrestha, A. and Irmak, S. (2015) 'Trade-offs between winter cover crop production and soil water depletion in the San Joaquin Valley, California', *Journal of Soil and Water Conservation*, 70(6), pp. 430–440. https://doi.org/10.2489/jswc.70.6.430

Obirikorang, K.A., *et al.* (2021) 'Aquaponics for improved food security in Africa: A review', *Frontiers in Sustainable Food Systems,* 5. https://doi.org/10.3389/fsufs.2021.705549

Patel, S.K., Sharma, A. and Singh, G.S. (2020) 'Traditional agricultural practices in India: An approach for environmental sustainability and food security', *Energy, Ecology and Environment*, 5(4), pp. 253–271. https://doi.org/10.1007/s40974-020-00158-2

Perotti, V.E., *et al.* (2020) 'Herbicide resistant weeds: A call to integrate conventional agricultural practices, molecular biology knowledge and new technologies', *Plant Science*, 290. https://doi.org/10.1016/j.plantsci.2019.110255

Phan, L.T.K., *et al.* (2021) 'Contamination of *Fusarium proliferatum* and *Aspergillus flavus* in the rice chain linked to crop seasons, cultivation regions, and traditional agricultural practices in Mekong Delta, Vietnam', *Foods*, 10(9). https://doi.org/10.3390/foods10092064

Rehman, A., *et al.* (2022) 'Sustainable agricultural practices for food security and ecosystem services', *Environmental Science and Pollution Research*, 29(56), pp. 84076–84095. https://doi.org/10.1007/s11356-022-23635-z

Russelle, M.P., Entz, M.H. and Franzluebbers, A.J. (2007) 'Reconsidering integrated crop-livestock systems in North America', *Agronomy Journal*, 99(2), pp. 325–334. https://doi.org/10.2134/agronj2006.0139

Sekaran, U., *et al.* (2021) 'Role of integrated crop-livestock systems in improving agriculture production and addressing food security – A review', *Journal of Agriculture and Food Research*, 5. Elsevier B.V. https://doi.org/10.1016/j.jafr.2021.100190

Shah, K.K., *et al.* (2021) 'Diversified crop rotation: An approach for sustainable agriculture production', *Advances in Agriculture*, 2021. https://doi.org/10.1155/2021/8924087

Singh, R. and Singh, G.S. (2017)'Traditional agriculture: A climate-smart approach for sustainable food production', *Energy, Ecology and Environment*, 2(5), pp. 296–316. https://doi.org/10.1007/s40974-017-0074-7

Strock, J.S., Porter, P.M. and Russelle, M.P. (n.d.) 'Cover cropping to reduce nitrate loss through subsurface drainage in the Northern U.S. Corn Belt', *Journal of Environmental Quality*, 33, pp. 1010–1016.

Teasdale, J.R. (2013) 'Contribution of cover crops to weed management in sustainable agricultural systems', Journal of Production Agriculture, 9(4), pp. 1010–1016.

Temesgen, B. (2021) 'Role and economic importance of crop genetic diversity in food security', *International Journal of Agricultural Science and Food Technology*, 7(1), pp. 164–169. https://doi.org/10.17352/2455-815x.000104

Thomson, A.M., *et al.* (2019) 'Sustainable intensification in land systems: Trade-offs, scales, and contexts', *Current Opinion in Environmental Sustainability*, 38, pp. 37–43. https://doi.org/10.1016/j.cosust.2019.04.011

Tyczewska, A., Twardowski, T. and Woźniak-Gientka, E. (2023) 'Agricultural biotechnology for sustainable food security', *Trends in Biotechnology*, 41(3), pp. 331–341. https://doi.org/10.1016/j.tibtech.2022.12.013

Wayman, S., *et al.* (2017). 'Organic and conventional farmers differ in their perspectives on cover crop use and breeding', *Renewable Agriculture and Food Systems*, 32(4), pp. 376–385. https://doi.org/10.1017/S1742170516000338

Yadav, N., *et al.* (2023) 'Recent advances in nanotechnology for the improvement of conventional agricultural systems: A review', *Plant Nano Biology*, 4, p. 100032. https://doi.org/10.1016/j.plana.2023.100032

19 Experiment Tools Used as a Biostimulant for Sustainable Crop Plants Improvement and Practices

Tushar Goyal, Deepak Kumar, Anand Kumar Gaurav, Arpan Mukherjee, Gowardhan Kumar Chouhan, and Jay Prakash Verma

19.1 INTRODUCTION

Biostimulants are formulations of certain organic or inorganic materials like seaweed extracts, humic acids and fulvic acids, animal- and vegetable-based components, and microorganisms (fungi and bacteria). Biostimulants' inoculation in small quantities to seeds, plants, or the rhizosphere induces natural processes that enhance nutrient uptake, nutrient efficiency, tolerance to stress, and improve crop quality and productivity in the US Farm Bill (Agriculture Act of 2018). Biostimulants can be used in combination with fertilizers, pesticides, and other growth regulators to improve growth and yield of plants. The ability of biostimulants to improve plant health and increase crop yield underlies their necessity for higher plant growth. Plants can benefit from biostimulants by using them to counter environmental stresses, including drought, salt, and extreme heat. They can also promote nutrient availability and soil quality, which again will lead to higher plant growth and production. Also, by stimulating the synthesis of plant hormones and accelerating nutrient absorption, biostimulants can promote plant development. Consequently, there will be greater root growth, flowering, and fruit production. To reduce the need for synthetic agrochemicals, biostimulants can promote more ecologically friendly and environmentally sustainable farming practices.

The importance of biostimulants for higher plant growth lies in their ability to improve plant nutrition, as they contain essential nutrients, such as nitrogen (N), phosphorus (P), and potassium (K), as well as other micronutrients, like iron (Fe), zinc (Zn), and manganese (Mn), which are required in smaller amounts but are critical for proper plant function. They can improve soil structure and fertility, increase microbial activity, and reduce soil-borne diseases. This, in turn, can improve plant growth and nutrient uptake.

Numerous varieties of chemically synthesized plant growth regulators are now utilized on an international market. Plant biostimulants, both microbial and non-microbial, are commonly employed in open field and greenhouse crops such as fruits, vegetables, berry crops, grapevines, ornamentals, turfs, and cereals (Basile *et al.*, 2020; Kaur *et al.*, 2021). Much scientific research has proven the effectiveness of man-made compounds in boosting crop yields and protecting plants from various biotic and abiotic stressors. For example, the potent germination-promoting activity of the man-made strigolactone GR24 has been recognized for a long time (Boyer *et al.*, 2014). Wheat (*T. aestivum*) and sunflowers (*H. annuus*) have both been shown to have increased salt tolerance after receiving exogenous applications of GR24 (Kausar and Shahbaz, 2017; Sarwar and Shahbaz, 2020). Despite the interest in chemical for agriculture, it is important to consider their potential detrimental impact on both the environment and human health (Zulfiqar *et al.*, 2020). In order

DOI: 10.1201/9781032630366-19

to achieve these goals, biostimulants are viewed as a cost-effective and environmentally friendly method of sustaining agriculture. In fact, Yasmeen *et al.* (2014) investigated the effectiveness of two substances, benzylaminopurine (BAP) and exogenously administered moringa leaf extract (MLE), which is cytokinin rich, in enhancing the growth of *Solanum lycopersicum* (tomato) plants. The study found that the foliar application of the MLE, a natural biostimulant, was more effective than synthetic BAP in promoting tomato growth and development, yield, and quality. Overall, biostimulants are a valuable tool for farmers and growers to improve plant growth and yield while also promoting sustainable agricultural practices.

This chapter's objective is to provide readers with a strong understanding of biostimulants, their multiple forms, and the potential uses of these substances in horticulture and agriculture. The action mechanisms of biostimulants as well as how they interact with plant physiology to enhance growth, yield, and stress tolerance are also covered. The chapter's overall goal is to provide readers with a thorough grasp of biostimulants and their potential applications in sustainable agriculture production.

19.2 TYPES OF BIOSTIMULANTS AND THEIR EFFECTS ON THE HOST PLANT GROWTH AND DEVELOPMENT

The different types of biostimulants include: humic substances (HS), seaweed extracts (SWE), protein-hydrolysates (PH), amino acids, and microbial-based plant-growth-promoting bacteria that promote plant development. Humic substances are made of organic matter (OM) and improve soil quality and health (Figure 19.1). Plant hormones and amino acids included in seaweed extracts promote plant development and improve tolerance to stress. Protein synthesis is improved, root growth is accelerated, and plant resistance to environmental stress is increased by amino acid-based biostimulants. Beneficial microorganisms like bacteria and fungus are present in microbial-based biostimulants, which help fix nitrogen, solubilize phosphorus, and enhance soil health (Figure 19.1). These biostimulants can be combined for best efficacy, and each one has certain advantages.

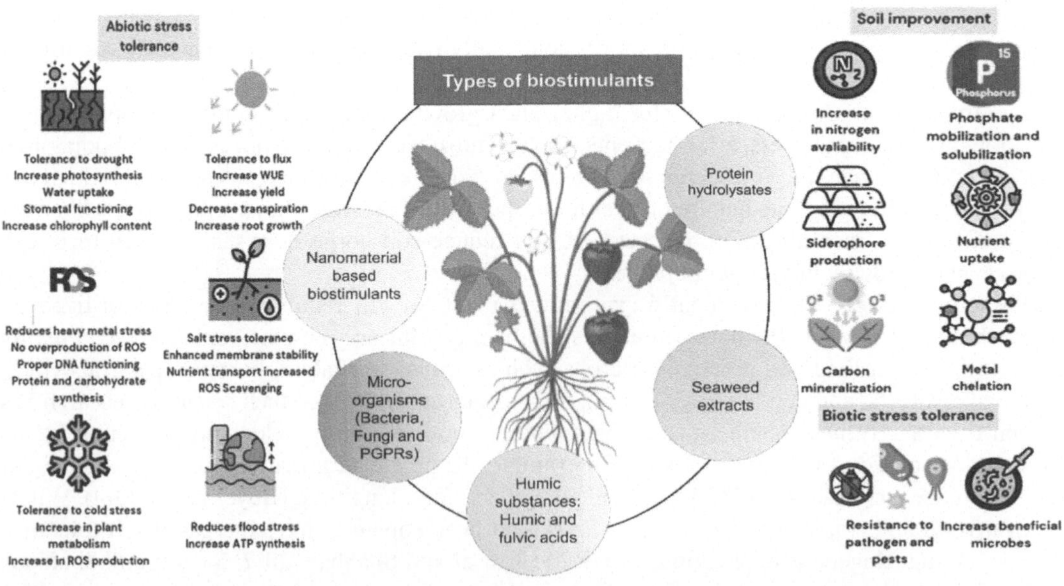

FIGURE 19.1 Diagrammatic representation of different types of biostimulants and their application.

19.2.1 Humic Substances/Components

Humic components (humic and fulvic acids), which makeup the majority (> 60%) of the organic carbon compounds on the planet, originate from the chemical and biological alteration of dead cell components as well as from microbial metabolism (Canellas *et al.*, 2015; Nardi *et al.*, 2009). HS are self-assembled associations made up of a mixture of comparatively low molecular mass components sustained by hydrogen bonding and hydrophobic interactions (Piccolo, 2002). Sprengel 1837 introduced the division of HS based on their solubility into humic acid (HA) and fulvic acid (FA), where HA can precipitate in acidic condition (~1–2 pH) and is soluble in alkaline conditions. Whereas FA remains soluble in acidic conditions, HS have been recognized as essential contributors to the regulation of many crucial ecological and environmental processes.

19.2.1.1 Humic Substances Have the Following Direct Impacts on the Growth and Development of Plants

The effect of HS on plant growth have been anticipated due to their positive effect on root architecture, soil fertility, and structure by acting on its microbial community (du Jardin, 2015). In support of HS as a biostimulant for higher plant growth, a study was done in which exogenous application of HS showed 22% increment of root and shoot dry weights of various plant species (Rose *et al.*, 2014). Many mechanisms impacting soil processes and plant physiology have been implicated in the biostimulation activity of HAs' nutrient availability on soil and absorption.

HAs directly improve soil nutrient availability and uptake by improving (1) soil structure, (2) cation exchange capacity and bringing down soil pH to neutral, (3) phosphorus solubilization by preventing Ca-P precipitation as well as improving the availability of micronutrients by preventing leaching, (4) enhancing lateral root and root hair growth because of auxin-like activity, which activates plasma membrane H^{+}-ATPase activity, and (5) promoting nitrate absorption by activating the key players (NR, GDH, and GER) in this (De Pascale *et al.*, 2017; Delgado *et al.*, 2002; García-Mina *et al.*, 2004; Halpern *et al.*, 2015; Pinton *et al.*, 1999; Schmidt *et al.*, 2007; Zandonadi *et al.*, 2016) (Table 19.1).

TABLE 19.1
Various Biostimulants and Their Effect on Crop Plants

Type of Biostimulant	Crop	Effect	Reference
Humic Substances	Maize	• Regulate plant hormone (IAA, ethylene, and jasmonic acid) activity by modifying plant receptors, phosphatase activity, and changing expression of different transcription factors.	(Souza *et al.*, 2022)
Humic Substances	Maize	• Root growth promotion (increase in fresh weight), proline accumulation, and catalase activity. • Modulation of genes encoding stress perception and cell signalling. • Increase plant stress tolerance against abiotic and abiotic stresses.	(Canellas *et al.*, 2020)
Protein Hydrolysates	Tomato	• Root growth promotion.	(Ceccarelli *et al.*, 2021)
Protein Hydrolysates	"Canasta" lettuce	• Improved yield, nutritional property such as total sugars, total phenolics, carotenoids, and nitrogen indices.	(Sabatino *et al.*, 2021)
Protein Hydrolysates	Cucumber	• Improved growth of cucumber seedlings. • Improved expression of nitrogen transporter genes, synthesis of amino acid, and helps in improving the xenobiotic components detoxification genes.	(Wilson *et al.*, 2015)

(Continued)

TABLE 19.1 ***(Continued)***
Various Biostimulants and Their Effect on Crop Plants

Type of Biostimulant	Crop	Effect	Reference
Seaweed Extracts	*Amaranthus polygamous*	• Increased plant growth, number of leaves, fresh weight of leaves, stem, and the whole plant. • Reduced recommended chemical fertilizer doses up to 50%.	(Senthuran *et al.*, 2019)
Seaweed Extracts	Sugarcane	• Yield enhancement of sugarcane up to 12.87%. • Improved water retention capacity and water content in sugar cane leaves. • Performed exceptionally well in insect-pest management.	(Chen *et al.*, 2019)
Seaweed Extracts	Tomato and sweet pepper	• Reduction in disease severity caused by *Xanthomonas compestris* and *Alternaria solani.* • Improved plant growth and yield. • Upregulation of IAA, gibberellin, and cytokinin synthesis genes.	(Ali *et al.*, 2019)
PGPR	C3 and C4 plants	• Improved drought tolerance of plants	(Zhao *et al.*, 2023)
PGPR	Tomato	• Enhance plant growth, development and yield and also improve the ISR (induction of systemic resistance). • Mitigated the negative effects of insect infestation via proline production, enhanced antioxidant enzymes synthesis, improve the activities of protease and polyphenol oxidases, phenolics, protein and chlorophyll, etc.	(Bano and Muqarab, 2017)
Silicon-Based Biostimulants	Potato	• Increased starch accumulation under water stress • Improved growth of early crop potatoes.	(Wadas, 2021, 2023)
PGPF	Lemon balm	• Drought stress was controlled by promoting photosynthesis, increasing the proline content, and increasing relative water content (RWC).	(Eshaghi Gorgi *et al.*, 2022)
PGPF	Rice and maize	• Increased the phytohormone production, defensive compounds, and enzymes that inhibit pathogen invasion.	(Murali *et al.*, 2021)

By significantly influencing primary metabolism, HS stimulates the metabolism of carbon and nitrogen, which in turn will lead to plant growth. Enzymes connected to N assimilation pathways, such as nitrate reductase (NR), glutamate dehydrogenase (GDH), and glutamine synthetase (GS), were induced by HS under various circumstances (Albuzio *et al.*, 1986; Canellas *et al.*, 2013; Hernandez *et al.*, 2015; Muscolo *et al.*, 1999). Moreover, relative to the addition of HS extracted from different components like peat, composts, or vermicomposts, plant physiological responses to brown coal (such as lignite, leonardite, and subbituminous coals) are less pronounced (Canellas and Olivares, 2014; Du Jardin, 2015).

HS not only affects plant primary metabolism and nutrient intake significantly, it may also have a substantial impact on secondary metabolism (Schiavon *et al.*, 2010). According to some recent studies, HS increased the enzyme expression of phenylalanine (tyrosine) ammonialyase (PAL/TAL), which catalyses the major step in the biosynthesis of phenolics by altering phenylalanine to the trans-cinnamic acid and the tyrosine to the p-coumaric acid. A significant phenol accumulation under stress condition in tomato and lettuce coincided with the expression of PAL/TAL (Hernandez *et al.*, 2015; Olivares *et al.*, 2015). Because of its recently documented impact in reducing drought stress, HS may also have other horticultural uses. In response to the administration of HS, plants under salinity stress (drought) showed the ability to osmotically respond by sustaining water uptake and cell turgor (Aydin *et al.*, 2012; Azevedo and Lea, 2011; García *et al.*, 2020; Paksoy *et al.*, 2010; Petrozza *et al.*, 2014; Türkmen *et al.*, 2004).

19.2.2 Protein Hydrolysates or Amino Acids Containing Products

The term "PH" refers to a mixture of amino acid, polypeptide, and oligopeptide synthesis by chemical and enzymatic hydrolyses of agroindustrial wastes, including both animal (leather, blood, feathers) and plant (vegetables and fruit byproduct) origins from specific biomass sources (legume crops, seeds, etc.) (Colla *et al.*, 2015; du Jardin, 2012; Halpern *et al.*, 2015; Maini, 2006; Schiavón *et al.*, 2008). The majority of the PHs biostimulants generally originate from animal-based or -derived proteins recovered by acid hydrolysis, with a fraction resulting through enzymatic hydrolysis of plant-derived proteins. In order to produce PHs from the byproducts of wet milling of corn and vegetables, more economical and sustainable methods are currently being used (Baglieri *et al.*, 2014; Pecha *et al.*, 2012).

19.2.2.1 Effect of PHs on Plant Metabolism and Physiology

For improved results, PHs are typically administered to plants as foliar sprays (Colla *et al.*, 2015; Colla *et al.*, 2017). PHs have been shown in numerous instances to act as biostimulants by modifying plant physiological and molecular processes which promote growth, increase crop production, and reduce the detrimental effects of stress on plants (Calvo *et al.*, 2014; Paradikovic *et al.*, 2011; Yakhin *et al.*, 2017). Direct impacts of PH biostimulation activity and the abiotic stress tolerance included (1) activation of the key enzyme that was involved in the nitrogen assimilation (NR, NiR, GS, and GOCAT) and carbon metabolism (citrate synthase, malate, and isocitrate dehydrogenase), (2) improved auxin- and gibberellin-like components activities, and (3) increased antioxidant enzyme activity, pigment biosynthesis property, and secondary metabolite production ability (Ertani *et al.*, 2013; Rouphael *et al.*, 2017, 2018; Schiavon *et al.*, 2008; Sestili *et al.*, 2018).

The applications of these compounds for seed treatments can provide some additional benefits. A animal-derived PH, gelatin, when used as capsules to the cucumber seeds shows a significant increase in the biomass, leaf area, and N content of new plants when evaluated with controlled or untreated seeds (Wilson *et al.*, 2015) (Table 19.1). These alterations in the plant biomass in response to gelatin capsules were correlated with the upregulation of both the amino acids and nitrate transporter genes and the xenobiotic compound detoxification system. Similar to this, a vegetal-PH based product containing a plant root hair-stimulating peptide has been utilized effectively as a seed treatment to increase root development, stress tolerance, and improve adherence of fungicides and micronutrients on the seed surface in maize (*Z. mays* L.) and soybean (*G. max*) plant (Bonini *et al.*, 2017)

It was observed in number of studies that PHs have the positive effects in secondary plant metabolism, and help plants to mitigate different stress conditions like salinity, drought, temperature, and the oxidative stress (Apone *et al.*, 2010; Ertani *et al.*, 2013; Kauffman *et al.*, 2007). Exogenous application of PHs to plants promoted vegetative plant growth and increased nutrient accumulation efficiency of plants, whether they are macro- or micronutrients (Halpern *et al.*, 2015). Researchers conducted a set of experiments on bananas in which a foliar application of the feather hydrolysed waste was given. This treatment results in shortening of the harvesting period by up to 28 days and increased hands number and weight of the bunch (Morales-Payan and Stall, 2003). For the above-mentioned reasons, it was clear that feather-degraded products could be bioenhancers to reduce the application of fertilizers and also reduce poultry waste.

In addition to modifying the primary and secondary metabolism, PHs stimulate production as well as accumulation of different antioxidant compounds, including carotenoids, flavonoids, and polyphenols (Ertani *et al.*, 2014; Gurav and Jadhav, 2013; Parrado *et al.*, 2007) (Table 19.1). These phytochemical compounds or natural antioxidants have various health benefits to prevent or reduce chronic diseases like type II diabetes, cardiovascular problems, inflammatory bowel disease, and cancer (Oude Griep *et al.*, 2011; Slavin and Lloyd, 2012). Keeping these points in mind, PH-based biostimulants are effective means of making crop production more economical and sustainable.

19.2.3 Seaweed Extracts (SWEs)

With roughly 10,000 species and an estimated 10% of the total oceans' productivity on the planet, seaweeds, also known as macroalgae, are a crucial part of the marine and coastal ecosystem. On the basis of their colour, seaweeds are distributed in three classes: pheophyceae, rhodophyta, and chlorophyta. While few in number, several species of seaweed are well-known for their use as food or supplements for both humans and animals, as well as in agriculture as manure/mulches and modified extracts (Craigie, 2011; Khan *et al.*, 2009; Rayorath *et al.*, 2008; Rönnbäck *et al.*, 2007). Nevertheless, from 1950s onward, the usage of entire seaweeds has typically been overtaken by the use of various types of extracts derived from different types of seaweed. Nowadays, the practice of employing SWEs as "plant biostimulants" is considerably more widespread. SWEs, when applied at a low concentration have the potential to induce plant physiological responses which in turn help improve plant growth, yield enhancement, nutritional content, and overall quality of the products. Both physical means like heat, pressure, and microwaves and chemical means such as solvents, acids, and alkalis are employed to extract seaweed. For the effectiveness of SWE, the commonly used method of seaweed extraction is alkaline-based extraction at high pressure. This methodology is effective for generating high-level extraction with slight degradation of the polysaccharides into oligomers. The most widely used method of application of SWE to plants is either by foliar application, root treatments, or a combination of both. Moreover, foliar application is more effective against disease control and yield enhancement (Ali *et al.*, 2016; Ali *et al.*, 2021) (Table 19.1).

19.2.3.1 Effect of SWEs on Plant Growth Promotions

SWEs are the most explored category for improving plant growth and yield of the last few decades, because of their positive response towards seeds germination and plant growth as well as the positive effects seen up to post-harvesting stages (Ali *et al.*, 2019) (Table 19.1). Application of SWEs onto seeds results in increased germination rates, root size, and root density (Rayorath *et al.*, 2008). In several studies, it was observed that after application of these compounds in tomatoes, cabbages, and marigolds, these plants have less transplantation shock when compared to untreated plants (Aldworth and Van Staden, 1987; Crouch and Van Staden, 1993). Polyuronides such as alginate and fucoidans are found in brown seaweeds, which have the property of soil conditioning and formation of high molecular weight polymers, which in turn improve the water-retention property of soil (Anderson, 2009; Hegazy *et al.*, 2009; Lattner *et al.*, 2003; Verkleij, 1992). This leads to improved growth of plant roots and soil, as well as flora and fauna activity (Lattner *et al.*, 2003; Verkleij, 1992). For example, *E. maxima* extracts show better root development in tomato plants as well as mug bean plants (Crouch and Van Staden, 1991; Finnie and Van Staden, 1985). In addition to increasing water and nutrient absorption, extracts of *A. nodosum* and *K. alvarezii* significantly promoted overall vigour and plant development (Senthuran *et al.*, 2019) (Table 19.1). Maize was able to absorb higher amounts of micronutrients (Zn, Fe, B, Cu, Mo, S, Mg, Ca, and Mn) than the untreated control plants when *A. nodosum* and *Laminaria* spp. extracts were applied (Ertani *et al.*, 2018). The absorption of potassium in cottonwood leaves was greatly boosted by the application of *A. nodosum* (Crouch and Van Staden, 1991). The potassium levels in the mustard leaves treated with *E. maxima* also increased in a similar way (Stasio *et al.*, 2017). Kahydrin (a vitamin K1 derivative), a commercially available SWE, modulates plasma membrane (PM) proton pumps and stimulates the release of H+ ions into the plant apoplast, which causes the rhizosphere area to become acidic (Lüthje and Böttger, 1995). Acidification altered the redox potential state of the soil and increased the solubility of different metal ions, releasing them for plant uptake.

Upon application, SWEs have a positive influence on phytohormonal activity of plants, which leads to higher plant growth and development. Commercial seaweed extracts had comparable physiological effects to phytohormones in terms of how they affected plant development, improving plant growth at small concentrations while inhibiting growth at the high concentrations (Khan *et al.*, 2009; Provasoli and Carlucci, 1974). For instance, lower concentrations of *E. maxima* extracts promoted root growth in tomatoes while inhibiting the root growth at higher concentrations (Finnie and Van Staden, 1985). Improved activity of plant endogenous cytokinis, isopentyladenine, cis-zeatin, and dihydozeatin in spinach was observed after application of *E. maxima* extracts, which have the property to regulate plant growth positively (Werner *et al.*, 2001). *A. nodosum* extracts betaine compounds when applied to tomatoes and sweet peppers inhibit degradation of chlorophyll contents, which results in improved plant height, leaf numbers, number of roots, length, and an increase in plant biomass (Digruber *et al.*, 2018; Ramkissoon *et al.*, 2017). In a published report, it was found that processed *E. maxima* extracts contained higher levels of plant growth hormones (ABA, GA, and brassinosteroids) than unprocessed extracts (Stirk *et al.*, 2014). Yet, in addition to chemical constituents in the extract, which are unrelated to phytohormones itself, seaweed extracts may also have phytohormone-like action. Using UPLC-ESI-MS/MS, Wally *et al.* (2013) evaluate the number of phytohormones in the commercially available SWEs and concluded that the phytohormone concentrations found in the extracts were insufficient to elicit different physiological activity in the host plants at the concentrations that were typically applied in the field conditions. Researchers also hypothesized that elicitor molecules included in a commercial *A. nodosum* extract cause endogenous phytohormone metabolism in the SWE-treated plants to be disturbed by the selective regulation of the phytohormome metabolic genes (Wally *et al.*, 2013).

Although nematode parasites critically infest plants and harm them, research on plants including *Arabidopsis thaliana* (Beckett and Van Staden, 1989), sunflowers (Vernieri *et al.*, 2006), and tomatoes (Mancuso *et al.*, 2006) have demonstrated that nematode infestation may be reduced by using SWEs. Moreover, sugar cane infestation brought on by borers, aphids, and thrips was considerably reduced by seaweed extracts, averting major economic loss (Carolina and Helena, 2019; Chen *et al.*, 2019) (Table 19.1). This decrease in infestation may be brought about by the pests' ovarian tissue cells' cytotoxicity, antifeedant actions, and growth suppression. For instance, a pink bollworm growth repellant effect was shown for an acyclic diterpenoid derived from *Sargassum* (Kubo *et al.*, 1985). Moreover, seaweed extracts work as triggers for plant defence mechanisms against damaging microorganism (bacterial, fungal, and viral) infections, shielding crops from significant commercial loss brought on by illnesses (Du Jardin, 2015; Fei *et al.*, 2017). SWEs have demonstrated the capacity to reduce the unpleasant sign of plant viruses and viroids, in addition to evoking resistance against bacterial and fungal diseases (Vera *et al.*, 2011). Pre-treatment with λ-carrageenan polysaccharide from the SWE reduced the symptoms of the tomato chlorotic dwarf viroid considerably, similar to how tobacco mosaic virus (TMV) in tobacco was substantially less severe when treated with the sulphated galactans, which make up a considerable portion of several SWE (Vera *et al.*, 2011). Also, the symptoms of the TMV in tobacco plants were reduced significantly when the plants treated with oligosaccharides from seaweeds (Jiménez *et al.*, 2011; Klarzynski *et al.*, 2003).

Many climatic pressures (drought, high temperatures, salt, and freezing conditions) reduce crop output in addition to strains brought on by pests and diseases. Reactive oxygen species (ROS) might accumulate because of these abiotic stressors, which can eventually harm the plant system (Jimenez *et al.*, 2011; Klarzynski *et al.*, 2003). It's very interesting to note that the plants exposed to the SWE (*A. nodosum* and *Sargassum* sp.) were able to survive the negative effects of different abiotic stressors. Many bacteria are activated in the rhizosphere by high exudate concentrations. The microbial population structure, size, and activity were influenced by the kind and pattern of root exudates. As anticipated, the microbiomes of the rhizosphere and phyllosphere were significantly impacted by the application of SWE to soil and foliage. The microbiome effect of the SWEs has also been hypothesized to affect the growth of plants. Consequently, rhizospheric microorganisms' plant growth promotion properties may be improved by seaweed extracts.

19.2.4 Plant Growth-Promoting Rhizobacteria (PGPR)

Plant growth-promoting rhizobacteria (PGPR) are a well-structured and regulated community of microorganisms that are always associated with plants growing under the field conditions (Chaparro *et al.*, 2014; Lundberg *et al.*, 2012; Turner *et al.*, 2013). The relationship of these tiny organisms is found with all eukaryotic organisms. All the primary plant structures, such as the root, shoot, leaf, flower, and seeds are connected with components of the phytomicrobiome such as the bacteria and fungus (Berg *et al.*, 2016). The circumstances at each of these structures, however, are very different, resulting in the presence of unique microbial communities in each. The most numerous and complex microbial community found in higher plants is the rhizomicrobiome, which is connected to the roots. The nitrogen-fixing rhizobia linked to legumes is the most well-known example (Gray and Smith, 2005). With the discovery of metagenomics and similar techniques, we are now able to determine how circumstances, plant genotypes, and plant growth affect membership in the phytomicrobiome, as many members of the phytomicrobiome cannot be cultured in lab conditions (Hirsch and Mauchline, 2012; Poli *et al.*, 2016; Wintermans *et al.*, 2016). This microbial community assisting early land plants faced several stressful conditions and nutrient availability.

The most prevalent subset of these organisms are PGPR, which dominate the rhizosphere area of host plants. These PGPR stimulate plant growth and development, reduce infections, enhance nutrient uptake, and boost stress tolerance property of all kinds, including biotic and abiotic stresses. The most well-studied major PGPR belong to *Rhizobium, Azotobacter* spp., *Azospirillum* spp., *Pseudomonas* spp., and *Bacillus* spp. groups and also work as plant biostimulants (Rouphael and Colla, 2020; Weller and Lugtenberg, 2015).

19.2.4.1 Beneficial Effects of PGPR on Plant

Plant growth-promoting bacteria (PGPB) are frequently found in a variety of habitats and are typically connected to many agricultural plant species, if not all of them. PGPR, which typically colonize the surfaces of the host plant roots and are tightly attached to the rhizosphere soil interface, are the type of PGPB that has been the subject of the most research. In much literature, it was suggested that PGPR can establish the endophytic relationships by entering the root interiors (Kabiraj *et al.*, 2020; Papik *et al.*, 2020; Woźniak *et al.*, 2019). To date, the mechanism by which PGPR benefits plants in their growth and development are not well-known. Because of the developing "omics" paradigms, the molecular foundation of association mechanisms between agricultural plants and bacteria that induce/stimulate physiological alterations are beginning to be understood. Different PGPR affect different aspects of the plant life cycle, including plant growth, development, and nutrition value; morphological and physiological growth and development; biotic and abiotic stress tolerance; interactions of agro-ecosystems with other species; and increased output.

Many scientific studies suggest that the PGPR produces hormones for providing plant protection as well as wall related-transcriptional changes, which include longer roots, more biomass, and a significant reduction in dimensions of stomata (Llorente *et al.*, 2016; Spaepen *et al.*, 2014). They are also responsible for activating auxin reaction genes, which in turn enhance plant development. *Rhizobium* (Faruque *et al.*, 2015), *Aeromonas, Azotobacter* (Dellagi, Quillere and Hirel, 2020), *Bacillus* (Ruzzi and Aroca, 2015), and *Pseudomonas* (Hernández-León *et al.*, 2015) represent a category of IAA producers PGPR. Cytokinins and giberellins producing PGPRs are also identified (Gupta *et al.*, 2012), but the mechanisms involved in how they synthesize these hormones are largely unknown. To maintain to rate of growth and development, PGPR can also regulate the amount of ethylene production by plants (Heydarian *et al.*, 2016).

The most important macronutrient for the plant growth and development of crops is nitrogen; however, because it is inert, most plants cannot access it. The PGPR converts atmospheric nitrogen to ammonia by the process of nitrogen fixation, and agricultural plants can use this source of nitrogen to increase yield (Bucao *et al.*, 2019). One of the most efficient and ecologically friendly techniques of replacing the usage of inorganic nitrogen fertilizers is the use of N_2-fixing bacteria as plant growth promoters (Kumar *et al.*, 2021) (Table 19.1). Free-living microbes (*Azotobacter, Azospirillum, Bacillus, Enterobacter, Pseudomonas, Burkholderia*, and *Serratia*), as well as symbiotic or associated microbes (*Rhizobium, Bradyrhizobium*, and some species of *Azospirillum* sp.) carry out biological nitrogen fixation (BNF) and provide nitrogen to the host plants (Singh *et al.*, 2016; Thilakarathna and Raizada, 2017). Moreover, diazotrophs from the *Frankia* sp. family may colonize a tiny subclass of non-legume plants, which can promote the growth of nitrogen-fixing root nodules. Another key macronutrient for plant metabolism and physiology is phosphorus, which is equally crucial for crop development and production. Phosphorus is necessary for cell division, biological oxidation, and other metabolic and physiological activities in plants, including photosynthesis (Gupta *et al.*, 2012). Because of loss of chemical phosphorous in the process of leaching, PGPR solubilize insoluble P by releasing some organic acids, which improves soil fertility and productivity (Basu *et al.*, 2021; Suleman *et al.*, 2018). Some PGPRs (*B. edaphicus, B. mucilaginosus, Acidothiobacillus* sp., *Pseudomonas* sp. *Burkholderia* sp., *Ferrooxidans* sp., and *Paenibacillus*) have the property to solubilize K, making them available to plants for proper growth and development (Bahadur *et al.*, 2017; Liu *et al.*, 2012; Meena *et al.*, 2017; Nath, Maurya and Meena, 2017; Sindhu *et al.*, 2016).

Global climate change (stresses, nutrient deficiency, salinity, drought, and heat stress) are responsible for yield reduction of up to 60–70% (Rouphael and Colla, 2018). Plant growth rhizobacteria can improve plant responses to stresses and boost physical, chemical, and biological activities by a variety of pathways (Fatima and Arora, 2021; Ilyas *et al.*, 2020; Silva *et al.*, 2020; Yuan *et al.*, 2020). Moreover, PGPR improve resistance to various diseases by suppression of plant pathogen and pest attacks. The mechanisms (direct or indirect) involve inhibiting pathogen growth by antibiosis, volatile organic compound (VOCs) production, siderophore-mediated inhibition, bacteriocin, and extracellular enzymatic lysis (Sayyed, 2019) (Table 19.1). Furthermore, by inducing systemic resistance or ISR and system-acquired resistance (SAR) pathways, PGPR are also capable of enhancing the defensive mechanism of their linked plants against pathogen attack (Maithani *et al.*, 2021). As a result, it is now crucial to apply multi-stress-resistant PGPR biostimulants to increase agricultural productivity, all while addressing issues of low yearly crop yields and global climate change.

19.2.5 Plant Growth-Promoting Fungi (PGPF)

The term "plant-growth-promoting fungus" (PGPF) refers to a diverse collection of nonpathogenic fungal species that live in the rhizosphere, on the host plant root surface, or inside the roots of agricultural plants, and have functions that aim to minimize the need for agrochemicals and promote benefits in plant growth and health (Malgioglio *et al.*, 2022). The PGPF were identified in studies that improved plant growth, immunity, and necessary secondary metabolite synthesis in tomatoes, oranges, apples, pears, cucumbers, and carrots, including *Gliocladium, Penicillium, Phoma, Phytophthora, Rhizoctonia, Talaromyces, and Trichoderma* (Attia *et al.*, 2022; Cantabella *et al.*, 2020; Kuzin *et al.*, 2020; Omomowo *et al.*, 2020; Pandit *et al.*, 2022; Syamsia *et al.*, 2021). PGPF have antagonistic and or biocontrol property against different pathogens (Omomowo *et al.*, 2020), improve plant ability to synthesize growth hormone (Akinola and Babalola, 2021), help in mineral solubilization, and improve defence system in plants. In addition to the aforementioned functions, PGPF also increase soil nutrient levels, restrict the invasion of phytopathogens on crop, and create 1-aminocyclopropane-1-carboxylate (ACC) deaminase and other phyto-hormones to inhibit the formation of ethylene in plants (Adedayo *et al.*, 2022).

19.2.5.1 Application and Impact of PGPF on Agricultural Plants

Soil microorganisms cohabit with the roots of agricultural plants in a network of interactions. In addition to stimulating the growth of agricultural plants, PGPR has been used to stimulate plant resistance against invading pathogens. In comparison to PGPR, little research has addressed the mechanisms through which PGPF is able to stimulate resistance to pathogen invasion (Hossain *et al.*, 2017; Muslim *et al.*, 2019) (Table 19.1). In addition to improving soil fertility and enhancing plant growth, PGPF are nonpathogenic saprophytes that exist naturally and feed on decaying organic materials (Malgioglio *et al.*, 2022) (Figure 19.1). They also strengthen the plant's ability to defend itself from pathogen attacks. The ability of PGPF to colonize roots is recognized as the primary mechanism necessary in avoiding phyto-pathogen invasion and also aids in nutrient digestion, therefore enhancing plant development (Malgioglio *et al.*, 2022). In spite of improving disease resistance, PGPF is potentially involved in P solubilization, auxin production, siderophore production, and enzyme production to boost plant development (Adedayo *et al.*, 2022).

Because of the positive significance of PGPF, researchers can use it as a booster to develop induced systemic resistance (ISR) and acquired systemic resistance (ASR) against different plant pathogens for plant growth and development in agriculture sectors (Dubey *et al.*, 2023; Yapa *et al.*, 2022). Apart from modifications in plant cell walls, PGPF promote a better collection of defense-related enzymes in plants; these are -1,3 glucanase, chitinase, peroxidases (POX), and phenylalanine ammonia-lyase (PAL) and are proportionately connected with phytopathogen protection (Zhou, Tang and Guo, 2018).

19.2.6 Silicon (Si) and Nanomaterial-Based Biostimulants

Numerus reports suggested the benefit of silicon (Si) based biostimulants on the plant health by mitigating environmental stresses. Si is a non-essential element for plants, but it has a biostimulatory effect on plant development that is well documented (Epstein, 2001; Ma and Yamaji, 2006). Application of Si-based biostimulants regulates various biochemical and physiological processes to alleviate environmental stress. It helps maintain plant water balance, enhance photosynthesis and nutrient absorption, control the production of genes that are stress-tolerant and antioxidants, and lessen oxidative stress (Ahire [13]*et al.*, 2021; Kovács *et al.*, 2022; Mir *et al.*, 2022). It was observed Si application benefits various crops such as wheat (Laane, 2017), cucumber (Ługowska, 2019; Shehata, 2018), tomato, and strawberry (Nada, 2020) by increasing the protein content, vitamin C, total phenol (TP), and total soluble solid (TSS) content (Figure 19.1).

In recent years, nanoparticles encapsulated silicon used widely for alleviating both biotic and abiotic stress caused by plant diseases and pest attacks, drought, temperature, salinity, nutrient imbalance, heavy metal toxicity, and wounds (Van Bockhaven *et al.*, 2013; Zhu and Gong, 2014). The processes underpinning Si's role in reducing abiotic stressors is currently not very well understood. The fundamental processes involved in the Si-mediated abiotic stress management in the higher plants are as described in the following: (1) anatomical alterations in the plant tissues owing to phytolith deposition that provide stiffness and modulate nutrients and water movement within the host plants, (2) augmentation of antioxidant defence systems in plants, and (3) increased nutrient as well as water mobility within the host plants, (4) harmful metal immobilization by co-precipitation or complexation with Si in the host plant tissues and soil, and (5) indirect control of the gene expression and signalling via plant growth hormones, according to current understanding. Si is being used in several commercial horticulture crops to increase tolerance to abiotic stressors, diseases, and pests, although its application is still very poorly understood. Because of that, Si is likely the sole element that confers resilience to many stressors while remaining non-toxic to environment and for the humans. Its usage as a biostimulant in horticulture is projected to rise significantly in the future.

19.3 CONCLUSION AND FUTURE CONSIDERATIONS

Biostimulants are becoming promising tools in agriculture for more economical and sustainable production. They have the capability of enhancing plant growth and development by improving plant abiotic and biotic stresses tolerance and nutrient uptake efficiency, improving the overall health of the host plant. Due to their multifaceted properties, they have been used extensively in agriculture in past decades. Biostimulants produced from diverse sources such as seaweed, microbes, and organic matter can be directly added to seeds, rhizosphere soil, foliar application, or a combination of these. Biostimulants have the potential ability to improve yield while minimizing the use of agro-chemical inputs and encourage more environmentally sound agricultural practices by fostering plant growth and enhancing crop production.

Although it has a favourable impact on agriculture, the combination of "omics" with some high-throughput technologies, including gas chromatography-mass spectrometry (GC-MS) and high-performance liquid chromatography (HPLC) allows scientists to better understand the molecular underpinnings and regulatory mechanisms of biostimulants, making them more advanced and highly effective. Even if they have the potential to boost crop yield, biostimulants might nevertheless confront a number of challenges. Some of the main challenges are: (1) Absence of a clear regulatory structure: Unlike fertilizers and pesticides, biostimulants lack a defined regulatory structure, making it challenging for businesses to bring products on the market and for farmers to identify those that are efficient and secure to use. (2) Inconsistent effectiveness: Since biostimulants are a complex mix of substances and microbes, ensuring consistent efficacy across diverse crops, habitats, and growth circumstances is challenging. This lack of consistency can also make it difficult to determine the best product for a farmer's unique needs. (3) Limited understanding of modes of action: The processes by which some biostimulants function are frequently poorly understood, despite the fact that they have been demonstrated to be effective in promoting host plant growth and stress (biotic and abiotic) tolerance. Developing new and better products and maximizing their utilization may be challenging as a result. (4) Limited knowledge of product interactions: Biostimulants are frequently used with additional input, including irrigation systems, herbicides, and fertilizers. Unfortunately, there is a lack of knowledge on how biostimulants interact with other input, which might result in unexpected effects and decreased efficacy. (5) Low consumer knowledge and education: Because biostimulants are new class of agricultural input, many farmers and growers might not be aware of their potential advantages or how to utilize them effectively. To enhance uptake and guarantee that biostimulants are used properly, better education and awareness campaigns are required.

In conclusion, future considerations for biostimulant development and usage include formulation optimization, integration with other sustainable agriculture techniques, study of mechanisms of action, and the creation of regulatory frameworks and standards to ensure that they are safe for human and environmental health. By addressing these considerations, biostimulants can be a promising approach for sustainable agriculture that can aid in global food security concerns.

ACKNOWLEDGMENT

The authors are highly thankful to SERB (EEQ/2021/001083), DST (DST/INT/SL/P-31/2021, DST/SEED/SCSP/STI/2020/426/G), and Banaras Hindu University, IoE (6031) incentives grants for providing support for applied research in plant-microbe interactions.

REFERENCES

Adedayo, A.A. *et al.* (2022) 'The application of plant growth-promoting rhizobacteria in *Solanum lycopersicum* production in the agricultural system: A review', *PeerJ*, 10, p. e13405.

Ahire, M.L. *et al.* (2021) 'Multifaceted roles of silicon in mitigating environmental stresses in plants', *Plant Physiology and Biochemistry*, 169, pp. 291–310.

Akinola, S.A. and Babalola, O.O. (2021) 'The fungal and archaeal community within plant rhizosphere: A review on their contribution to crop safety', *Journal of Plant Nutrition*, 44(4), pp. 600–618.

Albuzio, A., Ferrari, G. and Nardi, S. (1986) 'Effects of humic substances on nitrate uptake and assimilation in barley seedlings', *Canadian Journal of Soil Science*, 66(4), pp. 731–736.

Aldworth, S.J. and Van Staden, J. (1987) 'The effect of seaweed concentrate on seedling transplants', *South African Journal of Botany*, 53(3), pp. 187–189.

Ali, N. *et al.* (2016) 'The effect of *Ascophyllum nodosum* extract on the growth, yield and fruit quality of tomato grown under tropical conditions', *Journal of Applied Phycology*, 28, pp. 1353–1362.

Ali, O., Ramsubhag, A. and Jayaraman, J. (2019) 'Biostimulatory activities of *Ascophyllum nodosum* extract in tomato and sweet pepper crops in a tropical environment', *PLoS One*, 14(5), p. e0216710.

Ali, O., Ramsubhag, A. and Jayaraman, J. (2021) 'Phytoelicitor activity of *Sargassum vulgare* and *Acanthophora spicifera* extracts and their prospects for use in vegetable crops for sustainable crop production', *Journal of Applied Phycology*, 33, pp. 639–651.

Anderson, G. (2009) 'Seaweed extract shows improved fruit quality at McLaren Vale vineyard trial', *Australian and New Zealand Grapegrower and Winemaker*, 548, pp. 17–22.

Apone, F. *et al.* (2010) 'A mixture of peptides and sugars derived from plant cell walls increases plant defense responses to stress and attenuates ageing-associated molecular changes in cultured skin cells', *Journal of Biotechnology*, 145(4), pp. 367–376.

Attia, M.S. *et al.* (2022) 'Plant growth-promoting fungi as biocontrol tool against fusarium wilt disease of tomato plant', *Journal of Fungi*, 8(8), p. 775.

Aydin, A., Kant, C. and Turan, M. (2012) 'Humic acid application alleviate salinity stress of bean (*Phaseolus vulgaris* L.) plants decreasing membrane leakage', *African Journal of Agricultural Research*, 7(7), pp. 1073–1086.

Azevedo, R.A.D. and Lea, P.J. (2011) 'Research on abiotic and biotic stress–what next?', *Annals of Applied Biology*, 159(3), pp. 317–319.

Baglieri, A. *et al.* (2014) 'Fertilization of bean plants with tomato plants hydrolysates. Effect on biomass production, chlorophyll content and N assimilation', *Scientia Horticulturae*, 176, pp. 194–199.

Bahadur, I. *et al.* (2017) 'Mineral release dynamics of tricalcium phosphate and waste muscovite by mineral-solubilizing rhizobacteria isolated from Indo-Gangetic plain of India', *Geomicrobiology Journal*, 34(5), pp. 454–466.

Bano, A. and Muqarab, R.J.P.B. (2017) 'Plant defence induced by PGPR against *Spodoptera litura* in tomato (*Solanum lycopersicum* L.)', *Plant Biology*, 19(3), pp. 406–412.

Basile, B. *et al.* (2020) 'Appraisal of emerging crop management opportunities in fruit trees, grapevines and berry crops facilitated by the application of biostimulants', *Scientia Horticulturae*, 267, p. 109330.

Basu, A. *et al.* (2021) 'Plant growth promoting rhizobacteria (PGPR) as green bioinoculants: Recent developments, constraints, and prospects', *Sustainability*, 13(3), p. 1140.

Beckett, R.P. and Van Staden, J. (1989) 'The effect of seaweed concentrate on the growth and yield of potassium stressed wheat', *Plant and Soil*, 116, pp. 29–36.

Berg, G. *et al.* (2016) 'The plant microbiome explored: Implications for experimental botany', *Journal of Experimental Botany*, 67(4), pp. 995–1002.

Bonini, P. *et al.* (2017) 'Seed treatments with endophytic fungi and biostimulant compounds enhance yield of corn and soybean', in *The 3rd World Congress on the Use of Biostimulants in Agriculture*, Miami, FL. pp. 27–30.

Boyer, F.D. *et al.* (2014) 'New strigolactone analogs as plant hormones with low activities in the rhizosphere', *Molecular Plant*, 7(4), pp. 675–690.

Bucao, D.S., Yapit, R.H. and Gabriel, M.L.S. (2019) 'Biochemical characterization of microbials and their effects on the growth and yield of multiplier onion (*Allium ascalonicum* L.) in Northwestern Philippines', in Plant growth promoting rhizobacteria (PGPR): Prospects for sustainable agriculture. Springer, Singapore. pp. 75–91.

Calvo, P., Nelson, L. and Kloepper, J.W. (2014) 'Agricultural uses of plant biostimulants', *Plant and Soil*, 383, pp. 3–41.

Canellas, L.P. *et al.* (2013) 'A combination of humic substances and *Herbaspirillum seropedicae* inoculation enhances the growth of maize (*Zea mays* L.)', *Plant and Soil*, 366, pp. 119–132.

Canellas, L.P. and Olivares, F.L. (2014) 'Physiological responses to humic substances as plant growth promoter', *Chemical and Biological Technologies in Agriculture*, 1(1), pp. 1–11.

Canellas, L.P. *et al.* (2015) 'Foliar application of plant growth-promoting bacteria and humic acid increase maize yields', *Journal of Food Agriculture & Environment*, 13(1), pp. 131–138.

Canellas, L.P. *et al.* (2020) 'Plant chemical priming by humic acids', *Chemical and Biological Technologies in Agriculture*, 7(1), pp. 1–17.

Cantabella, D. *et al.* (2020) 'Inoculation of in vitro cultures with rhizosphere microorganisms improve plant development and acclimatization during immature embryo rescue in nectarine and pear breeding programs', *Scientia Horticulturae*, 273, p. 109643.

Carolina Feitosa de Vasconcelos, A. and Helena Garófalo Chaves, L. (2019) Edited by: Seyed Mahyar Mirmajlessi & Ramalingam Radhakrishnan. Publisher name & Location: Intechopen, London, United Kingdom. 'Biostimulants and their role in improving plant growth under abiotic stresses', in *Biostimulants in plant science*. pp. 3–16.

Ceccarelli, A.V. *et al.* (2021) 'Foliar application of different vegetal-derived protein hydrolysates distinctively modulates tomato root development and metabolism', *Plants*, 10(2), p. 326.

Chaparro, J.M., Badri, D.V. and Vivanco, J.M. (2014) 'Rhizosphere microbiome assemblage is affected by plant development', *The ISME Journal*, 8(4), pp. 790–803.

Chen, D. *et al.* (2019) 'Effects of seaweed extracts on promoting growth and improving stress resistance in sugarcane', *Asian Agricultural Research*, 11(4), pp. 69–76.

Colla, G. *et al.* (2015) 'Protein hydrolysates as biostimulants in horticulture', *Scientia Horticulturae*, 196, pp. 28–38.

Colla, G. *et al.* (2017) 'Foliar applications of protein hydrolysate, plant and seaweed extracts increase yield but differentially modulate fruit quality of greenhouse tomato', *HortScience*, 52(9), pp. 1214–1220.

Craigie, J.S. (2011) 'Seaweed extract stimuli in plant science and agriculture', *Journal of Applied Phycology*, 23, pp. 371–393.

Crouch, I.J. and Van Staden, J. (1991) 'Evidence for rooting factors in a seaweed concentrate prepared from *Ecklonia maxima*', *Journal of Plant Physiology*, 137(3), pp. 319–322.

Crouch, I.J. and Van Staden, J. (1993) 'Evidence for the presence of plant growth regulators in commercial seaweed products', *Plant Growth Regulation*, *13*, pp. 21–29.

De Pascale, S., Rouphael, Y. and Colla, G. (2017) 'Plant biostimulants: Innovative tool for enhancing plant nutrition in organic farming', *European Journal of Horticultural Science*, 82(6), pp. 277–285.

Delgado, A. *et al.* (2002) 'Phosphorus fertilizer recovery from calcareous soils amended with humic and fulvic acids', *Plant and Soil*, 245, pp. 277–286.

Dellagi, A., Quillere, I. and Hirel, B. (2020) 'Beneficial soil-borne bacteria and fungi: A promising way to improve plant nitrogen acquisition', *Journal of Experimental Botany*, 71(15), pp. 4469–4479.

Digruber, T. *et al.* (2018) 'Stimulation of energy willow biomass with triacontanol and seaweed extract', *Industrial Crops and Products*, 120, pp. 104–112.

Du Jardin, P. (2015) 'Bioestimulantes vegetais: Definição, conceito, principais categorias e regulação', *Scientia Horticulturae*, 196, pp. 3–14.

du Jardin, P. (2012) The science of plant biostimulants–A bibliographic analysis, Ad hoc study report.

Dubey, N. *et al.* (2023) 'Role of plant-microbial secondary metabolites in stress mitigation: Current knowledge and future directions', *Microbial based land restoration handbook*. CRC Press, Boca Raton. pp. 23–42.

Epstein, E. (2001) 'Silicon in plants: Facts vs. concepts', in *Studies in plant science*. Elsevier, Amesterdam, vol. 8, pp. 1–15.

Ertani, A. *et al.* (2009) 'Biostimulant activity of two protein hydrolyzates in the growth and nitrogen metabolism of maize seedlings', *Journal of Plant Nutrition and Soil Science*, 172(2), pp. 237–244.

Ertani, A. *et al.* (2013) 'Alfalfa plant-derived biostimulant stimulate short-term growth of salt stressed *Zea mays* L. plants', *Plant and Soil*, 364, pp. 145–158.

Ertani, A. *et al.* (2014) '*Capsicum chinensis* L. growth and nutraceutical properties are enhanced by biostimulants in a long-term period: Chemical and metabolomic approaches', *Frontiers in Plant Science*, 5, p. 375.

Ertani, A. *et al.* (2018) 'Evaluation of seaweed extracts from *Laminaria* and *Ascophyllum nodosum spp.* as biostimulants in *Zea mays* L. using a combination of chemical, biochemical and morphological approaches', *Frontiers in Plant Science*, 9, p. 428.

Eshaghi Gorgi, O. *et al.* (2022) 'Effect of plant growth promoting rhizobacteria (PGPR) and mycorrhizal fungi inoculations on essential oil in *Melissa officinalis* l. under drought stress', *Biologia*, 77, pp. 11–20.

Faruque, O.M. *et al.* (2015) 'Identification of *Bradyrhizobium* elkanii genes involved in incompatibility with soybean plants carrying the Rj4 allele', *Applied and Environmental Microbiology*, 81(19), pp. 6710–6717.

Fei, H. *et al.* (2017) 'Enhancing the productivity of hybrid poplar (Populus× hybrid) and switchgrass (*Panicum virgatum* L.) by the application of beneficial soil microbes and a seaweed extract', *Biomass and Bioenergy*, 107, pp. 122–134.

Fatima, T. and Arora, N.K. (2021) '*Pseudomonas entomophila* PE3 and its exopolysaccharides as biostimulants for enhancing growth, yield and tolerance responses of sunflower under saline conditions', *Microbiological Research*, 244, p. 126671.

Finnie, J.F. and Van Staden, J. (1985) 'Effect of seaweed concentrate and applied hormones on in vitro cultured tomato roots', *Journal of Plant Physiology*, 120(3), pp. 215–222.

García, M., Collado, S., Oulego, P., and Díaz M. (2020) 'The wet oxidation of aqueous humic acids', *Journal of hazardous materials*. Amsterdam, Elsevier, 396, p. 122402.

García-Mina, J.M., Antolin, M.C. and Sanchez-Diaz, M. (2004) 'Metal-humic complexes and plant micronutrient uptake: A study based on different plant species cultivated in diverse soil types', *Plant and Soil*, 258, pp. 57–68.

Gupta, M. *et al.* (2012) 'Isolation and identification of phosphate solubilizing bacteria able to enhance the growth and aloin-A biosynthesis of *Aloe barbadensis* Miller', *Microbiological Research*, 167(6), pp. 358–363.

Gurav, R.G. and Jadhav, J.P. (2013) 'A novel source of biofertilizer from feather biomass for banana cultivation', *Environmental Science and Pollution Research*, 20, pp. 4532–4539.

Gray, E.J. and Smith, D.L. (2005) 'Intracellular and extracellular PGPR: Commonalities and distinctions in the plant–bacterium signaling processes', *Soil Biology and Biochemistry*, 37(3), pp. 395–412.

Halpern, M. *et al.* (2015) 'The use of biostimulants for enhancing nutrient uptake', *Advances in Agronomy*, 130, pp. 141–174.

Hegazy, E.A. *et al.* (2009) 'Controlling of degradation effects in radiation processing of polymers', in *Controlling of degradation effects in radiation processing of polymers*. Vienna, Austria: International Atomic Energy Agency, p. 67.

Hernandez, O.L. *et al.* (2015) 'Humic substances from vermicompost enhance urban lettuce production', *Agronomy for Sustainable Development*, 35, pp. 225–232.

Hirsch, P.R. and Mauchline, T.H. (2012) 'Who's who in the plant root microbiome?', *Nature Biotechnology*, 30(10), pp. 961–962.

Hernández-León, R. *et al.* (2015) 'Characterization of the antifungal and plant growth-promoting effects of diffusible and volatile organic compounds produced by *Pseudomonas fluorescens* strains', *Biological Control*, 81, pp. 83–92.

Heydarian, Z. *et al.* (2016) 'Inoculation of soil with plant growth promoting bacteria producing 1-aminocyclopropane-1-carboxylate deaminase or expression of the corresponding acdS gene in transgenic plants increases salinity tolerance in *Camelina sativa*', *Frontiers in Microbiology*, 7, p. 1966.

Hossain, M.M., Sultana, F. and Hyakumachi, M. (2017) 'Role of ethylene signalling in growth and systemic resistance induction by the plant growth-promoting fungus *Penicillium viridicatum* in *Arabidopsis*', *Journal of Phytopathology*, 165(7–8), pp. 432–441.

Ilyas, N. *et al.* (2020) 'Rhizobacteria isolated from saline soil induce systemic tolerance in wheat (*Triticum aestivum* L.) against salinity stress', *Agronomy*, 10(7), p. 989.

Jiménez, E. *et al.* (2011) 'Anti-phytopathogenic activities of macro-algae extracts', *Marine Drugs*, 9(5), pp. 739–756.

Kabiraj, A. *et al.* (2020) 'Role of plant growth-promoting rhizobacteria (PGPR) for crop stress management' in *Sustainable agriculture in the era of climate change*. Springer Nature, Cham, Switzerland, pp. 367–389.

Kauffman, G.L., Kneivel, D.P. and Watschke, T.L. (2007) 'Effects of a biostimulant on the heat tolerance associated with photosynthetic capacity, membrane thermostability, and polyphenol production of perennial ryegrass', *Crop Science*, 47(1), pp. 261–267.

Kaur, H. *et al.* (2021) 'Biostimulants, the cinderella for plant development', in *Biostimulants for crops from seed germination to plant development*. New York City, NY: Academic Press, pp. 61–72.

Kausar, F. and Shahbaz, M. (2017) 'Influence of strigolactone (GR24) as a seed treatment on growth, gas exchange and chlorophyll fluorescence of wheat under saline conditions', *International Journal of Agriculture & Biology*, 19(2), pp. 321–327.

Khan, W. *et al.* (2009) 'Seaweed extracts as biostimulants of plant growth and development', *Journal of Plant Growth Regulation*, 28, pp. 386–399.

Klarzynski, O. *et al.* (2003) 'Sulfated fucan oligosaccharides elicit defense responses in tobacco and local and systemic resistance against tobacco mosaic virus', *Molecular Plant-Microbe Interactions*, 16(2), pp. 115–122.

Kovács, S., Kutasy, E. and Csajbók, J. (2022) 'The multiple role of silicon nutrition in alleviating environmental stresses in sustainable crop production', *Plants*, 11(9), p. 1223.

Kubo, I., Matsumoto, T. and Ichikawa, N. (1985) 'Absolute configuration of crinitol. An acyclic diterpene insect growth inhibitor from the brown algae *Sargassum tortile*', *Chemistry Letters*, 14(2), pp. 249–252.

Kumar, R. *et al.* (2021) 'Sustaining productivity through integrated use of microbes in agriculture', in *Role of microbial communities for sustainability*. Springer, Singapore, pp. 109–145.

Kuzin, A. *et al.* (2020) 'Soil fertility management in apple orchard with microbial biofertilizers', in *E3S Web of Conferences*. EDP Sciences, vol. 222, p. 03020.

Laane, H.M. (2017) 'The effects of the application of foliar sprays with stabilized silicic acid: An overview of the results from 2003–2014', *Silicon*, 9(6), pp. 803–807.

Lattner, D., Flemming, H.C. and Mayer, C. (2003) '13C-NMR study of the interaction of bacterial alginate with bivalent cations', *International Journal of Biological Macromolecules*, 33(1–3), pp. 81–88.

Liu, J. *et al.* (2012) 'Potential phosphorus leaching from sandy topsoils with different fertilizer histories before and after application of pig slurry', *Soil Use and Management*, 28(4), pp. 457–467.

Llorente, B.E., Alasia, M.A. and Larraburu, E.E. (2016) 'Biofertilization with *Azospirillum brasilense* improves in vitro culture of *Handroanthus ochraceus*, a forestry, ornamental and medicinal plant', *New Biotechnology*, 33(1), pp. 32–40.

Ługowska, M. (2019) 'Effects of bio-stimulants on the yield of cucumber fruits and on nutrient content', *African Journal of Agricultural Research*, 14(35), pp. 2112–2118.

Lüthje, S. and Böttger, M. (1995) 'On the function of a k-type vitamin in plasma membranes of maize (*Zea mays* L.) roots', *Mitteilungen aus dem Institut für allgemeine Botanik in Hamburg*, 25, pp. 5–13.

Lundberg, D.S., Lebeis, S.L., Paredes, S.H., Yourstone, S., Gehring, J., Malfatti, S., Tremblay, J., Engelbrektson, A., Kunin, V., Rio, T.G.D. and Edgar, R.C. (2012) 'Defining the core Arabidopsis thaliana root microbiome'. Nature, 488(7409), pp. 86–90.

Ma, J.F. and Yamaji, N. (2006) 'Silicon uptake and accumulation in higher plants', *Trends in Plant Science*, 11(8), pp. 392–397.

Maithani, D., Singh, H. and Sharma, A. (2021) 'Stress alleviation in plants using SAR and ISR: Current views on stress signaling network' in *Microbes and signaling biomolecules against plant stress: Strategies of plant-microbe relationships for better survival*. Singapore: Springer Singapore, pp. 7–36.

Maini, P. (2006) 'The experience of the first biostimulant, based on amino acids and peptides: A short retrospective review on the laboratory researches and the practical results', *Fertilitas Agrorum*, 1(1), pp. 29–43.

Malgioglio, G. *et al.* (2022) 'Plant-microbe interaction in sustainable agriculture: The factors that may influence the efficacy of PGPM application', *Sustainability*, 14(4), p. 2253.

Mancuso, S. *et al.* (2006) 'Marine bioactive substances (IPA Extract) improve foliar ion uptake and water stress tolerance in potted "Vitis vinifera" plants', *Advances in Horticultural Science*, 20(2), pp. 156–161.

Meena, V.S. *et al.* (2017) 'Plant beneficial rhizospheric microorganism (PBRM) strategies to improve nutrients use efficiency: A review', *Ecological Engineering*, 107, pp. 8–32.

Mir, R.A. *et al.* (2022) 'Multidimensional role of silicon to activate resilient plant growth and to mitigate abiotic stress', *Frontiers in Plant Science*, 13, p. 819658.

Morales-Payan, J.P. and Stall, W.M. (2003, December) 'Papaya (Carica papaya) response to foliar treatments with organic complexes of peptides and amino acids', *Proceedings of the Florida State Horticultural Society*, vol. 116, pp. 30–32.

Murali, M. *et al.* (2021) 'Bioprospecting of rhizosphere-resident fungi: Their role and importance in sustainable agriculture', *Journal of Fungi*, 7(4), p. 314.

Muscolo, A. *et al.* (1999) 'Earthworm humic matter produces auxin-like effects on Daucus carota cell growth and nitrate metabolism', *Soil Biology and Biochemistry*, 31(9), pp. 1303–1311.

Muslim, A. *et al.* (2019) 'Induction of systemic resistance in cucumber by hypovirulent binucleate *Rhizoctonia* against anthracnose caused by *Colletotrichum orbiculare*', *Tropical Life Sciences Research*, 30(1), p. 109.

Nada, M.M. (2020) 'Effect of foliar application with potassium silicate and Glycine betaine on growth and early yield quality of strawberry plants', *Journal of Plant Production*, 11(12), pp. 1295–1302.

Nardi, S. *et al.* (2009) 'Biological activities of humic substances', *Biophysico-Chemical Processes Involving Natural Nonliving Organic Matter in Environmental Systems*, 2(part 1), pp. 305–339.

Nath, D., Maurya, B.R. and Meena, V.S. (2017) 'Documentation of five potassium-and phosphorus-solubilizing bacteria for their K and P-solubilization ability from various minerals', *Biocatalysis and Agricultural Biotechnology*, 10, pp. 174–181.

Olivares, F.L. *et al.* (2015) 'Substrate biofortification in combination with foliar sprays of plant growth promoting bacteria and humic substances boosts production of organic tomatoes', *Scientia Horticulturae*, 183, pp. 100–108.

Omomowo, I.O., Adedayo, A.A. and Omomowo, O.I. (2020) 'Biocontrol potential of rhizospheric fungi from moringa oleifera, their phytochemicals and secondary metabolite assessment against spoilage fungi of sweet orange (*Citrus sinensis*)', *Asian Journal of Applied Sciences*, 8(1), pp. 35–48.

Oude Griep, L.M. *et al.* (2011) 'Colors of fruit and vegetables and 10-year incidence of stroke', *Stroke*, 42(11), pp. 3190–3195.

Paksoy, M., Türkmen, Ö. and Dursun, A. (2010) 'Effects of potassium and humic acid on emergence, growth and nutrient contents of okra (*Abelmoschus esculentus* L.) seedling under saline soil conditions', *African Journal of Biotechnology*, 9(33), pp. 5343–5346.

Pandit, M.A. *et al.* (2022) 'Major biological control strategies for plant pathogens', *Pathogens*, 11(2), p. 273.

Papik, J. *et al.* (2020) 'The invisible life inside plants: Deciphering the riddles of endophytic bacterial diversity', *Biotechnology Advances*, 44, p. 107614.

Parađiković, N. *et al.* (2011) 'Effect of natural biostimulants on yield and nutritional quality: An example of sweet yellow pepper (*Capsicum annuum* L.) plants', *Journal of the Science of Food and Agriculture*, 91(12), pp. 2146–2152.

Parrado, J. *et al.* (2007) 'Enzymatic vegetable extract with bio-active components: Influence of fertiliser on the colour and anthocyanins of red grapes', *Journal of the Science of Food and Agriculture*, 87(12), pp. 2310–2318.

Pecha, J. *et al.* (2012) 'Protein biostimulant foliar uptake modeling: The impact of climatic conditions', *AIChE Journal*, 58(7), pp. 2010–2019.

Petrozza, A. *et al.* (2014) 'Physiological responses to Megafol® treatments in tomato plants under drought stress: A phenomic and molecular approach', *Scientia Horticulturae*, 174, pp. 185–192.

Piccolo, A. (2002) 'The supramolecular structure of humic substances: A novel understanding of humus chemistry and implications in soil science.' In: *Advances in agronomy*. Elsevier, Amsterdam, pp. 57–134.

Pinton, R. *et al.* (1999) 'Modulation of uptake by water-extractable humic substances: Involvement of root plasma membrane H^+ ATPase', *Plant and Soil*, 215, pp. 155–161.

Poli, A. *et al.* (2016) 'Influence of plant genotype on the cultivable fungi associated to tomato rhizosphere and roots in different soils', *Fungal Biology*, 120(6–7), pp. 862–872.

Provasoli, L.A.A.F.C. and Carlucci, A.F. (1974) 'Vitamins and growth regulators. *Botanical monographs*', in *Algal physiology and biochemistry*. University of California Press, Berkeley, pp. 741–787.

Ramkissoon, A., Ramsubhag, A. and Jayaraman, J. (2017) 'Phytoelicitor activity of three Caribbean seaweed species on suppression of pathogenic infections in tomato plants', *Journal of Applied Phycology*, 29(6), pp. 3235–3244.

Rayorath, P. *et al.* (2008) 'Extracts of the brown seaweed *Ascophyllum nodosum* induce gibberellic acid (GA 3)-independent amylase activity in barley', *Journal of Plant Growth Regulation*, 27, pp. 370–379.

Rönnbäck, P. *et al.* (2007) 'Ecosystem goods and services from Swedish coastal habitats: Identification, valuation, and implications of ecosystem shifts', *AMBIO: a Journal of the Human Environment*, 36(7), pp. 534–544.

Rose, M.T. *et al.* (2014) 'A meta-analysis and review of plant-growth response to humic substances: Practical implications for agriculture', *Advances in Agronomy*, 124, pp. 37–89.

Rouphael, Y. and Colla, G. (2018) 'Synergistic biostimulatory action: Designing the next generation of plant biostimulants for sustainable agriculture', *Frontiers in Plant Science*, 9, p. 1655.

Rouphael, Y. and Colla, G. (2020) 'Biostimulants in agriculture', *Frontiers in Plant Science*, 11, p. 40.

Rouphael, Y. *et al.* (2017) 'Synergistic action of a microbial-based biostimulant and a plant derived-protein hydrolysate enhances lettuce tolerance to alkalinity and salinity', *Frontiers in Plant Science*, 8, p. 131.

Rouphael, Y. *et al.* (2018) 'High-throughput plant phenotyping for developing novel biostimulants: From lab to field or from field to lab?', *Frontiers in Plant Science*, 9, p. 1197.

Ruzzi, M. and Aroca, R. (2015) 'Plant growth-promoting rhizobacteria act as biostimulants in horticulture', *Scientia Horticulturae*, 196, pp. 124–134.

Sabatino, L. *et al.* (2021) 'Protein hydrolysates and mo-biofortification interactively modulate plant performance and quality of 'canasta' lettuce grown in a protected environment', *Agronomy*, 11(6), p. 1023.

Sarwar, Y. and Shahbaz, M. (2020) 'Modulation in growth, photosynthetic pigments, gas exchange attributes and inorganic ions in sunflower (*Helianthus annuus* L.) by strigolactones (GR24) achene priming under saline conditions', *Pakistan Journal of Botany* 52(1), pp. 23–31.

Sayyed, R. Z. (ed.) (2019) *Plant growth promoting rhizobacteria for sustainable stress management: Volume 2: rhizobacteria in biotic stress management*. Singapore: Springer Nature vol. 13.

Schiavon, M., Ertani, A. and Nardi, S. (2008) 'Effects of an alfalfa protein hydrolysate on the gene expression and activity of enzymes of the tricarboxylic acid (TCA) cycle and nitrogen metabolism in *Zea mays* L', *Journal of Agricultural and Food Chemistry*, 56(24), pp. 11800–11808.

Schiavon, M. *et al.* (2010) 'High molecular size humic substances enhance phenylpropanoid metabolism in maize (*Zea mays* L.)', *Journal of Chemical Ecology*, 36, pp. 662–669.

Schmidt, W. *et al.* (2007) 'Water-extractable humic substances alter root development and epidermal cell pattern in Arabidopsis', *Plant and Soil*, 300, pp. 259–267.

Senthuran, S. *et al.* (2019) 'Effect of seaweed extract (*Kappaphycus alvarezii*) on the growth, yield and nutrient uptake of leafy vegetable *Amaranthus polygamous', Tropical Agricultural Research*, 30(3), pp. 81–88.

Sestili, F. *et al.* (2018) 'Protein hydrolysate stimulates growth in tomato coupled with N-dependent gene expression involved in N assimilation', *Frontiers in Plant Science*, 9, p. 1233.

Shehata, S. A. (2018). Effect of foliar spray with potassium silicate on growth, yield, quality and storability of cucumber fruits. *Annals of Agricultural Science, Moshtohor*, 56(2), pp. 385–396.

Silva, L. *et al.* (2020). 'Biomass yield, nitrogen content and uptake, and nutritive value of alfalfa co-inoculated with plant-growth promoting bacteria', *IJIER*, 8, pp. 400–420.

Sindhu, S.S. *et al.* (2016) 'Potassium-solubilizing microorganisms (KSMs) and its effect on plant growth improvement', in *Potassium solubilizing microorganisms for sustainable agriculture.* Springer, New Delhi, pp. 171–185.

Singh, M. *et al.* (2016) 'Role of biofertilizers in conservation agriculture', in Conservation agriculture: An approach to combat climate change in Indian Himalaya. Singapore: Springer, pp. 113–134.

Slavin, J.L. and Lloyd, B. (2012) 'Health benefits of fruits and vegetables', *Advances in Nutrition*, 3(4), pp. 506–516.

Souza, A.C. *et al.* (2022) 'Plant hormone crosstalk mediated by humic acids', *Chemical and Biological Technologies in Agriculture*, 9(1), pp. 1–25.

Spaepen, S. *et al.* (2014) 'Phenotypical and molecular responses of *Arabidopsis* thaliana roots as a result of inoculation with the auxin-producing bacterium *Azospirillum brasilense*', *New Phytologist*, 201(3), pp. 850–861.

Stasio, E.D. *et al.* (2017) 'The influence of *Ecklonia maxima* seaweed extract on growth, photosynthetic activity and mineral composition of *Brassica rapa* L. subsp. sylvestris under nutrient stress conditions', *European Journal of Horticultural Science*, 82(6), pp. 286–293.

Stirk, W.A. *et al.* (2014) 'Abscisic acid, gibberellins and brassinosteroids in Kelpak®, a commercial seaweed extract made from *Ecklonia maxima*', *Journal of Applied Phycology*, 26, pp. 561–567.

Suleman, M. *et al.* (2018) 'Phosphate solubilizing bacteria with glucose dehydrogenase gene for phosphorus uptake and beneficial effects on wheat', *PLoS One*, 13(9), p. e0204408.

Syamsia, S. *et al.* (2021) Combination on endophytic fungal as the plant growth-promoting fungi (PGPF) on cucumber (*Cucumis sativus*), *Biodiversitas Journal of Biological Diversity*, 22(3), pp. 1194–1202.

Thilakarathna, M.S. and Raizada, M.N. (2017) 'A meta-analysis of the effectiveness of diverse rhizobia inoculants on soybean traits under field conditions', *Soil Biology and Biochemistry*, 105, pp. 177–196.

Türkmen, Ö. *et al.* (2004) 'Calcium and humic acid affect seed germination, growth, and nutrient content of tomato (*Lycopersicon esculentum* L.) seedlings under saline soil conditions', *Acta Agriculture Scandinavica, Section B-Soil & Plant Science*, 54(3), pp. 168–174.

Turner, T.R., James, E.K. and Poole, P.S. (2013) 'The plant microbiome', *Genome Biology*, 14(6), pp. 1–10.

Van Bockhaven, J., De Vleesschauwer, D. and Höfte, M. (2013) 'Towards establishing broad-spectrum disease resistance in plants: Silicon leads the way', *Journal of Experimental Botany*, 64(5), pp. 1281–1293.

Vera, J. *et al.* (2011) 'Seaweed polysaccharides and derived oligosaccharides stimulate defense responses and protection against pathogens in plants', *Marine Drugs*, 9(12), pp. 2514–2525.

Verkleij, F.N. (1992) 'Seaweed extracts in agriculture and horticulture: A review', *Biological Agriculture & Horticulture*, 8(4), pp. 309–324.

Vernieri, P.A.O.L.O. *et al.* (2006, October) 'Use of biostimulants for reducing nutrient solution concentration in floating system', *III International Symposium on Models for Plant Growth, Environmental Control and Farm Management in Protected Cultivation*, 718, pp. 477–484.

Wadas, W. (2021) 'Potato (*Solanum tuberosum* L.) Growth in response to foliar silicon application', *Agronomy*, 11(12), p. 2423.

Wadas, W. (2023) 'Nutritional value and sensory quality of new potatoes in response to silicon application', *Agriculture*, 13(3), p. 542.

Wally, O.S. *et al.* (2013) 'Regulation of phytohormone biosynthesis and accumulation in Arabidopsis following treatment with commercial extract from the marine macroalga *Ascophyllum nodosum*', *Journal of Plant Growth Regulation*, 32, pp. 324–339.

Weller, D. and Lugtenberg, B. (2015) *Principles of plant-microbe interactions—Microbes for sustainable agriculture.* Leiden: Springer.

Werner, T. *et al.* (2001) 'Regulation of plant growth by cytokinin', *Proceedings of the National Academy of Sciences*, 98(18), pp.10487–10492.

Wilson, H.T., Xu, K. and Taylor, A.G. (2015) 'Transcriptome analysis of gelatin seed treatment as a biostimulant of cucumber plant growth', *The Scientific World Journal*, 2015, p. 14.

Wintermans, P.C., Bakker, P.A. and Pieterse, C.M. (2016) 'Natural genetic variation in *Arabidopsis* for responsiveness to plant growth-promoting rhizobacteria', *Plant Molecular Biology*, 90, pp. 623–634.

Woźniak, M. *et al.* (2019) 'Endophytic bacteria potentially promote plant growth by synthesizing different metabolites and their phenotypic/physiological profiles in the Biolog GEN III MicroPlateTM Test', *International Journal of Molecular Sciences*, 20(21), p. 5283.

Yakhin, O.I. *et al.* (2017) 'Biostimulants in plant science: A global perspective', *Frontiers in Plant Science*, 7, p. 2049.

Yapa, N. *et al.* (2022) 'Biofertilizers: An emerging Trend in agricultural sustainability', *Chiang Mai Journal of Science*, 49(3), pp. 608–640.

Yasmeen, A. *et al.* (2014) 'Morphological and physiological response of tomato (*Solanum lycopersicum* L.) to natural and synthetic cytokinin sources: A comparative study', *Acta Physiologiae Plantarum*, 36, pp. 3147–3155.

Yuan, P. *et al.* (2020). 'Phenazine-producing rhizobacteria promote plant growth and reduce redox and osmotic stress in wheat seedlings under saline conditions', *Frontiers in Plant Science*, 11, p. 575314.

Zandonadi, D.B. *et al.* (2016) 'Plant proton pumps as markers of biostimulant action', *Scientia Agricola*, 73, pp. 24–28.

Zhao, X. *et al.* (2023) 'A meta-analysis on morphological, physiological and biochemical responses of plants with PGPR inoculation under drought stress', *Plant, Cell & Environment*, 46(1), pp. 199–214.

Zhou, L.S., Tang, K. and Guo, S.X. (2018) 'The plant growth-promoting fungus (PGPF) *Alternaria sp.* A13 markedly enhances *Salvia miltiorrhiza* root growth and active ingredient accumulation under greenhouse and field conditions', *International Journal of Molecular Sciences*, 19(1), p. 270.

Zhu, Y. and Gong, H. (2014) 'Beneficial effects of silicon on salt and drought tolerance in plants', *Agronomy for Sustainable Development*, 34, pp. 455–472.

Zulfiqar, F. *et al.* (2020) 'An overview of plant-based natural biostimulants for sustainable horticulture with a particular focus on moringa leaf extracts', *Plant Science*, 295, p. 110194.

Index

Note: Page numbers in **bold** and *italics* refer to tables and figures, respectively.

D

E

F

N

Q

R